Basic Structures
of Function Field Arit

Springer
*Berlin
Heidelberg
New York
Barcelona
Budapest
Hong Kong
London
Milan
Paris
Santa Clara
Singapore
Tokyo*

David Goss

Basic Structures of Function Field Arithmetic

 Springer

David Goss
Department of Mathematics
The Ohio State University
231 West 18th Avenue
Columbus, OH 43210-1174
USA
e-mail: goss@math.ohio-state.edu

```
Library of Congress Cataloging-in-Publication Data

Goss, David, 1952-
    Basic structures of function field arithmetic / David Goss.
      p.  cm.
    "Corrected second printing 1998 of the first edition 1996, which
  was originally published as volume 35 of the series Ergebnisse der
  Mathematik und ihrer Grenzgebiete. 3. Folge"--T.p. verso.
    Includes bibliographical references and index.
    ISBN 3-540-63541-6 (softcover : alk. paper)
    1. Fields, Algebraic.  2. Arithmetic functions.  3. Drinfeld
  modules.   I. Title.
  QA274.G686  1996b
  512'.74--dc21                                         97-35573
                                                            CIP
```

Corrected Second Printing 1998 of the First Edition 1996, which was originally published as Volume 35 of the series
Ergebnisse der Mathematik und ihrer Grenzgebiete. 3. Folge.

Mathematics Subject Classification (1991):
11G09, 11R58, 11T55, 11S40, 11S80,

ISBN 3-540-63541-6 Springer-Verlag Berlin Heidelberg New York

This work is subject to copyright. All rights are reserved, whether the whole or part of the material is concerned, specifically the rights of translation, reprinting, reuse of illustrations, recitation, broadcasting, reproduction on microfilms or in any other ways, and storage in data banks. Duplication of this publication or parts thereof is permitted only under the provisions of the German Copyright Law of September 9, 1965, in its current version, and permission for use must always be obtained from Springer-Verlag. Violations are liable for prosecution under the German Copyright Law.

© Springer-Verlag Berlin Heidelberg 1998
Printed in Germany

Typesetting: Camera-ready copy produced by the author's output file using a Springer T_EX macro package
SPIN 10645755   41/3143 - 5 4 3 2 1 0 – Printed on acid-free paper

In memory of
KURT J. EPPLER

Preface

Classical algebraic number theory concerns the properties satisfied by the rational numbers $\mathbb{Q}$ and those numbers α which satisfy a polynomial with rational coefficients. It has always been, and remains, a magical subject with the most wonderful and interesting structures one can imagine. The twists and turns of the theory are so subtle that they read like a masterful mystery story. Who in the nineteenth century, for instance, would ever have thought that a solution to Fermat's Last Theorem would arise from the study of elliptic modular functions? Yet this is precisely the way the well known solution due to G. Frey, J.-P. Serre, K. Ribet, R. Taylor and most importantly, A. Wiles, has proceeded. And it is clear that these fantastic results are by no means the end of the line. What ever else may happen, there will certainly be more wonderfully interesting mysteries and results in algebraic number theory in the years to come.

One of the tools essential to Wiles' proof [Wi1], and indeed essential to many of the great results in number theory of the late twentieth century, is the theory of arithmetic cohomology. Arithmetic cohomology arose via the realization that combinatorial questions about solutions to equations over finite fields could be attacked via *topological methods*. At first glance, of course, this seems nonsensical as all one seems to have over finite fields is a discrete set of solutions. However, it eventually became clear that the topology could be supplied by the *algebra* of the equations themselves. Like many advances in mathematics, the road to the construction of these topologies was opened up by the detailed study of a particular instance where one could work with ad hoc methods. For arithmetic cohomology theories, this "particular instance" was the theory of algebraic curves over finite fields.

Indeed, as curves are 1-dimensional, they are clearly the simplest nontrivial instances of all solution sets of polynomials (i.e., general varieties over fields). Beginning with E. Artin [Ar1], the study of solutions of curves over finite fields $\mathbb{F}_r$ (or their rational points) shifted to the study of the *zeta function* associated to the curve. This function was defined in analogy with the classical Riemann zeta function; however, it turned out to be a more elementary object in that it is *always* a rational function in r^{-s}. First Hasse, for elliptic curves, and then Weil [We2], [We4], for general curves, were able to prove the analog of the Riemann hypothesis for the zeta function of the curve through the use of the Jacobian of the curve; this is the "ad hoc method" for curves. The remarkable thing about the rational function decomposition of the zeta function a curve was that, via the analogy with complex curves, the factors appeared to arise from cohomology. That is, the numerator has degree $2g$ where g is the genus of the curve and the denominator splits into a product of two factors of degree 1, just as the 0 and 2 dimensional cohomology groups of a complex curve of genus g are 1-dimensional vector spaces and the 1-dimensional cohomology is a $2g$-dimensional space. This led to Weil's famous paper [We3] which marks the true origin of arithmetic cohomology.

At around the same time that the above "classical" theory of curves over finite fields (or, what is the same, function fields of dimension 1 over finite fields) was being discovered, another theory of curves was also being born. This theory has its origins in the paper [C1] of L. Carlitz. In this paper, Carlitz did something very different from the classical theory. Rather than attach to a curve a complex zeta function, he attached to it (for at least the rational curve, the polynomial ring $\mathbb{F}_r[T]$, and the "usual" infinite place ∞) an *exponential function* $e_C(z)$. This function, which was again constructed in an ad hoc fashion, has many of the same properties as the classical function e^z basic to all of mathematics. However, it also has the very peculiar property that it is an analytic function defined on a characteristic p space and *with values in characteristic p*. Indeed, whereas $e^{nz} = (e^z)^n$, Carlitz's function satisfies the functional equation $e_C(az) = C_a(e_C(z))$, for $a \in \mathbb{F}_r[T]$, where $C_a(u)$ is an *additive polynomial*. That is, $C_a(u+v) = C_a(u) + C_a(v)$; such polynomials may only be found non-trivially in finite characteristic. It was also found that $C_{ab}(u) = C_a(C_b(u)) = C_b(C_a(u))$ for polynomials a and b. The mapping $a \mapsto C_a(u)$ is called *the Carlitz module*.

Historically, mathematics has focused on complex analytic functions, both for their mathematical beauty and for their applicability to physics, engineering etc. Traditionally, therefore, mathematicians learn nothing about analysis in finite characteristic and, indeed, the common wisdom was that not much could be contained in this subject. However, through much cleverness, Carlitz was able to use his exponential function to obtain further objects of interest such as a "factorial" and "Bernoulli functions," and so on. These objects were developed somewhat quietly over the decades by Carlitz's students, L. Wade, H. Lee, C. Wagner, and most importantly D. Hayes. In fact, Hayes in [Ha4] established the basic fact that the division values of $e_C(z)$ generate good "cyclotomic" abelian extensions of the rational field in complete analogy with the division values of $e^{2\pi i z}$.

In 1974, the situation changed dramatically with the publication of V.G. Drinfeld's great paper *Elliptic Modules* [Dr1]. In this paper, Drinfeld began the theory of general exponential functions of arbitrary rank d (where $e_C(z)$ has rank 1) associated to an *arbitrary function field* with an *arbitrary* choice of place ∞. Drinfeld showed how such exponential functions give rise to *algebraic objects*, the elliptic modules (now called "Drinfeld modules" — the

Carlitz module is then the *simplest* of all Drinfeld modules), and constructed the moduli for such objects. Even more remarkably, Drinfeld was able to uniformize these moduli through the use of Tate's theory of rigid analytic spaces, all of which still takes place in characteristic p. Drinfeld's goal was to prove reciprocity laws à la Langlands, and indeed, his techniques, and their descendants, have proved very powerful; see, e.g., [Dr2], [Dr5], [Ka1], [HardK1], [La1], [LRS1].

On the other hand, now influenced by Drinfeld (the paper [Ha4] was not) Hayes went on to develop an explicit class field theory for general function fields. Whereas Drinfeld's original approach also gave class fields (as moduli, see our Subsection 7.1), Hayes' approach had the advantage that it was more elementary and "cyclotomic" and so easier to use in producing analogs of classical number theoretic objects à la Carlitz. Thus was born the two different branches of the theory of Drinfeld modules; the goal of one branch is to produce reciprocity laws, whereas the goal of the other is to develop an "arithmetic" in characteristic p. Of course, both branches influence each other and it seems, must ultimately belong to one single overarching theory.

This volume belongs to the second branch of the theory. It is dedicated to describing for the reader the fundamental properties of the arithmetic of function fields over finite fields. Thus we will describe the basic properties of Drinfeld modules, T-modules (a very important higher dimensional analog of Drinfeld modules introduced by G. Anderson in 1986, [A1] — if Drinfeld modules are analogous to elliptic curves, then T-modules are analogous to general abelian varieties), "shtukas," (a shtuka is a vector bundle version of a Drinfeld module), and so on. Our purpose here is to study these objects as being of interest *by themselves* in exactly the same manner one studies roots of unity, elliptic curves, or general abelian varieties.

Let **A** be the affine ring of elements of our fixed function field which are regular away from ∞. Thus **A** is a generalization of the ring $\mathbb{F}_r[T]$, and is a Dedekind domain with finite unit group and class number; but where this class number is almost always > 1. It is the "bottom ring" of the theory *by fiat* exactly as $\mathbb{Z}$ is the bottom ring of algebraic number theory. If one has a Drinfeld module ψ over a field L, then L becomes a *new* **A**-module via ψ. Using this **A**-module structure one can easily define the *Tate module* of ψ in analogy with elliptic curves. Again this Tate module lives in finite characteristic as a module over the completion of **A** at a prime. As usual one has functoriality, and, if L is finite, there is a Frobenius morphism. The characteristic polynomial of the Frobenius morphism at a prime has **A**-coefficients just as classically it has $\mathbb{Z}$-coefficients. Thus it makes sense to codify all these characteristic polynomials into an Euler product for a *characteristic p L-function* in complete analogy with L-functions of abelian varieties; indeed, these functions are basic to the theory and are presented in Section 8. A basic property of such functions is that they are *not* rational functions and are much more complex analytic objects; we describe them in great detail. Still the reader will quickly see that many basic ideas still remain to be discovered. For instance in Subsection 8.24, we will discuss some quite remarkable calculations that appear to be hinting at the appropriate type of *Riemann hypothesis* for these characteristic p L-functions. Yet, as of this writing, there is no clear indication of what arithmetic implications are contained in these calculations, etc.

In a similar way, the Carlitz factorial and Carlitz exponential lead to Γ-functions which are presented in Section 9. The reader will note that the theory of these Γ-functions is more advanced than that of the L-series. For instance, the Γ-functions of Section 9 all satisfy functional equations completely analogous to those of classical theory. Still, even here, there are many open questions that must eventually be answered.

From virtually the very beginning of classical number theory, it was understood that the Riemann zeta function satisfies its well known functional equation under $s \mapsto 1 - s$. And from the very beginnings of the theory of the L-functions of Section 8, it was known that these functions would *not* satisfy any *obvious* analog of this functional equation (the calculations of Subsection 8.24 mentioned above do not arise from any functional equation). Indeed, a simple count of the *two* types of Bernoulli objects that arise in the theory, show that they cannot be so easily matched up. Still, we have always felt that *something* must be underlying all of the structure that was appearing. While that "something" is not yet known, we are very much encouraged by the recent research of Anderson, described briefly in Subsection 10.6. With the constant discovery of such interesting phenomena, we are constantly encouraged to search for more.

This book is meant to be as self-contained as was possible to write in a reasonable amount of space. Still the reader must know basic number theory, such as in Weil's book of the same name [We1]. Moreover, it would definitely be of great help to the reader to be familiar with the basics of the theory of elliptic curves such as in [Sil1]. The reader will also, perhaps, notice that there are many loose ends in our presentation of the theory. This cannot be helped with such a young subject, and one that is undergoing such rapid growth (as attested to by the number of preprints, or papers recently published, listed in our references). In fact, the subject was advancing so rapidly that writing an exposition of it was both very challenging and exciting. Nothing would make the author happier than to know that some young mathematician has been spurred on by this book to help push the subject along. To that end, we have listed as many interesting open problems as was possible during the writing of this volume. For instance, in Subsection 8.25, the reader will learn that all of the theory developed for curves may ultimately be applicable in one form or another to *arbitrary* smooth varieties in finite characteristic. Indeed, only time, and the labors of many mathematicians, will eventually tell us how far we may go.

Due to space considerations we have not covered certain basic topics such as transcendence theory (though some important results are stated) or the theory of rigid analytic modular forms. For the former, we refer the reader to the papers of Jing Yu, L. Denis, M. Waldschmidt, J.-P. Allouche, Y. Hellegouarch, B. de Mathan, R. Tubbs, D. Brownawell, etc., which are listed in our references. For the latter, we refer the reader to [Go14] and to the very important new papers, [GeR2], [Ge29], etc., of E.-U. Gekeler and his school.

Before acknowledging the many people who have helped in the preparation of this volume, we need to mention three themes that appear consistently throughout this volume in the hope of avoiding possible future confusion by the reader. The first one is the relative difference in "size" between number fields and function fields. In algebraic number theory, one starts out with $\mathbb{Q}$, completes it to obtain the rational numbers $\mathbb{R}$ and then adjoins i to obtain the complex numbers $\mathbb{C}$, and then stops. But for function fields, when one completes, one ends up with a local field which not only has extensions of arbitrarily large degree, but also can have infinitely many distinct extensions of *fixed* degree. As such, the algebraic closure of this field is *huge*. In it, so to speak, number theory has "room to roam." Whereas in classical theory, objects may play many different roles all at once, for function fields, these different roles are played by *different* players. A very beautiful example of this is given by the theory of Thakur's Gauss sums for function fields (Subsection 9.8); because **A** can be so general, it turns out that generically (when **A** is *not* the polynomial ring) the "factorials" are distinct from the coefficients of the Taylor expansions of exponential functions — which is of course totally different from e^z. However, it is precisely the Taylor coefficients that define the *correct* general Γ-function that allows one to describe these Gauss sums in the manner of the classical Gross-Koblitz Theorem of algebraic number theory. Thus we see that the connection is between Gauss sums and exponentials and *not* Gauss sums and factorials; this is not seen in classical theory. By abusing language, and perhaps reality, we like to say that in the theory of function fields, number theoretic objects break up into little "arithmetic quarks." While the reader may not consider this analogy to be particularly apt, the reader *will* see this splitting-up phenomenon again and again throughout the book.

The second theme is what we call the "arithmetic/geometric dichotomy." For function fields there are *two* types of "cyclotomic extensions." The first type, obviously, are those obtained by adjoining roots of unity, which is the same as extending the field of constants. These constant field extensions are everywhere unramified. The second type is the ramified (geometric) extensions obtained by division values of Drinfeld modules. We call notions related to constant field extensions "arithmetic" and those related to Drinfeld modules "geometric." Of course, for $\mathbb{Q}$ both theories *collapse* into classical cyclotomic number fields (and so, again, we are dealing with a splitting-up type of phenomenon).

The third theme is what I call the "two T's." This arises in the following simple manner. Suppose, as very often happens, that we have a Drinfeld module for $\mathbb{F}_r[T]$ which is defined over $\mathbb{F}_r(T)$. Thus T is *both* an operator *and* a scalar. These are different roles and the "two T's" is just the idea that we need to keep them separate (so that, on occasion, when T appears as a scalar we will rename it θ). This idea is very similar to the classical (pre-schemes) notion of "generic points" of varieties and has surprising power, see Section 6 or Subsection 7.11. In fact, as became apparent during the writing of this book, our characteristic p L-functions and Γ-functions *arise precisely* from the different uses of T; L-functions arise from the use of T as an operator (via a Drinfeld module) whereas Γ-functions arise from T as scalar (via exponential functions). This realization may ultimately explain some of the surprising properties of our L-functions and Γ-functions; in any case, it is again not something seen classically (e.g., Remark 9.9.13). It is to be hoped that this phenomenon will be better understood in the near future.

I now have the happy task of acknowledging the many people whose help was essential for this volume. First of all, I would like to thank J.-P. Serre who not only asked me to write this volume, but was also kind enough to send me some unpublished conjectures on zeroes of classical Artin L-functions of number fields (Conjecture 8.24.1). I would also like to thank T. England for doing such a good job of typing this manuscript into TEX. I also thank, in no particular order, J. Roberts, B. Snyder, S. Mihalas, D. Shapiro, L. Guo, L. Denis, Y. Taguchi, D. Wan, M. Rosen, J.-P. Serre, M. Kapranov, D. Hayes, S. Sinha, D. Thakur, Jing Yu and B. Poonen for their help with various parts of this project. Special thanks are due to the referee for his careful reading of the manuscript.

I thank the National Science Foundation for its support during the preparation of this work.

I must also mention the names of G. Anderson, and D. Thakur who kindly communicated some of their own unpublished work for use in the volume, and who have contributed so much of fundamental value to the theory over the years. I am grateful to them for their help, and admiring of them for their mathematical prowess.

I thank Mike Rosen for his constant friendship and encouragement over the almost 20 years I have known him. His advice, input and contributions have been invaluable.

Finally, it is my pleasure to thank my wife Rita for putting up with all the difficulties of a modern author and his computer. I am deeply grateful for her support and understanding during the many years that it took to write this book. During this time my daughter Alyssa has gone from a squalling new-born to an opinionated new-person. Both Rita and Alyssa have made my life complete, and it is to both of them that I dedicate this work.

Columbus, Ohio
June, 1996

Table of Contents

1. **Additive Polynomials** 1
    1.1. Basic Properties 1
    1.2. Classification of Additive Polynomials 3
    1.3. The Moore Determinant 5
    1.4. The Relationship Between $k[x]$ and $k\{\tau\}$ 9
    1.5. The p-resultant 10
    1.6. The Left and Right Division Algorithms 12
    1.7. The τ-adjoint of an Additive Polynomial 15
    1.8. Dividing $\mathbb{A}^1$ by Finite Additive Groups 18
    1.9. Analogs in Differential Equations/Algebra 20
    1.10. Divisibility Theory 26
    1.11. The Semi-invariants of Additive Polynomials 31

2. **Review of Non-Archimedean Analysis** 35

3. **The Carlitz Module** 43
    3.1. Background 43
    3.2. The Carlitz Exponential 47
    3.3. The Carlitz Module 52
    3.4. The Carlitz Logarithm 56
    3.5. The Polynomials $E_d(x)$ 57
    3.6. The Carlitz Module over Arbitrary **A**-fields 59
    3.7. The Adjoint of the Carlitz Module 61

4. **Drinfeld Modules** 63
    4.1. Introduction 63
    4.2. Lattices and Their Exponential Functions 65
    4.3. The Drinfeld Module Associated to a Lattice 66
    4.4. The General Definition of a Drinfeld Module 69
    4.5. The Height and Rank of a Drinfeld Module 70
    4.6. Lattices and Drinfeld Modules over $\mathbf{C}_\infty$ 73
    4.7. Morphisms of Drinfeld Modules 79
    4.8. Primality in $\mathcal{F}\{\tau\}$ and **A** 86
    4.9. The Action of Ideals on Drinfeld Modules 86

4.10. The Reduction Theory of Drinfeld Modules 88
4.11. Review of Central Simple Algebra 93
4.12. Drinfeld Modules over Finite Fields 102
4.13. Rigidity of Drinfeld Modules 115
4.14. The Adjoint of a General Drinfeld Module 122

5. **T-Modules** 137
   5.1. Vector Bundles 137
   5.2. Sheaves and Differential Equations 139
   5.3. φ-sheaves 140
   5.4. Basic Concepts of T-modules 142
   5.5. Pure T-modules 148
   5.6. Torsion Points 151
   5.7. Tensor Products 156
   5.8. The Tensor Powers of the Carlitz Module 157
   5.9. Uniformization 159
   5.10. The Tensor Powers of the Carlitz Module redux 172
   5.11. Scattering Matrices 175

6. **Shtukas** 179
   6.1. Review of Some Algebraic Geometry 179
   6.2. The Shtuka Correspondence 182

7. **Sign Normalized Rank 1 Drinfeld Modules** 193
   7.1. Class-fields as Moduli 193
   7.2. Sign Normalization 196
   7.3. Fields of Definition of Drinfeld Modules 201
   7.4. The Normalizing Field 202
   7.5. Division Fields 205
   7.6. Principal Ideal Theorems 211
   7.7. A Rank One Version of Serre's Theorem 214
   7.8. Classical Partial Zeta Functions 218
   7.9. Unit Calculations 221
   7.10. Period Computations 223
   7.11. The Connection with Shtukas and Examples 227

8. **L-series** 235
   8.1. The "Complex Plane" S_∞ 236
   8.2. Exponentiation of Ideals 238
   8.3. v-adic Exponentiation of Ideals 243
   8.4. Continuous Functions on $\mathbb{Z}_p$ 245
   8.5. Entire Functions on S_∞ 248
   8.6. L-series of Characteristic p Arithmetic 254
   8.7. Formal Dirichlet Series 259

	8.8. Estimates	261
	8.9. L-series of Finite Characters	264
	8.10. The Question of Local Factors	269
	8.11. The Generalized Teichmüller Character	270
	8.12. Special-values at Negative Integers	272
	8.13. Trivial Zeroes	276
	8.14. Applications to Class Groups	279
	8.15. "Geometric" Versus "Arithmetic" Notions	283
	8.16. The Arithmetic Criterion for Cyclicity	284
	8.17. The "Geometric Artin Conjecture"	285
	8.18. Special-values at Positive Integers	290
	8.19. The Functional Equation of the Special-values	297
	8.20. Applications to Class Groups	301
	8.21. The Geometric Criterion for Cyclicity	307
	8.22. Magic Numbers	309
	8.23. Finiteness in Local and Global Fields	319
	8.24. Towards a Theory of the Zeroes	322
	8.25. Kapranov's Higher Dimensional Theory	341
9.	Γ-functions	349
	9.1. Basic Properties of the Carlitz Factorial	350
	9.2. Bernoulli-Carlitz Numbers	353
	9.3. The Γ-ideal	359
	9.4. The Arithmetic Γ-function	361
	9.5. Functional Equations	368
	9.6. Finite Interpolations	370
	9.7. Another v-adic Γ-function	374
	9.8. Gauss Sums	376
	9.9. The Geometric Γ-function	381
10.	Additional Topics	389
	10.1. The Geometric Fermat Equation	389
	10.2. Geometric Deligne Reciprocity and Solitons	390
	10.3. The Tate Conjecture for Drinfeld Modules	393
	10.4. Meromorphic Continuations of L-functions	394
	10.5. The Structure of the **A**-module of Rational Points	396
	10.6. Log-algebraicity and Special Points	397
	References	401
	Index	421

1. Additive Polynomials

1.1. Basic Properties

Let k be a field of finite characteristic p and let $\overline{k}$ be a fixed algebraic closure. Let $P(x) \in k[x]$ be a polynomial in x with coefficients in k. We say that $P(x)$ is *additive* if and only if $P(\alpha + \beta) = P(\alpha) + P(\beta)$ for $\{\alpha, \beta, \alpha + \beta\} \subseteq k$. We say that $P(x)$ is *absolutely additive* if and only if $P(x)$ is additive over $\overline{k}$.

Example 1.1.1. The polynomial $\tau_p(x) = x^p$ is absolutely additive for any k. Indeed
$$(x+y)^p = \sum_{i=0}^{p} \binom{p}{i} x^i y^{p-i},$$
and it is simple to see that $\binom{p}{i} \equiv 0 \,(p)$ for $i = 1, \ldots, p-1$. Thus $(x+y)^p = x^p + y^p$ which implies the result.

The above example is the well-known "p^{th}-power mapping." It is used everywhere in number theory and algebraic geometry.

Proposition 1.1.2. *Let $P(x)$ and $H(x)$ be additive polynomials. Then*
1. $P(x) + H(x)$ *is additive,*
2. $\alpha P(x)$ *is additive,* $\alpha \in k$,
3. $P(H(x))$ *is additive.*

Proof. This follows immediately from the definitions. □

Thus we see that the monomials $\tau_p^i(x) := x^{p^i}$, as well as all polynomials spanned by them, are absolutely additive.

Definition 1.1.3. We denote by $k\{\tau_p\}$ the subspace of $k[x]$ spanned by the linear combination of $\{\tau_p^i,\ i = 0, 1, 2, \ldots,\ \}$.

By 1.1.2, the space $k\{\tau_p\}$ forms a ring under composition. If $k \neq \mathbb{F}_p$, then $k\{\tau_p\}$ is *not* commutative. Indeed
$$\tau_p \alpha = \alpha^p \tau_p,$$
for $\alpha \in k$.

Not all additive polynomials are in $k\{\tau_p\}$, in general, as the next example makes evident.

Example 1.1.4. Let $k = \mathbb{F}_3$ and let
$$P(x) = x + (x^3 - x)^2 = x^6 + x^4 + x^2 + x.$$
Then for $\alpha \in \mathbb{F}_3$, $P(\alpha) = \alpha$ and so $P(x)$ is additive. Clearly $P(x) \notin \mathbb{F}_3\{\tau_p\}$.

The above example works as $k = \mathbb{F}_3$ is finite. We now assume that k is infinite where the situation is much different as shown by our next result.

Proposition 1.1.5. *Let k be an infinite field. Then a polynomial $P(x) \in k[x]$ is additive if and only if $P(x) \in k\{\tau_p\}$.*

Proof. The implication $\Leftarrow$ has already been established (Example 1.1.1). Thus we must establish $\Rightarrow$. To do so, we use the *formal derivative*, $P'(x)$, of $P(x)$ with respect to x. This is defined, as usual, by $P'(x) := \sum \alpha_i i x^{i-1}$ if $P(x) = \sum \alpha_i x^i$. It is a formal calculation to check that this operation satisfies the standard rules of differentiation.

We first establish that
$$P'(x) \equiv c$$
for some $c \in k$ if $P(x)$ is additive. Indeed, let $\alpha \in k$. Then the polynomial
$$P(x + \alpha) - P(x) - P(\alpha)$$
is 0 for all $x \in k$. Thus, as k is infinite, it must be identically 0. Now
$$P'(\alpha) = \frac{d}{dx}P(x + \alpha)\bigg|_{x=0} = \frac{d}{dx}(P(x) + P(\alpha))\bigg|_{x=0} = P'(0).$$
Thus, again by the infinitude of k, we see
$$P'(x) \equiv P'(0) \equiv c.$$

Consequently, we may write
$$P(x) = cx + \sum_{j=2}^{\nu} \alpha_j x^{n_j},$$
where $n_j \equiv 0 \ (p)$. We write
$$P(x) = P_0(x) + P_1(x)$$
where $P_0(x) = cx + \{\text{terms with } n_j \text{ a power of } p\}$, and $P_1(x) = \{\text{terms where } n_j \text{ is divisible by a prime} \neq p\}$. We must show that $P_1(x) \equiv 0$. Since we know that $P_0(x) \in k\{\tau_p\}$, we find $P_1(x) = P(x) - P_0(x)$ is also additive.

Let $\bar{k}$ be our fixed algebraic closure of k. It is clearly sufficient to show that $P_1(x) \equiv 0$ in $\bar{k}[x]$. Moreover, the mapping $\tau_p: \bar{k} \to \bar{k}$; $x \mapsto x^p$, is now an automorphism. Let p^e be the largest power of p dividing all n_j and let

$$P_2(x) = P_1(x)^{1/p^e} \in \bar{k}[x].$$

The mapping $x \mapsto x^{1/p^e}$ is easily seen to be additive (though *not* a polynomial) from $\bar{k}$ to itself; thus $P_2(x)$ is still additive, at least as a function from k to $\bar{k}$.

Using the additivity of $P_2(x)$ on k as above, we see that $P_2'(x)$ is identically 0. But, by the construction of $P_2(x)$, this can only happen if $P_2(x)$ vanishes, and the proof is complete. □

If k is stable under taking p-th roots, (as, for instance, $\bar{k}$ is) then the above argument can be much simplified by using the derivative to see that $P(x) = cx + H(x)^p$ and then using induction to conclude that $H(x)$ is of the desired form.

Corollary 1.1.6. *Let k be any field of characteristic p. Then the set of absolutely additive polynomials over k equals $k\{\tau_p\}$.*

Proof. The algebraic closure of any field is infinite. Thus we can apply the proposition. □

From now on we will drop "absolutely" and use the adjective "additive" to refer to an element of $k\{\tau_p\}$. We also fix a p-th power, $r = p^{m_0}$, and assume that $\mathbb{F}_r \subseteq k$. We set $\tau := \tau_p^{m_0}$ and we let $k\{\tau\}$ be the composition ring of polynomials in τ. It is now a simple matter to check that the ring $k\{\tau\}$ is the $\mathbb{F}_r$-algebra of $\mathbb{F}_r$-linear polynomials (i.e., $k\{\tau\}$ consists of those additive polynomials $P(x)$ such that $P(\zeta x) = \zeta P(x)$ for all $\zeta \in \mathbb{F}_r$).

Finally, we note that $k\{\tau_p\}$ is sometimes referred to as the "ring of Frobenius polynomials" or the "ring of p-polynomials."

1.2. Classification of Additive Polynomials

We continue with the ideas of the previous section.
Notation:
1. If $P(x)$ is additive, then we use the notation "$P(\tau_p)$" to refer to its representation in $k\{\tau_p\}$. Similarly, if $P(x)$ is $\mathbb{F}_r$-linear, then we use "$P(\tau)$" to denote its representation in $k\{\tau\}$. (**Warning:** With this convention it is *not* true that $P(\tau)$ is obtained from $P(x)$ by formally substituting τ for x!)
2. The multiplication, "$P(\tau) \cdot Q(\tau)$", etc., will always refer to multiplication inside $k\{\tau\}$.

3. We say that $P(\tau)$, etc., is *monic* if and only if $P(x)$ is monic.

4. Let $P(\tau) = \sum_{i=0}^{t} \alpha_i \tau^i$, $\alpha_t \neq 0$. We set $t = \deg P(\tau)$.

Notice that
$$r^{\deg P(\tau)} = \deg P(x).$$

Now let $k \supseteq \mathbb{F}_r$ be an algebraically closed field, and let $P(\tau) \in k\{\tau\}$. Clearly the roots of $P(x) \subset k$ form a subgroup. We shall see that, when $P(x)$ is a separable polynomial $\in k[x]$, the converse is also true. This is the so-called "Fundamental Theorem of Additive Polynomials."

Theorem 1.2.1. *Let $P(x) \in k[x]$ be a separable polynomial. Let*
$$\{w_1, \ldots, w_m\} \subset k$$
be the set of its roots. Then $P(x)$ is additive if and only if $\{w_1, \ldots, w_m\}$ is a subgroup.

Proof. What must be shown is the following: Let $W = \{w_1, \ldots, w_m\}$ be an additive subgroup of k and let
$$P(x) := P_W(x) := \prod_{i=1}^{m}(x - w_i).$$

Then $P(x)$ is additive.

Note that if $w \in W$, then $P(x+w) = P(x)$. Now let $y \in k$ and put
$$H(x) = P(x+y) - P(x) - P(y).$$

Clearly $\deg H(x) < \deg P(x)$. On the other hand, it is now trivial to see that $H(w) = 0$ for $w \in W$. As $m = \deg P > \deg H$, we conclude that $H(x) \equiv 0$.

Let y now be an arbitrary indeterminate and put
$$H_1(y) = P(x+y) - P(x) - P(y) \in k[x][y] = k[x, y].$$

We conclude that $H_1(\alpha) = 0$ for $\alpha \in k$. As k is infinite, we see that $H_1(y) \equiv 0$, which completes the proof. □

Corollary 1.2.2. *Let $P(x)$ be as in 1.2.1. Then $P(x)$ is $\mathbb{F}_r$-linear if and only if the roots W of $P(x)$ form an $\mathbb{F}_r$-vector subspace of k.*

Proof. Clearly, we need only show that if W is an $\mathbb{F}_r$-vector subspace of k, then $P(x)$ is $\mathbb{F}_r$-linear. Thus let $\zeta \in \mathbb{F}_r$ and put
$$H(x) = P(\zeta x) - \zeta P(x).$$

Let $|W| = r^j$; so $\deg P(x) = r^j$. As $\zeta^{r^j} = \zeta$, we conclude that $\deg H(x) < r^j$. On the other hand, $H(w) = 0$ for $w \in W$. Thus $H(x)$ must identically vanish. □

1.3. The Moore Determinant

We begin with recalling the classical *Vandermonde determinant*: Let A be the matrix

$$\begin{pmatrix} 1 & \cdots & 1 \\ x_1 & \cdots & x_n \\ \vdots & & \vdots \\ x_1^{n-1} & \cdots & x_n^{n-1} \end{pmatrix}.$$

Then

$$\det(A) = \prod_{j<i}(x_i - x_j).$$

In this subsection we present a characteristic p "τ-version" of this result. This is due to E. H. Moore [M1], and we follow the exposition given in [O1].

We will use the notation of the previous subsection: so k is a field with $\mathbb{F}_r \subseteq k$, τ is the r-th power mapping, etc. Let $W \subseteq k$ be an $\mathbb{F}_r$-subspace and let $\{w_1, \ldots, w_n\} \subseteq W$.

Lemma 1.3.1. *The set $\{w_1, \ldots, w_n\}$ is linearly independent over $\mathbb{F}_r$ if and only if, for every $i \geq 0$, the set*

$$\{\tau^i(w_1), \ldots, \tau^i(w_n)\}$$

is also.

Proof. Let $\{\alpha_1, \ldots, \alpha_n\} \subseteq \mathbb{F}_r$. Then

$$\sum \alpha_j w_j = 0 \iff \sum \alpha_j \tau^i(w_j) = 0. \qquad \square$$

Definition 1.3.2. We set

$$\Delta(w_1, \ldots, w_n) := \Delta_r(w_1, \ldots, w_n)$$

$$:= \det \begin{pmatrix} w_1 & \cdots & w_n \\ w_1^r & \cdots & w_n^r \\ \vdots & & \vdots \\ w_1^{r^{n-1}} & \cdots & w_n^{r^{n-1}} \end{pmatrix}$$

$$= \det \begin{pmatrix} \tau^0(w_1) & \cdots & \tau^0(w_n) \\ \vdots & & \vdots \\ \tau^{n-1}(w_1) & \cdots & \tau^{n-1}(w_n) \end{pmatrix}.$$

We call $\Delta(w_1, \ldots, w_n)$ the *Moore determinant*.

Lemma 1.3.3. *With the notation of Lemma 1.3.1. The set $\{w_1, \ldots, w_n\}$ is linearly independent over $\mathbb{F}_r$ if and only if $\Delta(w_1, \ldots w_n) \neq 0$.*

Proof. Suppose that $\Delta(w_1,\ldots,w_n) \neq 0$. We begin by establishing that this implies that $\{w_1,\ldots w_n\}$ must be linearly independent over $\mathbb{F}_r$. Indeed, let $\{\alpha_i\} \subseteq \mathbb{F}_r$ be chosen so that $\sum \alpha_i w_i = 0$. Then, by 1.3.1, we see that

$$\sum \alpha_i \begin{pmatrix} w_i \\ \tau(w_i) \\ \vdots \\ \tau^{n-1}(w_i) \end{pmatrix} = \begin{pmatrix} 0 \\ 0 \\ \vdots \\ 0 \end{pmatrix}.$$

As
$$\Delta(w_1,\ldots,w_n) \neq 0,$$
this immediately implies that $\alpha_i = 0$ for all i.

Conversely, assume that $\{w_1,\ldots,w_n\}$ is a linearly independent set over $\mathbb{F}_r$. We show that $\Delta(w_1,\ldots,w_n) \neq 0$. We work by induction; the case $n=1$ being easy. Thus let us assume the result for $n=t$ and assume that $\Delta(w_1,\ldots,w_{t+1}) = 0$. We can, therefore, find $\{\alpha_1,\ldots,\alpha_{t+1}\} \subseteq k$, not all zero, such that

$$\begin{array}{rcl} \alpha_1 w_1 + \cdots + \alpha_{t+1} w_{t+1} &=& 0 \\ \alpha_1 w_1^r + \cdots + \alpha_{t+1} w_{t+1}^r &=& 0 \\ \vdots & & \vdots \\ \alpha_1 w_1^{r^t} + \cdots + \alpha_{t+1} w_{t+1}^{r^t} &=& 0 \end{array}.$$

Without loss of generality, we may assume that $\alpha_1 = 1$. By raising the i-th equation to the r-th power, and subtracting it from the $(i+1)$-st equation, we find

$$\begin{array}{rcl} (\alpha_2 - \alpha_2^r)w_2^r + \cdots + (\alpha_{t+1} - \alpha_{t+1}^r)w_{t+1}^r &=& 0 \\ \vdots & & \vdots \\ (\alpha_2 - \alpha_2^r)w_2^{r^t} + \cdots + (\alpha_{t+1} - \alpha_{t+1}^r)w_{t+1}^{r^t} &=& 0 \end{array}.$$

On the other hand, $\{w_2^r,\ldots,w_{t+1}^r\}$ is a linearly independent set over $\mathbb{F}_r$. Induction then implies that

$$\Delta(w_2^r,\ldots,w_{t+1}^r) \neq 0;$$

thus our original equation had coefficients in $\mathbb{F}_r$. This implies that $\{w_1,\ldots,w_{t+1}\}$ is linearly dependent; a contradiction which gives the proof. $\square$

Corollary 1.3.4. *Let* $\dim_{\mathbb{F}_r} W = d$. *Then* $\{w_1,\ldots,w_d\} \subseteq W$ *is a basis over* $\mathbb{F}_r$ *if and only if* $\Delta(w_1,\ldots,w_d) \neq 0$. $\square$

Recall that, in Corollary 1.2.2, we characterized separable $\mathbb{F}_r$-linear polynomials as those separable polynomials whose roots form a finite dimensional $\mathbb{F}_r$-vector space. Using the Moore determinant, we can now be a bit more specific about the finite dimensional $\mathbb{F}_r$-subspace W and the associated $\mathbb{F}_r$-linear polynomial $P_W(x)$.

1.3. The Moore Determinant

Let $W \subseteq k$ be a finite dimensional $\mathbb{F}_r$-vector space. Let $\{w_1, \ldots, w_d\}$ be a chosen $\mathbb{F}_r$-basis for W and, for $1 \leq i \leq d$, let W_i be the span of $\{w_1, \ldots, w_i\}$. Set
$$P_i(x) := P_{W_i}(x) := \prod_{\alpha \in W_i} (x - \alpha);$$
so $P_W(x) = P_d(x)$. Recall that $P_i(\tau)$ denotes the element of $k\{\tau\}$ corresponding to $P_i(x)$.

Proposition 1.3.5.

1. *With multiplication in $k\{\tau\}$, we have*
$$P_i(\tau) = (\tau - P_{i-1}(w_i)^{r-1}\tau^0) \cdot P_{i-1}(\tau).$$

2. $P_i(x) = \Delta(w_1, \ldots, w_i, x)/\Delta(w_1, \ldots, w_i)$.
3. $P_W(\tau) = P_{\overline{W}_i}(\tau) \cdot P_i(\tau)$, *where $\overline{W}_i$ is the span of $\{P_i(w_{i+1}), \ldots, P_i(w_d)\}$.*

Proof. It is clear that
$$i = \deg P_i(\tau) = \deg\left((\tau - P_{i-1}(w_i)^{r-1}\tau^0) \cdot P_{i-1}(\tau)\right).$$
As both sides are monic in τ, we need only establish that they have the same set of roots ($= W_i$) in x.

Let $w \in W_i$ and write $w = cw_i + w_0$, where w_0 is spanned by the vectors $\{w_1, \ldots, w_{i-1}\}$. Then
$$(\tau - P_{i-1}(w_i)^{r-1}\tau^0) \cdot P_{i-1}(\tau)\big|_{x=w} = (\tau - P_{i-1}(w_i)^{r-1}\tau^0)\big|_{x=cP_{i-1}(w_i)}$$
$$= 0.$$
Thus the first part follows.

To see the second part, note that from 1.3.3 we know that x is a root of $\Delta(w_1, \ldots, w_i, x) = 0$ if and only if $x \in W_i$. Thus $P_i(x) = c\Delta(w_1, \ldots, w_i, x)$ for some $c \neq 0$. A moment's reflection shows that $c = 1/\Delta(w_1, \ldots, w_i)$.

The third part follows by checking that $P_W(\tau)$ and $P_{\overline{W}_i}(\tau)P_i(\tau)$ have the same roots. □

We shall often use 1.3.5 as follows: k will be Galois over some field k_1 and, for some i, both W_i and W are $\text{Gal}(k/k_1)$ stable. Then $P_W(\tau)$, $P_i(\tau)$, and $P_{\overline{W}_i}(\tau)$ all belong to $k_1\{\tau\}$ as is readily seen.

The next result follows readily from 1.3.5.2.

Lemma 1.3.6. *We have*
$$\frac{\Delta(w_1, \ldots, w_{d-1}, x)}{\Delta(w_1, \ldots, w_{d-1})} = \prod_{i=1}^{d-1} \prod_{k_i \in \mathbb{F}_r} (x - (k_1 w_1 + \cdots + k_{d-1} w_{d-1})).$$ □

1. Additive Polynomials

Corollary 1.3.7 (Moore). $\Delta(w_1,\ldots,w_d)$ equals

$$\prod_{i=1}^{d} \prod_{k_{i-1}\in\mathbb{F}_r} \cdots \prod_{k_1\in\mathbb{F}_r} (w_i + k_{i-1}w_{i-1} + \cdots + k_1w_1).$$

Proof. Let $x = w_d$ and apply induction to the formula

$$\Delta(w_1,\ldots,w_{d-1},w_d) = \Delta(w_1,\ldots,w_{d-1})$$
$$\cdot \prod_{i=1}^{d-1} \prod_{k_i\in\mathbb{F}_r} (w_d - (k_1w_1 + \cdots + k_{d-1}w_{d-1}))$$

which follows immediately from 1.3.6. $\square$

We also have a useful version of Wilson's Theorem as in [O1].

Corollary 1.3.8. *Let W, etc., be as above. Then*

$$\prod_{\substack{w\in W \\ w\neq 0}} w = (-1)^d \Delta(w_1,\ldots,w_d)^{r-1}.$$

Proof. From 1.3.6, we see that the coefficient of x in

$$\Delta(w_1,\ldots,w_d,x)/\Delta(w_1,\ldots,w_d)$$

is $\prod_{0\neq w\in W} w$. On the other hand, by expanding along the last column in $\Delta(w_1,\ldots,w_d,x)$, one finds that this term is

$$(-1)^d \frac{\Delta(w_1^r,\ldots w_d^r)}{\Delta(w_1,\ldots,w_d)} = (-1)^d \Delta(w_1,\ldots,w_d)^{r-1}. \quad \square$$

Formula 1.3.7 is the Moore Determinant Formula. Since $\{w_1,\ldots,w_d\}$ can be any linearly independent set, the reader will easily see that 1.3.7 is equivalent to the following: Let $\{x_1,\ldots x_d\}$ be algebraically independent variables. Then we have an equality of elements in $\mathbb{F}_r[x_1,\ldots,x_d]$

$$\Delta(x_1,\ldots,x_d) = \prod_{i=1}^{d} \prod_{k_{i-1}\in\mathbb{F}_r} \cdots \prod_{k_1\in\mathbb{F}_r} (x_i + k_{i-1}x_{i-1} + \cdots + k_1x_1).$$

Remark. 1.3.9. The right hand side of the above product contains exactly one representative, modulo $\mathbb{F}_r^*$, of each nonzero linear mapping over $\mathbb{F}_r$ of $x_1,\ldots,x_d$.

Finally, one can pass from the Moore determinant to the Vandermonde determinant (computed in k) via the formal substitution

$$x_i^{r^j} \mapsto x_i^j\,.$$

For instance,

$$\det\begin{pmatrix} x_1 & x_2 \\ x_1^r & x_2^r \end{pmatrix} = x_1 x_2^r - x_2 x_1^r \mapsto x_2 - x_1 = \det\begin{pmatrix} 1 & 1 \\ x_1 & x_2 \end{pmatrix}.$$

It would be amusing to know the mechanics of transforming the computation of the Moore determinant (i.e., Corollary 1.3.7) into the usual Vandermonde computation.

1.4. The Relationship Between $k[x]$ and $k\{\tau\}$

We continue with our previous notation. Let $f(x) \in k[x]$. Our goal here is to show the following result of Ore [O1].

Theorem 1.4.1. *There exists $g(\tau) \in k\{\tau\}$ such that $f(x)$ divides $g(x)$.*

The reader should note that the set of all $g(\tau)$ satisfying the condition of the theorem forms a left ideal in $k\{\tau\}$. Corollary 1.6.3 will imply that all such ideals are principal. The proof of Theorem 1.4.1 given just below will show that this ideal is non-trivial by exhibiting a non-trivial generator of it.

Proof of Theorem 1.4.1. We shall present two different proofs of this result.
Proof 1. Let $\bar{k}$ be a fixed algebraic closure of k and let $\{w_1, \ldots, w_d\} \subset \bar{k}$ be the roots of $f(x)$ chosen *without* multiplicity. Let $W = (w_1, \ldots, w_d)$ be the $\mathbb{F}_r$-span of $\{w_1, \ldots, w_d\}$. Put

$$P_W(x) = \prod_{w \in W}(x - w);$$

so by the results of §1.3, we know that $P_W(x)$ is $\mathbb{F}_r$-linear.

Let t be the largest multiplicity of any w_i and let e be the smallest positive integer so that

$$r^e \geq t\,.$$

Let $g(\tau) = \tau^e P_W(\tau) \in k\{\tau\}$. One then easily sees that $f(x)$ divides $g(x)$.
Proof 2. Let $d = \deg(f(x))$. For each $i \geq 0$, let $f_{(i)}(x)$ be the residue upon dividing x^{r^i} by $f(x)$. Let $v \leq d$ be the *smallest* integer such that $\{f_{(0)}(x), \ldots, f_{(v)}(x)\}$ are linearly dependent over k and let

$$\sum \beta_j f_{(j)}(x) = 0$$

be a non-trivial relation with $\{\beta_j\} \subseteq k$. Then $g(\tau) = \sum \beta_j \tau^j \in k\{\tau\}$ and $f(x)$ divides $g(x)$. □

As stated above, both proofs give the *same* polynomial provided that $g(\tau)$ in Proof 2 is normalized to be monic. Moreover, Proof 1 shows the important fact that as $\mathbb{F}_r \subseteq k$ every finite Galois extension of k can be obtained as the splitting field of an $\mathbb{F}_r$-linear polynomial.

Let $h(\tau)$ be any other additive polynomial such that $f(x)$ divides $h(x)$. It is clear that the $\mathbb{F}_r$-vector space W of Proof 1 *must* be contained in the set of roots of $h(x)$. Thus, by Proposition 1.3.5, we see that there exists $g_1(\tau) \in k\{\tau\}$ such that
$$h(\tau) = g_1(\tau)g(\tau).$$
This result also follows from the division algorithm in $k\{\tau\}$; see Section 1.6.

1.5. The p-resultant

We begin by recalling some of the standard material on resultants. For more, the reader may consult [L1, 206–211].

Let k be a field and let $\{f(x), g(x)\} \subset k[x]$. We write
$$f(x) = v_0 x^n + \cdots + v_n,$$
and
$$g(x) = w_0 x^m + \cdots + w_m.$$
The *resultant* $R(f,g)$ of $f(x)$ and $g(x)$ is defined to be $\det(A(f,g))$ where $A(f,g)$ is the $n+m$ square matrix

$$m\begin{cases} \\ \\ \end{cases} \quad n\begin{cases} \\ \\ \end{cases} \begin{pmatrix} v_0 \cdots v_n & & & \\ & v_0 \cdots v_n & & \\ & & \ddots & \\ & & & v_0 \cdots v_n \\ w_0 \cdots w_m & & & \\ & w_0 \cdots w_m & & \\ & & \ddots & \\ & & & w_0 \cdots w_m \end{pmatrix}$$

1.5. The p-resultant

Now let $\{t_1, \ldots, t_n\}$ be the roots of $f(x)$ counted *with* multiplicity and let $\{u_1, \ldots, u_m\}$ be the roots of $g(x)$ counted with multiplicity. One then calculates:

Proposition 1.5.1. $R(f,g) = v_0^m w_0^n \prod_{i=1}^{n} \prod_{j=1}^{m} (t_i - u_j).$ □

Corollary 1.5.2. *Suppose $v_0 w_0 \neq 0$. Then $R(f,g) = 0$ if and only if $f(x)$ and $g(x)$ have a root in common.* □

We now want to apply this theory in a suitable fashion to $\mathbb{F}_r$-linear polynomials. Let k be a field with $\mathbb{F}_r \subseteq k$. Let $\{f(\tau), g(\tau)\} \subset k\{\tau\}$. For simplicity, we assume that $f(x)$ and $g(x)$ are both monic and separable.

Let $\{w_1, \ldots, w_n\}$ be a basis for the $\mathbb{F}_r$-linear vector space of the roots of $f(x)$ and let $\{\psi_1, \ldots, \psi_m\}$ be the same for $g(x)$. Our next definition comes from [O1].

Definition 1.5.3. We set

$$R(f(\tau), g(\tau)) := \frac{\Delta(w_1, \ldots, w_n, \psi_1, \ldots, \psi_m)}{\Delta(w_1, \ldots, w_n)\Delta(\psi_1, \ldots, \psi_m)}.$$

$R(f,g)$ is the *p-resultant* of $f(\tau)$ and $g(\tau)$.

Proposition 1.5.4.
1. $R(f(\tau), g(\tau))^{r-1} = R(f(x)/x, g(x)/x).$

2. $$R(f(\tau), g(\tau)) = \frac{\Delta(f(\psi_1), \ldots, f(\psi_m))}{\Delta(\psi_1, \ldots, \psi_m)}$$
$$= (-1)^{mn} \frac{\Delta(g(w_1), \ldots, g(w_n))}{\Delta(w_1, \ldots, w_n)}.$$

Before turning to the proof of 1.5.4, we remark that it is clear a priori that there should be a relationship between $R(f(\tau), g(\tau))$ and $R(f(x)/x, g(x)/x)$. Indeed,

$R(f(x)/x, g(x)/x) = 0 \Leftrightarrow f(x), g(x)$ have a non $-$ trivial root in common,
$\Leftrightarrow \{w_1, \ldots, w_n, \psi_1, \ldots, \psi_m\}$ is a linearly,
dependent set,
$\Leftrightarrow \Delta(w_1, \ldots, w_n, \psi_1, \ldots, \psi_m) = 0.$

Proof of 1.5.4. 1. From Corollary 1.3.7, we know that $R(f(\tau),g(\tau))$ equals

$$\prod_{i=1}^{n} \prod_{(k_{i-1},\ldots,k_1)\in \mathbb{F}_r^{i-1}} \prod_{(\beta_1,\ldots,\beta_m)\in \mathbb{F}_r^m-\{(0,\ldots,0)\}} w_i + k_{i-1}w_{i-1} + \cdots + k_1 w_1 + \beta_1 \psi_1 + \cdots + \beta_m \psi_m.$$

(Note that the apparent lack of symmetry between $\{w_i\}$ and $\{\psi_j\}$ in the above formula is a consequence of Corollary 1.3.7.) Now fix i. Note that

$$\prod_{\zeta \in \mathbb{F}_r^*} \prod_{(k_{i-1},\ldots,k_1)\in \mathbb{F}_r^{i-1}} \prod_{(\beta_1,\ldots,\beta_m)\in \mathbb{F}_r^m-\{(0,\ldots,0)\}} \zeta w_i + k_{i-1}w_{i-1} + \cdots + \beta_m \psi_m$$

$$= \prod_{\zeta \in \mathbb{F}_r^*} \zeta^{r^{i-1}(r^m-1)} \prod_{(k_{i-1},\ldots,k_1)\in \mathbb{F}_r^{i-1}} \prod_{(\beta_1,\ldots,\beta_m)\in \mathbb{F}_r^m-\{(0,\ldots,0)\}} (w_i + k_{i-1}w_{i-1} + \cdots + \beta_m \psi_m)^{r-1}.$$

Note also that $\prod_{\zeta \in \mathbb{F}_r^*} \zeta^{r^{i-1}(r^m-1)} = 1$. The result now follows upon comparison with Proposition 1.5.1.

Part 2 follows upon using $f(\tau)$ and $g(\tau)$ to simplify the determinants. □

Example 1.5.5. Let $r = 3$. Let $n = m = 1$. Put $a := w_1$, $b := \psi_1$. Thus $f(x) = x^3 - a^2 x$ and $g(x) = x^3 - b^2 x$. One then finds easily that

$$R(f(x)/x, g(x)/x) = (a^2 - b^2)^2$$

and,

$$R(f(\tau), g(\tau)) = b^2 - a^2.$$

1.6. The Left and Right Division Algorithms

We continue to let k be a field containing $\mathbb{F}_r$. Let $\{f(\tau), g(\tau)\} \subset k\{\tau\}$. We begin by remarking that $f(\tau) \cdot g(\tau) = 0$ in $k\{\tau\}$ implies that $f(\tau)$ or $g(\tau)$ is 0. In particular, multiplication in $k\{\tau\}$ has *both* left and right cancelation properties; i.e., if $f(\tau)g(\tau) = f(\tau)h(\tau)$, then $g(\tau) = h(\tau)$, etc.

Definition 1.6.1. 1. We say that $f(\tau)$ is *right divisible* by $g(\tau)$ if there exists $h(\tau) \in k\{\tau\}$ such that
$$f(\tau) = h(\tau) \cdot g(\tau).$$

2. We say that $f(\tau)$ is *left divisible* by $g(\tau)$ if there exists $m(\tau) \in k\{\tau\}$ such that
$$f(\tau) = g(\tau) \cdot m(\tau).$$

1.6. The Left and Right Division Algorithms

The reader will easily see that if $f(\tau)$ is right divisible by $g(\tau)$, then $g(x) \mid f(x)$.

Proposition 1.6.2 (Right Division Algorithm). *Let $\{f(\tau), g(\tau)\} \subset k\{\tau\}$ with $g(\tau) \neq 0$. Then there exists $\{h(\tau), r(\tau)\} \subset k\{\tau\}$, with $\deg r(\tau) < \deg g(\tau)$, such that*
$$f(\tau) = h(\tau)g(\tau) + r(\tau).$$
Moreover, $h(\tau)$ and $r(\tau)$ are uniquely determined.

Proof. This follows in the same fashion as the classical division algorithm. □

Corollary 1.6.3. *Every left ideal of $k\{\tau\}$ is principal.*

Proof. Clear. □

The reader will note that $r(x)$ is the remainder upon applying the usual division algorithm to $\{f(x), g(x)\}$.

Definition 1.6.4. We say that k is *perfect* if and only if $\tau(k) = k$.

It is easy to see that every finite field is perfect (the mapping τ has trivial kernel and so a simple counting argument shows that it must be surjective). Moreover, every algebraically closed field is obviously perfect. Perfect fields have the property that *every* finite extension of them is separable. Let $\bar{k}$ be an algebraic closure of k. The *perfection* of $k \subseteq \bar{k}$, k^{perf}, is the subfield of all $x \in \bar{k}$ such that there exists $j \geq 0$ with $x^{p^j} \in k$. Clearly k^{perf}/k is totally inseparable and contains every totally inseparable extension of $k \subseteq \bar{k}$.

The following results are now clear.

Proposition 1.6.5 (Left Division Algorithm). *Let k be perfect and let $\{f(\tau), g(\tau)\}$ be as in 1.6.2. Then there exists $\{h(\tau), r(\tau)\} \subset k\{\tau\}$ with*
$$\deg r(\tau) < \deg g(\tau)$$
such that
$$f(\tau) = g(\tau) \cdot h(\tau) + r(\tau).$$
Furthermore, $r(\tau)$ and $h(\tau)$ are unique. □

Corollary 1.6.6. *If k is perfect, then every right ideal of $k\{\tau\}$ is principal.* □

Example 1.6.7. Let $k = \mathbb{F}_r(T)$. Put $f(\tau) = \tau^2 - \tau$ and $g(\tau) = \tau - T\tau^0$. Then
$$\tau^2 - \tau = (\tau + (T^r - 1)\tau^0)(\tau - T\tau^0) + T(T^r - 1)\tau^0,$$

and, in $\bar{k}\{\tau\}$,

$$\tau^2 - \tau = (\tau - T\tau^0)(\tau + (T-1)^{1/r}\tau^0) + T(T-1)^{1/r}\tau^0.$$

By using the *Euclidean Algorithm*, one can now compute the *right greatest common divisor* of $f(\tau)$, $g(\tau)$ defined as the monic generator of the *left* ideal generated by $f(\tau)$ and $g(\tau)$. We will denote this by "$(f(\tau),g(\tau))$" or "g.c.d.$(f(\tau),g(\tau))$." (See Example 1.10.3.)

Lemma 1.6.8. *Let k be a field of characteristic p and let $h(\tau) = (f(\tau), g(\tau))$. Then $h(x)$ is the greatest common divisor of $f(x)$ and $g(x)$.*

Proof. This follows from the fact that the remainders in the (right) division algorithms for $k\{\tau\}$ and $k\{x\}$ are the same. $\square$

Definition 1.6.9. We say that $f(\tau)$ and $g(\tau)$ are *right relatively prime* if and only if

$$(f(\tau), g(\tau)) = \tau^0.$$

Now let $k = \bar{k}$ and $f(\tau) = \sum_{j=0}^{n} a_j \tau^j$.

Lemma 1.6.10. *Let $0 \neq \alpha \in k = \bar{k}$.*
1. *$f(\tau)$ is right divisible by $\tau - \alpha\tau^0$ if and only if α is the $(r-1)$-st power of a root of $f(x) = 0$.*
2. *$f(\tau)$ is left divisible by $\tau - \alpha\tau^0$ if and only if α is a root of the equation*

$$a_0 + a_1^{1/r} y + a_2^{1/r^2} y^{(r+1)/r} + \cdots + a_{n-1}^{1/r^{n-1}} y^{(r^{n-1}-1)/((r-1)r^{n-2})}$$
$$+ a_n^{1/r^n} y^{(r^n-1)/((r-1)r^{n-1})} = 0.$$

Proof. These are exercises in long division. $\square$

Proposition 1.6.11. *Every $\mathbb{F}_r$-linear map from $\mathbb{F}_{r^n}$ to itself arises from some $P(\tau) \in \mathbb{F}_{r^n}\{\tau\}$.*

Proof. The dimension of the $\mathbb{F}_r$-vector space of $\mathbb{F}_r$-linear maps: $\mathbb{F}_{r^n} \to \mathbb{F}_{r^n}$ is n^2. On the other hand two $\mathbb{F}_r$-linear polynomials $f(\tau)$ and $g(\tau)$ give rise to the same linear map on $\mathbb{F}_{r^n}$ if and only if $f(\tau) - g(\tau)$ is right divisible by $\tau^n - \tau^0$. The maps $\{\tau^0, \ldots, \tau^{n-1}\}$ are, therefore, linearly independent over $\mathbb{F}_{r^n}$ as functions on $\mathbb{F}_{r^n}$. Their span over $\mathbb{F}_{r^n}$ is thus of dimension n^2 over $\mathbb{F}_r$. $\square$

1.7. The τ-adjoint of an Additive Polynomial

Let k be a field of characteristic p with $\mathbb{F}_r \subseteq k$. Let $f(\tau) \in k\{\tau\}$ with $\deg f(\tau) = n$. Let $W \subset \bar{k}$ be the $\mathbb{F}_r$-vector space of roots of $f(x) = 0$. We present here the "τ-adjoint" of $f(\tau)$. Our exposition is based on that of [O1].

Recall that $\tau\colon k^{\text{perf}} \to k^{\text{perf}}$ is an isomorphism. Thus we can also discuss "polynomials" $\sum \beta_i \tau^{-i}$ as operators on k^{perf}. We will denote this ring by $k^{\text{perf}}\{\tau^{-1}\}$.

Definition 1.7.1 (Ore). Let $f(\tau) = \sum_{i=0}^{n} \alpha_i \tau^i$, $\alpha_n \neq 0$.

1. We set $f^*(\tau) := \sum_{i=0}^{n} \alpha_i^{1/r^i} \tau^{-i} \in k^{\text{perf}}\{\tau^{-1}\}$; (this is the "$\tau^{-1}$-form" of the adjoint)
2. We set $f^{\text{ad}}(\tau) := \tau^n f^*(\tau) = \sum_{i=0}^{n} \alpha_i^{r^{n-i}} \tau^{n-i}$; (this is the "$\tau$-form" of the adjoint).

The reader will easily see that $f^{\text{ad}}(\tau) \in k\{\tau\}$ and $f^*(x) = \sum (\alpha_i x)^{1/r^i}$.

Example 1.7.2. Let $k = \mathbb{F}_r(T)$ and $f(\tau) = T^2 \tau^0 + (T^r + T)\tau + \tau^2$. Then

$$f^*(\tau) = T^2 \tau^0 + (T + T^{1/r})\tau^{-1} + \tau^{-2},$$

and

$$f^{\text{ad}}(\tau) = T^{2r^2} \tau^2 + (T^{r^2} + T^r)\tau + \tau^0.$$

The following lemma of Ore's is basic to the study of adjoints.

Lemma 1.7.3. *Let $f(\tau)$ and $g(\tau) \in k\{\tau\}$. Then $(f(\tau)g(\tau))^* = g^*(\tau)f^*(\tau)$.*

Proof. Let $\{a, b\} \subseteq k$. Note that

$$\begin{aligned}(a\tau^m \cdot b\tau^j)^* &= (ab^{r^m} \tau^{m+j})^* \\ &= a^{r^{-m-j}} b^{r^{-j}} \tau^{-m-j} \\ &= (b^{r^{-j}} \tau^{-j}) \cdot (a^{r^{-m}} \tau^{-m}).\end{aligned}$$

The proof now follows directly. $\square$

Thus, as one would expect, the formation of τ-adjoints (in τ^{-1}-form) anti-commutes. This leads to an anti-isomorphism of $\bar{k}\{\tau\}$ with $\bar{k}\{\tau^{-1}\}$. Note also that the roots of $f^*(x) = 0$ are precisely those of $f^{\text{ad}}(x) = 0$.

Lemma 1.7.4. *Let $h(\tau)$ and $g(\tau) \in \bar{k}\{\tau\}$. Set $h(\tau) = \sum_{i=0}^{t} \alpha_i \tau^i$, $\alpha_t \neq 0$, and $g(\tau) = \sum_{i=0}^{m} \beta_i \tau^i$, $\beta_m \neq 0$. Put $h_1(\tau) = \sum \alpha_i^{r^{-t-m}} \tau^{i-t}$. Then*

$$\tau^{-t-m} h(\tau) g(\tau) = h_1(\tau)(\tau^{-m} g(\tau)).$$

Proof. $\tau^{-t-m} h(\tau) = (h_1(\tau) \tau^{-m})$. □

Lemma 1.7.5. *$\beta \in \bar{k}$ is a nonzero root of $f^*(x) = 0$ if and only if $\beta^{(1-r)/r} \tau^0 - \tau^{-1}$ right divides $f^*(\tau)$.*

Proof. The if part is obvious. To see the only if part, note that β is a root of $f^{\text{ad}}(x) = 0$. Thus, by supplementing β to a basis for the roots of $f^{\text{ad}}(x) = 0$, and using 1.3.5, we see that $\tau - \beta^{r-1} \tau^0$ right divides $f^{\text{ad}}(\tau)$. By Lemma 1.7.4, we see that $\tau^0 - \beta^{(r-1)/r} \tau^{-1}$ right divides $f^*(\tau)$. But

$$\tau^0 - \beta^{\frac{r-1}{r}} \tau^{-1} = \beta^{\frac{r-1}{r}} \tau^0 (\beta^{\frac{1-r}{r}} \tau^0 - \tau^{-1}).$$ □

Lemma 1.7.6. *$\alpha \tau^0 - \tau$ left divides $f(\tau)$ if and only if $\alpha \tau^0 - \tau^{-1}$ right divides $f^*(\tau)$.*

Proof. This is obvious from 1.7.3. □

Corollary 1.7.7. *Let $0 \neq \alpha$. Then $\alpha \tau^0 - \tau$ left divides $f(\tau)$ if and only if $\alpha^{r/(1-r)}$ is a root of $f^*(x) = 0$.*

Proof. This follows from 1.7.5 and 1.7.6. □

Lemma 1.7.8. *Let $f(\tau) = \sum_{i=0}^{t} \alpha_i \tau^i$, $\alpha_0 \alpha_t \neq 0$. Then the double adjoint of $f(\tau)$, $(f^{\text{ad}}(\tau))^{\text{ad}}$, equals $\sum_{i=0}^{t} \alpha_i^{r^t} \tau^i$.*

Proof. We have $f^*(\tau) = \sum_{i=0}^{t} \alpha_i^{1/r^i} \tau^{-i} \Rightarrow$

$$f^{\text{ad}}(\tau) = \sum_{i=0}^{t} \alpha_i^{r^{t-i}} \tau^{t-i} \Rightarrow (f^{\text{ad}}(\tau))^* = \sum_{i=0}^{t} \alpha_i \tau^{i-t}$$

$$\Rightarrow (f^{\text{ad}}(\tau))^{\text{ad}} = \sum_{i=0}^{t} \alpha_i^{r^t} \tau^i.$$ □

1.7. The τ-adjoint of an Additive Polynomial

Corollary 1.7.9. *The roots of $(f^{\mathrm{ad}}(x))^{\mathrm{ad}} = 0$ are the r^t-th powers of the roots of $f(x) = 0$.* □

We now assume that $f(x)$ is separable. We will show that the roots of $f(x)$ and $f^*(x)$ generate the *same* algebraic extension of k.

Lemma 1.7.10. *Let $f(\tau) = cf_1(\tau)$, where $c \in k^*$ and $f_1(\tau)$ is monic. Then the roots of $f^*(x) = 0$ and $f_1^*(x) = 0$ generate the same extension of k.*

Proof. $f^*(\tau) = f_1^*(\tau) c \tau^0$. □

Theorem 1.7.11. *Let $f(\tau) = \sum_{i=0}^{t} \alpha_i \tau^i$, $\alpha_t \neq 0$. Assume that $\alpha_0 \neq 0$; so $f(x)$ is separable. Then the roots of $f(x)$ and $f^*(x)$ generate the same extension of k.*

Proof. Without loss of generality, we may assume that $f(\tau)$ is monic. Suppose that $0 \neq \beta$ is a root of $f^*(x) = 0$. By 1.7.7, we see that $\beta^{\frac{1-r}{r}} \tau^0 - \tau$ left divides $f(\tau)$; say
$$f(\tau) = (\beta^{\frac{1-r}{r}} \tau^0 - \tau) Q(\tau).$$
Let W be the roots of $f(x) = 0$ and $W_1 \subseteq W$ the subspace of roots of $Q(x) = 0$. Thus $Q(x) = -P_{W_1}(x) = -\prod_{w \in W_1}(x + w)$. By 1.3.5.1 we find that, as $f(\tau)$ is monic,
$$f(\tau) = (\tau - P_{W_1}(w)^{r-1} \tau^0) P_{W_1}(\tau)$$
for *any* $w \in W$ not in W_1. One easily concludes that
$$\beta^{\frac{1-r}{r}} = P_{W_1}(w)^{r-1},$$
or
$$\beta = c P_{W_1}(w)^{-r}$$
for some $c \in \mathbb{F}_r^*$.

Let L be the splitting field of $f(x)$; by assumption, L is separable over k. From above, we see that the splitting field of $f^{\mathrm{ad}}(x)$ is a subfield L_1 of L (as, obviously, $P_{W_1}(w) \in L$). Similarly, the splitting field of $(f^{\mathrm{ad}}(x))^{\mathrm{ad}}$ is a subfield $L_2 \subseteq L_1$. But, by 1.7.9, the roots of $(f^{\mathrm{ad}}(x))^{\mathrm{ad}} = 0$ are r^{th} powers of the roots of $f(x)$. This implies that L/L_2 is purely inseparable. But L/k is separable, so $[L : L_2] = 1$ or $L = L_1 = L_2$. □

We will give another proof of this result in Subsection 4.14 via a canonical Galois pairing between the roots of $f(x)$ and $f^*(x)$.

We can use the proof of the above result to express the roots of the adjoint in terms of the Moore Determinant.

Definition 1.7.12. Let $\{w_1, \ldots, w_t\}$ be a basis over $\mathbb{F}_r$ for $W :=$ roots of $f(x)$. Put $\overline{W} = \{\overline{w}_1, \ldots, \overline{w}_t\}$ where $\overline{w}_1 = \Delta(w_2, \ldots, w_t)/\Delta$, $\overline{w}_2 = \Delta(w_1, w_3, \ldots, w_t)/\Delta, \ldots, \overline{w}_t = \Delta(w_1, \ldots, w_{t-1})/\Delta$, $\Delta = \Delta(w_1, \ldots, w_t) \neq 0$.

Theorem 1.7.13 (Ore). *Let $f(\tau)$ be as in 1.7.11. Then $\{\overline{w}_1^r, \ldots, \overline{w}_t^r\}$ is an $\mathbb{F}_r$-basis for the roots of $f^*(x) = 0$.*

Proof. Let β be a non-zero root of $f^*(x) = 0$. As in the proof of 1.7.11, we see that $\beta = cP_{W_1}(w)^{-r}$, $c \in \mathbb{F}_r^*$, where $W_1 \subset W$ is an $(t-1)$-dimensional subspace and $w \in W \setminus W_1$. Let $\{v_1, \ldots, v_{t-1}\}$ be a basis for W_1. By 1.3.5.2, we conclude that
$$\beta = c\left(\frac{\Delta(v_1, \ldots, v_{t-1})}{\Delta(v_1, \ldots, v_{t-1}, w)}\right)^r.$$
Basic properties of determinants imply that
$$\left(\frac{\Delta(v_1, \ldots, v_{t-1})}{\Delta(v_1, \ldots, v_{t-1}, w)}\right)^r$$
may be expressed as linear combinations of $\{\overline{w}_1^r, \ldots, \overline{w}_t^r\}$. We therefore conclude that $\{\overline{w}_1^r, \ldots, \overline{w}_t^r\}$ span the roots of $f^*(x) = 0$. But this is a t-dimensional space. Thus $\{\overline{w}_1^r, \ldots, \overline{w}_t^r\}$ is an $\mathbb{F}_r$-basis. □

Corollary 1.7.14. *Let $\{w_1, \ldots, w_t\}$ be linearly independent over $\mathbb{F}_r$. Then so are*
$$\{\Delta(w_2, \ldots, w_t), \Delta(w_1, w_3, \ldots, w_t), \ldots, \Delta(w_1, \ldots, w_{t-1})\}. \quad \Box$$

1.8. Dividing $\mathbb{A}^1$ by Finite Additive Groups.

Let k, $\overline{k}$, etc., be as in the last section. Let $W \subseteq k$ be a finite $\mathbb{F}_r$-linear sub-vector space. Put $d = \dim_{\mathbb{F}_r}(W)$ and
$$f(\tau) := P_W(\tau);$$
so
$$f(x) = P_W(x) = \prod_{w \in W}(x - w) \in k[x].$$

We now regard W as a group of of operators on the affine line $\mathbb{A}^1$ over k. The action is simply
$$(w, x) \mapsto x + w.$$
Since $\mathbb{A}^1$ is affine and W is finite, one can form the quotient by forming rings of invariants (see [Mu1, §II.7] or [Sh1]). We denote the quotient by "$\mathbb{A}^1/k$."

1.8. Dividing $\mathbb{A}^1$ by Finite Additive Groups.

Proposition 1.8.1. *The W-invariant map, $x \mapsto f(x)$, gives an isomorphism $\widehat{\mathbb{A}^1/k} \xrightarrow{f} \mathbb{A}^1/k$.*

Proof. Let $k \subseteq k' \subseteq \overline{k}$. The proposition boils down to showing that, for any such k', and *any* W-invariant $g(x) \in k'[x]$, there exists $g'(x) \in k'[x]$ with

$$g(x) = g'(f(x)).$$

We do this by induction on the degree of $g(x)$.

The case where $\deg g(x) \leq r^d$ is clear.

Let $E = \{e_1, \ldots, e_t\}$ be the roots of $g(x)$ in $\overline{k}$ taken *without* multiplicity. Let $h_1(x) \in k'[x]$ be the minimal polynomial of e_1 over k' and let $E_1 \subseteq E$ be its set of roots. Let $E_2 \subseteq E$ be the union of orbits, under translation by W, of the elements of E_1. Note that all elements of E_2 have the same inseparability degree, say p^u, over k' as well as the same multiplicity, say m, in $g(x)$. Thus $\beta(x) = \prod_{e \in E_2}(x+e)^{p^u} \in k'[x]$ and $\beta(x)$ divides $g(x)$. Furthermore,

$$\prod_{e \in E_2}(x+e)^{p^u} = \prod_{e' \in E_3}(f(x)+f(e'))^{p^u}$$

where $E_3 \subseteq E_2$ is a set of representatives of W-orbits. But elementary field theory implies that

$$\prod_{e' \in E_3}(z+f(e'))^{p^u} \in k'[z];$$

the result follows by using induction on the degree of the quotient $g(x)/\beta(x) \in k'[x]$. □

Another proof of 1.8.1, more in line with what we shall do later on, may be given as follows: Let R be the fixed ring of $k'[x]$ under W. By standard theory, [Mu1, §7], one knows that R is finitely generated and normal; thus R is a Dedekind domain. Moreover, the map R to $k'[x]$, makes $k'[x]$ a finite R-module.

Now $f(x)$ gives rise to a W-invariant finite étale mapping

$$f: \mathbb{A}^1/k' \to \mathbb{A}^1/k'.$$

This mapping factors as

$$\begin{array}{ccc} \mathbb{A}^1/k' & \longrightarrow & \mathrm{Spec}\,(R) \\ & \searrow f & \downarrow \widehat{f} \\ & & \mathbb{A}^1/k' \end{array}.$$

The mapping $\widehat{f}$ must be finite, étale. It must also clearly have the property that every geometric fibre contains one reduced point. Therefore, $\widehat{f}$ is an isomorphism.

We finish this subsection by discussing a small variant of the above construction. Let W now be a finite dimensional $\mathbb{F}_r$-subspace of the separable closure, k^{sep}, of $k \subseteq \bar{k}$. We suppose that W is $\mathrm{Gal}(k^{\mathrm{sep}}/k)$-stable. Put

$$P_W(z) = \prod_{w \in W} (z - w),$$

as above. Clearly $P_W(z) \in k[z]$ is an $\mathbb{F}_r$-linear map and gives an isomorphism

$$W \backslash \mathbb{A}^1/k' \tilde{\to} \mathbb{A}^1/k',$$

where $k' = k(W)$. Moreover $k[P_W(z)]$ is *precisely* the ring of W-invariant elements of $k[z]$; the proof proceeds as above.

Suppose now that $W_0 \subseteq W$ is an $\mathbb{F}_r$-linear subspace which is, moreover, also Galois stable. Let $P_{W_0}(z)$ be defined in the obvious fashion.

Proposition 1.8.2. *$P_W(\tau)$ is right divisible by $P_{W_0}(\tau)$ in $k\{\tau\}$.*

Proof. The mapping

$$\mathbb{A}^1 \xrightarrow{P_W(z)} \mathbb{A}^1$$

factors as

$$\mathbb{A}^1 \xrightarrow{P_W(z)} W \backslash \mathbb{A}^1 = \mathbb{A}^1$$
$$\searrow P_{W_0}(z) \qquad \nearrow \widehat{P}(z)$$
$$W_0 \backslash \mathbb{A}^1 = \mathbb{A}^1$$

where $\widehat{P}(z)$ is $\mathbb{F}_r$-linear. The set of roots, W_1, of $\widehat{P}(z)$ is the image of W under $P_{W_0}(z)$. One sees that W_1 is also Galois stable and the result is established. $\square$

Proposition 1.8.2 also follows from Proposition 1.3.5.

1.9. Analogs in Differential Equations/Algebra

By now, the reader may have noticed a certain similarity between the results of the previous subsections and those coming from differential equations and algebra. In fact, this analogy is a source of inspiration for a number of results and is a basic theme of this volume.

Our purpose here is not to give any sort of complete review of differential equations/algebra. Rather, we recall enough material so that the reader may grasp the analogy.

Let k be any field of characteristic p with $\mathbb{F}_r \subseteq k$, and let $K = \mathbb{C}(t)$ where $\mathbb{C}$ is the complex numbers. On k we have the $\mathbb{F}_r$-linear operator

$$\tau: x \mapsto x^r\,;$$

while on K we have differentiation $D = \frac{d}{dt}$ which is obviously $\mathbb{C}$-linear.

1.9. Analogs in Differential Equations/Algebra

The ring of polynomials in τ, $k\{\tau\}$ is analogous to the ring of "polynomials" $K\{D\}$, i.e., expressions $\sum_{i=0}^{t} f_i(t)D^i$, $\{f_i(t)\} \subset K$. Indeed,

$$\tau \cdot a = a^r \tau,$$

for $a \in k$, as

$$Df = f' + fD,$$

for $f(t) \in \mathbb{C}(t)$ and $f'(t)$ is the derivative.

Moreover, elementary algebra implies that a "d^{th} order linear τ-equation", $f(z) = 0$, where $f(\tau) := \sum_{j=0}^{d} \alpha_j \tau^j$, $\{\alpha_j\} \subseteq k$, $\alpha_0 \alpha_d \neq 0$, has a basis of d linear independent solutions over the "τ-constants" $\mathbb{F}_r$. This is, of course, analogous to the well-known fact that an nth order differential equation

$$F(D)y = 0,$$

where $F(D) := \sum_{j=0}^{n} f_j(t)D^j$, $\{f_j\} \subset K$, $f_n \neq 0$ has n linearly independent solutions (on any connected domain where all $f_j(t)$ are regular and $f_n(t) \neq 0$) over the "D-constants" $\mathbb{C}$.

Let $\{w_1, \ldots, w_t\}$ be any t solutions to $f(z) = 0$. By 1.3.5, we know that $\{w_1, \ldots, w_t\}$ are linearly independent over $\mathbb{F}_r$ if and only if the Moore Determinant $\Delta(w_1, \ldots, w_t) \neq 0$. Similarly, let $\{y_1(t), \ldots, y_n(t)\}$ be any set of n solutions to $F(D)y = 0$. Then one knows that $\{y_1(t), \ldots y_n(t)\}$ is $\mathbb{C}$-linearly independent if and only if the *Wronskian*

$$W(t) := W(y_1, \ldots y_n)(t) := \det \begin{pmatrix} y_1(t) & \cdots & y_n(t) \\ y_1'(t) & \cdots & y_n'(t) \\ \vdots & & \vdots \\ y_1^{(n-1)}(t) & \cdots & y_n^{(n-1)}(t) \end{pmatrix}$$

is non-zero. Thus the Moore Determinant is analogous to the Wronskian as well as the Vandermonde Determinant.

This analogy can be pushed further to include the computation of *both* the Moore Determinant and the "computation" of the Wronskian. By the "computation" of the Wronskian, we mean, of course, Abel's formula

$$W'(t) + \frac{f_{n-1}(t)}{f_n(t)} W(t) = 0.$$

(So, among other things, the Wronskian is never zero or is *identically* zero.)

Let us review the derivation of Abel's result when $n = 2$. Thus we have an equation

$$f_2(t)y'' + f_1(t)y' + f_0(t)y = 0.$$

Assume that $f_2(t) \neq 0$ in the region of interest, and let us put $g_1(t) = f_1(t)/f_2(t)$ and $g_0(t) = f_0(t)/f_2(t)$. Thus

$$y'' + g_1(t)y' + g_0(t)y = 0.$$

Let $\{y_1, y_2\}$ be a $\mathbb{C}$-basis for the space of solutions; so

$$y_1'' + g_1(t)y_1' + g_0(t)y_1 = 0$$

and

$$y_2'' + g_1(t)y_2' + g_0(t)y_2 = 0.$$

Let us multiply the first equation by $-y_2$ and the second by y_1. Upon adding, we obtain

$$(y_1 y_2'' - y_2 y_1'') + g_1(t)(y_1 y_2' - y_1' y_2) = 0.$$

As $W'(t) = y_1 y_2'' - y_1'' y_2$, we obtain our objective

$$W'(t) + g_1(t) W(t) = 0.$$

Let us now carry this out in the case $n = 2$ in characteristic p. Our equation will be

$$\alpha_2 Z^{r^2} + \alpha_1 Z^r + \alpha_0 Z = 0; \qquad \alpha_0 \alpha_2 \neq 0.$$

Put $\beta_1 = \alpha_1/\alpha_2$ and $\beta_0 = \alpha_0/\alpha_2$; so

$$Z^{r^2} + \beta_1 Z^r + \beta_0 Z = 0. \qquad (*)$$

Let $\{w_1, w_2\}$ be an $\mathbb{F}_r$-basis for the solutions of $(*)$. So we have

$$w_1^{r^2} + \beta_1 w_1^r + \beta_0 w_1 = 0,$$

and

$$w_2^{r^2} + \beta_1 w_2^r + \beta_0 w_2 = 0.$$

Now multiply the top equation by $-w_2^r$, and the bottom by w_1^r. Upon adding, we find

$$(w_2^{r^2} w_1^r - w_1^{r^2} w_2^r) + \beta_0(w_2 w_1^r - w_1 w_2^r) = 0.$$

Thus $\Delta(w_1, w_2)^r - \beta_0 \Delta(w_1, w_2) = 0$, or $\Delta(w_1, w_2)$ satisfies the "first order" equation

$$Z^r - \beta_0 Z = 0;$$

clearly an analog of Abel's formula. In fact, this result generalizes and we have the next result.

Lemma 1.9.1. *Let* $f(\tau) = \sum\limits_{j=0}^{t} \alpha_j \tau^j$. *Let* $\{w_1, \ldots, w_t\}$ *be any t roots of* $f(z) = 0$. *Let* $\Delta = \Delta(w_1, \ldots, w_t)$. *Then we have*

$$\alpha_t \Delta^r + (-1)^{t+1} \alpha_0 \Delta = 0.$$

1.9. Analogs in Differential Equations/Algebra   23

Proof. If $\alpha_t = 0$, then $\{w_1, \ldots, w_t\}$ are linearly dependent, so $\Delta = 0$. Similarly if $\alpha_0 = 0$, we find $\Delta = 0$. Thus suppose $\alpha_0 \alpha_t \neq 0$ and put $\beta_0 = \alpha_0/\alpha_t$.

By definition,

$$\Delta = \det \begin{pmatrix} w_1 & \ldots & w_t \\ w_1^r & \ldots & w_t^r \\ \vdots & & \vdots \\ w_1^{r^{t-1}} & \ldots & w_t^{r^{t-1}} \end{pmatrix}.$$

Thus,

$$\Delta^r = \det \begin{pmatrix} w_1^r & \ldots & w_t^r \\ \vdots & & \vdots \\ w_1^{r^t} & \ldots & w_t^{r^t} \end{pmatrix}.$$

Now use $f(\tau)$ to express $w_i^{r^t}$ in terms of smaller powers for all i and substitute into the determinant. The result now follows from simple properties of determinants. □

Lemma 1.9.1 can also be deduced from the calculation (1.3.7) of the Moore Determinant. The above proof is modeled on the general classical proof of Abel's result and is due to M. Rosen. See also [He3, p. 129].

The reader should note that Abel's result uses the trace term of a polynomial whereas 1.9.1 uses the norm term.

Next we turn to the τ-adjoints of additive polynomials as discussed in Subsection 1.7. The main point here is the observation that our τ-adjoints are the analogs for additive polynomials of Lagrange's theory of adjoints for linear differential equations. For instance, if

$$F(D)y = f_0 y'' + f_1 y' + f_2 y,$$

then the adjoint is given by

$$F(D)^* y = (f_0 y)'' - (f_1 y)' + f_2 y.$$

This is clearly similar to $f^*(x)$ for $\deg f(\tau) = 2$; see the remark after Definition 1.7.1. For more, we refer the reader to [H1, p. 391] or [I1, §5.3].

There is also a remarkable and simple differential formalism associated to the r-th mapping τ as discussed in [He1]. The constants $\mathbb{C}$ of D are obviously obtained by solving the differential equation $Dy = 0$. However, the "constants" $\mathbb{F}_r$ of τ are obtained by solving $\tau y = y$. If one looks for an operator δ with $\mathbb{F}_r = \{y \mid \delta y = 0\}$, one is clearly led to the Artin-Schreier-like operator $\delta y := y^r - y = \tau(y) - y$. It turns out that this operator possesses many pleasing derivation-type formulae. For instance, one sees immediately that

$$\delta(xy) = x\delta y + \tau(y)\delta x,$$

(so δ is a "τ-derivation"). Moreover if one sets for $u \in k - \mathbb{F}_r$ and $y \in k$

$$\frac{\delta y}{\delta u} := \frac{\tau y - y}{\tau u - u},$$

then one has the "chain rule" for $x \in k - \mathbb{F}_r$

$$\frac{\delta y}{\delta x} = \frac{\delta y}{\delta u}\frac{\delta u}{\delta x}.$$

Moreover, it is easy to express the Moore determinant in terms of δ etc. Let $x \in k$ and let Ψ_x be the 2×2 matrix

$$\begin{pmatrix} x & \delta x \\ 0 & \tau x \end{pmatrix}.$$

Then the fact that δ is a τ-derivation is equivalent to having Ψ be a $\mathbb{F}_r$-algebra injection of k into the algebra $M_2(k)$ of 2×2 matrices. (The reader will easily deduce an analogous result for differentiation D.) For more, we refer the reader to [He1] where it is shown by a cohomological calculation that every τ-derivation is of the form $\lambda\delta$ where λ is a scalar.

We have mentioned that in Subsection 4.14 we will discuss a pairing that will give a different proof of Theorem 1.7.11. In fact, we will see that this pairing turns out to be analogous to a classical construction in the theory of differential equations.

Finally, we turn to the analogy with differential algebra. For classical ideas on differential algebra, we refer to [Kap1], [Beu1] and [Ma1]. For the "Tannakian" approach (i.e., based on tensor products), we refer to [K1], [K2] and [De1].

In differential algebra, one considers the Galois theoretic properties of solutions to differential equations. Thus, let K be a "differential field" with algebraically closed, subfield k of constants. This means that K is a field equipped with a derivation

$$a \mapsto a'$$

and $k = \{a \in K \mid a' = 0\}$. For example, let $K = \mathbb{C}(T)$, $k = \mathbb{C}$, equipped with

$$f(T) \mapsto \frac{d}{dT}f(T),$$

as usual. Let M be a differential over-field of K - so M has a derivation on it and K is stable with respect to this derivation. The *differential Galois group* of M/K is the group of all differential automorphisms (i.e., automorphisms of M/K which commute with the given derivation).

Now let

$$L(y) = y^{(n)} + a_1 y^{(n-1)} + \cdots + a_n y = 0 \tag{$*$}$$

be a differential equation with coefficients in K, and where "$y^{(i)}$" has the obvious meaning. A differential field M over K is a *Picard-Vessiot* extension [Kap1, p. 22] attached to the differential equation $(*)$ if and only if:

1.9. Analogs in Differential Equations/Algebra

1. M is generated over K by $\{u_1, \ldots, u_n\}$ such that $\{u_1, \ldots u_n\}$ satisfy (*) and are linearly independent over the constants of K.
2. M has the same field of constants as K.

One knows that, by our assumption that the base field is algebraically closed, one can always construct such extensions. The differential Galois groups of such extensions always turn out to be linear groups [Kap1, Theorem 5.9].

Example 1.9.2. As in [Kap1, p. 22], we exhibit an example where the differential Galois group is the full linear group. Let $M = \mathbb{C}\langle x_1, \ldots, x_n\rangle$ be the field generated by n differential indeterminates; that is, one first adjoins to $\mathbb{C}$ indeterminates $\{x_1, \ldots, x_n\}$, then one adjoins indeterminates $x'_1, \ldots, x'_n$ representing their derivatives, and so on. Finally, M is obtained as the quotient field. Let

$$\alpha = (\alpha_{ij}) \in GL_n(\mathbb{C}).$$

One sets
$$\alpha x_i^{(m)} = \sum \alpha_{ij} x_j^{(m)}, \quad \text{for } m = 0, \ldots.$$

In this fashion one obtains differential automorphisms of M.

Now let K be the fixed field of M under $GL_n(\mathbb{C})$. Define

$$L(y) = W(y, x_1, \ldots, x_n)/W(x_1, \ldots, x_n).$$

One sees that $L(y)$ is a linear differential equation with coefficients in K and that $\{x_1, \ldots, x_n\}$ satisfy $L(y) = 0$. Thus M/K is a Picard-Vessiot extension with differential Galois group equal to $GL_n(\mathbb{C})$.

We now return to additive polynomials. Here the differential Galois group gets replaced by the usual Galois group; so we have gone full circle – Galois groups → differential Galois groups → Galois groups. Let $f(\tau) \in k\{\tau\}$ be a separable polynomial where $\mathbb{F}_r \subseteq k$, and let K be a splitting field for $f(x)$. Let $\{w_1, \ldots, w_n\}$ be an $\mathbb{F}_r$-basis for the roots of $f(x)$. The fact that differential Galois groups are linear groups, now gets replaced by the obvious faithful representation of $\text{Gal}(K/k)$ into $GL_n(\mathbb{F}_r)$ arising from the action on $\{w_1, \ldots, w_n\}$.

Example 1.9.3. Following 1.9.3, we show here how to obtain the full group $GL_n(\mathbb{F}_r)$ as a Galois group by replacing the Wronskian with the Moore Determinant. Let $\{x_1, \ldots x_n\}$ be indeterminates and put $M = k(x_1, \ldots, x_n)$. Let $\alpha = (\alpha_{ij}) \in GL_n(\mathbb{F}_r)$ and put

$$\alpha x_i = \sum \alpha_{ij} x_j;$$

clearly α gives rise to an automorphism of M. We let K be the fixed field of $GL_n(\mathbb{F}_r)$. One now sees easily that M is the splitting field for the K-polynomial

$$\Delta(y, x_1, \ldots, x_n)/\Delta(x_1, \ldots x_n).$$

1.10. Divisibility Theory

We build here on Subsection 1.6, and discuss more properties of additive polynomials. As before, let k be a field of finite characteristic with $\mathbb{F}_r \subseteq k$, $\overline{k}$ a fixed algebraic closure, etc.

Definition 1.10.1. Let $f(\tau), g(\tau)$ be elements of $k\{\tau\}$. We define the (right) *least common multiple* of $f(\tau), g(\tau)$ to be the monic polynomial of smallest degree in $k\{\tau\}$ which is right divisible by both $f(\tau)$ and $g(\tau)$. We denote the least common multiple by l.c.m. $(f(\tau), g(\tau))$ or $\lceil f(\tau), g(\tau) \rceil$.

Lemma 1.10.2. *The least common multiple of $f(\tau)$ and $g(\tau)$ exists.*

Proof. The least common multiple is the monic generator of the left ideal which is the intersection of the left ideals generated by $f(\tau)$ and $g(\tau)$. This exists by Corollary 1.6.3. □

Example 1.10.3. Let us show how one can, in practice, find $\lceil f(\tau), g(\tau) \rceil$. Put $h(\tau) = \lceil f(\tau), g(\tau) \rceil$. For any $\beta(\tau) \in \overline{k}\{\tau\}$, let W_β be the $\mathbb{F}_r$-vector space of its roots in $\overline{k}$. Let us write

$$f(\tau) = f_0(\tau)\tau^a,$$

and

$$g(\tau) = g_0(\tau)\tau^b$$

for $a, b \in \mathbb{Z}_{\geq 0}$ and f_0, g_0 separable. Put

$$f_1(\tau) := \tau^{-a} f(\tau) = \tau^{-a} f_0(\tau) \tau^a,$$

and

$$g_1(\tau) := \tau^{-b} g(\tau) = \tau^{-b} g_0(\tau) \tau^b,$$

Thus, by perhaps passing to a purely inseparable extension k_1 of k, we see that f_1 and g_1 are separable. Let $c = \max\{a, b\}$; so

$$k_1^{r^c} \subseteq k.$$

Let W be the finite dimensional subspace of $\overline{k}$ generated by W_{f_1} and W_{g_1}. Let

$$h_1(x) = \prod_{w \in W} (x + w);$$

so $h_1(\tau) \in k_1\{\tau\}$. By Proposition 1.3.5 (for instance) one can find $a_1(\tau)$ and $b_1(\tau) \in k_1\{\tau\}$ with

$$h_1(\tau) = a_1(\tau) f_1(\tau)$$

and

$$h_1(\tau) = b_1(\tau) g_1(\tau).$$

1.10. Divisibility Theory   27

Therefore,
$$\tau^c h_1(\tau) = \tau^c a_1(\tau)\tau^{-c}\tau^{c-a}\tau^a f_1(\tau)$$
$$= \tau^c a_1(\tau)\tau^{-c}\tau^{c-a} f(\tau),$$

and
$$\tau^c h_1(\tau) = \tau^c b_1(\tau)\tau^{-c}\tau^{c-b} g(\tau).$$

Both $\tau^c a_1(\tau)\tau^{-c}$ and $\tau^c b_1(\tau)\tau^{-c}$ are in $k\{\tau\}$; thus $\tau^c h_1(\tau) \in k\{\tau\}$ and is right divisible by $f(\tau)$ and $g(\tau)$. It is now easy to see that $\tau^c h_1(\tau) = h(\tau)$.

In a similar fashion, the reader will see that
$$(f(\tau), g(\tau)) = \tau^t P_{W_1}(\tau),$$
where $t = \min\{a, b\}$ and $W_1 = W_{f_1} \cap W_{g_1}$.

Finally, one has
$$\deg\lceil f(\tau), g(\tau) \rceil = \deg f(\tau) + \deg g(\tau) - \deg(f(\tau), g(\tau)).$$

Example 1.10.4. Let $f(\tau) = \tau + \tau^0$, $g(\tau) = \tau$. Then $\lceil f(\tau), g(\tau)\rceil = \tau^2 + \tau$. Note that the least common multiple in $k[x]$ of $x^r + x$ and x^r is $x^{2r-1} + x^r \neq x^{r^2} + x^r$. So the notion of least common multiple differs in $k[x]$ and $k[\tau]$. However, let $f(\tau), g(\tau) \in k\{\tau\}$ and $h(\tau) = \lceil f(\tau), g(\tau)\rceil$, and let $h_0(x)$ be the least common multiple of $f(x)$ and $g(x)$ in $k[x]$. One then sees easily that $h_0(x) \mid h(x)$.

Lemma 1.10.5. *Let $f(\tau), g(\tau) \in k\{\tau\}$. Then there exists $d(\tau) \in k\{\tau\}$ with*
$$f(\tau)g(\tau) = d(\tau)f(\tau)$$
if and only if $\lceil f(\tau), g(\tau)\rceil$ right divides $f(\tau)g(\tau)$.

Proof. Let $h(\tau) = \lceil f(\tau), g(\tau)\rceil$. If $d(\tau)$ exists so that
$$f(\tau)g(\tau) = d(\tau)f(\tau),$$
then $f(\tau)g(\tau)$ is right divisible by *both* $g(\tau)$ and $f(\tau)$. Therefore, it must be divisible by $h(\tau)$.

Conversely, suppose that $f(\tau)g(\tau) = a(\tau)h(\tau)$. Then, as there exists $b(\tau) \in k\{\tau\}$ with $b(\tau)f(\tau) = h(\tau)$, so we see
$$f(\tau)g(\tau) = a(\tau)b(\tau)f(\tau).$$
Putting $d(\tau) = a(\tau)b(\tau)$ gives the result. □

We now suppose, for the moment, that $h(\tau) = \lceil f(\tau), g(\tau)\rceil$ actually equals $g(\tau)f(\tau)$. Thus,
$$g(\tau)f(\tau) = d(\tau)g(\tau),$$

for some $d(\tau) \in k\{\tau\}$ and $d(\tau) = g(\tau)f(\tau)g(\tau)^{-1}$. This provides *some* motivation for the next definition.

Definition 1.10.6. Let $f(\tau), g(\tau) \in k\{\tau\}$ with $\deg f(\tau) = d$ and $\deg g(\tau) = \delta$. Let α_d, β_δ be the coefficients of highest degree. Set
$$h(\tau) = \lceil f(\tau), g(\tau) \rceil;$$
so
$$h(\tau) = \xi(\tau)g(\tau),$$
for some $\xi(\tau) \in k\{\tau\}$. Set
$$(g*f)(\tau) = \alpha_d \beta_\delta^{r^c} \xi(\tau) = \alpha_d \beta_\delta^{r^c} h(\tau) g(\tau)^{-1},$$
where $c = \deg h(\tau) - \deg g(\tau)$. We call $(g*f)(\tau)$, the *transformation* of $f(\tau)$ by $g(\tau)$.

The reader will note that the constant involved in $(g*f)(\tau)$ is cooked up so that $(g*f)(\tau)$ and $f(\tau)$ have the same coefficient of highest degree.

Lemma 1.10.7. *Let $d(\tau)$ be the greatest common divisor of $f(\tau)$ and $g(\tau)$. Let $f(\tau) = f_0(\tau) d(\tau)$ and $g(\tau) = g_0(\tau) d(\tau)$. Then*
$$(g*f)(\tau) = (g_0 * f_0)(\tau).$$

Proof. We will establish, first of all, that $\lceil f(\tau), g(\tau) \rceil = \lceil f_0(\tau), g_0(\tau) \rceil d(\tau)$. From the degree formula given in Example 1.10.3, we need only show that $\lceil f_0(\tau), g_0(\tau) \rceil d(\tau)$ right divides $\lceil f(\tau), g(\tau) \rceil$. But suppose that $h(\tau) = \lceil f(\tau), g(\tau) \rceil$; then
$$h(\tau) = a(\tau) f(\tau) = b(\tau) g(\tau),$$
for some $a(\tau), b(\tau) \in k\{\tau\}$. Then
$$h(\tau) = a(\tau) f_0(\tau) d(\tau) = b(\tau) g_0(\tau) d(\tau);$$
thus,
$$h(\tau) d(\tau)^{-1} = c(\tau) \lceil f_0(\tau), g_0(\tau) \rceil,$$
for some $c(\tau) \in k\{\tau\}$. This then implies the above divisibility statement. Now
$$\lceil f(\tau), g(\tau) \rceil g(\tau)^{-1} = \lceil f_0(\tau), g_0(\tau) \rceil d(\tau) d(\tau)^{-1} g_0(\tau)^{-1}$$
$$= \lceil f_0(\tau), g_0(\tau) \rceil g_0(\tau)^{-1}.$$
The rest of the proof is now easy. □

Definition 1.10.8. Let $f(\tau)$ and $g(\tau)$ be relatively prime. We then say that $(g*f)(\tau)$ is a *special transformation* of $f(\tau)$ by $g(\tau)$.

1.10. Divisibility Theory

Thus Lemma 1.10.7 implies that $(g * f)(\tau)$ can always be obtained by a special transformation.

Definition 1.10.9. Let $f(\tau)$ and $g(\tau) \in k\{\tau\}$. If $f(\tau)$ and $g(\tau)$ are relatively prime, then we say that $(g * f)(\tau)$ is *similar* to $f(\tau)$.

Proposition 1.10.10. *Similarity is an equivalence relation.*

Proof. Reflexivity is obvious. We will show transitivity and leave the third condition – symmetry – to the reader. Suppose that

$$h_1(\tau) = (g_1 * f)(\tau), \qquad (f(\tau), g_1(\tau)) = \tau^0,$$

and

$$h_2(\tau) = (g_2 * h_1)(\tau), \qquad (g_2(\tau), h_1(\tau)) = \tau^0.$$

We need to show that

$$h_2(\tau) = (g_2 g_1 * f)(\tau),$$

and

$$\tau^0 = (g_2(\tau) g_1(\tau), f(\tau)).$$

As in the discussion before, 1.10.6, this *formally* looks like the obvious statement
"$(g_2(\tau) g_1(\tau)) f(\tau) (g_2(\tau) g_1(\tau))^{-1} = g_2(\tau)(g_1(\tau) f(\tau) g_1(\tau)^{-1}) g_2(\tau)^{-1}$,"
and may be established by the techniques of Proposition 1.3.5 and Example 1.10.3. □

For more properties of these transformations, we refer the reader to [O1].

Finally, we mention some results on "unique factorization" of elements in $k\{\tau\}$. For the proofs and more, we refer the reader to [O2], [O1].

Definition 1.10.11. Let $f(\tau) \in k\{\tau\}$ be monic. We say that $f(\tau)$ is *prime* or *irreducible* if it has no monic divisors in $k\{\tau\}$ except itself and τ^0.

Lemma 1.10.12. *Suppose that $f_0(\tau)$ and $f_1(\tau)$ are similar. Then $f_0(\tau)$ is prime if and only if $f_1(\tau)$ is.* □

Theorem 1.10.13. *Every monic polynomial has a decomposition into prime factors. Two different decompositions of the same polynomial will have the same number of factors. Through suitable orderings, the factors will be similar in pairs.* □

Definition 1.10.14. Let $f(\tau) \in k\{\tau\}$ be monic. We say that $f(\tau)$ is *completely reducible* if it is the least common multiple of prime polynomials.

Thus one finds primes $P_1(\tau), \ldots, P_t(\tau)$ such that
$$f(\tau) = \lceil P_1(\tau), \ldots, P_t(\tau) \rceil$$
and each $P_i(\tau)$ is relatively prime to the least common multiple of the others. Suppose now that $g(\tau) \in k\{\tau\}$ is arbitrary. We call the least common multiple of all primes which right divide $g(\tau)$, the *maximal completely reducible factor* of $g(\tau)$; let us denote it by $m^1(\tau)$. Thus
$$g(\tau) = g^1(\tau) \cdot m^1(\tau),$$
and the same argument applies to $g^1(\tau)$, etc. Therefore, we have shown the following.

Theorem 1.10.15. *Every polynomial has a unique representation as a product of maximal completely reducible factors.* □

Definition 1.10.16. Let $f(\tau) \in k\{\tau\}$. We say that $f(\tau)$ is *decomposable* if there exist $a(\tau), b(\tau) \in k\{\tau\}$ with
$$f(\tau) = \lceil a(\tau), b(\tau) \rceil.$$
Otherwise, $f(\tau)$ is *indecomposable*.

Theorem 1.10.17. *Let $f(\tau) \in k\{\tau\}$. Then $f(\tau)$ can be represented as the least common multiple*
$$f(\tau) = \lceil g_1(\tau), \ldots, g_t(\tau) \rceil$$
where $\{g_i(\tau)\} \subset k\{\tau\}$ and each $g_i(\tau)$ is indecomposable and relatively prime to the least common multiple of the others. Any two such representations will have the same number of components and these will be similar in pairs. □

Let k^{sep} be the separable closure of k.

Theorem 1.10.18 (Ore).
1. Let $f(\tau) \in k\{\tau\}$ be such that $f(x)$ is a separable polynomial. Let $V \subset \overline{k}$ be the $\mathbb{F}_r$-linear subspace of its roots and let θ be the corresponding representation of $\mathrm{Gal}(k^{\text{sep}}/k)$ on $GL(V)$. Then $f(\tau)$ is irreducible if and only if θ is irreducible.
2. Let $\{f(\tau), \theta\}$ be as in Part 1. Then $f(\tau)$ is indecomposable if and only if θ is indecomposable.

Proof.
1. Suppose θ is reducible. Let $V_0 \subseteq V$ be a non-trivial stable subspace. Let $P_{V_0}(x) = \prod_{v \in V_0}(x - v) \in k\{\tau\}$. Then $P_{V_0}(\tau)$ divides $f(\tau)$.
2. Follows in the same fashion. □

Theorem 1.10.18 is an analog of results of Loewy in differential equations. It illustrates, again, the closeness of additive polynomials and linear operators as discussed in Subsection 1.9.

1.11. The Semi-invariants of Additive Polynomials

Let k be a field with $\mathbb{F}_r \subseteq k$ and let $f(\tau) \in k\{\tau\}$.

Definition 1.11.1 (Ore). Let $g(\tau) \in k\{\tau\}$. Then $g(\tau)$ is said to be a *semi-invariant* of $f(\tau)$ if and only if $(g * f)(\tau)$ is a right divisor of $f(\tau)$.

Ore used the expression "invariant transformer" where we use "semi-invariant."

Proposition 1.11.2. *A necessary and sufficient condition that $g(\tau)$ be a semi-invariant of $f(\tau)$ is that*

$$f(\tau)g(\tau) = h(\tau)f(\tau),$$

for some $h(\tau) \in k\{\tau\}$.

Proof. Suppose that $f(\tau)g(\tau) = h(\tau)f(\tau)$ for some $h(\tau) \in k\{\tau\}$. Then, $f(\tau)g(\tau)$ is divisible by $f(\tau)$ and $g(\tau)$; thus

$$f(\tau)g(\tau) = a(\tau)\lceil f(\tau), g(\tau) \rceil$$

for some $a(\tau)$. We write this as

$$f(\tau)g(\tau) = a(\tau)\alpha(g * f)(\tau)g(\tau),$$

for some nonzero constant α. Since we can cancel right multiplication by $g(\tau)$, we conclude

$$f(\tau) = a(\tau)\alpha(g * f)(\tau),$$

which proves that $g(\tau)$ is a semi-invariant of $f(\tau)$.

Conversely, if $f(\tau) = a(\tau)(g * f)(\tau)$, then for some β

$$\begin{aligned} f(\tau)g(\tau) &= a(\tau)(g * f)(\tau)g(\tau) \\ &= a(\tau)\beta\lceil g(\tau), f(\tau)\rceil \\ &= a(\tau)\beta b(\tau)f(\tau), \end{aligned}$$

for some $b(\tau) \in k\{\tau\}$. Thus $f(\tau)g(\tau) = h(\tau)f(\tau)$ for $h(\tau) = a(\tau)\beta b(\tau)$. □

Corollary 1.11.3. *If $g(\tau)$ commutes with $f(\tau)$, then $g(\tau)$ is a semi-invariant of $f(\tau)$.*

Proof. This is obvious from Proposition 1.11.2. □

Corollary 1.11.4 (Ore). *The semi-invariants of $f(\tau)$ form a ring with unit element. This ring contains all left multiples of $f(\tau)$.*

Proof. This is a straightforward exercise using Proposition 1.11.2. □

Definition 1.11.5. We denote the ring of semi-invariants of $f(\tau)$ in $k\{\tau\}$ by $\mathrm{Sem}_k(f(\tau))$.

Let $\bar{k}$ be our chosen algebraic closure of k and let $W \subset \bar{k}$ be the $\mathbb{F}_r$-vector space of roots of $f(x)$. Let $g(\tau) \in \mathrm{Sem}_k(f(\tau))$. By 1.11.2, we see that $g(W) \subseteq W$. We thus obtain a representation

$$\rho \colon \mathrm{Sem}_k(f(\tau)) \to \mathrm{End}_{\mathbb{F}_r}(W);$$

we call ρ the *basic representation* of $\mathrm{Sem}_k(f(\tau))$.

Lemma 1.11.6. *Let $g_0(\tau) \in \mathrm{Sem}_k(f(\tau))$ and let $g_0(\tau) - g_1(\tau)$ be right divisible by $f(\tau)$. Then $g_1(\tau) \in \mathrm{Sem}_k(f(\tau))$ also.*

Proof. We have

$$f(\tau)g_1(\tau) = f(\tau)g_0(\tau) - f(\tau)(g_0(\tau) - g_1(\tau));$$

the right hand side can be written as $a(\tau)f(\tau)$ for some $a(\tau) \in k\{\tau\}$. □

Proposition 1.11.7. *Let $g_0(\tau)$ and $g_1(\tau)$ be as in 1.11.6. Suppose also that $f(\tau)$ is separable and that*

$$\deg f(\tau) > \deg g_0(\tau).$$

Then $g_0(\tau)$ is uniquely determined by the basic representation ρ.

Proof. Let $h_1(\tau)$ and $h_2(\tau)$ be any two elements of $k\{\tau\}$. If $h_1(w) = h_2(w)$ for all W, then one sees that $(h_1 - h_2)(\tau)$ is right divisible by $f(\tau)$. If $\deg h_i(\tau) < \deg f(\tau)$, $i = 1, 2$, then this implies that $h_2(\tau) - h_1(\tau) \equiv 0$. □

Proposition 1.11.8. *Let $f(\tau)$ be as in* Proposition 1.11.7. *Let W be its roots. Then $g(\tau) \in k\{\tau\}$ is a semi-invariant of $f(\tau)$ if and only if*

$$g(W) \subseteq W.$$

Proof. Let $g(\tau)$ be a semi-invariant. By 1.11.2, one sees that

$$f(\tau)g(\tau) = h(\tau)f(\tau)$$

for some $h(\tau) \in k\{\tau\}$. Let $w \in W$; one obtains

$$f(g(w)) = h(0) = 0,$$

so $g(w) \in W$.

1.11. The Semi-invariants of Additive Polynomials   33

Conversely, suppose $g(W) \subseteq W$. Then $f(\tau)g(\tau)$ is 0 on W. By 1.3.5, we see that $f(\tau)g(\tau)$ is right divisible by $f(\tau)$. Thus, by 1.11.2, we see that $g(\tau)$ is a semi-invariant of $f(\tau)$.   □

Let $f(\tau)$ continue to be chosen a separable element, and let $g(\tau) \in k\{\tau\}$ be a semi-invariant. We suppose that $g(\tau)$ is relatively prime to $f(\tau)$.

Lemma 1.11.9.
1. $(g * f)(\tau) = f(\tau)$.
2. Let $a(\tau)$ be chosen so that

$$a(\tau)g(\tau) + b(\tau)f(\tau) = \tau^0.$$

Then $a(\tau) \in \mathrm{Sem}_k(f)$ also.

Proof. Part 1 follows from a comparison of degrees. To see Part 2, note that $g(\tau)$ acts as an $\mathbb{F}_r$-linear automorphism of the roots W of $f(\tau)$. Thus $a(\tau)$ must also. By 1.11.8, we see it is a semi-invariant.   □

Lemma 1.11.10. *Let $\{W, f(\tau)\}$ be as above. Under the basic representation, those elements of $\mathrm{Sem}_k(f(\tau))$ which are relatively prime to $f(\tau)$ map to invertible elements in $\mathrm{End}_{\mathbb{F}_r}(W)$.*

Proof. This is immediate from 1.11.9.   □

Theorem 1.11.11 (Ore). *Let $\{W, f(\tau)\}$ be as above. Let $f(\tau)$ now be irreducible. Then the image in $\mathrm{End}_{\mathbb{F}_r}(W)$ of the semi-invariants of $f(\tau)$ is a finite field.*

Proof. Let $g(\tau)$ be a semi-invariant of $f(\tau)$ with non-zero image in $\mathrm{End}_{\mathbb{F}_r}(W)$. Then one sees that $g(\tau)$ must be prime to $f(\tau)$. By Lemma 1.11.10, we see that the image of $g(\tau)$ is invertible. Thus the image of the semi-invariants is a division ring which is also finite. The result now follows by Wedderburn's Theorem.   □

Finally we present another use of semi-invariants as suggested by N. Katz. Let us set $R := k\{\tau\}$ and let $f \in R$. Consider $M := R/Rf$ as a left R-module. Let $T: M \to M$ be an endomorphism of R-modules and let g be any element of R which projects to $T(1 + Rf) \in M$. One then sees that

$$T(r + Rf) = rT(1 + Rf) = r \cdot g + Rf;$$

thus T is given by right multiplication by g. However, not every $g \in R$ will work. One clearly needs $Rfg \subseteq Rf$; thus g must be a semi-invariant of f. In other words the elements of $\mathrm{End}_R(M)$ are represented by semi-invariants.

2. Review of Non-Archimedean Analysis

In this section we will give a rapid presentation of those properties of non-Archimedean analysis necessary for later sections. For more we refer the reader to [BGR1], [Bru1], [Kob1], [R1,144–159], [L1], and the first two chapters of [DGS1].

Let K be a field, of arbitrary characteristic, which is complete with respect to a real-valued *valuation* v. This means that there exists $v\colon K \to \mathbb{R} \cup \{\infty\}$ such that for all $x, y \in K$,

1. $v(x) = \infty \iff x = 0$,
2. $v(xy) = v(x) + v(y)$,
   and
3. $v(x+y) \geq \inf(v(x), v(y))$.

One then observes that v makes K a topological field: an element $x \in K$ is "small" if and only if $v(x)$ is large (in the usual sense on $\mathbb{R}$). Thus via v it makes sense to discuss Cauchy sequences in K. To say that K is "complete" means that, as with $\mathbb{R}$, every Cauchy sequence with elements in K converges to an element in K.

Let $\alpha \in \mathbb{R}$ with $0 < \alpha < 1$. One obtains an *absolute value*

$$|\ |_v \colon K \to \mathbb{R}_{\geq 0}$$

by setting $|x|_v = \alpha^{v(x)}$. It is easy to see that

1. $|x|_v = 0 \iff x = 0$,
2. $|xy|_v = |x|_v |y|_v$,
3. $|x+y|_v \leq \max\{|x|_v, |y|_v\}$.

Let

$$R := R_K := \{x \in K \mid |x|_v \leq 1\},$$

so R is the ring of "integers" of K, and

$$M := M_K := \{x \in K \mid |x|_v < 1\}.$$

It is a standard, and elementary, exercise to show that R is a local ring and $M \subset R$ is the maximal ideal.

There are two basic examples of this construction:

1. $K = \mathbb{Q}_p$, p a rational prime. Here one begins with the rational numbers $\mathbb{Q}$. If $x \in \mathbb{Q}^*$, then one decomposes x as

$$x = p^t \cdot x_0,$$

where $t \in \mathbb{Z}$ and neither the numerator nor denominator of x_0 involve p. One sets

$$v(x) := v_p(x) := t.$$

It is easy to see that v is a valuation. The field $\mathbb{Q}_p$ is the completion of $\mathbb{Q}$ with respect to this valuation, just as $\mathbb{R}$ is obtained from $\mathbb{Q}$. From the third property of valuations, and for any $\alpha \in \mathbb{R}$, $0 < \alpha < 1$, one has an equality of sets

$$|\mathbb{Q}_p|_v = |\mathbb{Q}|_v.$$

This is clearly not true about the classical absolute value on $\mathbb{Q}$ and $\mathbb{R}$.

2. $K = \mathbb{F}_r\left(\left(\frac{1}{T}\right)\right)$ = the field of formal Laurent series over $\mathbb{F}_r$ = finite field with r elements. The process of constructing K mirrors that of Part 1. One begins with $k = \mathbb{F}_r(T)$. One sets $v(0) = \infty$. If $x \in k^*$ is written

$$\left(\frac{1}{T}\right)^e x_0,$$

with the numerator and denominator of x_0 prime to T, then one sets

$$v(x) = e.$$

The reader may then easily check that K is the completion of k with respect to $v(x)$. (N.B., one normally works with the isomorphic (under $T \leftrightarrow 1/T$) field $\mathbb{F}_r((T))$. For our purposes, the formulation $\mathbb{F}_r\left(\left(\frac{1}{T}\right)\right)$ is best.) The valuation v extends continuously to K, etc.

Let K be as above and let L be a finite extension of K. One extends v uniquely to L as follows: Let $x \in L$ with $\alpha = N_K^L x \in K$. Then one sets

$$v(x) := \frac{1}{[L:K]} v(\alpha).$$

The reasoning behind this definition is illuminating and runs briefly as follows: Let $\overline{K}$ be a fixed algebraic closure of K containing L. *Suppose* that one had an extension of v to $\overline{K}$; we will also denote this extension by v. Let σ be any automorphism of $\overline{K}$ over K. The mapping $x \mapsto v(\sigma x)$ would also be a valuation extending v; standard results of valuation theory now imply that $v(\sigma x)$ must *equal* $v(x)$. Let p^t be the inseparability degree of $L \mid K$. Then

$$\alpha = N_K^L x = \prod_{\sigma \text{ distinct}} \sigma(x)^{p^t}.$$

The formula for $v(x)$ now follows easily. Conversely, one checks that this formula gives an extension of v to L, and that L is complete with respect to this extension. In this way, one extends v uniquely to all of $\overline{K}$. In terms of absolute values, one has $|x|_v = |\alpha|_v^{1/[L:K]}$.

A complete field K is called a *local field* if it is locally compact with respect to the topology induced by v. In the non-Archimedean case, this leads to a special choice of α (used above in the definition of $|\ |_v$) as follows: The ring of integers R of K is now *compact*; as $M \subset R$ is open, we deduce that the field R/M is *finite* with, say, q-elements. We then set $\alpha := 1/q$ and we say that the associated absolute value is *normalized*.

In practice, the types of fields K that are used are finite extensions of our basic examples, *or*, the completion of a fixed algebraic closure, $\overline{K}$, of K. (In the non-Archimedean case, while it is true that every finite extension is complete, it is *not* true that the algebraic closure is also complete and so one must pass to the completion.) It is also possible to work over $\overline{K}$ itself, provided our operations take Cauchy sequences converging in $\overline{K}$ to Cauchy sequences converging in $\overline{K}$. For instance, as we shall soon see, a power series with coefficients in K will have this property on those elements of $\overline{K}$ where it converges.

Proposition 2.1. *Let K be a complete field with valuation v. Let $\overline{K}$ be a fixed algebraic closure of K together with the canonical extension of v. Let $\widehat{\overline{K}}$ be its completion with respect to v. Then $\widehat{\overline{K}}$ remains algebraically closed.*

Proof. Let $P(x) = \sum_{j=0}^{t} \beta_j x^j \in \widehat{\overline{K}}[x]$ with $\beta_t = 1$. We need to show that $P(x)$ has a root in $\widehat{\overline{K}}$. Let L be a field containing $\widehat{\overline{K}}$ which contains a root α of $P(x)$, and equip L with the extension of v.

Pick $\delta > 0$ and let $P_1(x) = \sum_{j=0}^{t} \widehat{\beta}_j x^j \in \overline{K}[x]$ be chosen so that

$$\delta < \min_j \{v(\beta_j - \widehat{\beta}_j)\}, .$$

We have

$$P_1(\alpha) = P_1(\alpha) - P(\alpha) = \sum_{j=0}^{t} (\widehat{\beta}_j - \beta_j)\alpha^j .$$

Now let $\{\alpha_1, \ldots, \alpha_t\}$ be the roots of $P_1(x)$ in $\overline{K}$ so that

$$P_1(x) = \prod(x - \alpha_j) .$$

We conclude that

$$\prod(\alpha - \alpha_j) = P_1(\alpha) = \sum(\widehat{\beta}_j - \beta_j)\alpha^j .$$

Let $\xi_\alpha = \min_j \{v(\alpha^j)\} = \min_j \{jv(\alpha)\}$. Thus

$$\sum v(\alpha - \alpha_j) > \delta + \xi_\alpha.$$

In particular, for some j_0, we have

$$v(\alpha - \alpha_{j_0}) > \frac{\delta + \xi_\alpha}{t}.$$

We can thus construct a Cauchy sequence in $\overline{K}$ converging to α. This implies that $\alpha \in \widehat{\overline{K}}$. □

Let K be a complete field with valuation v. Let $\sum_{j=0}^{\infty} a_j$ be an infinite series with coefficients in K. One defines the convergence or divergence of this series by using partial sums *exactly* as in classical theory. However, *unlike* classical Archimedean theory, we have the following result.

Proposition 2.2. $\sum_{j=0}^{\infty} a_j$ *converges to an element of K if and only if* $\lim_{j \to \infty} a_j = 0$.

Proof. The only if statement follows as in classical theory. The if statement follows easily from the non-Archimedean property (Property 3) of valuation. □

Let $f(x) = \sum a_n x^n$ now be a power series with coefficients in K which we will assume complete and algebraically closed. Let $\alpha \in K$. In order for $f(x)$ to converge at α, we must have

$$\lim_{j \to \infty} a_j \alpha^j = 0,$$

or

$$\lim_{j \to \infty} v(a_j) + j v(\alpha) = \infty.$$

(The reader may easily work out a more classical looking statement using the absolute value $|\ |_v$ associated to v, should that be desired.)

Definition 2.3. We set

$$\rho(f) = -\lim_{j \to \infty} v(a_j)/j.$$

We call $\rho(f)$ the *order of convergence* of $f(x)$.

The next result is now easily seen.

Proposition 2.4. *Let $\alpha \in K$. Then $f(x)$ converges at α if $v(\alpha) > \rho(f)$ and diverges if $v(\alpha) < \rho(f)$.* □

2. Review of Non-Archimedean Analysis   39

Remark. 2.5. For the moment let us assume that K is complete but not necessarily algebraically closed. Let $f(x) = \sum a_j x^j$ be a power series with coefficients in K; as usual, we denote the set of such series by "$K[[x]]$". Suppose $\alpha \in \overline{K}$ with $v(\alpha) > \rho(f)$, and let $L = K(\alpha)$. Then clearly $f(\alpha)$ converges to an element of L; this is similar to what happens if $f(x)$ is actually a polynomial.

Therefore, given such a K, we can always pass to the completion of its algebraic closure should we desire to work with a complete field. However, from what we have just seen, one sees that this is often unnecessary.

Let $t \in \mathbb{R}$ and let K be as in Proposition 2.4. The set $\{x \in K \mid v(x) \geq t\}$ is a "closed disc;" the set $\{x \in K \mid v(x) > t\}$ is an "open disc" and the set $\{x \in K \mid v(x) = t\}$ is a "circle." (Please note that circles may be *empty* depending on our choice of t.) The reader may easily restate these definitions in the more familiar language of the associated absolute value.

Example 2.6. Let $f(x) = 1 + x + \cdots + x^n + \cdots$. Then $\rho(f) = 0$. Thus for $\alpha \in K$ with $v(\alpha) > 0$, $1 + \alpha + \cdots + \alpha^n + \cdots$ converges (and, of course, it converges to $1/(1-\alpha)$).

Definition 2.7. Let $f(x) = \sum_{j=0}^{\infty} a_j x^j \in K[[x]]$. Let $S = \{A_i\} \subset \mathbb{R}^2$, where $A_i = (i, v(a_i))$. The lower convex hull of S is the *Newton Polygon* of $f(x)$.

Let $(X, Y) \in \mathbb{R}^2$. Note that $A_i = (i, v(a_i))$ lies on the line
$$Y + v(x)X = v(a_i x^i)$$
as $v(a_i x^i) = v(a_i) + iv(x)$.

Proposition 2.8. *Let $\{m_i\}$ be the sequence of the slopes of the Newton Polygon of $f(x) = \sum_{j=0}^{\infty} a_j x^j$. Then $\{m_i\}$ is monotonically increasing and*
$$-\lim_{i \to \infty} m_i = \rho(f).$$

Proof. The monotonicity is easily seen. Let $b = \lim_i \{m_i\}$ and suppose that
$$v(\alpha) = -b' > -b.$$
We want to see that $f(x)$ converges at α. But notice that $v(a_i \alpha^i) = v(a_i) + iv(\alpha) > v(a_i) - ib$. On the other hand, it is clear that sufficiently far out, $(i, v(a_i))$ lies arbitrarily high above $(i, b'i)$ (which lies on the line of slope b' passing through the origin). Thus $v(a_i \alpha^i) \to \infty$.

Conversely, if $v(\alpha) < -b$, then one sees that the power series diverges at $x = \alpha$. □

We now study the zeros of $f(x)$. To simplify matters, and clearly without loss of generality, we will assume that $a_0 = 1$. Let t be a real number greater than $\rho(f)$. We are interested in the zeros of $f(x)$ on the circle $v(x) = t$. There are two possibilities:

1. There is no side of the Newton Polygon of $f(x)$ with slope $-t$. In this case, we claim that there is one and only one term of minimum valuation in $\Sigma a_j x^j$, $v(x) = t$. Indeed, if not, then suppose $i \neq j$ and

$$a = v(a_i x^i) = v(a_j x^j) = \inf_k \{v(a_k x^k)\}.$$

Then all the points A_k are above the line $Y + tX = a$ and the line from A_i to A_j is a side of the Newton Polygon of slope $-t$. This is a contradiction. Therefore, $v(f(x)) = v(a_i x^i)$ for some unique i and $v(x) = t$. This implies that there is no zero on $v(x) = t$.

2. Thus, let us suppose that there is a side from A_i to A_j, $i < j$, in the Newton Polygon of slope $-t$. We are then guaranteed, by the argument in Part 1, at least two terms of minimum valuation, and so there may be a zero on the circle of $v(x) = t$. Without loss of generality, let us denote these terms by i and j also. Pick $x_0 \in K$ with $v(x_0) = t$ and put $c = v(a_i x_0^i) = v(a_j x_0^j)$. We now consider the auxiliary power series

$$f^*(u) = \sum b_e u^e := a_i^{-1} x_0^{-i} f(x_0 u).$$

one sees without difficulty that $v(b_i) = v(b_j) = 0$, $v(b_e) \geq 0$ for $e \neq i, j$ and $v(u) = 0$ whenever $v(x) = t$.

Let us denote by R_K the local ring of K and by $M_K \subset R_K$ its maximal ideal. We see that $f^*(u)$ is a power series with coefficients in R_K. We can reduce $f^*(u)$ modulo M_K in the obvious fashion; we denote the reduction by $\overline{f}^*(u)$. We also denote the reduction of $a \in R_K$ by $\bar{a}$.

It is easy to see that there can be at most finitely many e with $v(b_e) = 0$. Thus, without loss of generality, we may assume that $v(b_e) > 0$ for $e < i$ and $e > j$. Consequently we find

$$\overline{f}^*(u) = u^j + \cdots + \overline{b}_i u^i$$
$$= u^i(u^{j-i} + \cdots + \overline{b}_i),$$

where $\overline{b}_i \neq 0$. Thus u^i and $u^{j-i} + \cdots + \overline{b}_i$ are relatively prime.

Hensel's Lemma now guarantees the existence of a monic polynomial $g(u)$ of degree $j - i$ and a power series $h(u) \in R_K[[u]]$ such that

$$\overline{g}(u) = u^{j-i} + \cdots + \overline{b}_i,$$
$$\overline{h}(u) = u^i,$$

and

$$f^*(u) = h(u)g(u).$$

Moreover, $h(u)$ converges for $v(u) \geq 0$ and $h(u) \neq 0$ whenever $v(u) = 0$.

Let $g(u) = \sum_{m=0}^{j-i} g_m u^m$. Then $g_0 \equiv \bar{b}_i \neq 0$ (M_K). As K is algebraically closed, $g(u)$ factors completely in K. As $v(g_0) = 0$, one must have $v(\alpha_w) = 0$ for any root α_w of $g(u)$.

Translating this back to $f(x)$, we see that $f(x)$ has exactly $j - i$ zeros on the circle $v(x) = t$ where $j - i$ is the length of the projection to the X-axis of the side of the Newton Polygon with slope $-t$.

We summarize this lengthy discussion in the following result.

Proposition 2.9. 1. *If there exists no side of the Newton Polygon of $f(x)$ with slope $-t$ ($t > \rho(f)$), then there are no zeros of $f(x)$ on the circle $v(x) = t$.*
2. *If there exists a side of the Newton Polygon of $f(x)$ of slope $-t$ ($t > \rho(f)$), then $f(x)$ has exactly m zeros on $v(x) = t$ where m is the length of the projection of the side of the Newton Polygon of slope $-t$ onto the X-axis.* □

Remarks. 2.10. 1. The above discussion shows that if $f(x)$ is a power series and λ is a root of $f(x)$ with $v(\lambda) > \rho(f)$, then

$$f(x)/(x - \lambda)$$

is also in $K[[x]]$ and

$$\rho(f(x)) = \rho(f(x)/(x - \lambda)).$$

2. Suppose that $f(x)$ has coefficients in a complete field L and let K be the completion of the algebraic closure of L. Then the above discussion establishes that the roots of $f(x)$ are actually algebraic over L. (Indeed, choose x_0 to be algebraic, etc.) Thus the theory really has a striking closeness to the theory of polynomials over L.

3. Suppose that K has characteristic $p > 0$, and suppose that K is the completion of the algebraic closure of L as above. Let $L^{\text{sep}} \subseteq K$ be the separable closure of L. It is not hard to see that K is also the completion of L^{sep}. Thus we may always assume that x_0 is separable algebraic over L. Now suppose that $f(x)$ is of the form $\sum a_i x^{p^i}$, $a_0 \neq 0$, $\{a_i\} \subset L$. Then the roots of $f(x)$ are in L^{sep}. Indeed one has

$$f'(x) \equiv a_0 \neq 0.$$

One now uses this observation, together with a separable choice of x_0, to force the above discussion to produce separable algebraic roots for $f(x)$.

4. Suppose that $f(x)$ has no zeros in the closed disc, $v(x) \geq t > \rho(f)$. Then $f(0) \neq 0$ and one can form the power series for $1/f(x)$. This power series converges in the region $v(x) > t$.

Let $t > \rho(f)$. As a small refinement of Proposition 2.9, we have the next result.

Proposition 2.11. *There exists only finitely many zeros of $f(x)$ in the disc $v(x) \geq t$.*

Proof. Let $t_1 > \rho(f)$. The ideas presented above show that $f(x)$ will have zeros on the circle $v(x) = t_1$ if and only if there is a side of the Newton Polygon of slope $-t_1$. Let $\{m_i\}$ be the sequence of the slopes of the Newton Polygon; thus
$$-\lim m_i = \rho(f).$$
Consequently there are only finitely many sides of the Newton Polygon of slope $L < -t$ which establishes the result. $\square$

Definition 2.12. Let $f(x) = \sum_{i=0}^{\infty} a_i x^i$. We say that $f(x)$ is *entire* if and only if $\rho(f) = -\infty$ (i.e., $f(x)$ converges for all x).

Proposition 2.13. *If $f(x)$ is an entire function with no zeros, then $f(x)$ is constant.*

Proof. Suppose that $f(x)$ is a nonconstant entire function. With a little reflection on the Newton Polygon, we see that we can always find a zero for $f(x)$. $\square$

The following very important theorem is an analog of the classical Weierstrass Factorization Theorem.

Theorem 2.14. *Let $f(x)$ be an entire function and let $\{\lambda_1, \ldots, \lambda_t, \ldots\}$ be its non-zero roots in K (which, we recall, is complete and algebraically closed). Then*
$$-\infty = \lim_t v(\lambda_t)$$
and
$$f(x) = cx^n \prod_t (1 - x/\lambda_t), \qquad n = \mathrm{ord}_{x=0} f(x),$$
for some constant c. Conversely if $\{\lambda_t\}$ is as above and $c \in K$, then the above product defines an entire function.

Proof. It is a straightforward exercise to see that such products define entire functions. Suppose now that $f(x)$ is any entire function. We form the function
$$f^*(x) = x^n \prod_t (1 - x/\lambda_t),$$
where $n = \mathrm{ord}_{x=0} f(x)$. We know that
$$g(x) = f(x)/f^*(x)$$
is entire with no zeros. Thus $g(x)$ is constant. $\square$

3. The Carlitz Module

We present here the details of the *Carlitz module*. This is the simplest of all Drinfeld modules and may be given in a concrete, elementary fashion. At the same time, most essential ideas about Drinfeld modules appear in the theory of the Carlitz module. Thus it is an excellent example for the reader to master and keep in mind when reading the more abstract general theory. Our basic reference is [C1], but see also [Go2].

3.1. Background

Let $\mathbf{A} = \mathbb{F}_r[T]$, $r = p^m$, and put $\mathbf{k} = \mathbb{F}_r(T)$. Let $v_\infty \colon \mathbf{k} \to \mathbb{R} \cup \{\infty\}$ be the valuation associated to $1/T$ as in our last section; so $v_\infty(1/T) = 1$. We denote the associated completion ("$\mathbf{k}_\infty$") of $\mathbf{k}$ by $\mathbf{K}$. The field $\mathbf{K}$ is, therefore, complete and is easily seen to be locally compact in the $1/T$-topology.

The reader will note that $\mathbf{k}$ is the field of functions on $\mathbb{P}^1/\mathbb{F}_r$, while $\mathbf{A}$ is the subring of those functions regular outside ∞.

Proposition 3.1.1. *The ring $\mathbf{A}$ is a discrete subring of $\mathbf{K}$. Moreover, $\mathbf{K}/\mathbf{A}$ is compact (i.e., $\mathbf{A}$ is "co-compact" in $\mathbf{K}$).*

Proof. Let $a \in \mathbf{A}$ with $v_\infty(a) > 0$. Then $a = 0$. Indeed, if $v_\infty(a) > 0$, then a has a zero at $\infty \in \mathbb{P}^1$; thus a is regular everywhere. Therefore, it is a constant function with zeros at ∞. The fact that $\mathbf{A}$ is discrete in $\mathbf{K}$ follows immediately.

To see the co-compactness of $\mathbf{A}$, one has but to observe that the "polar-part" of a Laurent series in $1/T$ is precisely a polynomial in T. Thus $\mathbf{K}/\mathbf{A}$ is isomorphic to $\frac{1}{T}\mathbb{F}_r[[\frac{1}{T}]]$. The ring $\mathbb{F}_r[[1/T]]$ is the inverse limit of the finite rings $\mathbb{F}_r[[1/T]]/(T^{-n})$, as $n \to \infty$, and is compact. Thus so is $1/T\mathbb{F}_r[[1/T]]$. □

Remark. 3.1.2. Henceforth, the reader should be aware of the following basic analogy:
$$\mathbf{A} \sim \mathbb{Z}, \quad \mathbf{k} \sim \mathbb{Q}, \quad \text{and} \quad \mathbf{K} \sim \mathbb{R}.$$

Indeed, both **A** and $\mathbb{Z}$ possess division algorithms and $\mathbb{Z}$ is discrete inside $\mathbb{R}$ (= "$\mathbb{Q}_\infty$"). Moreover, $\mathbb{R}/\mathbb{Z} \simeq S^1$ is compact. This analogy is at the heart of the theory presented in this book.

Thus **A** will be the "bottom ring" of the theory described here and **k** will be the "bottom field." Of course, in the classical approach to function fields there is no bottom.

Let $\overline{\mathbf{K}}$ be a fixed algebraic closure of **K** equipped with the canonical extension of v_∞. One thinks of $\overline{\mathbf{K}}$ as being analogous to $\mathbb{C}$ in that it is algebraically closed. However, it is neither locally compact *nor* complete. We let $\mathbf{C}_\infty$ be the completion of $\overline{\mathbf{K}}$. By Proposition 2.1, $\mathbf{C}_\infty$ is also algebraically closed and will be used in those occasions where a complete *and* algebraically closed field is needed.

For $d \geq 0$, we let $\mathbf{A}(d) := \{\alpha \in \mathbf{A} \mid \deg(\alpha) < d\}$; thus $\mathbf{A}(d)$ is the d-dimensional $\mathbb{F}_r$-vector space of polynomials of degree $< d$. Clearly

$$\mathbf{A} = \bigcup \mathbf{A}(d).$$

Definition 3.1.3. We set, $e_0(x) = x$, and for $d > 0$

$$e_d(x) := \prod_{\alpha \in \mathbf{A}(d)} (x - \alpha) = \prod_{\alpha \in \mathbf{A}(d)} (x + \alpha).$$

By Corollary 1.2.2, one sees that $e_d(x)$ is an $\mathbb{F}_r$-linear polynomial. Thus, in the notation of Section 1,

$$e_d(\tau) \in \mathbf{A}\{\tau\}.$$

We now want to give a closed form expression for the coefficients of $e_d(x)$.

Definition 3.1.4.

1. Let $i > 0$. We set
$$[i] := T^{r^i} - T \in \mathbf{A}.$$

2. We set $D_0 = 1$, and for $i > 0$,
$$D_i := [i][i-1]^r \cdots [1]^{r^{i-1}}.$$

3. We set $L_0 = 1$, and for $i > 0$,
$$L_i := [i][i-1] \cdots [1].$$

The numbers $[i]$, D_i and L_i are fundamental for the arithmetic of $\mathbb{F}_r[T]$. Their properties will be discussed at various points in this book. We note that $\deg[i] = r^i$, $\deg D_i = ir^i$ and $\deg L_i = r \cdot \frac{(r^i-1)}{(r-1)}$ and that their valuations at ∞ are the negatives of these numbers.

3.1. Background

Let L be some field extension of $\mathbf{k}$ containing an indeterminate x. Let $\{w_0, \ldots, w_d\}$ be $d+1$ elements of L which are linearly independent over $\mathbb{F}_r$. As in Section 1, we set

$$\Delta(w_0, \ldots, w_d) = \det \begin{pmatrix} w_0 & \cdots & w_d \\ w_0^r & \cdots & w_d^r \\ \vdots & & \vdots \\ w_0^{r^d} & \cdots & w_d^{r^d} \end{pmatrix}.$$

By Moore's formula (Corollary 1.3.7) we see that

$$\frac{\Delta(w_0, \ldots, w_d)}{\Delta(w_0, \ldots, w_{d-1})} = \prod_{\{\beta_j\} \subseteq \mathbb{F}_r} (w_d + \beta_0 w_0 + \cdots + \beta_{d-1} w_{d-1}).$$

We now substitute x for w_d and T^i for w_i, $i = 0, \ldots, d-1$. We obtain

$$\prod_{\alpha \in \mathbf{A}(d)} (x + \alpha) = \frac{\Delta(1, \ldots, T^{d-1}, x)}{\Delta(1, \ldots, T^{d-1})}.$$

Write

$$\Delta(1, \ldots, T^{d-1}, x) = \sum_{j=0}^{d} (-1)^{d-j} x^{r^j} M_j,$$

where M_j is the determinant of the minor obtained by crossing out the last column and j^{th} row. The reader will readily see that M_j is a determinant of Vandermonde type. Therefore,

$$M_j = \prod_{h > i} (T^{r^h} - T^{r^i}), \qquad h, i = 1, \ldots, d; \ h \neq j, \ i \neq j$$

$$= \frac{\prod_{h > i} (T^{r^h} - T^{r^i})}{\prod_{h > j} (T^{r^h} - T^{r^j}) \prod_{i < j} (T^{r^j} - T^{r^i})}$$

$$= \left(\prod_{i=1}^{d} D_i \right) \bigg/ D_j L_{d-j}^{r^j}.$$

In a similar fashion, one finds

$$\Delta(1, \ldots, T^{d-1}) = \prod_{i=1}^{d-1} D_i.$$

Thus, we obtain the following results of Carlitz.

3. The Carlitz Module

Theorem 3.1.5 (Carlitz). *We have*

$$e_d(x) = \prod_{\alpha \in A(d)} (x+\alpha) = \sum_{i=0}^{d} (-1)^{d-i} x^{r^i} \frac{D_d}{D_i L_{d-i}^{r^i}}.$$ □

The reader should immediately note that the coefficients $D_d/D_i L_{d-i}^{r^i}$ are actually integral, i.e., $\in \mathbf{A}$. Indeed, this is a trivial consequence of the product expansion for $e_d(x)$. We sometimes denote these coefficients by "$\begin{bmatrix} d \\ i \end{bmatrix}$" or "$\begin{bmatrix} d \\ i \end{bmatrix}_\mathbf{A}$".

Theorem 3.1.5 is so basic that we present a rather different derivation of it without using the Moore Determinant.

We begin by presenting some properties of D_i and L_i that are "factorial-like;" for more such formulas we refer the reader to Subsection 9.1.

Proposition 3.1.6.

1. $[i] = \prod\limits_{\substack{f \text{ monic prime} \\ \deg(f)|i}} f$.
2. $D_i = [i] D_{i-1}^r$.
3. $D_i = \prod\limits_{\substack{g \text{ monic} \\ \deg g = i}} g$.
4. L_i is the least common multiple of all polynomials of degree i.

Proof. 1. Note that

$$\frac{d}{dT}[i] = -1.$$

Thus $[i]$ is a separable polynomial. Part 1 is now an elementary exercise in the use of finite fields. Part 2 follows immediately from the definition. Parts 3 and 4 follow from Part 1 upon counting the number of times a given monic prime f divides the product of all monic polynomials of degree i (or, for Part 4, their least common multiplier). □

Corollary 3.1.7. $e_d(h) = D_d$ *for any monic polynomial h of degree d.*

Proof. A monic g of degree d can be uniquely written as $g = h + \alpha$, $\deg \alpha < d$. Thus the result follows from 3.1.6.3. □

Second Proof of Theorem 3.1.5. We claim that $e_d(x)$ may be written as

$$e_d(x) = e_{d-1}^r(x) - D_{d-1}^{r-1} e_{d-1}(x). \quad (*)$$

Both sides of the above equation are monic of the same degree. Thus we need only establish that they have the same set of roots $= \mathbf{A}(d)$. Thus let $\alpha \in \mathbf{A}(d)$. If $\alpha \in \mathbf{A}(d-1)$, then clearly α is a root of $*$. Thus assume that

$$\alpha = \zeta h,$$

with $\zeta \in \mathbb{F}_r^*$, and $\deg h = d - 1$. By 3.1.7, the right hand side of $*$ is

$$\zeta^r D_{d-1}^r - D_{d-1}^{r-1}\zeta D_{d-1} = 0,$$

as $\zeta^r = \zeta$. The result now follows by using induction and the properties of D_i and L_i. □

3.2. The Carlitz Exponential

In this subsection, we "pass to the limit" (as $d \to \infty$) in the formula we have obtained for $e_d(x)$ to obtain the *Carlitz exponential*.

From Theorem 3.1.5, we have

$$\prod_{\alpha \in \mathbf{A}(d)} (x + \alpha) = \sum_{j=0}^{d} (-1)^{d-j} x^{r^j} \frac{D_d}{D_j L_{d-j}^{r^j}}.$$

By 3.1.6.3, we have

$$D_i = \prod_{\substack{\deg g = i \\ g \text{ monic}}} g.$$

Thus

$$\prod_{\deg g = i} g = \prod_{\substack{\deg g = i \\ g \text{ monic}}} g^{r-1} \left(\prod_{\zeta \in \mathbb{F}_r^*} \zeta \right)$$

$$= \prod_{\substack{\deg g = i \\ g \text{ monic}}} -g^{r-1} = (-1)^{r^i} \prod_{\substack{\deg g = i \\ g \text{ monic}}} g^{r-1}$$

$$= -D_i^{r-1}.$$

Consequently,

$$\prod_{0 \neq \alpha \in \mathbf{A}(d)} \alpha = (-1)^d (D_0 \cdots D_{d-1})^{r-1}$$

$$= (-1)^d D_d / L_d.$$

We now divide the formula of Theorem 3.1.5 by $\prod_{0 \neq \alpha \in \mathbf{A}(d)} \alpha$. We obtain

$$x \prod_{0 \neq \alpha \in \mathbf{A}(d)} (1 + x/\alpha) = \sum_{j=0}^{d} (-1)^j \frac{x^{r^j}}{D_j} \frac{L_d}{L_{d-j}^{r^j}}.$$

48   3. The Carlitz Module

Let us put
$$\xi_d := \frac{[1]^{\frac{r^d-1}{r-1}}}{L_d}.$$

Lemma 3.2.1. *We have*
$$\xi_d = \prod_{j=1}^{d-1}(1-[j]/[j+1]).$$

Proof. Note that
$$\prod_{j=1}^{d-1}(1-[j]/[j+1]) = \prod_{j=1}^{d-1}\left(\frac{[j+1]-[j]}{[j+1]}\right).$$

Now $[j+1]-[j] = [1]^{r^j}$. Thus,
$$\prod_{j=1}^{d-1}\left(\frac{[j+1]-[j]}{[j+1]}\right) = \prod_{j=1}^{d-1}\frac{[1]^{r^j}}{[j+1]} = \frac{\prod_{j=0}^{d-1}[1]^{r^j}}{L_d}.$$

But $\sum_{j=0}^{d-1} r^j = \frac{r^d-1}{r-1}$ giving the result. □

Thus we see that the sequence $\{\xi_d\}_{d=1}^{\infty}$ has a limit in **K** which we denote by ξ_*. By Lemma 3.2.1 we see that
$$\xi_* = \prod_{j=1}^{\infty}\left(1-\frac{[j]}{[j+1]}\right).$$

Note that ξ_* is clearly a 1-unit in **K** (i.e., it is a unit in the ring $R_{\mathbf{K}} = \{x \in \mathbf{K} \mid v_\infty(x) \geq 0\}$ and is congruent to 1 modulo the maximal ideal $M_{\mathbf{K}} = \{x \in \mathbf{K} \mid v_\infty(x) > 0\}$). In particular, $v_\infty(\xi_*) = 0$.

Let $d > 0$.

Lemma 3.2.2.
1. $v_\infty(\xi_{d+1} - \xi_d) = r^d(r-1)$.
2. Set $\delta_d := \xi_d - \xi_*$. Then
$$v_\infty(\delta_d) = r^d(r-1).$$

Proof. Note that
$$\xi_{d+1} - \xi_d = -\frac{[d]}{[d+1]}\xi_d.$$

Thus
$$v_\infty(\xi_{d+1} - \xi_d) = r^{d+1} - r^d$$
giving Part 1.

To see Part 2, note that
$$-\delta_d = (\xi_{d+1} - \xi_d) + (\xi_{d+2} - \xi_{d+1}) + \cdots.$$
Thus, Part 2 follows from Part 1. □

Set $\beta_j := [1]^{\frac{r^j-1}{r-1}}$ and so $\xi_d = \frac{\beta_d}{L_d}$.

Lemma 3.2.3.
$$\frac{L_d}{L_{d-j}^{r^j}} = \frac{\beta_j \xi_{d-j}^{r^j}}{\xi_d}.$$

Proof. The right hand side of the statement of the lemma is equal to
$$\frac{[1]^{\frac{r^j-1}{r-1}} L_d [1]^{r^j\left(\frac{r^{d-j}-1}{r-1}\right)}}{[1]^{\frac{r^d-1}{r-1}} L_{d-j}^{r^j}}.$$
But $0 = r^j - 1 + r^j(r^{d-j} - 1) - (r^d - 1)$. □

Thus from Lemma 3.2.3, we see that
$$x \prod_{0 \neq \alpha \in A(d)} (1 + x/\alpha) = x \prod_{0 \neq \alpha \in A(d)} (1 - x/\alpha)$$
$$= \sum_{j=0}^d (-1)^j \frac{x^{r^j}}{D_j} \frac{L_d}{L_{d-j}^{r^j}}$$
$$= \sum_{j=0}^d (-1)^j \frac{x^{r^j}}{D_j} \beta_j \frac{\xi_{d-j}^{r^j}}{\xi_d}$$
$$= \frac{1}{\xi_d} \sum_{j=0}^d (-1)^j \frac{x^{r^j}}{D_j} \beta_j \xi_{d-j}^{r^j}.$$

Recalling that $\delta_j = \xi_j - \xi_*$, we find that the above equals
$$\frac{1}{\xi_d} \sum_{j=0}^d (-1)^j \frac{x^{r^j}}{D_j} \beta_j (\delta_{d-j}^{r^j} + \xi_*^{r^j}).$$

3. The Carlitz Module

Lemma 3.2.4. *For any $x \in \mathbf{C}_\infty$, as $d \to \infty$*

$$\sum_{j=0}^{d}(-1)^j \frac{x^{r^j}}{D_j} \beta_j \delta_{d-j}^{r^j} \to 0$$

in $\mathbf{C}_\infty$.

Proof. Note that
$$v_\infty(D_j) = -jr^j$$
and
$$v_\infty(\beta_j) = -r\frac{(r^j - 1)}{r - 1}.$$

Thus, by Lemma 3.2.2, we find

$$v_\infty\left((-1)^j \frac{x^{r^j}}{D_j}\beta_j \delta_{d-j}^{r^j}\right) = r^j v_\infty(x) + jr^j - r\frac{(r^j-1)}{r-1} + r^j r^{d-j}(r-1)$$

$$= \frac{r^j((v_\infty(x) + j + r^{d-j}(r-1))(r-1) - r) + r}{r-1}.$$

We now split
$$\sum_{j=0}^{d}(-1)^j \frac{x^{r^j}}{D_j}\beta_j \delta_{d-j}^{r^j}$$

into two sums, $\sum_1 + \sum_2$, where $\sum_1 := \sum_{2j<d}$, and $\sum_2 := \sum_{2j\geq d}$. With a little thought, one checks that both $\sum_1$ and $\sum_2$ tend to 0 as $d \to \infty$. This gives the lemma. □

Lemma 3.2.5. *For any $x \in \mathbf{C}_\infty$, the series*

$$\sum_{j=0}^{\infty}(-1)^j \frac{x^{r^j}}{D_j}\beta_j \xi_*^{r^j}$$

converges in $\mathbf{C}_\infty$.

Proof. This follows as in the proof of Lemma 3.2.4. □

Corollary 3.2.6. *Let $x \in \mathbf{C}_\infty$. Then*

$$x \prod_{0\neq\alpha\in A}(1 - x/\alpha) = \frac{1}{\xi_*}\sum_{j=0}^{\infty}\frac{x^{r^j}}{D_j}(-1)^j \beta_j \xi_*^{r^j}.$$

Proof. This follows from Lemma 3.2.4 and 3.2.5 and the expression for the left hand side given above. □

Definition 3.2.7. 1. Let λ be any $(r-1)$-st root of $-[1]$ in $\overline{\mathbf{K}}$. Then we set
$$\xi := \xi_C = \lambda \xi_*.$$

2. Let $x \in \mathbf{C}_\infty$. Then we set
$$e_C(x) = \sum_{j=0}^{\infty} \frac{x^{r^j}}{D_j}.$$

(This sum converges to an element of $\mathbf{C}_\infty$ as in the proof of Lemma 3.2.5.) The function $e_C(x)$ is the *Carlitz exponential*.

Summarizing, we have the following result due to Carlitz.

Theorem 3.2.8. *Let $x \in \mathbf{C}_\infty$. Then*
$$x \prod_{0 \neq \alpha \in \mathbf{A}} (1 - x/\alpha) = \frac{1}{\xi} \sum \frac{(\xi x)^{r^j}}{D_j} = \frac{1}{\xi} e_C(\xi x). \qquad \square$$

Corollary 3.2.9. *Put $L := \xi \mathbf{A} \in \overline{\mathbf{K}}$. Then for all $x \in \mathbf{C}_\infty$, we have*
$$x \prod_{0 \neq \alpha \in L} (1 - x/\alpha) = e_C(x).$$

Proof. If we substitute x/ξ for x in 3.2.8, we obtain
$$\frac{1}{\xi} e_C(x) = \frac{x}{\xi} \prod_{0 \neq \alpha \in L} (1 - x/\alpha),$$
which is equivalent to the statement of the corollary. $\square$

Remarks. 3.2.10. 1. Note that Corollary 3.2.9 gives the factorization of $e_C(x)$ guaranteed by Theorem 2.14.
2. Let $0 \neq \alpha \in L$, $\alpha \in L = \xi \mathbf{A}$. As $\xi_* \in \mathbf{K}$, we see that
$$\mathbf{K}(\alpha) = \mathbf{K}(\lambda),$$
and $\mathbf{K}(\lambda)$ is separable over $\mathbf{K}$. This is in keeping with 2.10.3.
3. By using the binomial theorem, for instance, we can choose an $(r-1)$-st root θ of $(1 - T^{1-r})$ in $\mathbf{K}$ which is a 1-unit. It is simple to see that θ is unique. We set
$$\xi_u = \theta \xi_*;$$
the element ξ_u is a 1-unit. Thus
$$\xi = \sqrt[r-1]{-T^r} \cdot \xi_u.$$

The reader will immediately see the analogy with

$$2\pi i = 2i\pi \, .$$

For more properties of ξ we refer the reader to Subsection 9.4.
4. We have normalized $e_C(x)$ in a slightly different fashion than in Carlitz's original paper. We do this in order to have a simpler time using $e_C(x)$ to describe abelian extensions.
5. The element ξ was shown to be transcendental over **k** by L.I Wade [Wad1].

3.3. The Carlitz Module

We now use $e_C(x)$ to describe a new module action of $\mathbf{A} = \mathbb{F}_r[T]$ on $\mathbf{C}_\infty$. This action is called the *Carlitz module*. It is the simplest example of a Drinfeld module.

Proposition 3.3.1. *Let $x \in \mathbf{C}_\infty$. Then*

$$e_C(Tx) = Te_C(x) + e_C(x)^r \, .$$

Proof. We have

$$e_C(x) = \sum_{i=0}^{\infty} \frac{x^{r^i}}{D_i} \, .$$

So

$$e_C(Tx) = \sum_{i=0}^{\infty} T^{r^i} \frac{x^{r^i}}{D_i} \, ,$$

or

$$e_C(Tx) - Te_C(x) = \sum_{i=0}^{\infty} (T^{r^i} - T) \frac{x^{r^i}}{D_i} \, .$$

But $D_i = (T^{r^i} - T)D_{i-1}^r$; so

$$e_C(Tx) - Te_C(x) = \sum_{i=1}^{\infty} \frac{x^{r^i}}{D_{i-1}^r} = \left(\sum_{i=0}^{\infty} \frac{x^{r^i}}{D_i}\right)^r \, . \qquad \square$$

Let $a \in \mathbf{A}$ with $a = \sum_{j=0}^{d} a_j T^j$, $\{a_j\} \subseteq \mathbb{F}_r$, $a_d \neq 0$.

Corollary 3.3.2. *Let $x \in \mathbf{C}_\infty$. Then*

$$e_C(ax) = ae_C(x) + \sum_{j=1}^{d} C_a^{(j)} e_C(x)^{r^j}$$

where $\{C_a^{(j)}\} \subset \mathbf{A}$ and $C_a^{(d)} = a_d$.

3.3. The Carlitz Module

Proof. Note that for $i \geq 1$,
$$e_C(T^i x) = e_C(T(T^{i-1}x)).$$

Thus $\{C_{T^i}^{(j)}\}$ can be computed via recursion. For instance,
$$\begin{aligned}
e_C(T^2 x) &= Te_C(Tx) + e_C(Tx)^r \\
&= T(Te_C(x) + e_C(x)^r) + (Te_C(x) + e_C(x)^r)^r \\
&= T^2 e_C(x) + (T^r + T)e_C^r(x) + e_C(x)^{r^2}.
\end{aligned}$$

Thus the coefficients for $e_C(ax)$ can now be computed using $\mathbb{F}_r$-linearity. The result follows easily. □

As in Section 1, let $\tau \colon \mathbf{C}_\infty \to \mathbf{C}_\infty$ be the rth power mapping, $\tau(x) = x^r$, and, for any subfield M of $\mathbf{C}_\infty$, we let $M\{\tau\}$ be the composition ring of $\mathbb{F}_r$-linear polynomials.

Definition 3.3.3. Let $\{C_a^{(j)}\}$ be as in Corollary 3.3.2. Then we set
$$C_a(\tau) = a\tau^0 + \sum_{j=1}^{d} C_a^{(j)} \tau^j,$$
where $d = \deg a$.

Thus, we have the fundamental functional equation for $e_C(x)$
$$e_C(ax) = C_a(e_C(x)).$$

Theorem 3.3.4. *The mapping from $\mathbf{A}$ to $\mathbf{k}\{\tau\}$, $a \mapsto C_a$, is an injection of $\mathbb{F}_r$-algebras.*

Proof. It is clear that the map $a \mapsto C_a$ is $\mathbb{F}_r$-linear and injective. Thus we need only show that it is a mapping of algebras; i.e., for $a, b \in \mathbf{A}$
$$C_{ab} = C_a \cdot C_b,$$
where $C_a \cdot C_b$ is the multiplication in $\mathbf{k}\{\tau\}$ (and so is multiplication of additive polynomials). But
$$C_{ab}(e_C(x)) = e_C(abx) = e_C(a(bx)) = C_a(e_C(bx)) = C_a(C_b(e_C(x))),$$
which gives the result. □

Definition 3.3.5. We call the mapping $\mathbf{A} \mapsto \mathbf{k}\{\tau\}$, $a \mapsto C_a$, the *Carlitz module*. It is denoted by "C."

3. The Carlitz Module

Remarks. 3.3.6. 1. In fact, with the obvious definitions, $C_a \in \mathbf{A}\{\tau\}$ for all $a \in \mathbf{A}$.

2. What is "really" going on with the Carlitz module is the following: from Theorem 2.14, one knows that every non-constant entire function (in non-Archimedean analysis) is *surjective*. Let $L = \mathbf{A}\xi$ be the zeros of $e_C(x)$; we then have an isomorphism

$$\mathbf{C}_\infty / L \xrightarrow{\sim} \mathbf{C}_\infty$$

via $e_C(x)$. Now the group on the left is obviously an **A**-module; thus by transport of structure, we obtain a new **A**-action on $\mathbf{C}_\infty$. This action *is* the Carlitz module.

3. If one recalls that $D_i = \prod_{\substack{g \text{ monic} \\ \deg g = i}} g$, one sees immediately that $e_C(x)$ is analogous to the classical exponential function

$$e^x = \sum_{n=0}^\infty \frac{x^n}{n!}$$

except that the expansion for $e_C(x)$ only involves the monomials $\{x^{r^i}\}$. In particular, the Carlitz module is an **A**-analog of the multiplicative group $\mathbb{G}_m$ (as algebraic group) with the usual $\mathbb{Z}$-action. This analogy will be used very often in this work.

Definition 3.3.7. The *division values* (or *division points*) of the Carlitz module (or exponential) are the values $\{e_C(a\xi) \mid a \in \mathbf{k}\} \subset \mathbf{C}_\infty$.

Let $a = b/f \in \mathbf{k}$, $\{b, f\} \subset \mathbf{A}$ with $f \neq 0$. Then $e_C(a\xi)$ is a root of $C_f(x) = 0$; thus it belongs to the algebraic closure of **k** in $\mathbf{C}_\infty$.

For now, we establish the following weak, but very important, result. The reader should immediately see the analogy with cyclotomic fields.

Proposition 3.3.8. *Let $L \subseteq \mathbf{C}_\infty$ be an extension of **k**. Let $a \in \mathbf{k}$ and let*

$$L_1 = L(e_C(a\xi)).$$

Then L_1 is an abelian extension of L.

Proof. Let $a = b/f$ be an irreducible rational function. Then

$$e_C\left(\frac{b}{f}\xi\right) = C_b(e_C(\xi/f));$$

thus $L_1 \subseteq L(e_C(\xi/f))$. Thus, by the standard arguments of Galois theory, we may assume that $L_1 = L(e_C(\xi/f))$. Put $\rho := e_C(\xi/f)$.

Since the coefficients of $C_g(\tau)$ are in **A** for all g, we see immediately that L_1 contains *all* values

$$e_C\left(\frac{g}{f}\xi\right);$$

i.e., L_1 contains all f-division points. As an **A**-module, the Carlitz exponential assures us that the **A**-module of f-division points is isomorphic to **A**$/(f)$.

Since L_1 contains all f-division points, it is easy to see that it is Galois over L. Let G be the Galois group. As $C_g(\tau) \in \mathbf{A}\{\tau\}$ for all g, we see that the action of G on the f-division points *commutes* with the action of **A**. Thus, for $\sigma \in G$, we see that $\sigma(\rho)$ is also an **A**-module generator of the f-division points. We, therefore, obtain an injection $G \hookrightarrow \mathbf{A}/(f)^*$ giving the result. $\square$

Definition 3.3.9. Let $g \in \mathbf{A}$. We set

$$C[g] := \left\{ e_C\left(\frac{b}{g}\xi\right) \mid b \in \mathbf{A} \right\} \subset \mathbf{C}_\infty.$$

We call $C[g]$ the *module of g-division points*. It is **A**-module isomorphic to $\mathbf{A}/(g)$. A generator of $C[g]$ as an **A**-module is called a *primitive g-th division point*.

We have seen that, as **A**-module, $C[g] \simeq \mathbf{A}/(g)$. Note also that if $\zeta \in \mathbb{F}_r^*$, then

$$C[g] = C[\zeta g].$$

Thus, $C[g]$ depends *only* on the *ideal* in **A** generated by g. Consequently let $I \subseteq \mathbf{A}$ be an ideal. We set

$$C[I] := C[i],$$

for *any* generator i of I.

Finally we finish this section with a formula for the coefficients $\{C_a^{(j)}\}$ of $C_a(\tau)$ as in [Go3]. Our next section will present another formula due to Carlitz.

Let $a \in \mathbf{A}$ and, as above, put

$$C_a(\tau) = a\tau^0 + \sum_{j=1}^{d} C_a^{(j)} \tau^j.$$

For the moment and for the sake of simplicity let us put

$$a_j := C_a^{(j)}.$$

Proposition 3.3.10. *Let a, a_j, etc. be as above. Then we have*

$$a_1 = \frac{a^r - a}{T^r - T}, \quad a_2 = \frac{a_1^r - a_1}{T^{r^2} - T}, \ldots, a_i = \frac{a_{i-1}^r - a_{i-1}}{T^{r^i} - T}, \ldots.$$

Moreover, if $a = \zeta f$, for $\zeta \in \mathbb{F}_r^*$ and f monic of degree d, then $a_d = \zeta$.

Before turning to the proof, we remark that once $a_d = \zeta$, then $a_{d+1} = a_{d+2} = \cdots = 0$.

Proof. Write $C_a = a\tau^0 + \chi_a$, where $\chi_a \in \mathbf{A}\{\tau\}$. Thus, $\chi_T = \tau$. Now $C_a C_T = C_T C_a$ in $\mathbf{k}\{\tau\}$. Therefore,

$$(a\tau^0 + \chi_a)C_T = C_T(a\tau^0 + \chi_a),$$

or

$$C_T a\tau^0 - a\tau^0 C_T = \chi_a C_T - C_T \chi_a. \qquad (*)$$

The result now follows upon equating coefficients of τ^j on both sides of the above equation. $\square$

Remarks. 3.3.11. 1. One sees that $a_i \neq 0$ for $i = 1, \ldots, d$. Moreover, $\deg(a_i)$ can be easily found from the proposition.
2. Let $u, v \in \mathbf{k}\{\tau\}$ and set, as usual,

$$[u, v] = uv - vu.$$

One sees easily that, as $[u,v]$ is the commutator, the map $v \mapsto [u,v]$ is a derivation of $\mathbf{k}\{\tau\}$. (Indeed: $u(v_1 v_2) - (v_1 v_2)u = (uv_1 - v_1 u)v_2 + v_1(uv_2 - v_2 u)$.) Moreover, the equation $(*)$ given just above can be written

$$[C_T, a\tau^0] = -[C_T, \chi_a].$$

One thinks of this equation as being a *derivation equation* which defines the Carlitz module.
3. Proposition 3.3.10 is different from the one in [Go3] due to our normalization of $e_C(x)$.

3.4. The Carlitz Logarithm

Recall that

$$e_C(x) = \sum_{i=0}^{\infty} \frac{x^{r^i}}{D_i},$$

$D_0 = 1$. Thus the derivative $e'_C(x)$ is identically 1. Consequently, we may formally derive an *inverse* for $e_C(x)$ about the origin with a non-trivial radius of convergence. We call this function "$\log_C(x)$." It is clearly $\mathbb{F}_r$-linear, as $e_C(x)$ is.

By definition, as formal power series,

$$e_C(\log_C(x)) = \log_C(e_C(x)) = x,$$

(see, [C1], Theorem 6.1). Now, $e_C(x)$ has the functional equation,

$$e_C(Tx) = Te_C(x) + e_C(x)^r.$$

Thus
$$\log_C(e_C(Tx)) = Tx = \log_C(Te_C(x)) + \log_C(e_C(x)^r).$$

Substituting $\log_C(x)$ for x, we obtain
$$T\log_C(x) = \log_C(Tx) + \log_C(x^r).$$

As $\log'_C(x)$ is also identically 1, one finds
$$\log_C(x) = x + \frac{x^r}{-[1]} + \frac{x^{r^2}}{[1][2]} + \frac{x^{r^3}}{-[1][2][3]} + \cdots$$
$$= \sum_{i=0}^{\infty} (-1)^i \frac{x^{r^i}}{L_i}.$$

Recall that in Definition 2.3 we discussed the order of convergence of a non-Archimedean power series – the order of convergence being the valuation theoretic version of the standard radius of convergence.

Proposition 3.4.1. *The order of convergence of* $\log_C(x)$, $\rho(\log_C(x))$, *is* $-\frac{r}{r-1} = \frac{r}{1-r}$.

Proof. This follows immediately once one notices that
$$v_\infty(L_i) = r\frac{(1-r^i)}{(r-1)}.$$
$\square$

The reader should note that
$$v_\infty(\xi) = -\frac{r}{r-1}.$$
Thus the logarithm converges "up to the smallest non-zero period of $e_C(x)$."

3.5. The Polynomials $E_d(x)$

In this subsection, we present another formula for the Carlitz module due to Carlitz [C1]. Our exposition is modeled on that of [AT1].

Let $\log(x)$, e^x be the usual complex-valued functions. Simple calculus gives the identity
$$(1+t)^x = e^{x\log(1+t)},$$
which one expands about $t = 0$ as
$$(1+t)^x = \sum_{n=0}^{\infty} \binom{x}{n} t^n.$$

58   3. The Carlitz Module

Of course $\{\binom{x}{n}\}$ are the binomial polynomials

$$\binom{x}{n} = \frac{x(x-1)\cdots(x-n+1)}{n!}.$$

We now do a very similar thing with the Carlitz module. Let $e_C(x)$ and $\log_C(x)$ be the Carlitz exponential and logarithm as in the previous subsections.

Definition 3.5.1. We set

$$e_C(z\log_C(x)) = \sum_{j=0}^{\infty} E_j(z)x^{r^j}.$$

Note that, as $e_C(x)$ is entire, $e_C(z\log(x))$ converges (at least) for all $\{(z,x)\}$ with $v_\infty(x) > \frac{r}{1-r}$ by Proposition 3.4.1.

Proposition 3.5.2.
1. $E_j(z)$ is an $\mathbb{F}_r$-linear polynomial of degree r^j.
2. $E_j(a) = 0$ for all $a \in \mathbf{A}(j)$.
3. $E_j(T^j) = 1$.

Proof. Part 1 follows from the power series definition of $E_j(z)$. To see Part 2, put $z = a \in \mathbf{A}(j)$. Then

$$e_C(a\log_C(x)) = C_a(e_C(\log_C(x))) = C_a(x)$$

is a polynomial in x of degree $r^{\deg(a)} < r^j$. Thus $E_j(a) = 0$. Now set $a = T^j$. Then the above formula and our knowledge of $C_{T^j}(x)$ imply that $E_j(T^j) = 1$ giving Part 3. □

Corollary 3.5.3. We have $E_j(z) = \dfrac{e_j(z)}{D_j}$.

Proof. The polynomial $e_j(z)$ has degree r^j. Its zero set is $\mathbf{A}(j)$. Finally by 3.1.7 we see that

$$\frac{e_j(T^j)}{D_j} = 1.$$

We therefore obtain the corollary. □

Corollary 3.5.4 (Carlitz). Let $\{C_a^{(j)}\}$ be as in Definition 3.3.2. Then, for all j,

$$C_a^{(j)} = E_j(a) = \frac{e_j(a)}{D_j}.$$

Proof. We have seen that

$$e_C(a \log_C(x)) = C_a(x) = \sum_{j=0}^{\infty} E_j(a) x^{r^j}.$$

Thus the result follows from 3.5.3. □

Corollary 3.5.5 (Carlitz). *Let $a \in \mathbf{A}$. Then $e_j(a)/D_j \in \mathbf{A}$ also.*

Proof. The coefficients of $C_a(x)$ are in **A**. □

Corollary 3.5.4 gives another description of the coefficients of C_a. It should be compared with Proposition 3.3.10. In fact, it would be interesting to know the exact relationship between the two. Moreover, Corollary 3.5.5 shows how close the analogy between $E_j(z)$ and $\binom{x}{j}$ really is.

3.6. The Carlitz Module over Arbitrary A-fields

We now want to study the Carlitz module over arbitrary fields, not just those containing **k**. Let L be a field containing $\mathbb{F}_r$. As with **k**, it is reasonable to expect that the Carlitz module over L will give rise to a map **A** to $L\{\tau\}$. If we compose this map with the derivative map $L\{\tau\} \to L$ (i.e., the ring homomorphism $L\{\tau\} \to L$ given by taking an $\mathbb{F}_r$-linear polynomial to its coefficient of τ^0) we obtain our first definition.

Definition 3.6.1. Let L be a field. We say that L is an **A**-field if and only if there is a morphism $\iota \colon \mathbf{A} \to L$. Let $\wp = \ker(\iota)$. We call $\wp$ the *characteristic* of L. We say that L has *generic characteristic* if and only if $\wp = (0)$.

Thus if L has generic characteristic, then L contains **k** as a subfield. Let $\overline{L}$ be a fixed algebraic closure of L with the **A**-structure coming from ι.

The procedure for considering the Carlitz module over L is now clear: One simply applies ι to the coefficients of $C_a(\tau)$, for $a \in \mathbf{A}$ — which are elements of **A** — to obtain elements in $L\{\tau\}$.

Let $a \in \mathbf{A}$. Note that

$$C'_a(x) = \iota(a).$$

Thus if $a \notin \wp$, then $C_a(x)$ is still a separable polynomial. Note also the similarity between the above normalization and the standard normalization in the theory of complex multiplication of elliptic curves.

Via C, the field $\overline{L}$ now becomes an **A**-module: Let $a \in \mathbf{A}$ and $\alpha \in \overline{L}$. Then we have

$$(a, \alpha) \mapsto C_a(\alpha).$$

One says that α is an "a-torsion point" if and only if $C_a(\alpha) = 0$ and so on. One sets $C[a] \subset \overline{L}$ to be the roots of $C_a(x) = $ the module of a-torsion points.

As before, one sees that $C[a]$ depends only on the ideal generated by a. And, if $I = (i)$ is an **A**-ideal, then we set

$$C[I] := C[i].$$

In general, if K is any **A**-field, we will use the notation "$C(K)$" to denote K viewed as **A**-module via C.

Our main goal in this subsection is to describe the torsion points of C in $C(\overline{L})$ as **A**-modules. The reader will note the similarity between our arguments and those standard ones used classically for roots of unity.

Theorem 3.6.2.
1. Let $a \notin \wp = \ker(\iota)$. Then $C[a] \subset C(\overline{L})$ is isomorphic to $\mathbf{A}/(a)$.
2. Let $(f) = \wp$. Then $C[f^i] = \{0\} \subset \overline{L}$.

Proof. 1. We know that $C[a]$ is a finite **A**-module of order $r^{\deg(a)}$. As **A** is a principal ideal domain, we can decompose

$$C[a] \simeq \oplus \mathbf{A}/(f_i)^{e_i},$$

where f_i is prime and $e_i > 0$. The elements in **A** which are prime to a act as automorphisms of $C[a]$. Thus $f_i \mid a$ for all i and so $f_i \notin \wp$ for all i. Moreover, the number of elements in $\mathbf{A}/(f_i)$ is $r^{\deg f_i}$. Thus by simply counting f_i-division points, one sees that $f_i \neq f_j$ for $i \neq j$ implying that $C[a]$ is cyclic. One now sees easily that $C[a]$ is **A**-module isomorphic to $\mathbf{A}/(a)$.

To see Part 2 look first at $C[f]$ = roots of $C_f(x)$. Since $C'_f(x) = \iota(f) = 0 + (\wp)$, we see that $C_f(x)$ is no longer separable. Thus it has $< r^{\deg(f)}$ roots in $\overline{L}$. However, the arguments in Part 1 imply that $C[f]$ must be **A**-module isomorphic to $\mathbf{A}/(f^i)$ for some i. Counting again implies that $i = 0$ or

$$C_f(x) \equiv x^{r^{\deg f}} \ (\wp \mathbf{A}[x]).$$

The result follows.

Alternatively, Part 2 can be seen directly through Proposition 3.3.10. □

Of course, $C[f^i]$, etc., should actually be viewed as a *finite group scheme* with the induced **A**-action. In this case, $C[f^i]$ is isomorphic to

$$\overline{L}[x]/(x^{r^{i \deg f}}).$$

Recall that $\mathbb{Z}/(p)^* \simeq \mathbb{Z}/(p-1)$ as abelian groups, etc. We now present the analogous result for the Carlitz module. Let f be a monic prime of **A**. Set $\wp = (f)$ and $\mathbb{F}_\wp = \mathbf{A}/\wp$. Let $\mathbb{F}_{\wp^n}$ be the unique (up to isomorphism) extension of $\mathbb{F}_\wp$ of degree n. Thus $\mathbb{F}_{\wp^n}$ is an **A**-field via the map $\mathbf{A} \to \mathbf{A}/\wp \hookrightarrow \mathbb{F}_{\wp^n}$, and is an **A**-module via C.

Theorem 3.6.3. *Via C, $\mathbb{F}_{\wp^n}$ is **A**-module isomorphic to $\mathbf{A}/(f^n - 1)$.*

Proof. $\mathbb{F}_{\wp^n}$ is a finite **A**-module and must be cyclic by Theorem 3.6.2. Now by Part 2 of Theorem 3.6.2 we see that

$$C_{f^n}(x) \equiv x^{r^{n \deg f}} \pmod{\wp \mathbf{A}[x]}.$$

Thus C_{f^n-1} annihilates $\mathbb{F}_{\wp^n}$. A simple count now implies the result. □

Theorem 3.6.3 illustrates again the remarkable similarities between C and the multiplicative group $\mathbb{G}_m$. One can try to push this similarity in a number of directions and we present one such direction here.

Thus, let a be an integer $\neq -1, 0$ or 1. We assume that a is *square free*. Let $N_a(x)$ be the number of primes $p \leq x$ for which a is primitive modulo p, i.e., a generates $\mathbb{Z}/p^*$. Then, in 1927, E. Artin conjectured that $N_a(x)$ had a positive density (of a certain given form). This result is now known modulo certain generalized Riemann Hypotheses. For a good reference, we refer the reader to [Le1].

We now try to do an analogous thing with **A**, C, etc. For the moment, let M be the set of all *monic* primes of **A**. Standard arguments on primes in arithmetic progression imply that the greatest common divisor of

$$\{f - 1 \mid f \in M \text{ and } \deg f > 1\}$$

is 1 *unless* $r = 2$ in which case it is easily seen to be $T(T + 1)$.

As will be seen in later sections, the classical proofs of cyclotomic theory also work for division points of the Carlitz module. One finds that $C(\mathbf{k})$ has non-trivial **A**-torsion only when $r = 2$. In this case, the torsion submodule is

$$C[T(T+1)] = \{0, T, T+1, 1\}.$$

The reader may profit by comparing what we have seen for $T(T + 1)$ and what happens classically for 2, the odd primes, and $\{\pm 1\}$.

Thus let $a \in \mathbf{A}$ be nonzero. If $r = 2$, we also require $a \notin \{0, T, T+1, 1\}$ and a is *not* of the form $C_T(b)$ or $C_{T+1}(b)$ for some $b \in \mathbf{A}$. For example, a might be a prime of degree > 1. Let $N_a(x)$ be the set of primes $\wp$ of **A** where the residue of a generates $C(\mathbb{F}_\wp)$. The analogy with classical theory leads one to expect an analog of Artin's conjecture for the Carlitz module and, in fact, this has been established in [Hsu1].

3.7. The Adjoint of the Carlitz Module

Let $\mathbf{k}^{\text{perf}} \subset \mathbf{C}_\infty$ be the perfection of **k**. Our goal in this short subsection is to show how Section 1.7 allows us to deduce the existence of the "τ-adjoint," or "adjoint," to the Carlitz module C.

3. The Carlitz Module

Recall that
$$C_a(\tau) = a\tau^0 + \sum_{i=1}^{\deg(a)} C_a^{(i)} \tau^i \in \mathbf{k}\{\tau\}.$$

By using Definition 1.7.1, we are led to our next definition.

Definition 3.7.1. We set
$$C_a^*(\tau) = \tau^0 + \sum_{i=1}^{\deg(a)} (C_a^{(i)})^{1/r^i} \tau^{-i} \in \mathbf{k}^{\text{perf}}\{\tau^{-1}\},$$
where $\mathbf{k}^{\text{perf}}\{\tau^{-1}\}$ is the ring of Frobenius polynomials in τ^{-1}.

Lemma 3.7.2. $C_{ba}^*(\tau) = C_a^*(\tau)C_b^*(\tau) = C_b^*(\tau)C_a^*(\tau) = C_{ab}^*(\tau).$

Proof. We have
$$C_{ba}(\tau) = C_b(\tau)C_a(\tau) = C_a(\tau)C_b(\tau) = C_{ab}(\tau).$$

Thus the result follows by applying Lemma 1.7.3. □

Remarks. 3.7.3. 1. Just as C is generated by C_T, so C_T^* is generated by $C_T^* = T\tau^0 + \tau^{-1}$. Thus
$$\begin{aligned} C_{T^2}^* &= (T\tau^0 + \tau^{-1})(T\tau^0 + \tau^{-1}) \\ &= T^2\tau^0 + (T^{1/r} + T)\tau^{-1} + \tau^{-2}. \end{aligned}$$

In general, one obtains C_a^* for C_a by formally treating τ as a variable and then replacing τ^i by τ^{-i}.

2. Proposition 3.3.10 can be adapted to C^*. We leave this as an exercise for the reader.

3. All the ideas of torsion points etc., make sense for C^*. We note, however, that the torsion points of C^* are actually algebraic over $\mathbf{k}$ (just raise C_a^* to the r^d-th power for $d = \deg a$). By 1.7.11 we see that the I-torsion points of C and C^* generate the *same* extension of $\mathbf{k}$.

4. Drinfeld Modules

4.1. Introduction

In this section, we begin discussing *Drinfeld modules*, our basic objects of study. Drinfeld modules are generalizations of the Carlitz module of our last section. Moreover, most of the salient features of Drinfeld modules are seen on the Carlitz module. Among these are: 1. The "lattice" $\mathbb{F}_r[T]\xi$; 2. The exponential of $\mathbb{F}_r[T]\xi$ (= the Carlitz exponential $e_C(x)$); 3. The multiplication law of $e_C(x)$ (= the Carlitz module C); 4. The algebraicity of C and its reductions to fields of "finite" characteristic; and 5. The action of $\mathbb{F}_r[T]$ on rational points via C.

The basic reference for this chapter is, of course, Drinfeld's original paper [Dr1]. Other sources are [Dr2], [DeH1], [Ha1], [Ha2] and [Ge2].

Drinfeld modules exist in remarkable generality. So we now fix some notation that will be used throughout this book. Let X be a smooth, projective, geometrically connected curve over the finite field $\mathbb{F}_r$, $r = p^{m_0}$. Let $\infty \in X$ be a *fixed* closed point of degree d_∞ over $\mathbb{F}_r$. We set $\mathbf{k}$ to be the function field of X and $\mathbf{A} \subset \mathbf{k}$ to be the ring of functions regular *outside* ∞. Let v_∞ be the valuation associated to the prime ∞ and let $\mathbf{K} = \mathbf{k}_\infty$ be the associated completion. Let $\overline{\mathbf{K}}$ be a fixed algebraic closure of $\mathbf{K}$ and let $\mathbf{C}_\infty$ be the completion of $\overline{\mathbf{K}}$ coming from the canonical extension of v_∞ to $\overline{\mathbf{K}}$; we also use v_∞ to denote this extension to $\overline{\mathbf{K}}$ and thus to $\mathbf{C}_\infty$. We note again that $\mathbf{C}_\infty$ is algebraically closed.

For divisors D of X, we set $\deg D$ to be the degree of D over $\mathbb{F}_r$. We put $d_\infty = \deg \infty$. For $x \in \mathbf{K}$, we set $\deg x = -d_\infty v_\infty(x)$. Thus if $0 \neq a \in \mathbf{A}$, then $\deg a = \log_r(\#\mathbf{A}/(a)) =$ the degree of the *finite part* of the divisor of a.

We let $|x|_\infty$ be the normalized absolute value associated to ∞. Thus

$$|x|_\infty = r^{\deg x} = r^{-d_\infty v_\infty(x)}.$$

Note that $|?|_\infty$ and $\deg(?)$ have canonical extensions to $\mathbf{C}_\infty$ arising from that of $v_\infty(?)$.

It is well-known, and directly seen, that $\mathbf{A}$ is a Dedekind domain. In fact, one has $\mathrm{Spec}(\mathbf{A}) = X - \infty$. The unit group of $\mathbf{A}$ is $\mathbb{F}_r^*$ as every non-constant $a \in \mathbf{A}$ must have zeros in order to balance out the pole at ∞ (the degree of a principal divisor *on* X must be 0).

4. Drinfeld Modules

Proposition 4.1.1. *Let $I \subset \mathbf{k}$ be a non-trivial $\mathbf{A}$-fractional ideal. Then $I \subset \mathbf{K}$ is discrete and co-compact.*

Proof. By definition, there is an $\alpha \in \mathbf{k}^*$ with $\alpha I \subseteq \mathbf{A}$. The discreteness follows from this. The compactness of $\mathbf{K}/I$ follows, for example, from the Riemann-Roch Theorem. □

Thus, Proposition 4.1.1 is the generalization to general $\mathbf{A}$ of Proposition 3.1.1.

Let $\mathrm{Jac}_X(\mathbb{F}_r)$ be the finite group of divisors of degree 0 on X modulo divisors of elements of $\mathbf{k}^*$. We set $h(\mathbf{k})$ to be the order of this group; we call $h(\mathbf{k})$ the "class number of (the field) $\mathbf{k}$."

Let D be a divisor of degree 0 defined over $\mathbb{F}_r$. We write

$$D = D_0 + c\infty$$

where $c \in \mathbb{Z}$ and D_0 is a divisor involving closed points in $X - \infty$. Thus

$$\deg D_0 = -cd_\infty .$$

We set

$$\pi(D) = D_0 ;$$

it is clear that π gives rise to a homomorphism from $\mathrm{Jac}_X(\mathbb{F}_r)$ to the usual class group, $\mathrm{Cl}(\mathbf{A})$, of $\mathbf{A}$ as a Dedekind domain.

Lemma 4.1.2. *The following sequence is exact*

$$0 \longrightarrow \mathrm{Jac}_X(\mathbb{F}_r) \xrightarrow{\pi} \mathrm{Cl}(\mathbf{A}) \xrightarrow{\deg} \mathbb{Z}/(d_\infty) \longrightarrow 0 .$$

Proof. It is clear that π is injective. From above we see that

$$\deg(\pi(D)) = \deg(D_0) \in (d_\infty) .$$

Conversely, let D_0 be a fractional ideal of $\mathbf{A}$ viewed as a divisor on X – so ideals of $\mathbf{A}$ will correspond to effective divisors – with $\deg D_0 \in (d_\infty)$. It is clear that we can complete D_0 to a divisor D of degree 0 on X as above.

Finally, it is well-known that we can choose on X a divisor D_1 of degree 1 only involving closed points in $X - \infty$. Thus D_1 comes from a fractional ideal of $\mathbf{A}$ and so establishes that the map to $\mathbb{Z}/(d_\infty)$ is surjective. □

Let $h(\mathbf{A})$ be the class number of $\mathbf{A}$ as a Dedekind domain.

Corollary 4.1.3. *We have $h(\mathbf{A}) = d_\infty h(\mathbf{k})$.* □

Thus the set-up that we are using is the natural generalization of that used for the Carlitz module: $\mathbf{A}$ plays the role of $\mathbb{Z}$, $\mathbf{k}$ the role of $\mathbb{Q}$, $\mathbf{K}$ the role of $\mathbb{R}$, etc. However, it is apparent that the analogies are now somewhat weaker than in the $\mathbb{F}_r[T]$-case. Indeed, **A** can now have non-principal ideals, etc. This will lead eventually to the construction of certain fields (i.e., certain Hilbert class fields) that *also* play some of the role classically played by $\mathbb{Q}$, etc. In other words, in the function field theory, one needs *many* fields to play the roles classically played by the one field $\mathbb{Q}$. An important aspect of the function field theory is that this "splitting" allows us to pinpoint which properties are at the core of various well-known results (e.g., sometimes, the role of $\mathbb{Q}$ will be played by **k** and other times by a Hilbert class field — this can *not* be seen in classical theory as $\mathbb{Q}$ is its own Hilbert class field).

This tendency of the function field theory to illustrate the "decomposition" of classical number theoretic objects into their "arithmetic components" is a general phenomenon. In this sense, one obtains insight of a "metamathematical" nature from function fields.

4.2. Lattices and Their Exponential Functions

We can now begin to define Drinfeld modules. As we did with the Carlitz module, we will begin analytically over **K**.

We continue with the notation of Subsection 4.1. Let **M** be a complete extension of $\mathbf{K} \subseteq \mathbf{C}_\infty$.

Definition 4.2.1. An **A**-submodule $L \subset \mathbf{C}_\infty$ (with the usual multiplication of **A**) is called an **M**-*lattice* (or *lattice*, or **A**-*lattice*, if **M** is not specified) if and only if
1. L is finitely generated as **A**-module;
2. L is discrete in the topology of $\mathbf{C}_\infty$;
3. Let $\mathbf{M}^{\mathrm{sep}} \subseteq \mathbf{C}_\infty$ be the separable closure of **M**. Then L should be contained in $\mathbf{M}^{\mathrm{sep}}$ and be stable under $\mathrm{Gal}(\mathbf{M}^{\mathrm{sep}}/\mathbf{M})$.

The *rank* of L is its rank as a finitely generated torsion-free (=finitely generated projective) submodule of $\mathbf{C}_\infty$.

The discreteness of L is equivalent to having any finite ball in $\mathbf{C}_\infty$ contain only finitely many elements of L.

Example 4.2.2. Let $\mathbf{A} = \mathbb{F}_r[T]$, etc. Let ξ be as in Definition 3.2.7. It is simple to see that $L = \mathbf{A}\xi$ is a **K**-lattice.

Definition 4.2.3. Let L be as in Definition 4.2.1. Then we set
$$e_L(x) = x \prod_{\substack{\alpha \in L \\ 0 \neq \alpha}} (1 - x/\alpha).$$

4. Drinfeld Modules

Proposition 4.2.4. *The product for $e_L(x)$ converges for all x in $\mathbf{C}_\infty$. The resulting entire function has Taylor expansion about $x = 0$ with coefficients in $\mathbf{M}$.*

Proof. The first statement is a simple consequence of the discreteness of L. The second statement is a simple consequence of the $\mathrm{Gal}(\mathbf{M}^{\mathrm{sep}}/\mathbf{M})$ stability of $L \subset \mathbf{M}^{\mathrm{sep}}$ as well as the continuous nature of the Galois action on $\mathbf{M}^{\mathrm{sep}}$. □

Proposition 4.2.5. *The entire function $e_L(x)$ is $\mathbb{F}_r$-linear.*

Proof. Write $L = \cup L_i$ where L_i is a finite dimensional $\mathbb{F}_r$-vector space. Thus

$$e_L(x) = \lim_i e_{L_i}(x)$$

where $e_{L_i}(x) = x \prod_{0 \neq \alpha \in L_i} (1 - x/\alpha)$. The polynomial $e_{L_i}(x)$ is $\mathbb{F}_r$-linear by Corollary 1.2.2; the result follows directly. □

Corollary 4.2.6. *The function $e_L(x)$ gives rise to an isomorphism of abelian groups*

$$\mathbf{C}_\infty/L \stackrel{e_L(x)}{\simeq} \mathbf{C}_\infty.$$

Proof. The kernel of $e_L(x)$ as an $\mathbb{F}_r$-linear map is just L. Moreover, $e_L(x)$ is surjective on $\mathbf{C}_\infty$ as is *every* non-constant entire function. The result is now easy. □

4.3. The Drinfeld Module Associated to a Lattice

We continue with the notation of the last subsection. Thus L is a lattice associated to an extension $\mathbf{M}$ of $\mathbf{K}$ inside $\mathbf{C}_\infty$. We set $d = \mathrm{rank}_\mathbf{A}(L)$.

Let $0 \neq a \in \mathbf{A}$. The following result, due to Drinfeld, is fundamental to the theory.

Theorem 4.3.1. *We have the following equality of entire functions*

$$e_L(ax) = ae_L(x) \prod_{0 \neq \alpha \in a^{-1}L/L} (1 - e_L(x)/e_L(\alpha)).$$

Proof. Set

$$P(x) = x \prod_{0 \neq \alpha \in a^{-1}L/L} (1 - x/e_L(\alpha)).$$

As $e_L(x)$ is $\mathbb{F}_r$-linear, $\{e_L(a^{-1}L/L)\}$ is a finite $\mathbb{F}_r$-vector space; thus $P(x)$ is an $\mathbb{F}_r$-linear polynomial. Therefore, $P'(x) \equiv 1$.

Now the entire functions $e_L(ax)$ and $aP(e_L(x))$ have the same divisor. By above, they have the same derivative. Thus by Theorem 2.14, they are equal. $\square$

We will use the notation "ϕ_a^L", or just "ϕ_a" if L is understood, for the polynomial $ax \prod_{0 \neq \alpha \in a^{-1}L/L} (1 - x/e_L(\alpha))$. Therefore, from the proof of Theorem 4.3.1, we see that
$$\phi_a \in \mathbf{M}\{\tau\},$$
where τ is the rth power mapping and $\mathbf{M}\{\tau\}$ is the composition ring of Frobenius polynomials in τ. As $\mathbf{A}$ is a Dedekind domain and L is finitely generated and torsion-free, one knows [CuR1, 22.5, 22.11] that there is an isomorphism of $\mathbf{A}$-modules
$$L \simeq \mathbf{A}^{d-1} \oplus I,$$
where I is a non-zero ideal of $\mathbf{A}$. Thus
$$a^{-1}L/L \simeq \bigoplus_{i=1}^{d} \mathbf{A}/(a),$$
and, in particular, has $r^{d \deg(a)}$ elements. Consequently,
$$\deg \phi_a(\tau) = d \deg(a).$$

Proposition 4.3.2. *The mapping $a \in \mathbf{A} \mapsto \phi_a \in M\{\tau\}$ has the following properties:*
1. *It is $\mathbb{F}_r$-linear;*
2. *If $a \in \mathbb{F}_r \subset \mathbf{A} \Rightarrow \phi_a = a\tau^0$;*
3. *$\phi_{ab}(\tau) = \phi_a(\tau)\phi_b(\tau) = \phi_b(\tau)\phi_a(\tau) = \phi_{ba}(\tau)$.*

Proof. 1. This follows immediately from the definitions. 2. If $a \in \mathbb{F}_r^*$, then $a^{-1}L = L$. 3. $\phi_{ab}(e_L(x)) = e_L(abx) = \phi_a(e_L(bx)) = \phi_a(\phi_b(e_L(x)))$, etc. $\square$

Definition 4.3.3. The injection $\mathbf{A} \to \mathbf{M}\{\tau\}$, $a \mapsto \phi_a$, associated to L is called the *Drinfeld module associated to L*. Its *rank* is $d = \text{rank}_\mathbf{A}(L)$.

Definition 4.3.4. Let L_1, L_2 be two $\mathbf{A}$-lattices of the same rank. A *morphism* from L_1 to L_2 is an element $c \in \mathbf{C}_\infty$ with $cL_1 \subseteq L_2$. If the ranks of L_1 and L_2 are different, then we only allow $0 \in \mathbf{C}_\infty$ to be a morphism. If, further, L_1, L_2 are M-lattices for some complete field $\mathbf{K} \subseteq \mathbf{M} \subseteq \mathbf{C}_\infty$, then an $\mathbf{M}$-*morphism* is a *morphism* from L_1 to L_2 given by some $c \in \mathbf{M} \subseteq \mathbf{C}_\infty$.

The reader should note that if L_1 and L_2 are contained in a field $\mathcal{F}$, then so is c.

Proposition 4.3.5. *Let ϕ and ψ be two Drinfeld modules associated to lattices L_1 and L_2, respectively, of the same rank. Let $c \in \mathbf{C}_\infty$ be a morphism from L_1 to L_2. Then via the isomorphisms*

$$e_{L_1}: \mathbf{C}_\infty/L_1 \xrightarrow{\sim} \mathbf{C}_\infty, \quad e_{L_2}: \mathbf{C}_\infty/L_2 \xrightarrow{\sim} \mathbf{C}_\infty,$$

the element $c \in \mathbf{C}_\infty$ corresponds to a polynomial $P(\tau) := P_c(\tau) \in \mathbf{C}_\infty(\tau)$ with

$$P\phi_a = \psi_a P$$

for all $a \in \mathbf{A}$.

Proof. Look at the function $e_{L_2}(cx)$. By definition, it is zero on the lattice $c^{-1}L_2$ which contains L_1 by assumption. As L_1 and $c^{-1}L_2$ have the same rank, we deduce that

$$c^{-1}L_2/L_1$$

is finite. Now put

$$P(x) := P_c(x) = cx \prod_{0 \neq \alpha \in c^{-1}L_2/L_1} (1 - x/e_{L_1}(\alpha)).$$

As usual, $P(x)$ is $\mathbb{F}_r$-linear. Moreover, the function

$$P(e_{L_1}(x))$$

has a simple zero at each point of $c^{-1}L_2$ with derivative c. Thus

$$P(e_{L_1}(x)) = e_{L_2}(cx)$$

and the result is now easily established. □

Remarks. 4.3.6. 1. Let L be a lattice as above with associated Drinfeld module ϕ. The action of $a \in \mathbf{A}$ via ϕ_a can be summarized by the commutative diagram:

$$\begin{array}{ccc} \mathbf{C}_\infty/L & \xrightarrow{a} & \mathbf{C}_\infty/L \\ e_L \downarrow \sim & & e_L \downarrow \sim \\ \mathbf{C}_\infty & \xrightarrow{\phi_a} & \mathbf{C}_\infty \end{array}$$

Similarly if L_1 and L_2 are lattices with Drinfeld modules ϕ, ψ respectively, and if $c \in \mathbf{C}_\infty$ is a morphism from L_1 to L_2, then the morphism from ϕ to ψ associated to c is summarized by:

$$\begin{array}{ccc}
\mathbf{C}_\infty/L_1 & \xrightarrow{c} & \mathbf{C}_\infty/L_2 \\
e_{L_1} \downarrow \tilde{\rightarrow} & & e_{L_2} \downarrow \tilde{\rightarrow} \\
\mathbf{C}_\infty & \xrightarrow{P_c} & \mathbf{C}_\infty
\end{array}$$

2. Let L and ϕ be as above. Suppose that L is an **M**-lattice for some complete subfield $\mathbf{M} \subseteq \mathbf{C}_\infty$. Let $\mathbf{M}_1 \subseteq \mathbf{M}^{\text{sep}}$ be a finite dimensional extension of **M** which contains L. Then $\mathbf{M}_1$ also contains *all* points

$$\{e_L(b\alpha) \mid b \in \mathbf{k} \text{ and } \alpha \in L\}.$$

Note that these are *precisely* the torsion points of ϕ, i.e., those points $\lambda \in \mathbf{C}_\infty$ such that there exists $a \in \mathbf{A}$ with $\phi_a(\lambda) = 0$. Finally, $\phi_a \in \mathbf{M}\{\tau\}$ for all a.

4.4. The General Definition of a Drinfeld Module

We now continue along the path previously taken with the Carlitz module — the analytic theory via lattices dictates the general definitions over arbitrary fields. We also continue with previous notation.

Definition 4.4.1. An **A**-*field* $\mathcal{F}$ is a field $\mathcal{F}$ equipped with a fixed morphism $\imath \colon \mathbf{A} \to \mathcal{F}$. The prime ideal $\wp$ which is the kernel of $\imath$ is called the *characteristic* of $\mathcal{F}$. We say $\mathcal{F}$ has *generic characteristic* if and only if $\wp = (0)$; otherwise we say that $\wp$ is *finite* and $\mathcal{F}$ has *finite characteristic*.

Definition 4.4.1 obviously agrees with our previous definition in the case $\mathbf{A} = \mathbb{F}_r[T]$.

As in Section 1, over $\mathcal{F}$ we have the ring $\mathcal{F}\{\tau\}$, with τ the rth power mapping. Let $f(\tau) = \sum_{i=0}^{v} a_i \tau^i \in \mathcal{F}\{\tau\}$. We set

$$Df := a_0 = f'(\tau).$$

It is clear that the mapping $\mathcal{F}\{\tau\} \to \mathcal{F}$, $f \mapsto Df$, is a morphism of $\mathbb{F}_r$-algebras.

We then have the following fundamental definition.

Definition 4.4.2. Let $\phi \colon \mathbf{A} \to \mathcal{F}\{\tau\}$ be a homomorphism of $\mathbb{F}_r$-algebras. Then ϕ is a *Drinfeld module* over $\mathcal{F}$ if and only if
1. $D \circ \phi = \imath$;
2. For some $a \in \mathbf{A}$, $\phi_a \neq \imath(a)\tau^0$.

The normalization in 4.4.2.1 is analogous to the normalization used in complex multiplications of elliptic curves. The condition 4.4.2.2 is obviously a non-triviality condition.

70   4. Drinfeld Modules

Note that if $a \notin \wp$ = the characteristic of $\mathcal{F}$, then ϕ_a is separable. Note also that we could have used $\mathcal{F}\{\tau_p\}$ in Definition 4.4.2 instead of $\mathcal{F}\{\tau\}$. Indeed, it is trivial to see that the image of $\mathbb{F}_r$ under any homomorphism to $\mathcal{F}\{\tau_p\}$ must be $\mathbb{F}_r \tau_p^0$. Thus the image of ϕ inside $\mathcal{F}\{\tau_p\}$ must be contained in $\mathcal{F}\{\tau\}$ by commutativity.

Via ϕ, any extension $\mathcal{L}$ of $\mathcal{F}$ becomes an **A**-module. We denote this module by $\phi(\mathcal{L})$.

Proposition 4.3.5 leads to the next definition.

Definition 4.4.3. Let ϕ and ψ be two Drinfeld modules over an **A**-field $\mathcal{F}$. A *morphism* from ϕ to ψ over $\mathcal{F}$ is a polynomial $P(\tau) \in \mathcal{F}\{\tau\}$ with

$$P\phi_a = \psi_a P$$

for all $a \in \mathbf{A}$. Nonzero morphisms are called *isogenies*.

Let ϕ be a Drinfeld module over $\mathcal{F}$, and let $I \subseteq \mathbf{A}$ be an ideal. As **A** is a Dedekind domain, one knows that I may be generated by (at most) two elements $\{i_1, i_2\} \subset I$. Since $\mathcal{F}\{\tau\}$ has a right division algorithm (cf. Subsection 1.6), there exists a right greatest common divisor in $\mathcal{F}\{\tau\}$. It is the *monic* generator of the left ideal of $\mathcal{F}\{\tau\}$ generated by ϕ_{i_1}, ϕ_{i_2}.

Definition 4.4.4. We set ϕ_I to be the monic generator of the left ideal of $\mathcal{F}\{\tau\}$ generated by ϕ_{i_1} and ϕ_{i_2}.

Now let $\overline{\mathcal{F}}$ be a fixed algebraic closure of $\mathcal{F}$.

Definition 4.4.5. We define $\phi[I] \subset \phi(\overline{\mathcal{F}})$ to be the finite subgroup given by the roots of ϕ_I. If $a \in \mathbf{A}$ then we set $\phi[a] := \phi[(a)]$.

Remarks. 4.4.6. 1. The group $\phi[I]$ is clearly stable under $\{\phi_a\}_{a \in \mathbf{A}}$. Thus it is *also* a finite **A**-module.
2. If the characteristic, $\wp$, divides I, then ϕ_I will not be separable. As such, ϕ_I gives rise to a finite **A**-stable *subgroup scheme* of $\mathbb{G}_a$ whose $\overline{\mathcal{F}}$-points equals $\phi[I]$. We will use the same notation for this group scheme when no confusion will result.

4.5. The Height and Rank of a Drinfeld Module

We continue with the notation and set-up of the previous subsections. Thus ϕ is a Drinfeld module over the **A**-field $\mathcal{F}$.

For each $a \in \mathbf{A}$, we set

$$\mu_\phi(a) := \mu(a) := -\deg \phi_a(\tau);$$

(so $\mu(0) = \infty$).

4.5. The Height and Rank of a Drinfeld Module

Lemma 4.5.1. *There exists a nonzero rational number d such that*
$$\mu(a) = dd_\infty v_\infty(a) = -d\deg(a).$$

Proof. The mapping μ satisfies $\mu(ab) = u(a) + \mu(b)$, and $\mu(a+b) \geq \min\{\mu(a), \mu(b)\}$. By our non-triviality assumptions on ϕ, μ gives rise to a place of $\mathbf{k}$ which can only correspond to ∞. This gives the result. $\square$

Corollary 4.5.2. *Let $\phi\colon \mathbf{A} \to \mathcal{F}\{\tau\}$ be a Drinfeld module. Then ϕ is injective.* $\square$

Proposition 4.5.3. *The number d of Lemma 4.5.1 is a positive integer.*

Proof. Let $\mathfrak{B}$ be a prime of $\mathbf{A}$ different from the characteristic of $\mathcal{F}$. Let h be the class number of $\mathbf{A}$ as a Dedekind domain and let $(a) = \mathfrak{B}^h$. Let $\overline{\mathcal{F}}$ be a fixed algebraic closure of $\mathcal{F}$. Consider
$$\phi[a] \subset \overline{\mathcal{F}};$$
this is a finite $\mathbf{A}$-module of precisely $r^{d\deg a}$ elements by Lemma 4.5.1. On the other hand, one sees that all $b \in \mathbf{A}$, b prime to $\mathfrak{B}$, must act as *automorphisms* of $\phi[a]$. The general theory of modules over Dedekind domains [CuR1, 22.16] thus gives
$$\phi[a] \simeq \bigoplus_{i=1}^{t} \mathbf{A}/(\mathfrak{B}^{e_i}),$$
for some integers t and $\{e_i\}$. If one uses this result to count the number of elements in $\phi[\mathfrak{B}^j]$, for $j \leq h$, one sees that $t = d$ and $e_i = h$ for all i finishing the proof. $\square$

The above proof is very similar to the classical argument which shows that a finite multiplicative group of roots of unity is cyclic. From 4.5.3 one sees easily that if $b \in \mathbf{A}$ is prime to the characteristic of $\mathcal{F}$, then $\phi[b] \subset \phi(\overline{\mathcal{F}})$ is isomorphic to $(\mathbf{A}/(b))^d$.

Definition 4.5.4. The integer of Proposition 4.5.3 is the *rank* of the Drinfeld module ϕ.

Remarks. 4.5.5. 1. If $a \in \mathbf{A}$ is prime to the characteristic of $\mathcal{F}$, we see that $\phi[a] \simeq (\mathbf{A}/(a))^{\text{rank }\phi}$ as $\mathbf{A}$-modules.
2. Suppose $\mathcal{F} \subseteq \mathbf{C}_\infty$ and that ϕ comes from a lattice L as in Subsection 4.3. Using Part 1, it is very easy to see that ϕ and L have the *same* rank. Thus Definition 4.5.4 agrees with Definition 4.3.3.
3. It is now simple to see that one can have a non-zero morphism only between two Drinfeld modules of the same rank.

We now turn to the definition of the "height" of ϕ. For this, we assume that $\text{char}(\mathcal{F}) = \wp \neq (0)$. We let $v_\wp : k \to \mathbb{Z}$ be the normalized valuation associated to $\wp$; thus, e.g., if $a \in \mathbf{k}$ has a zero at $\wp$ of order t, one has $v_\wp(a) = t$, etc.

For each $a \in \mathbf{A}$, let $\omega(a)$ be the smallest integer $t \geq 0$ with τ^t occurring in ϕ_a with *nonzero* coefficient.

Lemma 4.5.6. *There exists a positive rational number h with*
$$\omega(a) = h v_\wp(a) \deg \wp,$$
for all $a \in \mathbf{A}$.

Proof. The mapping $a \mapsto \omega(a)$ satisfies $\omega(ab) = \omega(a) + \omega(b)$, and $\omega(a+b) \geq \min\{\omega(a), \omega(b)\}$. As we have seen before, one can now extend ω to a valuation on $\mathbf{k}$ which must correspond to $\wp$. This gives the existence of h. $\square$

Proposition 4.5.7. *Let h be as in Lemma 4.5.6. Then h is a positive integer.*

Proof. Let $\phi[\wp]$ be the $\mathbf{A}/\wp$-module of $\wp$-division points in a fixed algebraic closure $\overline{\mathcal{F}}$ of $\mathcal{F}$. The points of $\phi[\wp]$ are the roots of the equation $\phi_\wp(x) = 0$ where $\phi_\wp$ is given in Definition 4.4.4. We write
$$\phi_\wp(\tau) = \alpha \tau^{\omega(\wp)} + \{\text{higher terms}\}$$
with $\alpha \neq 0$. Thus clearly
$$\#\phi[\wp] = r^{\deg \phi_\wp(\tau) - \omega(\wp)}.$$
On the other hand, if t is the dimension of $\phi[\wp]$ as an $\mathbf{A}/\wp$-vector space, then
$$\#\phi[\wp] = r^{t \deg \wp}.$$

As $\mathbf{A}$ has a finite class number, one can find e (e.g., e = class number of $\mathbf{A}$) with $\wp^e$ principal, say
$$\wp^e = (a).$$
By using the filtration $\phi[\wp^i] \subseteq \phi[\wp^{i+1}]$, $i = 0, \ldots$, one sees that
$$\#\phi[a] = \#\phi[\wp^e] = r^{e(\deg \phi_\wp(\tau) - \omega(\wp))}$$
$$= r^{et \deg \wp}.$$

Using Lemma 4.5.6, one computes
$$\#\phi[a] = r^{d \deg(a) - \omega(a)}$$
$$= r^{ed \deg \wp - he \deg \wp}.$$

Thus $h = d - t$ and the result follows from 4.5.3. $\square$

Definition 4.5.8. The natural number h of Proposition 4.5.7 is called the *height* of ϕ.

The reader will note that the genesis of the above definition lies in the standard notion of the height of a formal group.

Example 4.5.9. Let $\mathbf{A} = \mathbb{F}_r[T]$, $\mathbf{k} = \mathbb{F}_r(T)$. Let $\mathcal{F} = \mathbf{k}$ and let C be the Carlitz module $C\colon \mathbf{A} \to \mathbf{k}\{\tau\}$, $C_T = T\tau^0 + \tau$. Then C is clearly of rank 1 as a Drinfeld module over any $\mathbf{A}$-field. Moreover, if $(f) = \wp$ is a prime ideal of $\mathbf{A}$, f monic, then C is of height 1 at $\wp$. That is, consider C as a Drinfeld module over $\mathbf{A}/\wp$ by reducing the coefficients mod $\wp$. Then C has height 1 over $\mathbf{A}/\wp$ as $C_f(x)$ now is an inseparable polynomial and so an easy counting argument implies that, modulo $\wp$, $C_f(\tau) = \tau^{\deg f}$.

4.6. Lattices and Drinfeld Modules over $\mathbf{C}_\infty$

We begin by presenting some material from [DeH1,§3] on discrete modules in vector spaces over local fields. For this, we first abstract a bit from the situation of the last subsection and consider a more general set-up. So we let K now be a local field and $A \subset K$ a *discrete* subring with K/A compact. We let k be the quotient field of A. The basic examples are:

1. $A = \mathbf{A}$, $K = \mathbf{K}$, $k = \mathbf{k}$;
2. $A = \mathbb{Z}$, $K = \mathbb{R}$, $k = \mathbb{Q}$;
3. k is a complex quadratic field, $A =$ the ring of integers and $K = \mathbb{C}$.

Let V be a finite dimensional vector space over K. We equip V with the usual sup-norm based on an absolute value $|?|$ on K. It is well-known that *all* K-norms on V compatible with $|?|$ are equivalent to each other, and thus to the sup-norm. A subgroup $H \subset V$ is said to be *discrete* if and only if there is a non-trivial neighborhood N of 0 with $H \cap N = \{0\}$.

Definition 4.6.1. Let H be a finitely generated A-module. The *rank* of H is the dimension of $k \otimes_A H$. We denote it by "$\mathrm{rank}_A(H)$."

Note that our chosen examples of A are Dedekind domains. Thus, in these cases, one is able to decompose H up to isomorphism as

$$H \simeq T \oplus \underbrace{A \oplus \cdots \oplus A}_{t-\text{times}} \oplus I,$$

where T is a finitely generated torsion module, $I \subseteq A$ is a non-vanishing ideal and $t+1$ is the rank of A.

Proposition 4.6.2. *Let V be a finite dimensional K-vector space of dimension d. Let $H \subset V$ be a discrete A-module. Then H is finitely generated over A and $\mathrm{rank}_A(H) \leq d$.*

Proof. Let $W = K \cdot H \subseteq V$. So W is a finitely generated vector space and we let $\{w_1, \ldots, w_n\}$ be a basis for W. Set

$$L = Aw_1 + \cdots + Aw_n \subseteq H \subseteq W \subseteq V.$$

As H is discrete in V, there is a neighborhood N^1 of 0 in V with $H \cap N^1 = \{0\}$.

One can readily find a neighborhood N of 0 in V with $N + N = N^1$. Thus one sees that for $x, y \in H$,

$$\{N + x\} \bigcap \{N + y\} \neq \phi \iff x = y.$$

Thus $N + L$ is a neighborhood of 0 in V/L whose intersection with H/L is $\{0\}$. Thus H/L is a discrete subgroup of V/L and also of the compact abelian group

$$W/L = (Kw_1 + \cdots + Kw_n)/(Aw_1 + \cdots + Aw_n).$$

As such H/L is finite; therefore it is finitely generated and

$$\dim_k(k \otimes_A L) = \dim_k(k \otimes_A H) = \dim_K(W)$$

giving the result. □

Proposition 4.6.3. *Let H be a projective A-module contained in a finite dimensional K-vector space V. Then H is discrete in V if and only if the mapping $\theta \colon K \otimes_A H \to V$ is injective.*

Proof. Assume that H is discrete in V. By 4.6.2, we know that it is then finitely generated. Let $\mathrm{im}(\theta)$ be the image of θ. Moreover, $H \subset \mathrm{im}(\theta) \subseteq V$ so that H must be discrete in $\mathrm{im}(\theta)$. As such, 4.6.2 gives

$$\mathrm{rank}_A(H) \leq \dim_K(\mathrm{im}(\theta)).$$

On the other hand, for any finitely generated H, we have

$$\mathrm{rank}_A(H) = \dim_k(k \otimes_A H) = \dim_K(K \otimes_A H) \geq \dim_K(\mathrm{im}(\theta)).$$

Thus $\dim_K(K \otimes_A H) = \dim_K(\mathrm{im}(\theta))$ and θ is injective.

Conversely, let us now assume that θ is injective; we thus deduce that H has finite rank (and is therefore also finitely generated as it is torsion free). As H is projective, we can find a finitely generated projective H^1 with $H \oplus H^1$ free over A of rank $d_0 < \infty$. Upon adding $K \otimes H^1$ to $K \otimes H$, we now have the situation of a free module L of rank d_0 inside K^{d_0} and such that $K \cdot L = K^{d_0}$. Thus L is discrete and therefore so is $H \subset K \otimes H$. As θ is injective, H is also discrete inside V. □

We can now present the equivalence between the categories of lattices, with morphisms coming from elements of $\mathbf{C}_\infty$, and Drinfeld modules, with morphisms as in 4.4.3. We also now revert to the notation of the previous subsection.

Our exposition is modeled on that of [Ha1] and [Ha2].

Let **M** be a *complete* subfield of $\mathbf{C}_\infty$ which contains **K**. (So for instance, **M** is a finite extension of **K**.) Thus **M** is an **A**-field in the obvious fashion. Let $\varphi: \mathbf{A} \to \mathbf{M}\{\tau\}$ be a Drinfeld module of fixed rank d. Our goal is to establish that ψ comes from an **M**-lattice in the manner of Subsection 4.3. Moreover, if ψ and ϕ are Drinfeld modules of rank d over **M** and $P \in \mathbf{M}\{\tau\}$ is a morphism from ϕ to ψ, then we will show that P arises from a morphism of the corresponding **M**-lattices.

Let $0 \neq a \in \mathbf{A}$. From the definition of ψ, we see that $\psi_a(\tau) = a\tau^0 + \{\text{higher terms}\}$. Thus, we can formally expand $\tau^0/\psi_a(\tau)$ into a "power series"

$$\tau^0/\psi_a(\tau) = \frac{1}{a}\tau^0 + \sum_{i=1}^{\infty} a_i \tau^i \in \mathbf{M}\{\{\tau\}\},$$

where $\mathbf{M}\{\{\tau\}\}$ is the ring of "Frobenius power series" over **M** with the obvious definitions. Therefore, ψ extends uniquely to an injection $\psi: \mathbf{k} \to \mathbf{M}\{\{\tau\}\}$. This motivates our next definition.

Definition 4.6.4. Let $\psi: \mathbf{k} \to \mathbf{M}\{\{\tau\}\}$ be a morphism such that $D \circ \psi_a = a$ (where D of a power series is the coefficient of τ^0). We call ψ a *formal* **k**-*module* over **M**. We say that ψ is *non-trivial* if and only if $\psi_a \neq a\tau^0$ for some $0 \neq a \in \mathbf{k}$.

Therefore a Drinfeld module ψ gives rise to a unique formal **k**-module, which we will also denote by "ψ."

Lemma 4.6.5. *Let* $f(\tau) = \sum_{i=0}^{\infty} a_i \tau^i \in \mathbf{M}\{\{\tau\}\}$ *with* $Df = \alpha := a_0$. *Suppose* α *is transcendental over* $\mathbb{F}_r$. *Then there exists a unique power series*

$$\lambda_f = \sum_{i=0}^{\infty} c_i \tau^i \in \mathbf{M}\{\{\tau\}\}$$

with $c_0 = 1$, *and*

$$\lambda_f \cdot \alpha\tau^0 = f\lambda_f$$

(with the multiplication in $\mathbf{M}\{\{\tau\}\}$*).*

Proof. The equation

$$\lambda_f \cdot \alpha\tau^0 = f\lambda_f$$

is equivalent to the recurrence,

$$(\alpha^{r^i} - \alpha)c_i = \sum_{j=1}^{i} a_j c_{i-j}^{r^j}$$

for all $i \geq 1$. By assumption $\alpha^{r^i} - \alpha \neq 0$ for all i. Thus upon setting $c_0 = 1$, we can find c_i, $i > 0$, uniquely by recursion. □

Corollary 4.6.6. *Let $f(\tau)$, etc., be as in 4.6.5. Then $\lambda_f \mathbf{M} \lambda_f^{-1}$ is the centralizer of $f(\tau)$ in $\mathbf{M}\{\{\tau\}\}$.*

Proof. In $\mathbf{M}\{\{\tau\}\}$ the centralizer of $\alpha\tau^0$ is clearly $\mathbf{M} \cdot \tau^0$. Moreover, by the lemma,
$$\lambda_f \alpha \lambda_f^{-1} = f.$$
The result now follows immediately. □

Proposition 4.6.7. *Let $\psi \colon \mathbf{k} \to \mathbf{M}\{\{\tau\}\}$ be a formal module over $\mathbf{M}$. Then there exists a unique power series $e_\psi \in \mathbf{M}\{\{\tau\}\}$ with $De_\psi = 1$ and*
$$\psi_a = e_\psi a e_\psi^{-1},$$
for all $a \in \mathbf{k}$.

Proof. If ψ is trivial, then we set $e_\psi = \tau^0$. If ψ is non-trivial, let $\alpha \in \mathbf{k}$ be chosen with $\psi_\alpha \neq \alpha\tau^0$, so α is transcendental over $\mathbf{k}$. Let
$$e_\psi = \lambda_{\psi_\alpha}.$$
Choose $x \in \mathbf{k}$. Then $\psi_x \psi_\alpha = \psi_\alpha \psi_x$ by definition. Thus, by 4.6.6 we deduce the existence of $t \in \mathbf{M}$ with
$$\psi_x = e_\psi t e_\psi^{-1}.$$
Upon looking at the coefficient of τ^0, in the above equation, we find $t = x$.
The uniqueness follows from 4.6.5. □

Remarks. 4.6.8. 1. Let τ_p be the pth power morphism; there is thus an injection $\mathbf{M}\{\{\tau\}\} \hookrightarrow \mathbf{M}\{\{\tau_p\}\}$. A formal k-module $\psi \colon \mathbf{k} \to \mathbf{M}\{\{\tau\}\}$ thus gives rise to a formal k-module $\psi \colon \mathbf{k} \to \mathbf{M}\{\{\tau_p\}\}$. It is important to note that by the uniqueness properties (Lemma 4.6.5) of e_ψ, we obtain the *same* power series whether we work in $\mathbf{M}\{\{\tau\}\}$ or in $\mathbf{M}\{\{\tau_p\}\}$.
2. When $\psi = C$ is the Carlitz module, one sees easily that e_ψ is just the exponential of C, e_C, as discussed in Section 3.

We can now prove the main result of this section. It is the *analytic uniformization theorem* for Drinfeld modules.

Theorem 4.6.9. *Let ψ be a Drinfeld module over $\mathbf{M}$ of rank $d > 0$. Then there is an $\mathbf{M}$-lattice $L := L_\psi$ of rank d such that ψ is the associated Drinfeld module in the manner of Theorem 4.3.1. Moreover, the association $\psi \mapsto L_\psi$ gives rise to an equivalence of categories between the category of Drinfeld*

4.6. Lattices and Drinfeld Modules over $\mathbf{C}_\infty$

modules of rank d over **M** and the category of **M**-lattices of rank d (equipped with **M**-morphisms of **M**-lattices).

Proof. We know that ψ gives rise to a non-trivial formal **k**-module and thus to an element $e_\psi \in \mathbf{M}\{\{\tau\}\}$. From Proposition 4.6.7, we have

$$e_\psi a \tau^0 = \psi_a e_\psi$$

for all $a \in \mathbf{A}$.

Our first order of business is to show that e_ψ is entire. We write

$$e_\psi = \sum_{i=0}^{\infty} c_i \tau^i, \quad c_0 = 1.$$

The reader will easily note that e_ψ is entire if and only if $c_i^{1/r^i} \to 0$ as $i \to \infty$. Let $a \in \mathbf{A}$ with $\deg a > 0$; so a is clearly transcendental over $\mathbb{F}_r$. Put

$$\psi_a = a\tau^0 + \sum_{i=1}^{t} a_i \tau^i.$$

Let $n \geq t$. From the recursive formula used in the proof of Lemma 4.6.5, we find

$$(a^{r^n} - a)c_n = \sum_{j=1}^{t} a_j c_{n-j}^{r^j}. \tag{$*$}$$

Let v_∞ be the valuation at ∞ on **k** which we have extended to $\mathbf{C}_\infty$. We find from $*$ that

$$v_\infty(a) + \frac{v_\infty(c_n)}{r^n} \geq \min_{i \leq j \leq t} \{v_\infty(a_j)/r^n + r^{j-n} v_\infty(c_{n-j})\},$$

or

$$\frac{v_\infty(c_n)}{r^n} \geq \min_{1 \leq j \leq t} \left\{ \frac{v_\infty(a_j)}{r^n} + r^{j-n} v_\infty(c_{n-j}) \right\} - v_\infty(a).$$

Let θ be chosen so that $\theta < v_\infty(a) < 0$. Then for n sufficiently large, say $n \geq n_0$, we have

$$\min_{i \leq j \leq t} \left\{ \frac{v_\infty(a_j)}{r^n} \right\} < v_\infty(a) - \theta.$$

Thus,

$$\frac{v_\infty(c_n)}{r^n} \geq \min_{1 \leq j \leq t} \left\{ \frac{v_\infty(c_{n-j})}{r^{n-j}} \right\} - \theta.$$

By using this formula recursively, we see that $\frac{v_\infty(c_n)}{r^n} \to \infty$ as $-n\theta$ does. Thus, e_ψ is entire.

78   4. Drinfeld Modules

Now let L_ψ be the kernel of e_ψ = its set of zeros. From 2.10.3, we see that L_ψ is contained in the separable closure of $\mathbf{M} \subseteq \mathbf{C}_\infty$ and is Galois stable and discrete. Moreover, for $a \in \mathbf{A}$, we have

$$e_\psi(ax) = \psi_a(e_\psi(x))$$

(now viewing e_ψ as a function of x). Thus, one sees that L_ψ is a discrete $\mathbf{A}$-module. To see that L_ψ is a lattice, we need only show that it is finitely generated.

To see the finite generation of L_ψ, look at the $\mathbf{K}$-vector space $V = \mathbf{K}L_\psi$. We first establish that this vector space is finitely generated. Suppose not, and let $\{m_1, \ldots, m_t, \ldots\}$ be an infinite sequence of elements of L_ψ which are linearly independent over $\mathbf{K}$. Let $V_i = \mathbf{K}m_1 + \cdots + \mathbf{K}m_i$, and let $L_i := L_\psi \cap V_i$. We see that L_i is now an $\mathbf{A}$-lattice in a finite dimensional $\mathbf{K}$-vector space. By Proposition 4.6.2, we see that L_i is finitely generated. Moreover, we see that

$$a^{-1}L_i/L_i \simeq (A/(a))^i.$$

But $a^{-1}L_i/L_i \subseteq a^{-1}L_\psi/L_\psi \simeq \psi[a] \simeq (A/(a))^d$. We therefore obtain a contradiction once $i > d$. Thus we deduce that L_ψ is finitely generated of rank d.

Let ϕ and ψ be two Drinfeld modules over $\mathbf{M}$ of rank d with $\mathbf{M}$-lattices L_ϕ and L_ψ. Let $P \in \mathbf{M}\{\tau\}$ be a morphism from ϕ to ψ. So

$$P\phi_a = \psi_a P$$

for all $a \in \mathbf{A}$. One then checks that

$$e_\psi^{-1} P e_\phi$$

commutes with $x\tau^0$ for all $x \in \mathbf{k}$. It thus corresponds to an element $c \in \mathbf{M}$ which is easily seen to be a morphism from L_ϕ to L_ψ. □

Thus we see a very strong similarity between Drinfeld modules of a *fixed* rank over $\mathbf{C}_\infty$ and elliptic curves over $\mathbf{C}$. Let ψ be one such Drinfeld module with exponential function $e(z)$. We let $\log(z) := \log_\psi(z)$ be the inverse function to $e(z)$ (as obtained by formally inverting the power series for $e(z)$). One sees easily that $\log(z)$ has a non-trivial radius of convergence; the exact domain of convergence is given by Proposition 4.14.2.

Finally we finish this subsection by quoting a fundamental result due to Jing Yu [Yu1]. Let $\mathcal{F} \subset \mathbf{C}_\infty$ be a finite extension of $\mathbf{k}$ and let ψ be a Drinfeld module defined over $\mathcal{F}$ (that is, $\psi_a(\tau) \in \mathcal{F}\{\tau\}$ for all $a \in \mathbf{A}$). Then Yu establishes the following analog of the classical Hermite-Lindemann Theorem.

Theorem 4.6.10. *Let $e_\psi(x)$ be the exponential function of ψ with associated lattice L_ψ. Then $e_\psi(\alpha)$ is transcendental over $\mathbf{k}$ for all $0 \neq \alpha$ in the algebraic* *closure of* **k** *contained in* $\mathbf{C}_\infty$. *In particular, all* non-zero *elements of* L_ψ *are transcendental over* **k**. □

The elements of L_ψ are sometimes called the *periods* of ψ. In the special case of the Carlitz module, 4.6.10 was established by L.I. Wade [Wad1]. For more on the transcendence theory associated to Drinfeld modules, etc., we refer the reader to [Yu2]. General theory on the fields of definition of Drinfeld modules is presented in Subsection 7.3.

4.7. Morphisms of Drinfeld modules

We will present here the basic material on morphisms of Drinfeld modules. The reader will note that, as a quite general rule, the results are exactly what one would expect from the analogy with elliptic curves.

Let $\mathcal{F}$ be an **A**-field and let $\overline{\mathcal{F}}$ be a fixed algebraic closure. Let ϕ and ψ be two Drinfeld modules over $\mathcal{F}$ of fixed rank $d > 0$. Recall that a morphism from ϕ to ψ over $\mathcal{F}$ is an element $P(\tau) \in \mathcal{F}\{\tau\}$ with

$$P\phi_a = \psi_a P$$

for all $a \in \mathbf{A}$. Such morphisms form an **A**-module denoted "$\mathrm{Hom}_\mathcal{F}(\phi, \psi)$." If $\phi = \psi$, then we denote $\mathrm{Hom}_\mathcal{F}(\phi, \psi)$ by "$\mathrm{End}_\mathcal{F}(\phi)$;" it is a subring of $\mathcal{F}\{\tau\}$ under composition. The modules $\mathrm{Hom}(\phi, \psi)$, $\mathrm{End}(\phi)$ are always understood to be considered over $\overline{\mathcal{F}}$.

Proposition 4.7.1. *Let* $P \in \mathcal{F}\{\tau\}$ *be a morphism from* ϕ *to* ψ. *Then* P *is an isomorphism if and only if* $\deg P(\tau) = 0$.

Proof. P is an isomorphism if and only if there exists $Q \in \mathcal{F}\{\tau\}$ with $P \cdot Q = \tau^0$. □

Let P be as in the above proposition. Then the *scheme-theoretic* kernel of P is the finite subgroup scheme of $\mathbb{G}_a$ given by

$$\mathrm{Spec}(\mathcal{F}[x]/(P(x))).$$

The points of this subgroup scheme in $\overline{\mathcal{F}}$ form the kernel of the mapping of **A**-modules

$$\phi(\overline{\mathcal{F}}) \xrightarrow{P} \psi(\overline{\mathcal{F}}).$$

Let $a \in \mathbf{A}$.

Proposition 4.7.2. *Let* $\alpha \in \overline{\mathcal{F}}$ *be an* a-*division point of* ϕ. *Then* $P(\alpha)$ *is an* a-*division point of* ψ.

Proof. We have
$$P(\phi_a(\alpha)) = P(0) = 0 = \psi_a(P(\alpha)).$$
□

Now let $I \subseteq \mathbf{A}$ be an ideal.

Corollary 4.7.3. *Let $\alpha \in \phi[I]$. Then $P(\alpha) \in \psi[I]$.* □

Let $\mathcal{G}$ be any algebraically closed overfield of $\overline{\mathcal{F}}$.

Proposition 4.7.4. *The natural inclusion*
$$\mathrm{Hom}_{\overline{\mathcal{F}}}(\phi, \psi) \hookrightarrow \mathrm{Hom}_{\mathcal{G}}(\phi, \psi)$$
is an equality.

Proof. We will show that any $P \in \mathrm{Hom}_{\mathcal{G}}(\phi, \psi)$ must have coefficients which are algebraic over $\mathcal{F}$. Let $a \in \mathbf{A}$ be prime to the characteristic of $\mathcal{F}$ with
$$\deg \phi_a(\tau) = \deg \psi_a(\tau) > \deg P(\tau).$$
By 4.7.2, we see that P maps $\phi[a]$ to $\psi[a]$. The result now follows by applying the Lagrange Interpolation Theorem. □

Remark. 4.7.5. The reader will see that, in fact, all elements of $\mathrm{Hom}(\phi, \psi)$ are defined over the separable closure of $\mathcal{F} \subseteq \overline{\mathcal{F}}$.

Suppose for the next result that $\mathcal{F}$ has generic characteristic, so $\mathbf{k} \subseteq \mathcal{F}$.

Proposition 4.7.6. $\mathrm{End}_{\mathcal{F}}(\phi)$ *is commutative.*

Proof. ϕ is defined over some finitely generated subfield of $\mathcal{F}$ which we can embed in $\mathbf{C}_\infty$. So, without loss of generality, we may consider $\mathcal{F} = \mathbf{C}_\infty$. Now 4.7.4 and 4.6.9 assure us that $\mathrm{End}_{\mathcal{F}}(\phi)$ is a subring of the ring of endomorphisms of the lattice associated to ϕ. As this ring is obviously commutative, the result follows. □

We have noted that, via ϕ, $\mathrm{End}_{\mathcal{F}}(\phi)$ becomes an **A**-module. As $\mathrm{End}_{\mathcal{F}}(\phi) \subseteq \mathcal{F}\{\tau\}$, this module is obviously torsion-free. Our next goal is to establish that this module is finitely generated. Our proof of this follows [DeH1].

For $P \in \mathrm{End}_{\mathcal{F}}(\phi)$ we set
$$\delta(P) = -\deg(P),$$
(e.g., see Subsection 4.5).

Lemma 4.7.7. *The mapping* $P \mapsto \delta(P)$ *satisfies:*
1. $\delta(P) = \infty$ *if and only if* $P = 0$;
2. $\delta(PQ) = \delta(P) + \delta(Q)$;
3. $\delta(P + Q) \geq \min\{\delta(P), \delta(Q)\}$;
4. *Let* $a \in \mathbf{A}$. *Then*

$$\delta(aP) = \delta(\phi_a P) = -d \deg(a) + \delta(P).$$

Proof. This is now a familiar exercise. □

The mapping δ thus gives rise to a *norm* over $\mathbf{K}$ on the vector space $V = \mathrm{End}_\mathcal{F}(\phi) \otimes_{\mathbf{A}} \mathbf{K}$. Note also that $\mathrm{End}_\mathcal{F}(\phi)$ is a *discrete* submodule of V just as $\mathbf{A}$ is discrete $\mathbf{K}$.

Theorem 4.7.8. $\mathrm{End}_\mathcal{F}(\phi)$ *is a projective* $\mathbf{A}$*-module of rank* $\leq d^2$.

Proof. We know that $M := \mathrm{End}_\mathcal{F}(\phi)$ is a discrete submodule of V. Let $W \subseteq V$ be a finite dimensional $\mathbf{K}$-subspace. By 4.6.2, $M_W := W \cap M$ is a finitely generated $\mathbf{A}$-module.

Let $0 \neq a \in \mathbf{A}$. Note that $a^{-1}M_W/M_W$ injects into $a^{-1}M/M$. (Indeed, suppose that $m \in a^{-1}M_W$ and $m \in M$; so $m \in W \cap M = M_W$.)

Now suppose that M does not have finite rank. Then we can find an infinite sequence of elements $\{m_i\} \subseteq M$ which are linearly independent over $\mathbf{k}$ and $\mathbf{K}$. Let $V_i = \mathbf{K}m_1 + \cdots + \mathbf{K}m_i$. Then $M_i := V_i \cap M$ is finitely generated, projective, and of rank i by 4.6.2. From above, we see that $a^{-1}M_i/M_i$ injects into $a^{-1}M/M$. Note that $a^{-1}M_i/M_i \simeq \mathbf{A}/(a)^i$.

On the other hand, let a be prime to the characteristic of $\mathcal{F}$. Then the natural map

$$\mathrm{End}_\mathcal{F}(\phi) \otimes_{\mathbf{A}} \mathbf{A}/(a) \to \mathrm{End}_{\mathbf{A}}(\phi[a])$$

is injective. Indeed, suppose that $P \in \mathrm{End}_\mathcal{F}(\phi)$ gives rise to the trivial endomorphism on $\phi[a]$. By Subsection 1.8 (for instance) we see that

$$P = P_0 \phi_a,$$

for some $P_0 \in \mathcal{F}\{\tau\}$. One now checks readily that $P_0 \in \mathrm{End}_\mathcal{F}(\phi)$. As $\mathrm{rank}_{\mathbf{A}/(a)} \mathrm{End}(\phi[a]) = d^2$, the result follows. □

Through the use of Proposition 4.7.13, the reader will have no difficulty extending Theorem 4.7.8 to morphisms between different Drinfeld modules of fixed rank.

Definition 4.7.9. Let ϕ and ψ be two Drinfeld modules over a field $\mathcal{F}$ with P an isogeny (i.e., a non-zero morphism) over $\mathcal{F}$ from ϕ to ψ.
1. We say that P is a *separable* isogeny if and only if $P(\tau)$ is separable.

82   4. Drinfeld Modules

2. We say that P is *purely inseparable* if and only if $P(\tau) = \tau^j$ for some $j > 0$.

Proposition 4.7.10. *Let $P(\tau)$ be an inseparable morphism from ϕ to ϕ over $\mathcal{F}$. Then $\mathcal{F}$ must have finite characteristic $\wp$. Moreover, let $\deg \wp = d$. Then j, as defined in 4.7.9.2, must be a multiple of d.*

Proof. Let $P(\tau) = c\tau^e + \{\text{higher terms}\}$, where $c \neq 0$. Let $\imath \colon \mathbf{A} \to \mathcal{F}$ be the structure map. We find that
$$\imath(a)^{r^e} = \imath(a)$$
for all a. Thus the characteristic must be finite. The rest of the result follows easily. $\square$

Let G be a finite group scheme over $\mathcal{F}$ [Mul], [Sh1]. We let G^0 be the connected component of the identity ($=$ the "infinitesimal" part of G); G^0 is a subgroup scheme of G. We let $G^{\text{ét}}$ be the étale quotient of G by G^0. We then have the well-known exact sequence
$$0 \to G^0 \to G \to G^{\text{ét}} \to 0.$$

Proposition 4.7.11. *Let $H \subset \mathbb{G}_a/\mathcal{F}$ be a finite subgroup scheme. Let ϕ be a Drinfeld module of rank $d > 0$ over $\mathcal{F}$. Then H is the scheme-theoretic kernel of an isogeny $P \colon \phi \to \psi$ if and only if H is $\mathbf{A}$-invariant (via ϕ) and*
$$H^0 = \begin{cases} \{0\} & \text{if characteristic } \mathcal{F} = (0) \\ \mathcal{F}[x]/(x^{r^{td}}), \ d = \deg \operatorname{char} \mathcal{F}, \ t \geq 0, & \text{otherwise} \end{cases}.$$

Proof. The necessity follows as in 4.7.10. Let us now show the sufficiency; this is modeled on the material in Section 1.8.

Let $\overline{\mathcal{F}}$ be a fixed algebraic closure of $\mathcal{F}$ and let $H_1 = \{\alpha_1, \ldots, \alpha_m\}$ be the points of H in $\overline{\mathcal{F}}$. By our $\mathbf{A}$-stability assumption, H_1 is a finite dimensional $\mathbb{F}_r$-subspace of $\overline{\mathcal{F}}$.

Let $P_{H_1}(x) = x \prod_{0 \neq \alpha \in H_1} (1 - x/\alpha)$. So $P_{H_1}(x)$ is $\mathbb{F}_r$-linear (and is a nonzero multiple of the function used in Section 1.8). We set
$$P_H(\tau) = \tau^{td} P_{H_1}(\tau)$$
$$= P_{H_2}(\tau)\tau^{td},$$
where $H_2 = \tau^{td} H_1$ and $P_{H_2}(x)$ is defined in the obvious way.

We now show that $P_H(\tau)$ gives an isomorphism $\mathbb{G}_a/H \to \mathbb{G}_a$ such that the resulting $\mathbf{A}$-action on $\mathbb{G}_a$ is a Drinfeld module ψ of rank d. (The reader should think of the analytic uniformization of Drinfeld modules.) This is easily done in two steps:

1. Dividing by H^0. Here we obtain the quotient Drinfeld module ϕ^0 with
$$\tau^{td}\phi_a = \phi_a^0 \tau^{td},$$
by simply applying τ^{td} to the *coefficients* of ϕ_a for all $a \in \mathbf{A}$. (Note that $\tau^{td}\phi_a\phi_b = \phi_a^0 \tau^{td}\phi_b = \phi_a^0\phi_b^0\tau^{td}$, etc.)
2. We now divide ϕ^0 by the finite $\mathbf{A}$-module H_2 to get ψ. (We note that H_2 is the $\overline{\mathcal{F}}$-valued points of $H^{\text{ét}}$.) This operation causes no problems, as we are able to work at the level of geometric points. □

Remarks. 4.7.12. 1. Let $P: \phi \to \psi$ be an isogeny and let H be the scheme-theoretic kernel of P. We will sometimes write "ϕ/H" for ψ. This notation is obviously very suggestive and is in line with standard notation in the theory of abelian varieties.

2. Let $P_H(\tau)$ be as in the proof of 4.7.11 and let $\widetilde{P}_H(\tau) = cP_H(\tau)$ for some nonzero $c \in \mathcal{F}$. Then one can also use $\widetilde{P}_H(\tau)$ to give an isomorphism $\mathbb{G}_a/H \to \mathbb{G}_a$. In this case $\phi/H \simeq c\psi c^{-1}$ and is isomorphic to ψ. Thus the isomorphism class of ϕ/H is independent of the choice used.

Proposition 4.7.13. *Let $P: \phi \to \psi$ be an isogeny. Then there exists an isogeny $\widehat{P}: \psi \to \phi$ such that*
$$\widehat{P}P = \phi_a$$
for some nonzero $a \in \mathbf{A}$.

Proof. Let H be the scheme-theoretic kernel of P. As H is finite, one can find $0 \neq a \in \mathbf{A}$ with a annihilating H. Thus $H \subseteq \phi[a]$ and $\phi/H = \psi$. let H_1 be the image of $\phi[a]$ under the map $\phi \to \psi$. Then one has an isogeny $\psi \to \psi_1 := \psi/H_1$. However, it is clear that the map $\phi \to \psi \to \psi_1$ gives an isomorphism $\phi/\phi[a] \xrightarrow{\sim} \psi_1$. Thus $\psi_1 \xrightarrow{\sim} \phi$ giving the result. □

Corollary 4.7.14. 1. $P\widehat{P} = \psi_a$.
2. *Isogeny gives rise to an equivalence relation on Drinfeld modules over $\mathcal{F}$.*

Proof. 1. We have $\widehat{P}P = \phi_a$. Thus $P\widehat{P}P = P\phi_a = \psi_a P$. Canceling gives $P\widehat{P} = \psi_a$. Part 2 follows easily. □

Corollary 4.7.15. $\text{End}_{\mathcal{F}}(\phi) \otimes_{\mathbf{A}} \mathbf{k}$ *is a finite dimensional division algebra over $\mathbf{k}$.*

Proof. This now follows from 4.7.8 and 4.7.13. □

Corollary 4.7.16. *Let $P: \phi \to \psi$ be an isogeny. Then $\text{End}_{\mathcal{F}}(\phi)$ and $\text{End}_{\mathcal{F}}(\psi)$ have the same rank over $\mathbf{A}$.*

Proof. Let $\widehat{P}$ be as in 4.7.13. Then the map from $\operatorname{End}_{\mathcal{F}}(\psi)$ to $\operatorname{End}_{\mathcal{F}}(\phi)$, $\alpha \mapsto \widehat{P}\alpha P$, is an injection of **A**-modules. Thus $\operatorname{rank}_{\mathbf{A}} \operatorname{End}_{\mathcal{F}}(\psi) \leq \operatorname{rank}_{\mathbf{A}} \operatorname{End}_{\mathcal{F}}(\phi)$. Now reverse ϕ and ψ. □

Thus, in generic characteristic, Corollary 4.7.15 tells us that the endomorphisms of a Drinfeld module form an order in some finite dimensional (commutative) extension of **k**. In fact, as with complex multiplication of elliptic curves, one can say more about these extensions. This is contained in our next result. The proof we use was shown to us by Jiu-Kang Yu.

Proposition 4.7.17. $\operatorname{End}_{\mathcal{F}}(\phi) \otimes_{\mathbf{A}} \mathbf{K}$ *is a finite dimensional division algebra over* **K**.

Proof. From 4.7.15, we know that $U = \operatorname{End}_{\mathcal{F}}(\phi) \otimes_{\mathbf{A}} \mathbf{k}$ is a division algebra. Let $\alpha \in U$. As **k** is contained in the center of U, we see that α is contained in some finite field extension $E/\mathbf{k}$ with $\mathbf{k} \subseteq E \subseteq U$. It thus suffices to show that $E \otimes_{\mathbf{k}} \mathbf{K}$ remains a field. Set $t = [E\colon \mathbf{k}]$.

As we have seen, the map $\alpha \mapsto \deg \alpha$, $\alpha \in \mathcal{F}\{\tau\}$, induces a valuation ∞_1 on E which lies over the valuation ∞ on **k**. Put $\mathcal{O} = E \cap \mathcal{F}\{\tau\}$; the ring $\mathcal{O}$ is an order above **A** in E. We are free to develop a theory of Drinfeld modules over $\mathcal{O}$ exactly as in the complex multiplication of elliptic curves, see [Ha1]. The natural injection of $\mathcal{O}$ into $\mathcal{F}\{\tau\}$ gives such an object which we denote by ψ.

Recall that d is the rank of ϕ and we set d_1 to be the rank of ψ.

Let $\widetilde{\mathcal{O}}$ be the ring of **A**-integers in E. The rings $\mathcal{O}$ and $\widetilde{\mathcal{O}}$ are equal upon inverting some nonzero element $f \in \mathcal{O}$. Let $a \in \mathbf{A}$ be prime to the characteristic of $\mathcal{F}$ and also prime to f. Note that the a-division points of ϕ are the same as those of ψ. We conclude that

$$(\mathbf{A}/(a))^d = (\mathcal{O}/(a))^{d_1}$$

as **A**-modules. As $\mathcal{O}$ is **A**-projective of rank t by assumption, we conclude that $d = d_1 t$.

Let $\deg_1(a)$, for $a \in \mathbf{A}$, be the degree *over* $\mathbb{F}_r$ of a at ∞_1. One then finds that

$$\deg_1(a) = \deg(a) t.$$

This is enough to force E to have only one prime above ∞ and thus for $E \otimes_{\mathbf{k}} \mathbf{K}$ to remain a field. □

Remark. 4.7.18. Let E be a finite extension of $\mathbf{k} \subset \mathbf{C}_\infty$. We say that E is a CM_∞-*field* if and only if it contains one and only one prime above ∞. One can now easily construct Drinfeld modules ψ with "complex multiplication by the maximal order $\mathcal{O} = \mathbf{A}$-integers in E," i.e., such that $\mathcal{O}$ injects into $\operatorname{End}(\psi)$. Indeed let L be a lattice for $\mathcal{O}$ inside $\mathbf{C}_\infty$ of rank d_1. Then L is an **A**-lattice of rank $d = d_1[E\colon \mathbf{k}]$. The analytic theory now gives a Drinfeld module ψ associated to L. It clearly has the correct properties. We say that ψ has "sufficiently many complex multiplications" if and only if $d_1 = 1$; i.e., $L \simeq I$ where I is an $\mathcal{O}$-ideal.

Finally, the use of non-maximal orders in the proof of 4.7.17 can be avoided. That this is so will be the last result of this subsection.

Let E be a field over **k** and let $\mathcal{O} \subset E$ be an order above **A**. Let $\widetilde{\mathcal{O}}$ now be the maximal order = the ring of **A**-integers. The *conductor* of $\mathcal{O}$ is the largest ideal $\mathfrak{c}$ of $\widetilde{\mathcal{O}}$ which is also an ideal of $\mathcal{O}$. (The reader can easily show the non-triviality of such an ideal using the theory of modules over **A**.)

Proposition 4.7.19. *Let $\mathcal{F}$ be an **A**-field and ϕ a Drinfeld module over $\mathcal{F}$. Let $\mathcal{O}$ inject into $\mathrm{End}_{\mathcal{F}}(\phi)$ over **A**. Then there is a Drinfeld module ψ over $\mathcal{F}$ which is isogenous to ϕ and such that $\widetilde{\mathcal{O}} \xrightarrow{\sim} \mathrm{End}_{\mathcal{F}}(\psi)$.*

Proof. Let $\phi[\mathfrak{c}]$ be the group scheme of $\mathfrak{c}$-division points of ϕ. Let $\psi = \phi/\phi[\mathfrak{c}]$ and let π be the projection map $\phi \to \psi$. We will show that $\widetilde{\mathcal{O}}$ is isomorphic to $\mathrm{End}_{\mathcal{F}}(\psi)$.

Let $\alpha \in \widetilde{\mathcal{O}}$ and let $0 \neq c \in \mathfrak{c}$. Thus $\alpha c \in \mathfrak{c} \subseteq \mathcal{O}$, and so αc gives rise to an endomorphism of ϕ. We have

$$\mathrm{Ker}(\pi \circ \alpha c \colon \phi \to \psi) = \phi[\alpha c \mathfrak{c}]$$
$$\mathrm{Ker}(\pi \circ c \colon \phi \to \psi) = \phi[c\mathfrak{c}] \subseteq \phi[\alpha c \mathfrak{c}].$$

Thus we can find a morphism $\widetilde{\alpha} \colon \psi \to \psi$ so that the following diagram is commutative:

$$\begin{array}{ccc} \phi & \xrightarrow{\alpha c} & \phi \\ {\scriptstyle c}\downarrow & & \\ \phi & & \downarrow{\scriptstyle \pi} \\ {\scriptstyle \pi}\downarrow & & \\ \psi & \xrightarrow{\widetilde{\alpha}} & \psi \end{array}$$

The morphism $\widetilde{\alpha}$ is now seen to correspond to $\alpha \in \mathcal{O}$ giving the result. $\square$

We thank K. Rubin for his help with the above argument – we note that it works just as well for elliptic curves. (In [Ha1] this result is also established by D. Hayes using the $*$ operation of Subsection 4.9. Indeed, once one applies $*$ with the conductor, the isogenous module can be seen to extend to the maximal order. This is in essence also the proof given above.)

4.8. Primality in $\mathcal{F}\{\tau\}$ and A

Let $\mathcal{F}$ be an **A**-field and let ϕ be a Drinfeld module over $\mathcal{F}$ of rank $d > 0$. Let $f \in \mathcal{F}\{\tau\}$. Recall that in Definition 1.10.11 we called f *prime* if and only if it has no monic divisors in $\mathcal{F}\{\tau\}$ save itself and τ^0. We call $a \in \mathbf{A}$ *prime* if and only if (as usual) the ideal (a) is prime. Our purpose here is to show how the existence of ϕ allows us to relate these two notions.

We note first that if a factors as bc with $\deg(b) > 0$ and $\deg(c) > 0$, then ϕ_a is clearly *not* prime.

Theorem 4.8.1. *Let $\mathcal{F}$, ϕ, etc., be as above. Suppose that $\phi_a \in \mathcal{F}\{\tau\}$ is prime. Then $a \in \mathbf{A}$ is prime also.*

Proof. First of all, if ϕ_a is prime, then it is obviously separable and a must be prime to the characteristic of $\mathcal{F}$. Let $W := \phi[a]$ be the $\mathbb{F}_r$-linear vector space of roots of ϕ_a in some fixed algebraic closure of $\mathcal{F}$. Then we have the injection of $\mathbf{A}/(a)$ into $\text{End}_{\mathbb{F}_r}(W)$ in the obvious manner.

On the other hand, since the image of $\mathbf{A}$ under ϕ is commutative, *all* elements ϕ_b, $b \in \mathbf{A}$, are semi-invariants of ϕ_a (Corollary 1.11.3). Thus the image of $\mathbf{A}/(a)$ lies in the image of the semi-invariants of ϕ_a. But by Ore's Theorem (Theorem 1.11.11) this image is a finite field. Thus $\mathbf{A}/(a)$ is a finite integral domain and so also a field. □

Let $I \subseteq \mathbf{A}$ be an ideal. In the same fashion, one can show that if ϕ_I is prime, then I is prime also.

4.9. The Action of Ideals on Drinfeld Modules

Let $\mathcal{F}$ be an **A**-field via $\imath \colon \mathbf{A} \to \mathcal{F}$. Let ϕ be a fixed Drinfeld module over $\mathcal{F}$ of rank $d > 0$. Let $I \subseteq \mathbf{A}$ be an ideal and let $\phi_I \in \mathcal{F}\{\tau\}$ be as in Definition 4.4.4. Thus the polynomial ϕ_I describes the finite group scheme $\phi[I]$.

Write $\phi_I(\tau) = \widetilde{\phi}_I(\tau)\tau^{t_I}$, with $\widetilde{\phi}_I(\tau)$ separable and t_I a non-negative integer. Set $D_I :=$ coefficient of τ^0 in $\widetilde{\phi}_I(\tau) =$ derivative with respect to x of $\widetilde{\phi}_I(x)$. Thus $D_I \neq 0$ by assumption and we put

$$\widehat{\phi}_I(\tau) = \frac{1}{D_I}\phi_I(\tau).$$

In the proof of 4.7.11, the function $\widehat{\phi}_I(\tau)$ was used to form the quotient Drinfeld module $\phi/\phi[I]$. As will be seen later on in this subsection, the function $\widehat{\phi}_I(\tau)$ is good from the viewpoint of lattices. However, it is not optimal from an algebraic view-point as we have divided by D_I.

4.9. The Action of Ideals on Drinfeld Modules

Therefore, we will use here the function $\phi_I(\tau)$ itself to form quotient Drinfeld modules. We will denote the quotient $\phi/\phi[I]$, constructed via $\phi_I(\tau)$, as "ϕ^I" or "$I * \phi$." If we denote by $\widehat{\phi}^I$ the quotient formed by $\widehat{\phi}_I$, then, as in Remark 4.7.12.2, we see that

$$\phi^I = D_I \widehat{\phi}^I D_I^{-1}.$$

By definition, $\phi_I(\tau)$ is the monic generator in $\mathcal{F}\{\tau\}$ of the left ideal I_ϕ generated by $\{\phi_i\}_{i \in I}$. This ideal is obviously stable under multiplication on the right by ϕ_a for any $a \in \mathbf{A}$. Therefore, we deduce that

$$\phi_I \phi_a = \phi_a^1 \phi_I$$

for some $\phi_a^1 \in \mathcal{F}\{\tau\}$.

Lemma 4.9.1. *We have $\phi_a^1 = \phi_a^I$ for all $a \in \mathbf{A}$.*

Proof. By definition, we have

$$\phi_I \phi_a = \phi_a^I \phi_I$$

for $a \in \mathbf{A}$. Thus

$$\phi_a^1 \phi_I = \phi_a^I \phi_I,$$

and canceling on the right gives the result. □

Lemma 4.9.2. *Let I, J be ideals of $\mathbf{A}$. Then, we have*
1. $\phi_{IJ} = (J * \phi)_I \phi_J$.
2. $I * (J * \phi) = (IJ) * \phi$.

Proof. These follow directly from the definitions. □

4.9.2.1 can also be used to give an elementary proof of 4.8.1.

Suppose that $I = (i)$ is principal and let $0 \neq h_i$ be the coefficient of *highest* degree in ϕ_i. Thus

$$\phi_I = h_i^{-1} \phi_i.$$

Clearly ϕ_i is also an endomorphism of ϕ over $\mathcal{F}$; thus $I * \phi = h_i^{-1} \phi h_i$ is isomorphic to ϕ. We therefore have our next result.

Proposition 4.9.3. *Let $\mathrm{Isom}_{\mathcal{F}}(d)$ be the isomorphism classes over $\mathcal{F}$ of Drinfeld modules of rank $d > 0$. Then there is a natural action of $\mathrm{Cl}(\mathbf{A})$ on $\mathrm{Isom}_{\mathcal{F}}(d)$ induced by $I * \phi$.* □

Suppose now that $\mathcal{F} = \mathbf{C}_\infty$. Thus ϕ corresponds to a lattice L of rank d. Let $e_L(z)$ be the exponential function of L.

Proposition 4.9.4. *$\widehat{\phi}_I(e_L(z))$ is the exponential function of $\widehat{\phi}^I$.*

Proof. By definition we have

$$e_L(az) = \phi_a(e_L(z)).$$

Thus

$$\widehat{\phi}_I(e_L(az)) = \widehat{\phi}_I(\phi_a(e_L(z)))$$
$$= \widehat{\phi}_a^I(\widehat{\phi}_I(e_L(z))).$$

The result now follows upon noting that the derivative of $\widehat{\phi}_I(e_L(z))$ is identically 1. $\square$

Corollary 4.9.5. 1. $\widehat{\phi}^I$ *corresponds to the lattice* $I^{-1}L$.
2. ϕ^I *corresponds to the lattice* $D_I I^{-1} L$.

Proof. The first part is obvious. The second part follows from the equality

$$\phi^I = D_I \widehat{\phi}^I D_I^{-1}.$$

$\square$

The elegant formalism developed in this subsection is due to David Hayes (see, e.g., [Ha2]).

4.10. The Reduction Theory of Drinfeld Modules

Let $\mathcal{F}$ be an **A**-field equipped with a non-trivial discrete valuation v. *We assume that* $v(\mathbf{A}) \geq 0$. In this subsection we will discuss the reduction theory of Drinfeld modules at the maximal ideal of $\mathcal{F}$ associated to v. Our sources are [Dr1] and [Tak1]. The reader who is knowledgeable in the theory of elliptic curves will find much that is familiar here.

Let $\mathcal{O}_v \subset \mathcal{F}$ be the valuation ring of v; thus $\mathcal{O}_v = \{\alpha \in \mathcal{F} \mid v(\alpha) \geq 0\}$. Let $\imath\colon \mathbf{A} \to \mathcal{F}$ be the structure map; so, by assumption, $\imath(\mathbf{A}) \subset \mathcal{O}_v$. We let $M_v \subset \mathcal{O}_v$ be the maximal ideal, $M_v := \{\alpha \in \mathcal{O}_v \mid v(\alpha) > 0\}$. Finally set $F_v := \mathcal{O}_v/M_v$.

Let ϕ be a Drinfeld module over $\mathcal{F}$ of fixed rank $d > 0$.

Definition 4.10.1. 1. We say that ϕ has *integral coefficients* if and only if ϕ_a has coefficients in $\mathcal{O}_v$ for all $a \in \mathbf{A}$ *and* the reduction modulo M_v of these coefficients defines a Drinfeld module (of *some* rank $0 < d_1 \leq d$) over F_v. The reduced Drinfeld module will be denoted by "ϕ^v."
2. We say that ϕ has *stable reduction* at v if and only if there exists a Drinfeld module ψ over $\mathcal{F}$, with ψ isomorphic to ϕ over $\mathcal{F}$, and ψ has integral coefficients.
3. We say that ϕ has *good reduction* at v if and only if it has stable reduction at v *and*, in addition, ϕ^v has rank d. (Thus "ϕ gives rise to a Drinfeld module over the *ring* $\mathcal{O}_v$.")

4.10. The Reduction Theory of Drinfeld Modules

4. We say that ϕ has *potential stable* (resp. *potential good*) reduction at v if and only if there exists an extension $(\mathcal{G}, w)$ of $(\mathcal{F}, v)$ such that ϕ has stable (resp. good) reduction at w.

Let $f(\tau) = \sum_{j=0}^{t} c_j \tau^j \in \mathcal{F}\{\tau\}$. We set

$$v(f(\tau)) := \min\{v(c_j)/(r^j - 1) \mid j > 0\}.$$

Lemma 4.10.2. *Let $u \in \mathcal{F}^*$. Then the Drinfeld module $u\phi u^{-1}$ has integral coefficients at v if and only if*

$$v(u) = \min\{v(\phi_a) \mid a \in \mathbf{A} - \mathbb{F}_r\}.$$

Proof. Let $a \in \mathbf{A}$ be non-constant and set

$$\phi_a = \sum_{j=0}^{t} \phi_j(a) \tau^j, \quad \phi_0(a) = \iota(a), \; \phi_t(a) \neq 0.$$

Thus

$$u\phi_a u^{-1} = \sum_{j=0}^{t} u^{1-r^j} \phi_j(a) \tau^j.$$

In order for $u\phi u^{-1}$ to have integral coefficients it is needed that

$$u^{1-r^j} \phi_j(a) \in \mathcal{O}_v$$

for all $\{j, a\}$, *and* such that for some $j_0 > 0$, and some $a_0 \in \mathbf{A}$, $u^{1-j_0} \phi_{j_0}(a_0) \in \mathcal{O}_v^*$. (Indeed, this follows from the fact that Drinfeld modules gives valuations at ∞ as in Subsection 4.5.) The result now follows. $\square$

The next result is due to Drinfeld.

Proposition 4.10.3. *Let ϕ be a Drinfeld module over $\mathcal{F}$ as above. Then there is a natural number $e_v(\phi)$ which is prime to p such that the following two properties are equivalent for a finite extension $(\mathcal{G}, w)$ of $(\mathcal{F}, v)$:*

1. *ϕ has stable reduction at w;*
2. *The index of ramification of w over v is divisible by $e_v(\phi)$.*

Proof. Note that $\mathbf{A}$ is finitely generated over $\mathbb{F}_r$. Thus $\min\{v(\phi_a) \mid a \in \mathbf{A} - \mathbb{F}_r\}$ exists as an element of $\mathbb{Q}$. The result now follows. $\square$

Corollary 4.10.4. 1. *Every ϕ (as above) has potential stable reduction.*
2. *If $d = 1$, then ϕ has potential good reduction.*

90   4. Drinfeld Modules

Proof. The first part follows from the proposition. To see the second, notice that for rank 1 modules, stable and good reduction are the same. □

Now let $\mathcal{F}^{\text{sep}}$ be the separable closure of $\mathcal{F}$ in a fixed algebraic closure $\overline{\mathcal{F}}$. As we saw in Section 2, we can extend v to $\overline{\mathcal{F}}$. Let $\overline{v}$ be one such extension. As we have *not* assumed that $\mathcal{F}$ is complete, $\overline{v}$ may *not* be unique. However, any two such extensions are conjugate via an automorphism of $\overline{\mathcal{F}}$ over $\mathcal{F}$.

Let $\mathcal{F}^{\text{sep}}$ be the separable closure of $\mathcal{F}$ in $\overline{\mathcal{F}}$. Let $G := \text{Gal}(\mathcal{F}^{\text{sep}}/\mathcal{F})$ and let M be a G-module. Let $I_{\overline{v}}$ be the inertia subgroup of G at $\overline{v}$. We say that M is *unramified* at v if and only if $I_{\overline{v}}$ acts trivially on M. By our remarks above, this notion is independent of our choice of $\overline{v}$.

Our next result, due to T. Takahashi [Tak1], is an analog for Drinfeld modules of the classical result of *Ogg-Néron-Shafarevich* in the theory of abelian varieties.

Theorem 4.10.5. *Let $\wp \in \text{Spec}(\mathbf{A})$ be different than the characteristic of $F_v = \mathcal{O}_v/M_v$. Let ϕ be a Drinfeld module over $\mathcal{F}$. Then ϕ has good reduction at v if and only if the G-module $\phi[\wp^\infty] := \bigcup_{m \geq 1} \phi[\wp^m]$ is unramified at v.*

Proof. Suppose first that ϕ has good reduction at v. It is then easy to see that $\phi[\wp^\infty]$ is unramified at v.

Thus let us assume that $\phi[\wp^\infty]$ is unramified at v as a G-module. As $\text{Cl}(\mathbf{A})$ has finite order, say h, we have $\wp^h = (b)$ is principal for some $b \in \mathbf{A}$. As we saw in Section 2, we can extend v to a valuation $\overline{v}$ on $\overline{\mathcal{F}}$ and thus on $\mathcal{F}^{\text{sep}}$.

We will establish, first of all, that ϕ has stable reduction at v.

Claim 4.10.6. *Let $\alpha \in \phi[b]$. Then $\overline{v}(\alpha)$ is an integer and*

$$\max_{0 \neq \alpha \in \phi[b]} \{\overline{v}(\alpha)\} = -v(\phi_b).$$

Proof of 4.10.6. The first part is an immediate consequence of the non-ramification at v. To see the second part, let

$$\phi_b = \sum b_j \tau^j,$$

where τ is the rth power morphism. Thus

$$\phi_b(x) = \sum b_j x^{r^j} = x \sum b_j x^{r^j - 1}.$$

The theory of Newton polygons, as given in Section 2, implies that the maximum value M of $\overline{v}(\alpha)$ is given by

$$\max\{(v(b_0) - v(b_j))/(r^j - 1) \mid j > 0\}.$$

4.10. The Reduction Theory of Drinfeld Modules

As b is prime to the characteristic of $\mathcal{F}$, we see that $v(b_0) = 0$. Therefore, the second part follows from the definition of $v(\phi_b)$. This completes the proof of 4.10.6.

Thus, as $\phi[\wp^\infty]$ is unramified, we conclude that $v(\phi_b)$ is an integer. Now let $(\mathcal{G}, w)$ be a finite extension of $(\mathcal{F}, v)$ where ϕ has stable reduction as in 4.10.4. Let $u \in \mathcal{G}^*$ be chosen so that $u\phi u^{-1}$ has integral coefficients at w.

Claim 4.10.7. $e_v(\phi) = 1$.

Proof of 4.10.7. Let F_w be the residue field of $(\mathcal{G}, w)$ at w. We know that $u\phi u^{-1}$ is a Drinfeld module over F_w of some positive rank. Thus, if $a \in \mathbf{A} - \mathbb{F}_r^*$, then the reduction of $u\phi_a u^{-1}$ must have positive degree in τ. Arguing as before (e.g., the proof of 4.10.2), we see that $w(u) = w(\phi_a)$. We conclude that $v(\phi_a)$ is independent of a *and* that $v(\phi_a) = v(\phi_b)$ is integral (i.e., in $\mathbb{Z}$). It is now easy to see that this forces $e_v(\phi) = 1$ and establishes 4.10.7.

Therefore, we may assume that ϕ has integral coefficients at v, and we now establish that the reduction of ϕ at v is, in fact, good. For this it is enough to show that the leading coefficient of ϕ_b is a unit of v. We will assume that this is *not* so and arrive at a contradiction.

By our choice of b, we see that it is a v-unit. As such, there must be an element α_1 of $\phi[b]$ with $\overline{v}(\alpha_1) < 0$.

Claim 4.10.8. There exists a root α_2 of the equation in x, $\phi_b(x) = \alpha_1$, such that $\overline{v}(\alpha_1) < \overline{v}(\alpha_2) < 0$.

Proof of 4.10.8. Suppose that $\overline{v}(\alpha) \leq \overline{v}(\alpha_1)$ for all roots of $\phi_b(x) = \alpha_1$. An easy computation then implies that the coefficients of $\alpha_1^{-1}\phi_b\alpha_1$ are $\overline{v}$-integers. This forces $\overline{v}(\alpha_1^{-1}) \leq v(\phi_b) = 0$ (since ϕ has integral reduction at v). Thus we find $\overline{v}(\alpha_1) \geq 0$ contradicting the assumption that $\overline{v}(\alpha_1) < 0$. This contradiction now establishes 4.10.8.

As $\phi_b(\alpha_2) = \alpha_1$ and ϕ has integral coefficients, we deduce that $\overline{v}(\alpha_2) < 0$. In a similar fashion, we can find $\alpha_n \in \mathcal{F}^{\text{sep}}$, for $n \geq 1$, such that

$$\phi_b(\alpha_{n+1}) = \alpha_n, \quad \overline{v}(\alpha_n) < \overline{v}(\alpha_{n+1}) < 0.$$

As each $\alpha_n \in \phi[b^n] \subset \phi[\wp^\infty]$, we deduce – from our ramification assumption on $\phi[\wp^\infty]$ – that $\overline{v}(\alpha_n)$ is an integer for all n. As $\overline{v}(\alpha_n) < \overline{v}(\alpha_{n+1}) < 0$, for $n \geq 1$, we deduce a contradiction establishing the theorem. □

Definition 4.10.9. Let $\wp$ be a prime of $\mathbf{A}$ and let $\mathbf{k}_\wp$, $\mathbf{A}_\wp$ be the respective completions of these rings at $\wp$. We set

$$T_\wp(\phi) = \text{Hom}_{\mathbf{A}}(\mathbf{k}_\wp/\mathbf{A}_\wp, \phi[\wp^\infty]).$$

We call $T_\wp(\phi)$ the $\wp$-*adic Tate module* of ϕ.

It is easy to see that $T_\wp(\phi)$ is an $\mathbf{A}_\wp$-module and gives rise to a covariant functor on the category of Drinfeld modules. If $\wp$ is as in 4.10.5, then $T_\wp(\phi)$ is isomorphic to $\mathbf{A}_\wp^d$. Moreover, under this hypothesis, morphisms are distinguished by their action on Tate modules, (see Proposition 4.12.11).

Note that
$$T_\wp(\phi) \simeq \varprojlim \phi[\wp^m]$$
$$\simeq \varprojlim \phi[b^n],$$
where b is as above. Note further that $\mathrm{Aut}_\mathbf{A}(\phi[\wp^\infty]) \simeq \mathrm{Aut}_{\mathbf{A}_\wp}(T_\wp(\phi))$. Thus $T_\wp(\phi)$ comes equipped with a (continuous) G-action. Theorem 4.10.5 can then be restated as the following result.

Corollary 4.10.10. *Let ϕ, etc., be as in* Theorem 4.10.5. *Then ϕ has good reduction at v if and only if $T_\wp(\phi)$ is an unramified G-module.* □

Let $\mathcal{F}_v$ be the completion of $\mathcal{F}$ at v and let $\mathcal{F}_v^{nr}$ be its maximal unramified extension in some fixed algebraic closure.

Corollary 4.10.11. *ϕ has potential good reduction at v if and only if the image of $I_{\bar{v}}$ in $\mathrm{Aut}_{\mathbf{A}_\wp}(T_\wp(\phi))$ is finite. When this is the case, the extension $\mathcal{F}_v^{nr}(\phi[\wp^\infty])/\mathcal{F}_v^{nr}$ is independent of $\wp$ and cyclic of degree $e_v(\phi)$.*

Proof. This follows from Theorem 4.10.5 and Proposition 4.10.3. □

Corollary 4.10.12. *Let ϕ have potential good reduction at v. Let $I \subseteq \mathbf{A}$ be a non-trivial ideal prime to v.*
1. *The extension $\mathcal{F}_v^{nr}(\phi[I])/\mathcal{F}_v^{nr}$ is independent of I and tamely ramified of degree $e_v(\phi)$.*
2. *The Galois module $\phi[I]$ is unramified at v if and only if ϕ has good reduction at v.*

Proof. Let $\wp$ be a prime dividing I. The extension $\mathcal{F}_v^{nr}(\phi[\wp^\infty])/\mathcal{F}_v^{nr}(\phi[\wp])$ is tamely ramified by 4.10.11 and 4.10.2. Moreover, the Galois group of this extension injects into the kernel of the natural map
$$\mathrm{Aut}_\mathbf{A}(\phi[\wp^\infty]) \to \mathrm{Aut}_\mathbf{A}(\phi[\wp])$$
which is a pro-p-group. Thus the extension is trivial and we deduce that $\mathcal{F}_v^{nr}(\phi[\wp^\infty]) = \mathcal{F}_v^{nr}(\phi[\wp])$. We thus see that $\mathcal{F}_v^{nr}(\phi[\wp])$ is independent of $\wp$ and, therefore, that $\mathcal{F}_v^{nr}(\phi[I]) = \mathcal{F}_v^{nr}(\phi[\wp])$ is independent of $\wp$. The result follows. □

Our next result is a version for Drinfeld modules of a basic result of G. Faltings. It is due to Y. Taguchi [Tag1]. As the proof would take us too far afield, we refer the reader to [Tag1] for details.

Let $\mathcal{F}$ now be a finite extension of $\mathbf{k}$ and let $\mathcal{F}^{\text{sep}} \subset \overline{\mathbf{k}}$ be its separable closure. Let $G := \text{Gal}(\mathcal{F}^{\text{sep}}/\mathcal{F})$ and let ϕ be a Drinfeld module over $\mathcal{F}$. Let $\wp \in \text{Spec}(\mathbf{A})$ be a prime and, as before, let $T_\wp(\phi)$ be the $\wp$-adic Tate module; thus $T_\wp(\phi)$ has a continuous G-action.

Theorem 4.10.13 (Taguchi). $T_\wp(\phi)$ *is a semi-simple G-module.* □

In [Go4, §3.8] the reader will find a treatment of the *Brauer-Nesbitt Theorem* which is particularly relevant to the above representations. Also, in Subsection 10.3 we discuss the *Tate Conjecture for Drinfeld modules*.

4.11. Review of Central Simple Algebra

Let L be an $\mathbf{A}$-field and let ϕ be a Drinfeld module over L of rank $d > 0$. If $\mathbf{k} \subseteq L$, then, by Proposition 4.7.6, we see that $\text{End}(\phi)$ is *always* a commutative domain. (Indeed, it is always isomorphic to the endomorphism ring of the lattice associated to ϕ.) However, if L has characteristic $\wp \neq 0$, then this no longer need be true, exactly as with elliptic curves over finite fields.

The theory of Drinfeld modules over finite fields, and their rings of endomorphisms, is intimately connected with the theory of central simple algebras. Indeed, the theory offers truly beautiful examples of many of the standard results on such algebras. As such, for the convenience of the reader, we will present here a short general review, with few proofs, of central simple algebra as well as some non-commutative ring theory. The knowledgeable reader may skip this review with no difficulty and pass to the next subsection where a few more technical facts will also be recalled.

The first remark to be made is that one *cannot* take for granted many of the constructions familiar in commutative algebra. For instance, let R be a commutative domain with unit; so R has no non-trivial zero divisors. Then, of course, one can easily embed R into its field of fractions. If, however, R is a non-commutative domain with unit (so R also has no non-trivial zero divisors), then R need not be embeddable in a division ring of fractions, (see [FD1, Supp. Ex. 5b]). However, there are conditions that allow the formation of fractions which are fortunately satisfied by the rings $L\{\tau\}$ of Section 1.

Thus, let $f(\tau), g(\tau) \in L\{\tau\}$ be nonzero elements. By example 1.10.3, one can find the right least common multiple $h(\tau) \neq 0$ of $f(\tau)$ and $g(\tau)$. So, by definition
$$h(\tau) = a(\tau)f(\tau) = b(\tau)g(\tau)$$
for some nonzero $a(\tau), b(\tau) \in L\{\tau\}$. This is precisely what is needed!

Definition 4.11.1. Let R be a non-commutative domain with unit. Then R satisfies the *left Ore condition* if and only if given nonzero $a, b \in R$, there exist nonzero $a', b' \in R$ with $a'a = b'b$.

Suppose for the moment that R satisfies the left Ore condition and that R is contained in a division ring D. Let a, b, c, d be four nonzero elements in R. Choose nonzero b', d' with

$$\beta = b'b = d'd$$

as we are now guaranteed. Suppose further that, in D,

$$b^{-1}a = d^{-1}c.$$

Left multiplication by β gives

$$b'a = d'c,$$

which has the distinct *advantage* of being an equation in R. We are thus led to our next definition.

Definition 4.11.2. Let $\widehat{D}$ be the set of ordered pairs $\{(a,b) \mid a, b \in R, \ b \neq 0\}$. We define the relation "$\equiv$" on $\widehat{D}$ by

$$(a,b) \equiv (c,d) \iff b'a = d'c$$

where $b'b = d'd$ as above.

One checks that $\equiv$ is an equivalence relation. Set $D := D(R) =$ the set of equivalence classes of $\widehat{D}$ under $\equiv$. The equivalence class of (a,b) is denoted by "$b^{-1}a$." We give D an addition by defining

$$b^{-1}a + d^{-1}c = (b'b)^{-1}(b'a + d'c);$$

so we have "common denominators." Multiplication is given by

$$(b^{-1}a)(d^{-1}c) = (a'b)^{-1}(d'c)$$

where $a'a = d'd$. One then checks that D is a division ring called the (left) division ring of fractions.

Conversely, suppose that R can be embedded in a division ring D with the property that *all* elements of D can be written in the form $x^{-1}y$, for $0 \neq x$ and $y \in R$. Then given $a, b \in R$, with $b \neq 0$, we are guaranteed the existence of b', a' in R such that in D

$$ab^{-1} = (a')^{-1}b';$$

thus $a'a = b'b$ in R. In other words, one obtains the left Ore conditions in R.

In particular, we are thus always guaranteed the existence of a left division ring of fractions for $L\{\tau\}$.

Suppose now that L is perfect; we therefore also obtain the existence of a left division algorithm. We leave it to the reader to check that, in this case, $L\{\tau\}$ also satisfies the right Ore condition (with the obvious definition). We can, therefore, also form the right division ring of fractions for $L\{\tau\}$. We also leave it to the reader to check that, when L is perfect, both left and right rings of fractions are isomorphic.

The reader may well wonder if it is always possible to embed a domain R *some way* into a division ring. The answer is no; see [Be1] for example.

Definition 4.11.3. 1. Let R be a ring with unit. Then R is *simple* if and only if it has no non-trivial two sided ideals.
2. The center, $Z(R)$, of R is defined to be $\{\alpha \in R \mid \alpha r = r\alpha, \ \forall r \in R\}$.

Thus $Z(R)$ is a commutative subring of R.

Definition 4.11.4. Let R be as above. Then R is (left) *Artinian* if and only if every descending chain of left ideals stabilizes.

We then have the following basic result.

Theorem 4.11.5 (Wedderburn). *Let R be simple and Artinian. Then R is isomorphic to the ring $M_n(D)$ of $n \times n$ matrices over a division ring D.*

Proof. [FD1, Theorem 1.15]. □

Corollary 4.11.6. *Under the above assumptions, $Z(R)$ is a field.* □

Definition 4.11.7. Let R be a ring and L an arbitrary field. Then R is an *L-algebra* if and only if there is a ring homomorphism of L into the center of R.

The basic example is $M_n(L)$ over L.

From now on, all rings R will be finite dimensional L-algebras where L is as in 4.11.7; thus they are automatically Artinian.

Definition 4.11.8. We say that an L-algebra R is *central simple* over L if and only if R is simple, and $Z(R) = L$ (so R is *central over L*).

Thus $R \approx M_n(D)$ where D is a division ring and $Z(R) = Z(D) = L$.

Given two L-algebras A and B, we can form the tensor product $A \otimes_L B$ in the usual fashion. This is also an L-algebra and one checks the following result.

Proposition 4.11.9. *If A and B are central simple over L, then so is $A \otimes_L B$.*

Proof. [FD1, Cor. 3.6]. □

Suppose now that L_1 is an overfield of L and A is central simple over L. We then have the change of base functor $A \mapsto A_{L_1} := A \otimes_L L_1$.

Proposition 4.11.10. A_{L_1} *is central simple over* L_1.

Proof. [FD1, Th. 3.5]. □

Suppose for the moment that R is again an arbitrary ring. We define the *opposite ring* R° as follows. The additive group of R° is the additive group of R. However, multiplication $a \circ b$ of elements a and b in R° is defined by $a \circ b := ba$ where ba is calculated in R.

The following results illustrates the relationship between rings and their opposites: let D be a division ring and let V be an n-dimensional *left* vector space over D. Then
$$\mathrm{Hom}_D(V, V) \simeq M_n(D^\circ).$$
Conversely, if V is an n-dimensional *right* vector space over D, then
$$\mathrm{Hom}_D(V, V) \simeq M_n(D).$$

If A is central simple over L, then so is A° and $A \otimes_L A^\circ$. There is an algebra mapping $\Pi: A \otimes_L A^\circ \to \mathrm{End}_L(A)$ given by
$$a \otimes b \mapsto (x \mapsto axb).$$
As $A \otimes_L A^\circ$ is simple, the map must be injective and a dimension count gives surjectivity. Thus we have our next result.

Proposition 4.11.11. *Let* A, A° *be as above. Let* $n = \dim_L(A)$. *Then*
$$A \otimes_L A^\circ \simeq M_n(L).$$
□

Let $\alpha \in A$ be an invertible element. Then α gives rise to an automorphism of A given by $x \mapsto \alpha x \alpha^{-1}$. Such an automorphism is said to be *inner*.

Theorem 4.11.12 (Skolem, Noether). *Let A be central simple over L and let B be a simple L-algebra. Let $\sigma_1, \sigma_2: B \to A$ be two L-algebra morphisms. Then there exists an inner automorphism φ of A such that $\sigma_2 = \varphi \circ \sigma_1$.*

Proof. [FD1, Th. 3.14]. □

Corollary 4.11.13. *Every L-linear automorphism of A is inner.* □

Suppose now that $S \subseteq A$ is any subset. We define the *centralizer of S in A* to be
$$\{a \in A \mid as = sa, \ \forall s \in S\}.$$
The centralizer of S is denoted by "$Z_A(S)$;" it is a subalgebra of A.

4.11. Review of Central Simple Algebra

Theorem 4.11.14 (Centralizer Theorem). *Let A be central simple over L and let $B \subseteq A$ be a simple subalgebra of A. Then:*
1. $Z_A(B)$ *is simple;*
2. $[A:L] = [B:L][Z_A(B):L]$;
3. $Z_A(Z_A(B)) = B$.

Proof. [FD1, Th. 3.15]. □

Corollary 4.11.15. *Let A be central simple over L. Then $\dim_L(A)$ is a square, say d^2. If $A = D$ is a division ring, then the maximal commutative subfields of D are exactly d-dimensional and are equal to their centralizers.*

Proof. We know that $A \simeq M_n(D)$ for some central division ring D. Thus
$$\dim_L(A) = n^2 \dim_L(D).$$
Let $K \supseteq L$ be a maximal subfield; so $K \subseteq Z_D(K)$. Moreover, $K = Z_D(K)$ since any element of $Z_D(K)$ can be adjoined to K to obtain a bigger commutative field and K is maximal. Thus
$$\dim_L(D) = [K:L]^2. \qquad \square$$

The integer d is the *degree* of D and d^2 is its *rank*.

Definition 4.11.16. *Let K be an extension of L. We say that K is a splitting field for a central simple algebra if and only if $A \otimes_L K \simeq M_n(K)$ for some n.*

Arguments similar to those given in Corollary 4.11.15 show the following important result.

Theorem 4.11.17. *Every maximal subfield of a finite dimensional central division algebra D is a splitting field for D.* □

Theorem 4.11.18. *Let D be a central division algebra over L. Then there exists a maximal subfield of D which is also separable over L.*

Proof. [Sc1, Th. 8.5.5]. □

Corollary 4.11.19. *There are no non-trivial central division rings over a separably closed field.* □

Our next definition gives the fundamental *Brauer group* of a field L. This group classifies the central simple algebras and central division algebras over L and is denoted "Br(L)."

98   4. Drinfeld Modules

Definition 4.11.20. 1. Let A, B be two central simple algebras over L. We say that A is *similar* to B (and write $A \sim B$) if and only if they are both isomorphic to matrix rings over the *same* division ring.
2. The similarity classes of central simple algebras forms the *set* $\mathrm{Br}(L)$.
3. The tensor product gives a commutative product on $\mathrm{Br}(L)$.
4. By 4.11.11, the opposite algebra acts as the inverse in $\mathrm{Br}(L)$. Thus $\mathrm{Br}(L)$ becomes a group under tensor product.

Splitting fields can now be characterized through the use of Brauer equivalence.

Theorem 4.11.21. *Let D be a finite dimensional central division algebra over L. A finite extension K of L is a splitting field for D if and only if there exists a central simple algebra A similar to D (so $A \simeq M_t(D)$, some t) such that K can be embedded in A and $Z_A(K) = K$. In this case, we have $[K:L] = td$ and $[D:L] = d^2$.*

Proof. [Jac1, Th. 4.8]. □

Definition 4.11.22. Let A be a finite dimensional algebra over a field L. Let $a \in A$. The element a gives an L-linear endomorphism $\hat{a}$ of A through left multiplication. The determinant of this endomorphism is called the *norm* of a and denoted "$N_{A/L}(a)$." The trace of $\hat{a}$ is called the *trace* of a and denoted "$\mathrm{Tr}_{A/L}(a)$."

Suppose now that A is central simple over L. Let K be a splitting field of A and choose an isomorphism ι of $A \otimes K$ with $M_n(K)$. For $a \in A$, we then have the determinant and trace of $\iota(a)$ as an element of $M_n(K)$. The Skolem-Noether Theorem immediately implies that this trace and determinant are independent of the choice of ι. In fact, much more is true.

Lemma 4.11.23. *The determinant and trace of $\iota(a)$ are in L. They are independent of the choice of splitting field.*

Proof. [Sc1, Lemma 8.5.7]. □

Definition 4.11.24. The determinant and trace given in Lemma 4.11.23 are called the *reduced norm* and *reduced trace* of a respectively. They are denoted by "$N^{\mathrm{red}}_{A/L}(a)$" and "$\mathrm{Tr}^{\mathrm{red}}_{A/L}(a)$."

By passing to a splitting field K, one is able to do computations in $M_n(K)$. One then sees readily that

$$N_{A/L}(a) = (N^{\mathrm{red}}_{A/L}(a))^n,$$

and

$$\mathrm{Tr}_{A/L}(a) = n\mathrm{Tr}^{\mathrm{red}}_{A/L}(a).$$

4.11. Review of Central Simple Algebra

Suppose now that L is a local non-Archimedean field; so L is a finite extension of $\mathbb{Q}_p$ or $\mathbb{F}_p\left(\left(\frac{1}{T}\right)\right)$ for some non-constant T. In particular L is complete with respect to a valuation v. Let D be a finite dimensional division ring over L and let $\mathcal{O} \subset L$ be the ring of integers of L with respect to v (so $\mathcal{O} = \{\alpha \in L \mid v(\alpha) \geq 0\}$). Let $\beta \in D$. As D is finite dimensional over L, we know that β satisfies a polynomial equation with coefficients in L (indeed, the span of $\{\beta^0, \beta, \beta^1, \ldots\}$ is finite dimensional). We say that β is *integral with respect to* v if and only if β satisfies a monic polynomial with coefficients in $\mathcal{O}$ (as is the standard definition).

For $\beta \in D$ we set

$$v(\beta) := \frac{1}{[D:L]} v(N_{D/L}(\beta)),$$

which agrees with our earlier definition in Section 2 when D is commutative. One checks that this definition is compatible with towers $D_0 = D \subseteq D_1 \subseteq D_2 \subseteq \cdots$.

Theorem 4.11.25. *The valuation v prolongs to D. The set $\mathcal{O}_D$ of all integers of D is a subring of D. It is the valuation ring of v ($= \{\beta \in D \mid v(\beta) \geq 0\}$).*

Proof. [Sch1, Chapter 2.7, Th. 9] or [Re1, Th. 12.8]. □

As D is finite dimensional over L it is complete with respect to v. Moreover $\mathcal{O}_D$ is *the* maximal order of D; i.e., it is maximal among those unital, $\mathcal{O}$ subalgebras (obvious definition) of D which are finitely generated and which contain an L-basis for D.

Let L be arbitrary again; we need one last set of definitions before we present the famous computation of the Brauer groups in arithmetic. Let K/L be a Galois extension of degree n with group $G := \text{Gal}(K/L)$. We assume that G is *cyclic* with generator σ. Let $a \in L^*$. We now construct an algebra (called a *cyclic algebra*) A as follows: A contains K and an element e such that $\{1, e, \ldots, e^{n-1}\}$ is a basis for A over K. We further require that in A:
1. $e^n = a \cdot 1$, and
2. $e\lambda = \sigma(\lambda) \cdot e$ for $\lambda \in K$.

One checks readily that these definitions do, in fact, give an algebra. The algebra A is denoted "$(K/L, \sigma, a)$."

Theorem 4.11.26. *The algebra $A = (K/L, \sigma, a)$ is a central simple algebra over L.*

Proof. [Sc1, Th. 8.12.1]. □

From dimensional considerations, we see that K is a maximal commutative subfield of A, and, as in Theorem 4.11.17, one checks that K is a splitting field for A.

4. Drinfeld Modules

Theorem 4.11.27. 1. *A central simple algebra A of dimension n^2 over L is isomorphic to a cyclic algebra if A contains a commutative subfield K of degree n over L such that K/L is a cyclic extension.*
2. *$(K/L, \sigma, a) \simeq (K/L, \sigma, b)$ if and only if $a/b \in N_{K/L}(K^*)$.*
3. *$(K/L, \sigma, a)$ splits if and only if $a \in N_{K/L}(K^*)$.*
4. *$(K/L, \sigma, a) \otimes (K/L, \sigma, b)$ is equivalent to $(K/L, \sigma, ab)$ in the Brauer group.*

Proof. [Sc1, Th 8.12.2, Th, 8.12.4, Lemma 8.12.6 and Th. 8.12.7]. □

Remark. 4.11.28. The above definition is quite common. However, Deuring [Deu1] and Weil [We1] *transpose* the relation 2 in the above definition of a cyclic algebra. Thus they obtain the Brauer *inverse* of our definition.

We are now in a position to discuss the computation of Brauer groups in arithmetic. We begin with the local theory. By definition, the local fields are $\mathbb{R}$, $\mathbb{C}$, finite extensions of $\mathbb{Q}_p$ or finite extensions of $\mathbb{F}_p\left(\left(\frac{1}{T}\right)\right)$. The first two are Archimedean and the others non-Archimedean. As $\mathbb{C}$ is algebraically closed, there are no non-trivial division algebras over it. Over $\mathbb{R}$ there is, up to isomorphism, one non-commutative central division algebra. This algebra is *Hamilton's Quaternions* given as $(\mathbb{C}/\mathbb{R}, ^-, -1)$. It is denoted "$\mathbb{H}$." Thus $\mathrm{Br}(\mathbb{R}) \simeq \frac{1}{2}\mathbb{Z}/\mathbb{Z}$.

Now let L be a finite extension of $\mathbb{Q}_p$ or $\mathbb{F}_p\left(\left(\frac{1}{T}\right)\right)$.

Theorem 4.11.29. 1. *Every central simple algebra over L has an unramified (thus cyclic) splitting field. Every central simple algebra is cyclic.*
2. *There is a canonical isomorphism (called the "invariant isomorphism")*

$$\mathrm{inv}_L \colon \mathrm{Br}(L) \xrightarrow{\sim} \mathbb{Q}/\mathbb{Z}.$$

3. *Under the isomorphism of Part 2, the fraction k/n corresponds to the cyclic algebra $(K/L, \sigma, \pi^k)$ where K/L is the cyclic unramified extension of degree n, σ is the Frobenius automorphism and $\pi \in L$ is a uniformizing element (= generator of the maximal ideal in the ring of integers). If $(k, n) = 1$, then $(K/L, \sigma, \pi^k)$ is a division ring.*
4. *In general, the invariant of $(K/L, \sigma, a)$ depends only on $\mathrm{ord}_\pi(a)$, where π is as in Part 3.*
5. *Let K/L be a finite extension of degree n and let $r_{K/L} \colon \mathrm{Br}(L) \to \mathrm{Br}(K)$ be the "restriction map" given by $A \mapsto A \otimes_L K$. Then the following diagram is commutative:*

$$\begin{array}{ccc} \mathrm{Br}(L) & \xrightarrow{\mathrm{inv}_L} & \mathbb{Q}/\mathbb{Z} \\ {\scriptstyle r_{K/L}}\big\downarrow & & \big\downarrow {\scriptstyle x \mapsto nx} \\ \mathrm{Br}(K) & \xrightarrow{\mathrm{inv}_K} & \mathbb{Q}/\mathbb{Z} \end{array}$$

6. *Suppose D is a central division algebra with $[D:L] = d^2$. Any extension K/L with $[K:L] = d$ can be embedded in D as a maximal commutative subfield.*

Proof. See [Deu1] or [We1] for Parts 1-5 (keeping Remark 4.11.28 in mind). For Part 6 notice that Part 5 immediately implies that K is a splitting field for D. Thus the result follows from Theorem 4.11.21. $\square$

Finally, we let L be a global field; so L is a finite extension of $\mathbb{Q}$ or $\mathbb{F}_p(T)$ for some non-constant T. Let M_L be the set of normalized places of L (as in [We1]), and, if $\mathbb{Q} \subseteq L$, let M_L^∞ be the set of Archimedean places. A place in $M_L - M_L^\infty$ is *discrete*. For $w \in M_L$ we have the completion L_w and the restriction $\operatorname{Br}(L) \to \operatorname{Br}(L_w)$ given by $A \mapsto A_w := A \otimes L_w$.

Theorem 4.11.30. 1. *Every central simple algebra over L is cyclic.*
2. *For almost all $w \in M_L$ (i.e, outside of finitely many), A_w splits over L_w.*
3. *The local invariant homomorphisms induce a canonical exact sequence*

$$0 \to \operatorname{Br}(L) \to \bigoplus_{w \text{ discrete}} \mathbb{Q}/\mathbb{Z} \oplus \bigoplus_{w \text{ real}} \frac{1}{2}\mathbb{Z}/\mathbb{Z} \xrightarrow{\beta} \mathbb{Q}/\mathbb{Z} \to 0$$

where β is the sum of all local invariants.

Proof. See [Re1] or [We1]. $\square$

Definition 4.11.31. 1. Let L be an arbitrary field with A a central simple algebra over L. Suppose $A \sim M_t(D)$ where D has degree d. We set $\operatorname{ind}(A) = d$; it is the *Schur index* of A.
2. We set $\exp(A)$ to be the order of the class of A in $\operatorname{Br}(L)$.

Lemma 4.11.32. *Let A, L, etc., be as in Definition 4.11.31.*
1. $\exp(A) \mid \operatorname{ind}(A)$.
2. *Every prime divisor of $\operatorname{ind}(A)$ is a prime divisor of $\exp(A)$.*

Proof. [FD1, Prop. 4.16 and Lemma 4.17]. $\square$

Theorem 4.11.33. *Let L be a global field and A a central simple algebra over L. For $w \in M_L$ set $i_w := \operatorname{ind}(A_w)$. Let K be a finite extension of L.*
1. *K is a splitting field for A if and only if for each prime W of K we have*

$$i_w \mid [K_W:L_w]$$

where w is the restriction of W to L.
2. *We have*

$$\operatorname{ind}(A) = \exp(A) = \operatorname{l.c.m.}\{i_w\}.$$

102   4. Drinfeld Modules

Proof. [Re1, Th. 32.15 and Th. 32.17]. □

Finally we remark that by Theorem 4.11.18, every central simple algebra A over L splits over a separable extension K/L. Thus A also splits over the Galois closure of K. One can therefore reinterpret $\text{Br}(L)$ in terms of *Galois cohomology*, see [Se1] for example.

4.12. Drinfeld Modules over Finite Fields

In this subsection we return to Drinfeld modules. We let **A**, **k**, etc., be as in Subsection 4.1. Let $\mathbb{F}_q$ be the finite field with q-elements where we suppose that $q = r^s$ for some integer $s > 0$. We further suppose that there exists a morphism $\imath \colon \mathbf{A} \to \mathbb{F}_q$ making $\mathbb{F}_q$ into an **A**-field. Clearly the kernel of $\imath$ is a non-trivial prime $\wp$ of **A**.

Let ϕ be a fixed Drinfeld module of rank d over $\mathbb{F}_q$. By definition ϕ gives an injection of **A** into $\mathbb{F}_q\{\tau\}$, where τ is the rth power morphism. From the discussion at the beginning of Section 4.11, we see that the domain $\mathbb{F}_q\{\tau\}$ can be imbedded in its division ring of fractions; we call this division ring "$\mathbb{F}_q(\tau)$." As $\mathbb{F}_q$ is perfect, we may use either the left *or* right division ring of quotients.

Set $F := \tau^s \in \mathbb{F}_q\{\tau\}$. Note that F fixes $\mathbb{F}_q$ and is thus an endomorphism of ϕ. It is therefore integral over **A**.

Lemma 4.12.1. $\mathbb{F}_q(\tau)$ *is a central division algebra over* $\mathbb{F}_r(F)$ *(in the sense of* Definition 4.11.8*) of dimension* s^2.

Proof. This is a straightforward exercise. □

Our next task is to compute the invariants (see 4.11.29.2) of $\mathbb{F}_q(\tau)$ at the primes of $\mathbb{F}_r(F)$. We begin by noting that $\mathbb{F}_q(\tau)$ is a *cyclic algebra* over $\mathbb{F}_r(F)$ (see the discussion before Theorem 4.11.26 for definitions). Indeed, the reader will easily see that

$$\mathbb{F}_q(\tau) \simeq (\mathbb{F}_q(F)/\mathbb{F}_r(F), \sigma, F),$$

where σ is the Frobenius automorphism of the constant field extension $\mathbb{F}_q(F)/\mathbb{F}_r(F)$. Thus, we can use the calculations in Theorem 4.11.29 *once* we have reduced ourselves to the local situation. For this we recall some facts from Deuring [Deu1] (but always with 4.11.28 in mind!). Thus let L be a field and let K be a cyclic extension of L of degree n with generator σ.

Proposition 4.12.2. 1. *Let m be prime to n. Then*

$$(K/L, \sigma, \alpha) = (K/L, \sigma^m, \alpha^m).$$

2. Let E be an extension of L. Let $M = KE$ be the compositum of K and E. Let j be the smallest power > 0 so that σ^j lies in $\mathrm{Gal}(M/E)$ (viewed as a subgroup of $\mathrm{Gal}(K/L)$). Then

$$(K/L, \sigma, \alpha) \otimes E = (M/E, \sigma^j, \alpha).$$

Proof. Part 1 is readily seen. For Part 2, see [Deu1, Chapter V, §5, Satz 4]. □

Corollary 4.12.3. *Let K/L be a cyclic extension of global fields and let w be a prime of L. Then*

$$(K/L, \sigma, \alpha) \otimes L_w = (K_W/L_w, \sigma^{n/n_w}, \alpha)$$

where W is any extension of w to K and n_w is the local degree $[K_W : L_w]$. □

Theorem 4.12.4. *Let L be a global function field with constant field $\mathbb{F}_r$ (thus L is a finite extension of $\mathbb{F}_r(T)$ for some nonconstant $T \in L$). Let K be an extension of L obtained by adjoining a root of unity (thus K is a constant field extension of L). Let $n = [K:L]$ and let σ be the automorphism of K/L which induces the rth power morphism on $\mathbb{F}_r$. Finally let $\alpha \in L^*$. Then for each place w of L, the invariant of $(K/L, \sigma, \alpha) \otimes L_w$ is*

$$\frac{\mathrm{ord}_w(\alpha) \deg_{\mathbb{F}_r}(w)}{n} \pmod{1}.$$

Proof. Let W be a place of K extending w. From 4.12.3 we see that

$$(K/L, \sigma, \alpha) \otimes L_w = (K_W/L_w, \sigma^{n/n_w}, \alpha).$$

Set $\delta := \deg_{\mathbb{F}_r}(w)$. One see that $n_w = n/(n, \delta) \implies n/n_w = (n, \delta)$. As $\delta/(n, \delta)$ is prime to $n/(n, \delta)$, by 4.12.2.1 we see that

$$(K_W/L_w, \sigma^{(n,\delta)}, \alpha) = (K_W/L_w, \sigma^\delta, \alpha^{\delta/(n,\delta)}).$$

On the other hand, σ^δ is the Frobenius of the constant field extension K_W/L_w. Thus, by 4.11.29 we see that the invariant is

$$\frac{\delta/(n,\delta)\mathrm{ord}_w(\alpha)}{n/(n,\delta)} = \frac{\mathrm{ord}_w(\alpha) \deg_{\mathbb{F}_r}(w)}{n} \pmod{1}. \qquad \square$$

The above elegant computation was shown to us by M. Rosen. It is our pleasure to thank him here.

As a corollary to 4.12.4, we see that the *product formula* immediately implies that the sum of the invariants of $(K/L, \sigma, \alpha)$ is 0 in agreement with 4.11.30.3.

Corollary 4.12.5. $\mathbb{F}_q(\tau)$ *is the central division algebra over* $\mathbb{F}_r(F)$ *with invariant* $1/s$ *at the zero of* F, $-1/s$ *at the pole of* F, *and* 0 *elsewhere.* □

Recall that an *order* in $\mathbb{F}_q(\tau)$ over $\mathbb{F}_r[F]$ is a unital subring which is finite over $\mathbb{F}_r[F]$ and has maximal rank. It is *maximal* if and only if it cannot be strictly imbedded in a larger order.

Lemma 4.12.6. $\mathbb{F}_q\{\tau\}$ *is a maximal* $\mathbb{F}_r[F]$-*order in* $\mathbb{F}_q(\tau)$.

Proof. We need only check maximality. Suppose R is an order with $\mathbb{F}_q\{\tau\} \subseteq R$. As R is finite over $\mathbb{F}_r[F]$, one sees that there exists $0 \neq a \in \mathbb{F}_r[F]$ with $Ra \subseteq \mathbb{F}_q\{\tau\}$. But Ra is then a left ideal of $\mathbb{F}_q\{\tau\}$; thus by Corollary 1.6.3 it is principal. So $Ra = \mathbb{F}_q\{\tau\}b$ for some b or

$$R = \mathbb{F}_q\{\tau\}x,$$

$x = ba^{-1}$. Thus $x \in R$ and $x^2 \in R$ can be written rx for some $r \in \mathbb{F}_q\{\tau\}$. As R is a domain, we see that $x = r$. Thus $R \subseteq \mathbb{F}_q\{\tau\}$ giving the result. □

Note that 4.12.5 implies that the central simple $\mathbb{F}_r[F]$-algebra,

$$\mathbb{F}_q(\tau) \otimes_{\mathbb{F}_r[F]} \mathbb{F}_r((F))$$

is actually a central division algebra with invariant $1/s$. In fact, one can see directly that

$$\mathbb{F}_q(\tau) \otimes_{\mathbb{F}_r[F]} \mathbb{F}_r((F)) \simeq \mathbb{F}_q((\tau)),$$

where $\mathbb{F}_q((\tau))$ is the ring of "finite-tailed Laurent series in τ." One can also see directly $\mathbb{F}_q((\tau))$ is a division ring through the use, for instance, of the geometric series. Similar remarks hold for the completion of $\mathbb{F}_q(\tau)$ at the pole of F.

Let $\mathbb{F}_q\{\{\tau\}\} \subset \mathbb{F}_q((\tau))$ be defined in the obvious fashion. One can check (either directly, or by general theory; see e.g., 4.12.9.2) that $\mathbb{F}_q\{\{\tau\}\}$ is *the* (see 4.11.25) maximal $\mathbb{F}_r[[F]]$-order in $\mathbb{F}_q((\tau))$.

Sources for the next set of results are [Dr2], [Ge3] and [Yub1]. I also wish to acknowledge very useful communication with the authors of [Ge3] and [Yub1].

Set $R := \mathbb{F}_q\{\tau\}$ and consider $\mathbf{A}$ as being embedded in R via our Drinfeld module ϕ. By definition, the centralizer $Z_R(\mathbf{A})$ of $\mathbf{A}$ in R is

$$\{\alpha \in R \mid \alpha\phi_a = \phi_a\alpha, \ \forall a \in \mathbf{A}\}.$$

Thus the reader will immediately see that

$$Z_R(\mathbf{A}) = \mathrm{End}_{\mathbb{F}_q}(\phi).$$

The embedding of $\mathbf{A}$ by ϕ extends to an embedding of $\mathbf{k}$ in $\mathbb{F}_q(\tau)$. We define D to be the centralizer of $\mathbf{k}$ in $\mathbb{F}_q(\tau)$ (obvious definition). Clearly D is a division ring with $\mathbf{k}$ contained in its center and one checks that $D = \mathrm{End}_{\mathbb{F}_q}(\phi) \otimes \mathbf{k}$.

Lemma 4.12.7. *D is central over* $\mathbf{k}(F)$.

Proof. One sees easily that $\mathbf{k}(F)$ is contained in the center of D and that $D = Z_{\mathbb{F}_q(\tau)}(\mathbf{k}(F))$. Thus the result follows by 4.11.14.3. □

Put $E = \mathbf{k}(F)$ and $t^2 = \dim_E(D)$. Set v equal to the place of $\mathbf{k}$ associated to $\wp =$ the characteristic of $\mathbb{F}_q$ as an $\mathbf{A}$-field. Since D and E are simple subalgebras of $\mathbb{F}_q(\tau)$, we are free, for example, to use the powerful techniques of 4.12.4.

In our next result, we view ϕ as having complex multiplication by an $\mathbf{A}$-order in E. By 4.7.19 we can pass to an isogenous Drinfeld module and have complex multiplication by the maximal order $\mathcal{O} =$ ring of $\mathbf{A}$-integers. To be independent of the choice of isogenous Drinfeld module, as with elliptic curves, we pass to a category where isogenies are *invertible*. This is the category of *Drinfeld modules up to isogeny*. As every isogeny divides an element of $\mathbf{A}$, we need only invert the elements of $\mathbf{A} - \{0\}$. Note that the Tate modules are now also "tensored with $\mathbf{k}$" in this new category. These rational Tate spaces will be denoted "$V_w(\phi)$," etc., as vector spaces over the completion $\mathbf{k}_w$.

For the convenience of the reader, recall that $D = \text{End}_{\mathbb{F}_q}(\phi) \otimes \mathbf{k}$, $q = r^s$, $F = \tau^s$ is the Frobenius morphism of ϕ over $\mathbb{F}_q$, $E = \mathbf{k}(F)$ and $t^2 = \dim_E(D)$.

Theorem 4.12.8. 1. *There is a unique place v_E of E which is a zero of F, and a unique place ∞_E of E which is the pole of F. The place v_E lies above v and ∞_E is the unique place over the infinite prime of $\mathbf{k}$.*
2. *The central division algebra D over E is characterized by $\text{inv}_{v_E}(D) = 1/t$, $\text{inv}_{\infty_E}(D) = -1/t$ and $\text{inv}_w(D) = 0$ for any other place w of E.*
3. *In the category of Drinfeld modules up to isogeny, we have $V_{v_E}(\phi) = 0$. For any place w of $E \neq v_E, \infty$, the module $V_w(\phi)$ is a vector space over $E_w = \mathcal{O}_w \otimes_\mathbf{A} \mathbf{k}$ of dimension t.*
4. *We have $\text{rank}\,\phi = t[E:\mathbf{k}]$.*
5. *Let $|?|_\infty$ be the unique extension to E of the normalized valuation of $\mathbf{k}$ corresponding to ∞. Then*
$$|F|_\infty = q^{1/d} \qquad (d = \text{rank}\,\phi).$$

Proof. One sees readily that the index of $\mathbb{F}_q(\tau)$ is s. Similarly, at the pole or zero of F, the index is also s. Now let E' be a maximal commutative subfield containing E. By 4.11.17, E' is a splitting field for $\mathbb{F}_q(\tau)$ and by 4.11.15 $[E':\mathbb{F}_r(F)] = s$. Thus 4.11.33.1 implies that there is a unique place of E' lying above the zero of F and a unique place above the pole. The same is obviously also true for E.

The zero and pole of F correspond to valuations on $\mathbb{F}_r(F)$ while v and ∞ are places of $\mathbf{k}$. Thus we need some words to explain why v_E lies over v, etc.

106   4. Drinfeld Modules

Consider the completion $E_{v_E} \subseteq \mathbb{F}_q((\tau))$; by 4.11.25 the valuation on $\mathbb{F}_q((\tau))$ (measuring divisibility by τ) induces v_E on E. For $a \in \wp$ (= the characteristic of $\mathbb{F}_q$) one sees immediately that $\phi_a \equiv 0 \pmod{\tau}$. Thus v lies above v_E. Similarly for ∞, giving Part 1.

To see Part 2 note that, by [Jac1, Sec. 4.7, Ex.1 or Th. 4.1], the class of D in $\mathrm{Br}(E)$ is the same as $\mathbb{F}_q(\tau) \otimes E$. By 4.11.29.5, we see that

$$\mathrm{inv}_{v_E}(D) = [E : \mathbb{F}_r(F)] \cdot \mathrm{inv}_{F=0}(\mathbb{F}_q(\tau)).$$

By [Jac1, Th. 4.11], we see that $[E : \mathbb{F}_r(F)] = s/t$. Thus 4.12.5 gives

$$\mathrm{inv}_{v_E}(D) = 1/t.$$

The rest of Part 2 follows in a similar fashion (compare 4.7.17).

To see Part 3, set $\tilde{\mathcal{O}} = E \cap \mathbb{F}_q\{\tau\}$. One sees that $\tilde{\mathcal{O}}$ is an order in E and that the inclusion of $\tilde{\mathcal{O}}$ in $\mathbb{F}_q\{\tau\}$ gives a Drinfeld module extending ϕ. By passing to an isogenous Drinfeld module $\tilde{\phi}$, we can get complex multiplication by $\mathcal{O}$ as in Proposition 4.7.19. Note that as F lies in the center of $\mathbb{F}_q(\tau)$, $\tilde{\phi}_F = F$.

To compute the rank of $\tilde{\phi}$, we find that $\deg_\tau F = s$ and $\deg_{\infty_E} F = [E : \mathbb{F}_r(F)]$. Thus

$$\mathrm{rank}\,\tilde{\phi} = s/[E : \mathbb{F}_r(F)]$$
$$= s \cdot t/s$$
$$= t.$$

It is clear that the characteristic of $\tilde{\phi}$ is the ideal associated to v_E. So, for all $w \neq v_E, \infty$, the space $T_w(\tilde{\phi})$ is free of rank t. Moreover,

$$T_{v_E}(\tilde{\phi}) = \varprojlim \tilde{\phi}[F^j] = \{0\}.$$

Part 3 follows directly as does Part 4.

Finally, we turn to Part 5. For a place w of a global field L, let $|?|_w$ denote here the normalized absolute value at w. The following are now readily seen to be equivalent:

1. $|F|_\infty = q^{1/d}$,
2. $|F|_{\infty_E} = q^\alpha$, $\alpha = [E : \mathbf{k}]/d$,
3. $|F|_{v_E} = q^{-\alpha}$, $\alpha = [E : \mathbf{k}]/d$
4. $|F|_{F=0} = q^{-\beta}$, $\beta = [E : \mathbf{k}]/(d[E : \mathbb{F}_r(F)])$.

But, of course, $|F|_{F=0} = 1/r$. Thus statement 4 is equivalent to $s[E : \mathbf{k}] = d[E : \mathbb{F}_r(F)]$. By Part 4, this is equivalent to $s = t[E : \mathbb{F}_r(F)]$ which was established in the proof of Part 2.   □

To proceed further we need to recall, once again, some techniques from the theory of central simple algebras and their orders. Let A be an arbitrary Dedekind domain and let K be its quotient field. Let H be a central simple algebra over K and let R be an A-order in H.

4.12. Drinfeld Modules over Finite Fields   107

Proposition 4.12.9. 1. R is contained in a maximal order R_1.
2. R is a maximal order if and only if $R \otimes A_w$ is a maximal A_w-order in $H \otimes K_w$ for all primes w of A.

Proof. Part 1 is [Re1, Cor. 10.4]. Part 2 is [Re1, Cor. 11.6]. □

Now let R be a maximal order of H. A *left R-ideal* Λ of H is a finitely generated torsion-free left R-submodule of H such that $K \cdot \Lambda = H$. Let α be an invertible element of H. Then clearly $\Lambda \alpha$ is also a left R-ideal. The number of equivalence classes under multiplication by invertible elements is called the *class number* of R. It is finite whenever K is a global field [Re1, Sec. 26]. It is independent of the maximal order used and may also be obtained by using right ideals, [Re1, Sec. 26, Ex. 7].

Let R, α be as above. Then $\alpha R \alpha^{-1}$ is maximal whenever R is. The number of equivalence classes of maximal orders is called the *type number* of H. It is known that the type number is *smaller* than the class number (see [Deu1, Sec. VI, §8] or use [Re1, Th. 21.6]).

The importance for us of this discussion lies in our next result.

Lemma 4.12.10. *Every maximal $\mathbb{F}_r[F]$-order in $\mathbb{F}_q(\tau)$ is conjugate to $\mathbb{F}_q\{\tau\}$.*

Proof. As $\mathbb{F}_q$ is perfect, we have both left and right division algorithms in $\mathbb{F}_q\{\tau\}$. One thus easily sees that the class number is 1 and the result follows from the previous discussion. □

For the moment, now let L be an arbitrary **A**-field and let w be a prime of **A** not equal to the characteristic of L. Let ϕ and ψ be two Drinfeld modules over L of rank $d > 0$. As in Definition 4.10.9, we can form the w-adic Tate modules $T_w(\phi)$ and $T_w(\psi)$. Both of these are free of rank d over $\mathbf{A}_w$. A morphism $f: \phi \to \psi$ gives an $\mathbf{A}_w$-morphism

$$f_*: T_w(\phi) \to T_w(\psi).$$

Proposition 4.12.11. *The natural map*

$$\mathrm{Hom}_L(\phi, \psi) \otimes \mathbf{A}_w \to \mathrm{Hom}_{\mathbf{A}_w}(T_w(\phi), T_w(\psi))$$

is injective with torsion free cokernel.

Proof. Suppose $f \in \mathrm{Hom}_L(\phi, \psi)$ is nonzero. By 4.7.13, we are thus guaranteed a non-trivial isogeny $\widehat{f}: \psi \to \phi$. The map $\alpha \mapsto \widehat{f} \circ \alpha$ then gives an injection of $\mathrm{Hom}_L(\phi, \psi)$ into $\mathrm{End}_L(\phi)$. As such, by Theorem 4.7.8, we see that $M = \mathrm{Hom}_L(\phi, \psi)$ is projective and finitely generated; say

$$M \simeq \mathbf{A}^t \oplus I,$$

for some $t \geq 0$ and some ideal $I \subseteq \mathbf{A}$.

Suppose $m \in M$ is such that m_* is zero. Write $m = \alpha_1\rho_1+\cdots+\alpha_t\rho_t+i\rho_{t+1}$ where $\{\rho_1,\ldots,\rho_{t+1}\} \subset M$ give the above decomposition, $i \in I$ and $\{\alpha_e\} \subset \mathbf{A}$. Since m_* is 0, m is trivial on $\phi[w^j]$ for all $j \geq 0$. Suppose $w^{j_0} = (b)$, where j_0 is the class number of $\mathbf{A}$ and $b \in \mathbf{A}$. As in 4.7.13, we see that m_* factors through b^e for all $e \geq 0$. In other words one can find $\lambda_e \in \mathrm{Hom}_L(\phi,\psi)$ such that
$$b^e \cdot \lambda_e = m$$
for all e. This shows that α_i, $i = 1,\ldots t$, and i are divisible by b^e, $e \geq 1$; thus they vanish. This proves the first part of the result.

For the second part, suppose that
$$\alpha f = m$$
for $f \in \mathrm{Hom}_{\mathbf{A}_w}(T_w(\phi), T_w(\psi))$, $\alpha \in \mathbf{A}_w$ and m is written as above. Let π_w be a uniformizer at w and let $\alpha = \pi_w^h \beta$, where $h \geq 0$ and β is a unit. As above we see that $\alpha_1,\ldots,\alpha_t$, i are all divisible by π_w^h in $\mathbf{A}_w$. Thus the result follows. $\square$

The reader should note that the injectivity of the map of 4.12.11 essentially follows because an additive polynomial can have only finitely many roots.

Let us now return to the $\mathbf{A}$-field $L := \mathbb{F}_q$. Let ϕ and ψ be two Drinfeld modules of rank d over $\mathbb{F}_q$ and let w be a place of $\mathbf{A}$ not equal to v (which is the place associated to the characteristic of $\mathbb{F}_q$ as an $\mathbf{A}$-field). Let $\overline{\mathbb{F}}_q$ be an algebraic closure of $\mathbb{F}_q$ and put $G := \mathrm{Gal}(\overline{\mathbb{F}}_q/\mathbb{F}_q)$. The Tate modules are obviously $\mathbf{A}_w[G]$-modules and the map of 4.12.11 is obviously a morphism of $\mathbf{A}_w[G]$-modules.

Theorem 4.12.12. 1. *The map*
$$\mathrm{Hom}_{\mathbb{F}_q}(\phi,\psi) \otimes \mathbf{A}_w \to \mathrm{Hom}_{\mathbf{A}_w[G]}(T_w(\phi), T_w(\psi))$$
is bijective.
2. *As above, let F be the q-Frobenius endomorphism of ϕ. Let $m_\phi(u)$ be its minimal polynomial over $\mathbf{k}$ as a field element and $f_\phi(u)$ its characteristic polynomial as an endomorphism of $T_w(\phi)$. Then, in the notation of Theorem 4.12.8, $f_\phi = m_\phi^t$. Consequently $f_\phi(u)$ is independent of w.*
3. *The modules ϕ and ψ are isogenous over $\mathbb{F}_q \iff m_\phi = m_\psi \iff f_\phi = f_\psi$.*

Proof. Part 2 of the theorem follows directly from Theorem 4.12.8.3.

From Proposition 4.12.11 one sees readily that the map given in Part 1 injective and has torsion free kernel. Therefore Part 1 is equivalent to establishing that
$$\mathrm{Hom}_{\mathbb{F}_q}(\phi,\psi) \otimes \mathbf{k}_w \to \mathrm{Hom}_{\mathbf{k}_w}(V_w(\phi), V_w(\psi))^G$$
is bijective. The $\mathbf{k}_w$-vector space $\mathrm{Hom}_{\mathbf{k}_w}(V_w(\phi), V_w(\psi))^G$ can also be written

$$\mathrm{Hom}_{\mathbf{k}_w[F]}(V_w(\phi), V_w(\psi)),$$

where $V_w(\psi)$ and $V_w(\psi)$ are viewed as $\mathbf{k}_w[F]$-modules. By 4.12.8.3,

$$V_w(\phi) \simeq (\mathbf{k}_w[x]/m_\phi(x))^t$$

for some indeterminate x. Thus, if ϕ and ψ are isogenous, one has $m_\phi = m_\psi$ and $f_\phi = f_\psi$. If ϕ and ψ are not isogenous, then

$$\mathrm{Hom}_{\mathbf{k}_w[F]}(V_w(\phi), V_w(\psi)) = \{0\}.$$

Conversely, if $f_\phi = f_\psi$, then $m_\phi = m_\psi$ by 4.12.8.3. If $m_\phi = m_\psi$, then $\mathrm{End}_{\mathbb{F}_q}(\phi) \otimes \mathbf{k}$ is isomorphic to $\mathrm{End}_{\mathbb{F}_q}(\psi) \otimes \mathbf{k}$ by Theorem 4.12.8.2 (as, indeed, they are both central over isomorphic fields with the same invariants). By the theorem of Skolem-Noether (Theorem 4.11.12), this isomorphism is induced by an inner automorphism $f \mapsto zfz^{-1}$ of $\mathbb{F}_q(\tau)$. As

$$\mathbb{F}_q(\tau) = \mathbb{F}_q\{\tau\} \otimes_{\mathbb{F}_r[F]} \mathbb{F}_r(F)$$

we may assume that $z \in \mathbb{F}_q\{\tau\}$. One then checks that z is an isogeny from ϕ to ψ. This completes the proof of the theorem. $\square$

The connection between $f_\phi(u)$ and the reduced norm on D is given in [Ge3, §3].

Theorem 4.12.13. *Suppose that the Frobenius $F \in \mathbf{A}$. Then $\mathrm{End}_{\mathbb{F}_q}(\phi)$ is a maximal order in $D = \mathrm{End}_{\mathbb{F}_q}(\phi) \otimes \mathbf{k}$.*

Proof. We know that D splits locally into matrix algebras at the primes of $\mathbf{k}$ not equal to v or ∞. Let w be a prime of $\mathbf{A}$ not equal to v. Then the proof of Theorem 4.12.12 shows that $\mathrm{End}_{\mathbb{F}_q}(\phi) \otimes \mathbf{A}_w$ is a matrix algebra over $\mathbf{A}_w$ and thus a maximal order in the split algebra $D \otimes \mathbf{k}_w$.

Next we focus on the place v. From our computation of invariants and 4.11.32.1, we see that $D \otimes \mathbf{k}_v$ is *still* a division algebra. It is a division subring of $\mathbb{F}_q((\tau))$. Theorem 4.11.25 implies that the intersection of the maximal order $\mathbb{F}_q\{\{\tau\}\}$ of $\mathbb{F}_q((\tau))$ with $D \otimes \mathbf{k}_v$ is *the* maximal order of $D \otimes \mathbf{k}_v$. But this intersection is $\mathrm{End}_{\mathbb{F}_q}(\phi) \otimes_{\mathbb{F}_r[F]} \mathbb{F}_r[[F]]$.

Therefore $\mathrm{End}_{\mathbb{F}_q}(\phi) \otimes \mathbf{A}_w$ is a maximal order for all primes of $\mathbf{A}$. The result now follows from 4.12.9.2. $\square$

Let $\overline{\mathbf{k}}$ be a fixed algebraic closure of $\mathbf{k}$.

Definition 4.12.14. An element $F \in \overline{\mathbf{k}}$ is called a *Weil number of rank d* for the $\mathbf{A}$-field $\mathbb{F}_q$ if and only if

1. F is integral over $\mathbf{A}$.
2. There is only one place of $\mathbf{k}(F)$ which is a zero for F. This place lies above the characteristic v.

3. There is only one place of $k(F)$ lying above ∞.
4. $|F|_\infty = q^{1/d}$, where $|?|_\infty$ is the unique extension to $k(F)$ of the normalized absolute value of k corresponding to ∞.
5. $[k(F):k]$ divides d.

The set of Weil numbers of rank d for $\mathbb{F}_q$ is acted on by the group of automorphisms of $\bar{k}$ over k. Let W_d be the set of conjugacy classes of Weil numbers of rank d under this action. From Theorem 4.12.8 and Theorem 4.12.12, we obtain an injection

$$\{\text{isogeny classes of Drinfeld } \mathbf{A}\text{-modules of rank } d \text{ over } \mathbb{F}_q\} \hookrightarrow W_d.$$

This injection takes the isogeny class of a given Drinfeld module ϕ to the roots of the polynomial $m_\phi(u)$ as defined in 4.12.12.2.

Theorem 4.12.15. *The above map is a bijection.*

Proof. Let F be a Weil number of rank d. Set $E := k(F)$ and let v_E be the unique zero of F above v; similarly let ∞_E be the infinite prime of E.

Consider the extension $E/\mathbb{F}_r(F)$. One sees that v_E is the only zero of F above the zero of F in $\mathbb{F}_r(F)$ and similarly for the pole. Let $q = r^s$. Then Property 4 implies that

$$[E:\mathbb{F}_r(F)] = s[E:k]/d.$$

Put $t = d/[E:k]$ and let D be the central division algebra of degree t^2 over E with $\text{inv}_{v_E}(D) = 1/t$, $\text{inv}_{\infty_E}(D) = -1/t$ and all other invariants equal to 0. Let L be a maximal subfield of D such that $E \subseteq L$. As before, the places v_E and ∞_E are inert in L. Moreover, one computes

$$s = [L:\mathbb{F}_r(F)].$$

Using the obvious isomorphism

$$\mathbb{F}_r(F) \tilde{\to} \mathbb{F}_r(\tau^s)$$

of rational function fields, we view L as an $\mathbb{F}_r(\tau^s)$-algebra. We now form the tensor product

$$\widetilde{D} = \mathbb{F}_q(\tau) \otimes_{\mathbb{F}_r(F)} L.$$

A simple calculation using 4.12.5, 4.11.29.5, and the results of the last paragraph, shows that *all* local invariants of $\widetilde{D}$ vanish. Thus L is a splitting field of $\mathbb{F}_q(\tau)$.

By Theorem 4.11.21, we see that L embeds in $\mathbb{F}_q(\tau)$. Let $\beta: E \to \mathbb{F}_q(\tau)$ be the restriction of this embedding to E. Let $\mathcal{O}$ be the ring of functions of E regular away from ∞_E. By [Re1, Sec. 8, Ex. 1] we can embed $\beta(\mathcal{O})$ in some $\mathbb{F}_r[\tau^s]$-order R, and by Proposition 4.12.9.1, we may assume that R is *maximal*.

Lemma 4.12.10 now assures us that we may assume $R = \mathbb{F}_q\{\tau\}$. One now verifies that
$$\phi := \beta \mid_{\mathbf{A}}$$
is a Drinfeld module of rank d over $\mathbb{F}_q$ with Weil number F. □

The reader should note that if F is a Weil number of rank d associated to $\mathbb{F}_q$ and the isogeny class of ϕ, then F^n is associated to the isogeny class of ϕ over $\mathbb{F}_{q^n}$, etc.

The above results constitute the analog of the classical theory of *Tate-Honda* for Drinfeld modules. These were announced in [Dr2] and we have closely followed the exposition in [Yub1].

Recall that $\wp$ is the characteristic of $\mathbb{F}_q$ as an **A**-field. Let $\overline{\mathbb{F}}_q$ be an algebraic closure of $\mathbb{F}_q$.

Definition 4.12.16. Let ϕ be a Drinfeld module of rank d over $\mathbb{F}_q$. We say that ϕ is *supersingular* if and only if $\phi[\wp] = \phi[\wp](\overline{\mathbb{F}}_q)$ (= the module of $\wp$-division points in $\overline{\mathbb{F}}_q$) is trivial.

In fact, the supersingularity of ϕ clearly depends only on the isogeny class of ϕ. Thus there is a good notion of *supersingular Weil numbers*, etc.

Proposition 4.12.17. *Let ϕ be as in 4.12.16. Then the following are equivalent:*

1. *There is a finite extension $\mathbb{F}_{q^n}$ of $\mathbb{F}_q$ such that the division ring*
$$\mathrm{End}_{\mathbb{F}_{q^n}}(\phi) \otimes \mathbf{k}$$
has dimension d^2 over $\mathbf{k}$.
2. *Some power of the Frobenius endomorphism F of ϕ over $\mathbb{F}_q$ lies in $\mathbf{A}$.*
3. *ϕ is supersingular.*
4. *The field $\mathbf{k}(F)$ has only one prime above $\wp$.*

Proof. The equivalence of 1 and 2 follows from Theorem 4.12.8. Thus, suppose 2. Let $\alpha = F^m \in \mathbf{A}$ for some m. Then ϕ_α is purely inseparable. Moreover (α) must be a power of $\wp$. Thus ϕ is supersingular. Conversely let ϕ be supersingular and let $\wp^j = (\beta)$ for some β. Thus ϕ_β is purely inseparable, i.e., $\phi_\beta = cF^w$, for some $c \in \mathbb{F}_q^*$ and $w > 0$. Therefore,
$$\phi_{\beta^{q-1}} = F^{(q-1)w}$$
giving Part 2.

The equivalence of 3 and 4 comes from Theorem 4.12.8. Indeed 4.12.8.3 gives the existence of $\wp$-division points if and only if $\wp$ is *not* inert in $\mathbf{k}(F)$.

One can also see easily that 4 $\Longrightarrow$ 2. Indeed, by 4.12.8 (or 4.7.17) $\mathbf{k}(F)$ has only one prime above ∞. Without loss, we can assume that ϕ has complex multiplication by the ring of **A**-integers, $\mathcal{O}$, in $\mathbf{k}(F)$. Now the unit group of $\mathcal{O}$ is finite by assumption, and 4 implies that

$$(F) = \wp_E^j$$

where $\wp_E$ lies above $\wp$. Let σ be any automorphism of $E/\mathbf{k}$ (which may not be separable). Then

$$(\sigma(F)) = \wp_E^j$$

or $\sigma(F) = uF$, where u is a unit in $\mathcal{O}$. This clearly gives 2. □

Let ϕ, n be as in 4.12.17.1. We remark that by 4.7.8, $\text{End}_{\mathbb{F}_{q^n}}(\phi)$ has maximal rank over **A**. Moreover by 4.12.13, $\text{End}_{\mathbb{F}_{q^n}}(\phi)$ is a maximal order in $\text{End}_{\mathbb{F}_{q^n}}(\phi) \otimes \mathbf{k}$.

Example 4.12.18. Let $p = 2$; so r is a 2-power. Set $\mathbf{A} = \mathbb{F}_r[T]$, $\wp = (T)$ and $\mathbb{F}_q = \mathbf{A}/\wp \simeq \mathbb{F}_r$ (so $q = r$). Let ϕ be the rank 2 Drinfeld **A**-module over $\mathbb{F}_q$ given by $\phi_T = \tau^2$. Clearly ϕ is supersingular. The q-Frobenius F equals τ. Thus $F^2 = \phi_T$. Put $E = \mathbf{k}(F)$; so $[E:\mathbf{k}] = 2$ and E is purely inseparable over $\mathbf{k}$. By Part 4 of Theorem 4.12.8, we have $\text{End}_{\mathbb{F}_q}(\phi) = \mathbf{A}$ and $\text{End}_{\mathbb{F}_{q^2}}(\phi)$ is a maximal order in a division ring of rank 4.

Suppose now that $\mathbb{F}_q$ has dimension m over the field $\mathbf{A}/\wp$. As above, let ϕ be a rank d Drinfeld module over $\mathbb{F}_q$. Let $f(u) := f_\phi(u) \in \mathbf{A}[u]$ be the characteristic polynomial of the q-Frobenius morphism F acting on $T_w(\phi)$ for some prime $w \neq \wp$. From 4.12.12, we see that $f(u)$ is independent of w.

Proposition 4.12.19. *We have*

$$(f(0)) = \wp^m.$$

Proof. This follows from Parts 1, 4 and 5 of Theorem 4.12.8 and Theorem 4.12.12. □

Corollary 4.12.20. $\wp^m$ *is a principal ideal.* □

The reader will note that 4.12.20 puts very strong restrictions on which extensions $\mathbb{F}_{q^m}$ may have Drinfeld modules over them. For instance, if $\wp$ is itself not principal, then there are *no* Drinfeld modules over $\mathbf{A}/\wp$.

For the next set of results, we follow Gekeler, [Ge3]. Let M be a finite **A**-module. We define the *Euler-Poincaré characteristic* (or *Fitting ideal*), $\chi(M)$, of M as follows: It is the ideal of **A** defined by

1. $\chi(M) = \wp$, if $M \simeq \mathbf{A}/\wp$, for $\wp$ prime;
2. Let $0 \to M_1 \to M_2 \to M_3 \to 0$ be exact. Then

$$\chi(M_2) = \chi(M_1)\chi(M_3).$$

4.12. Drinfeld Modules over Finite Fields

Proposition 4.12.21. *Let ϕ, $f(u)$, etc., be as above. Then*
$$(f(1)) = \chi(\mathbb{F}_q),$$
where $\mathbb{F}_q$ is viewed as a finite $\mathbf{A}$-module via ϕ.

Proof. Let $|?|_\infty$ be the normalized valuation at ∞ on $\mathbf{k}$. From 4.12.8.5 and the fact that $|1|_\infty = 1$, it readily follows that
$$|f(1)|_\infty = q.$$
Now let w be a prime of $\mathbf{A}$ not equal to $\wp$. One checks that
$$T_w(\phi)/(F-1)T_w(\phi) \simeq \mathbb{F}_q^{(w)},$$
where $\mathbb{F}_q^{(w)}$ is the submodule of $\mathbb{F}_q$ of w-power division points. Standard module theory thus implies that $(f(1))$ and $\chi(\mathbb{F}_q)$ agree at all primes of w except, perhaps, $\wp$.

To finish, we note that $\mathbb{F}_q$ is the direct sum over all its w-power torsion elements where w ranges over *all* $\mathbb{F}_q^{(w)}$, $w \in \operatorname{Spec}(\mathbf{A})$. As $|f(1)|_\infty = q$ and $\chi(\mathbb{F}_q)/f(1)$ is a w-unit at all $w \neq \wp$, we see that the product formula implies that it is a $\wp$-unit also. This gives the result. □

Corollary 4.12.22. *With the above notation, $\chi(\mathbb{F}_q)$ is principal.* □

For the moment, fix $w \neq \wp$ in $\operatorname{Spec}(\mathbf{A})$. Set $V := V_w(\phi) = T_w(\phi) \otimes \mathbf{k}$ as a $\mathbf{k}_w$-vector space. Thus
$$f(u) = \det(1 - uF \mid V).$$
Define
$$f_i(u) := \det(1 - uF \mid \wedge^i V).$$
As $f(u) = f_1(u)$ is independent of $w \neq \wp$, so are all $f_i(u)$, $i \geq 0$.

Let $h(u)$ be any function in u. As usual set
$$\frac{d}{du} \log h(u) = h'(u)/h(u).$$

Proposition 4.12.23. *We have*
$$\sum_{k \geq 1} \det(1 - F^k \mid V) u^k = u \frac{d}{du} \log \prod_{0 \leq e \leq d} f_e(u)^{(-1)^{e+1}}.$$

Proof. This is standard linear algebra; see [Ge3, Lemma 5.6]. □

Definition 4.12.24 (Gekeler). We set
$$Z(u) := Z(\phi, u) := \prod_{0 \leq e \leq d} f_e(u)^{(-1)^{e+1}}.$$

114   4. Drinfeld Modules

By 4.12.23 and 4.12.21, one can derive $\chi(\mathbb{F}_{q^n})$, $n \geq 1$, from $Z(u)$.

Due to Gekeler [Go4, Th. 3.2.8], one knows that ϕ is determined up to isogeny by its rank *and* the numbers

$$\{\det(1 - F^k \mid V) \mid k \geq 1\}.$$

The following example, due to Thakur, implies that this result is *false* without knowledge of the rank.

Example 4.12.25. Set $r = 2$. Put

$$Z_1(u) = \frac{1 - Tu}{1 - u}$$

and

$$Z_2(u) = \frac{1 - Tu^2}{(1 - u)(1 - Tu)}.$$

One sees that $Z_1(u)$ and $Z_2(u)$ are the Z-functions of Drinfeld modules of ranks 1 and 2, respectively. However, they both have the same log-derivatives.

Suppose that rank $\phi = 1$. Then by Part a of [Go4, Prop. 3.2.14] one knows that ϕ is determined up to isogeny by the *ideals* $\chi(\mathbb{F}_{q^n})$, $n \geq 1$. If rank $\phi > 1$, one only knows, so far, that ϕ is determined up to $\overline{\mathbb{F}}_q$-isogeny by $\chi(\mathbb{F}_{q^n})$, [Go4, Prop. 3.2.14.b].

Question 4.12.26. Is ϕ determined up to $\mathbb{F}_q$-isogeny by the ideals $\chi(\mathbb{F}_{q^n})$, $n \geq 1$?

Remark. 4.12.27. Let A be an abelian variety over $\mathbb{F}_q$ where q is a p-power. Let ℓ be a prime of $\mathbb{Z}$ not equal to p and set

$$P(u) := \det(1 - Fu \mid T_\ell(A)) \in \mathbb{Z}[u]$$

where $T_\ell(A)$ is the ℓ-adic Tate module. Let $T_p(A)$ be the p-adic Tate module of A. Then one knows that

$$\det(1 - Fu \mid T_p(A)) = \prod_\alpha (1 - \alpha u) \in \mathbb{Z}_p[u]$$

where α runs over the *unit roots* of $P(u) \in \mathbb{Z}_p[u]$. A similar result for Drinfeld modules can be readily obtained from Theorem 4.12.8.

See also the paper [Pot3] for interesting restatements of some of the results of this subsection.

4.13. Rigidity of Drinfeld Modules

For the moment, let k be a field and X a complete (e.g., projective) variety over k. Let Y and Z be *any* two varieties over k. Let $p_2: X \times Y \to Y$ be the projection on the second factor. Let $f: X \times Y \to Z$ be a morphism (as k-algebraic varieties) with the following property: there exists some $y_0 \in Y$ such that $f(X \times \{y_0\})$ is a single point $z_0 \in Z$.

Lemma 4.13.1 (Rigidity for Varieties). *Let X, Y, Z, etc., be as above. Then there exists a morphism $g: Y \to Z$ such that $f = g \circ p_2$.*

Proof. ([Mu1,§4]) Let $x_0 \in X$ be any point and define $g: Y \to Z$ by $g(y) = f(x_0, y)$. As $X \times Y$ is a variety, in order to show that $f = g \circ p_2$ one need only show that they are equal on some non-trivial open subset of $X \times Y$. Let $U \subseteq Z$ be an affine open subset with $z_0 \in U$. Set $W = Z - U$ and $G = p_2(f^{-1}(W))$. As X is complete, one knows that p_2 is proper; thus G is *closed*. Moreover, as $f(X \times \{y_0\}) = \{z_0\}$, one sees that $y_0 \notin G$. Set $V = Y - G$; thus V is nonempty and open.

Now let $y \in V$. By construction, the complete variety $X \times \{y\}$ maps to U; thus $X \times \{y\}$ maps to a single point of U. Consequently, if $x \in X$, $y \in V$, then
$$f(x, y) = f(x_0, y) = g \circ p_2(x, y)$$
completing the proof. □

Corollary 4.13.2. *Let X and Y be two abelian varieties over k and let $f: X \to Y$ be any morphism. Then $f(x) = h(x) + \alpha$, where $h: X \to Y$ is a homomorphism of abelian varieties and $\alpha \in Y$.*

Proof. ([Mu1,§4]) Set $\alpha = f(0)$ and let $g(x) = f(x) - f(0)$; so $g(0) = 0$. Let $\phi: X \times X \to Y$ be defined by
$$\phi(x, y) = g(x + y) - g(x) - g(y).$$
We see that $\phi(X \times \{0\}) = \phi(\{0\} \times X) = 0$. Thus, by 4.13.1 we deduce that $\phi \equiv 0$ on $X \times X$. However,
$$\phi \equiv 0 \iff g \text{ is a homomorphism},$$
and the result is established. □

The reader will notice the similarity between the proof of 4.13.2 and the proof of Theorem 1.2.1.

Corollary 4.13.3. *Let X and Y be as in 4.13.2.*
1. Let $f: X \to Y$ be a morphism (as varieties) such that for all $n > 0$, the n-division points of X are mapped to the n-division points of Y set-theoretically. Then f is a homomorphism of abelian varieties.

116   4. Drinfeld Modules

2. Let $f\colon X \to Y$ be a morphism (as varieties) such that f takes the set of all torsion points of X to the set of all torsion points of Y. Then $f(x) = h(x) + \alpha$, where $h(x)$ is a homomorphism of abelian varieties and α is a torsion point of Y.

Proof. These are obvious from 4.13.2. □

In this subsection we will discuss the Drinfeld modules analog of 4.13.3 due to B. Poonen [Po1], with the aid of A. Tamagawa. We will concentrate on establishing the results for endomorphisms while making a few remarks on what holds for arbitrary morphisms of Drinfeld modules. As the underlying variety of a Drinfeld module is $\mathbb{G}_a$, it is clear that no rigidity of the form 4.13.1, 4.13.2 can hold. Thus all proofs of the analogs of 4.13.3 must somehow use the **A**-action of the Drinfeld module.

Let **A**, **k**, etc., be as in Subsection 4.1. Let L be an **A**-field and let ϕ be a rank d Drinfeld module over L.

Theorem 4.13.4. *Let $g(x) \in L[x]$ have the property that g maps $\phi[a]$ to itself set-theoretically for all $a \in \mathbf{A}$. Then $g \in \mathrm{End}_L(\phi) \subseteq L\{\tau\}$.*

Proof. First of all, as $g(\phi[1]) \subseteq \phi[1]$, we deduce that $g(0) = 0$. Secondly, it is clear that we may work over *any* extension field of L.

We now suppose that L has generic characteristic; thus $\mathbf{k} \subseteq L$. As before, we let $\mathbf{C}_\infty$ be the completion of the algebraic closure of $\mathbf{k}_\infty$. Note that ϕ is defined over a finitely generated subfield of L and that this subfield may be injected into $\mathbf{C}_\infty$. Thus we may assume that $L = \mathbf{C}_\infty$.

Let $\Lambda \subset \mathbf{C}_\infty$ be the **A**-lattice of rank d associated to ϕ and let $e(z) := e_\Lambda(z)$ be the exponential associated to ϕ. Let $\log(z)$ be the composition inverse to $e(z)$. Clearly, $\log(z)$ is also $\mathbb{F}_r$-linear and, as $e(z)$ is entire, $\log(z)$ has a non-trivial radius of convergence. Thus the function

$$h(z) := \log(g(e(z)))$$

has $h(0) = 0$ and also converges in some non-trivial neighborhood U of 0.

The analytic theory of Drinfeld modules, Subsection 4.6, gives an isomorphism of $a^{-1}\Lambda/\Lambda$ with $\phi[a]$ for all $a \in \mathbf{A}$. If $z \in U \cap a^{-1}\Lambda$, then our assumption on g implies that $h(z) \in a^{-1}\Lambda$.

Now fix $\lambda \in \Lambda - \{a\}$ and let $\{a_j\} \subseteq \mathbf{A}$ be a sequence with $|a_j|_\infty \to \infty$ as $j \to \infty$. Thus $h(a_j^{-1}\lambda) \in a_j^{-1}\Lambda$ for $j \gg 0$. In particular,

$$h(a_j^{-1}\lambda) \cdot a_j \in \Lambda$$

for $j \gg 0$ is a sequence in Λ which converges to $h'(0)\lambda$. However, Λ is discrete; thus $h(a_j^{-1}\lambda)a_j$ must eventually be constant. Consequently, for $t \gg 0$

$$h(a_t^{-1}\lambda) = a_t^{-1}\lambda^1$$

4.13. Rigidity of Drinfeld Modules   117

for some $\lambda^1 \in \Lambda$. As $h(z)$ is analytic near 0 and agrees with a linear monomial on a sequence approaching 0, we conclude that $h(z)$ *is* a linear monomial; thus $g(z)$ is also $\mathbb{F}_r$-linear. Moreover, $h(z)$ commutes with scalar multiplication by $a \in \mathbf{A}$ and so $g(z)$ commutes with ϕ_a, at least in *some* non-trivial neighborhood of 0. But, as $g(z)$ and ϕ_a are polynomials, we conclude that they commute; i.e., $g \in \mathrm{End}(\phi)$. This concludes the proof in generic characteristic.

Suppose now that L has finite characteristic $\wp \neq 0$ in $\mathrm{Spec}(\mathbf{A})$. We begin here by assuming first that $g(x)$ is *additive*; an argument given later on will allow us to remove this restriction. Recall that we set τ_p to be the p-power map; $\tau_p(x) = x^p$. Thus $g(x)$ has a representation as a polynomial $g(\tau_p) \in L\{\tau_p\}$.

Set $m = \deg_{\tau_p}(g(\tau_p))$. We will show, first of all, that for *any* $a \in \wp - \{0\}$, g commutes with ϕ_{a^n+1}, $n > m$. Now if $a \in \wp$, then $a^n + 1 \notin \wp$; thus $\phi_{a^n+1}(\tau_p)$ is separable. As g takes $\phi[\alpha]$ to itself for all $\alpha \in \mathbf{A}$ by assumption, we conclude that $\phi_{a^n+1}(g(\tau_p))$ annihilates $\phi[a^n+1]$. As $\phi_{a^n+1}(\tau_p)$ is separable, the arguments of Subsection 1.3 (or 1.8) show that we may write $\phi_{a^n+1}(g(\tau_p))$ as $h(\phi_{a^n+1}(\tau_p))$ for some $h(\tau_p) \in L\{\tau_p\}$ with $\deg_{\tau_p} h = \deg_{\tau_p} g$.
As $\phi_{a^n+1}(g(\tau_p)) = h(\phi_{a^n+1}(\tau_p))$, we see that

$$\phi_{a^n}(g(\tau_p)) + g(\tau_p) = h(\phi_{a^n}(\tau_p)) + h(\tau_p). \qquad (*)$$

As $a \in \wp$, the coefficient of τ_p in $\phi_a(\tau_p)$ vanishes; thus $\phi_{a^n}(\tau_p)$ contains no monomial of the form τ_p^j, $j \leq n$. Therefore, Equation $*$ implies that $g(\tau_p) = h(\tau_p)$ showing that g and ϕ_{a^n+1} commute.

As $g(\tau_p)$ also commutes with τ_p^0, we conclude that g commutes with $\phi_{a^n} = \phi_a^n$ also. For such n, the argument given above also implies that g commutes with $\phi_{a^{n+1}} = \phi_a^{n+1}$. In order to see that g commutes with ϕ_a, we pause to establish a short lemma.

Lemma 4.13.5. *Let R be a domain (i.e., a unital ring with no non-trivial divisors). Let $x \in R$ be nonzero and suppose that $g \in R$ commutes with x and xy for some $y \in R$. Then g also commutes with y.*

Proof. We have

$$x(gy) = (xg)y = g(xy) = (xy)g = x(yg).$$

Thus canceling by x implies that $gy = yg$. $\square$

Returning now to the proof of 4.13.4, we see that Lemma 4.13.5 shows that g commutes with ϕ_a as it commutes with ϕ_a^n and ϕ_a^{n+1}. Similarly for any $b \in \mathbf{A}$, g commutes with ϕ_a and ϕ_{ab} (as $ab \in \wp$). Thus g commutes with ϕ_b giving the result for g additive.

To finish the proof we now present the argument, due to A. Tamagawa, that allows us to remove the additivity assumption. We may, without loss of generality, assume that L is algebraically closed. We let T be the set of

$g(x) \in L[x]$ for which there exists a nonzero $c \in \mathbf{A}$ with the property that $\phi_c \circ g(x)$ maps $\phi[a]$ to itself for all $a \in \mathbf{A}$.

Note that $T \cap L\{\tau_p\} \subseteq \text{End}_L(\phi)$. Indeed, we have seen that $\phi_c(g(\tau_p))$ must be an element of $\text{End}_L(\phi)$. Thus $\phi_c(g(\tau_p))$ commutes with ϕ_a, for $a \in \mathbf{A}$. But ϕ_a also commutes with ϕ_c; thus 4.13.5 implies that ϕ_a commutes with $g(\tau_p)$.

We now use induction on the degree of $g(x)$ to show that $T \subseteq \text{End}_L(\phi)$. Let $0 \neq b \in \mathbf{A}$ and $u \in \phi[b]$. Set

$$g_u(x) = g(x+u) - g(x) - g(u).$$

If $\phi_c(g(x))$ maps $\phi[a]$ to itself for all $a \in \mathbf{A}$, then one sees easily that $\phi_c(g_u(x))$ does the same. Thus $g_u(x) \in T$. As $\deg g_u(x) < \deg g(x)$ the induction hypothesis implies that

$$g_u \in \text{End}_L(\phi).$$

As we vary b, we find infinitely many possible elements u to use as above. On the other hand, as in Subsection 4.7, we know that $\text{End}_L(\phi)$ is discrete under the valuation given by taking degrees. Thus there are only finitely many endomorphisms of degree $< \deg g$. By the pigeonhole principle, there exists $j(x) \in \text{End}_L(\phi)$ such that

$$g_u(x) = g(x+u) - g(x) - g(u) = j(x)$$

for infinitely many u's. Thus, as a polynomial in variables x, y we conclude that

$$g(x+y) - g(x) - g(y) = j(x).$$

Interchanging x and y gives

$$g(x+y) - g(x) - g(y) = j(y);$$

thus $j(x) = j(y)$. In other words, $j(x)$ is constant, and as $g(0) = 0$, we conclude that $j(x) \equiv 0$. Thus g is additive and the proof is now complete. $\square$

Recall that in Subsection 1.11, we defined the semi-invariant of $f(\tau) \in L\{\tau\}$. By 1.11.2 we know that $g(\tau)$ is a semi-invariant of $f(\tau)$ if and only if there exists $h(\tau) \in L\{\tau\}$ with

$$f(\tau)g(\tau) = h(\tau)f(\tau). \tag{4.13.6}$$

Thus the semi-invariants of $f(\tau)$ are easily seen to form a *subring* of $L\{\tau\}$ which we denote "$\text{Sem}_L(f(\tau))$."

Let ϕ be our Drinfeld module over L as before. We define the *semi-invariants of ϕ*, $\text{Sem}_L(\phi)$, to be

$$\bigcap_{a \in \mathbf{A}} \text{Sem}_L(\phi_a).$$

It is easy to see that $\text{Sem}_L(\phi)$ contains $\text{End}_L(\phi)$.

Corollary 4.13.7. *We have the equality* $\mathrm{Sem}_L(\phi) = \mathrm{End}_L(\phi)$.

Proof. From the characterization, Equation 4.13.6, just given, it is clear that $g(\tau) \in \mathrm{Sem}_L(\phi)$ stabilizes $\phi[a]$ for all $a \in \mathbf{A}$. Thus the result follows immediately from 4.13.4. □

It is remarkable and satisfying that the obvious analog of Theorem 4.13.4 is *also* valid for $\mathbb{G}_m$ in any characteristic; we refer the reader to [Po1] for the details.

There is a version of Theorem 4.13.4 which works for morphisms of Drinfeld modules. Thus let ψ be another Drinfeld module over L; we specifically do *not* make any assumptions on its rank.

Theorem 4.13.8. *Let $g(x) \in L[x]$ have the property that for all $a \in \mathbf{A}$, g maps $\phi[a]$ set-theoretically to $\psi[a]$. Then g is a morphism of Drinfeld modules from ϕ to ψ.*

Proof. We leave this as an exercise for the reader. □

Theorem 4.13.8 gives a very beautiful characterization of morphisms of Drinfeld modules. For instance, if $\mathrm{rank}\,\phi \neq \mathrm{rank}\,\psi$, then $g \equiv 0$ is the only polynomial satisfying the assumptions of 4.13.8.

We now turn to a version for endomorphisms of Drinfeld modules of 4.13.3.2.

Theorem 4.13.9. *Suppose that L has generic characteristic. Let $g(x) \in L[x]$ and suppose that at least one of the following conditions hold:*
1. *There is an infinite $\mathbf{A}$-submodule S of the torsion module of ϕ (over an algebraic closure of L) such that g maps S to S.*
2. *There exist infinitely many $a \in \mathbf{A}$ for which g maps $\phi[a]$ to itself.*

Then there exists a torsion point c of ϕ and an endomorphism j of ϕ, both defined over L, so that
$$g(x) = j(x) + c.$$

Before turning to the proof of Theorem 4.13.9, we present some remarks and a lemma.

Remarks. 4.13.10. 1. Theorem 4.13.9 may fail if L has non-zero characteristic. Indeed, suppose L is *finite* and let $\overline{L}$ be an algebraic closure of L. Let ϕ be a Drinfeld module over L; it is then clear that $\overline{L}$ is *precisely* the set of torsion points of ϕ. But *any* $g(x) \in L[x]$ maps $\overline{L}$ to itself. Thus the first part of the theorem is violated.

2. Now let $\mathbf{A} = \mathbb{F}_r[T]$, etc., and let C be the Carlitz module. Let $L = \mathbf{A}/(T) \simeq \mathbb{F}_r$ and let $\phi = \widetilde{C}$ be the reduction of C at (T). Thus

120   4. Drinfeld Modules

$$\phi_T = \tau \in L\{\tau\}.$$

As ϕ has rank 1, $\text{End}(\phi) = \mathbf{A}$. On the other hand, let $g \in \mathbb{F}_{r^2}\{\tau\}$, where $\mathbb{F}_{r^2}$ is viewed as the extension of degree two of L. It is clear that g commutes with ϕ_{T^2} and thus with ϕ_a for $a \in \mathbb{F}_r[T^2]$. Moreover, g maps $\phi[a]$ to itself for $a \in \mathbb{F}_r[T^2]$. Thus we see that Theorem 4.13.9 can fail with its second hypothesis when the characteristic is finite.

Lemma 4.13.11. *Let ϕ be a Drinfeld module over an $\mathbf{A}$-field L of generic characteristic. Let $R \subseteq L$ be the $\mathbf{k}$-algebra generated by the coefficients of ϕ_a and the inverses of the leading coefficients of ϕ_a, for all nonzero $a \in \mathbf{A}$. Then R is finitely generated over $\mathbf{k}$.*

Proof. (A. Tamagawa) As is well-known, the algebra $\mathbf{A}$ is finitely generated over $\mathbb{F}_r$; say by $\{\alpha_1, \ldots, \alpha_j\}$. Thus the $\mathbf{k}$-algebra T generated by *all* coefficients of ϕ_a is generated by the coefficients of $\{\phi_{\alpha_1}, \ldots, \phi_{\alpha_j}\}$.

Now fix $\alpha_0 \in \mathbf{A} - \mathbb{F}_r$. We claim that R is generated over T by the inverse of the leading coefficient of ϕ_{α_0}. Indeed, let $0 \neq a \in \mathbf{A}$,

$$\phi_{\alpha_0} = c_0 \tau^{n_0} + \{\text{lower order terms}\},$$

and

$$\phi_a = c\tau^n + \{\text{lower order terms}\}.$$

One knows that $\phi_a \phi_{\alpha_0} = \phi_{\alpha_0} \phi_a$; thus equating higher terms yields

$$1/c = c^{r^{n_0}-2}/c_0^{r^n-1} \in T[c_0^{-1}]$$

giving the result. $\square$

Proof of Theorem 4.13.9. Without loss of generality, we can assume that $g(0) = 0$. Thus we need to show that g is an endomorphism of ϕ. Now, if there are infinitely many $a \in \mathbf{A}$ for which g maps $\phi[a]$ to itself, then the absolute values of these a's tend to infinity. We can, therefore, use the first part of the proof of Theorem 4.13.4 to complete the proof.

Therefore we may assume from now on that it is the first hypothesis of the statement of the theorem which holds. The proof now involves a number of steps.

Step 1: We reduce to the case where L is a finite extension of $\mathbf{k}$. Thus suppose that we have a counterexample g over an arbitrary field L with $g(S) \subseteq S$, but $g \notin \text{End}_L(\phi)$. Thus there exists $b \in \mathbf{A}$ with

$$\phi_b(g(x)) - g(\phi_b(x)) \neq 0.$$

Let α be a nonzero coefficient of a monomial in this difference. By 4.13.11 we know that the $\mathbf{k}$-algebra B generated by all coefficients of ϕ_a, $0 \neq a \in \mathbf{A}$, the coefficients of $g(x)$, and the inverse of the leading coefficient in ϕ_a is finitely generated over $\mathbf{k}$.

Let $\overline{\mathbf{k}}$ be an algebraic closure of $\mathbf{k}$. By Proposition 5.23 in [AM1] (with $A = \mathbf{k}$, $B = B$, $v = \alpha$, $\Omega = \overline{\mathbf{k}}$) one can find a $\mathbf{k}$-algebra morphism $\rho\colon B \to \overline{\mathbf{k}}$ with $\rho(\alpha) \neq 0$. The image $\rho(B)$ is a finitely generated subalgebra of $\overline{\mathbf{k}}$ and so is a *finite* extension of $\mathbf{k}$. The standard "going up" theorem allows us to extend ρ to a homomorphism from the integral closure $\overline{B}$ of B in L to $\overline{\mathbf{k}}$. Note that the *entire* torsion submodule of ϕ must lie in $\overline{B}$; thus by applying ρ we obtain a Drinfeld module $\widetilde{\phi}$ over $\rho(B)$ of the same rank (indeed, all leading coefficients are invertible in B). Moreover, as $\rho(\alpha) \neq 0$, we see immediately that $\rho(g)$ is *not* an endomorphism of $\widetilde{\phi}$. As ϕ and $\widetilde{\phi}$ have the same rank and we are in generic characteristic, we see that ρ gives rise to an isomorphism between the torsion submodules of ϕ and $\widetilde{\phi}$. In particular, $\rho(S)$ is an infinite torsion submodule which is mapped onto itself by $\rho(g)$. Thus we now have a counterexample over a finite extension of $\mathbf{k}$.

Step 2: Let L and its algebraic closure $\overline{L}$ be embedded in $\mathbf{C}_\infty$. Then we want to show that under $|?|_\infty$ the module S is not discrete.

Let Λ be the lattice corresponding to ϕ and $e(z) := e_\Lambda(z)$ its exponential function. We have the isomorphism

$$e\colon \mathbf{C}_\infty/\Lambda \mapsto \mathbf{C}_\infty.$$

Thus $S \subset e(\mathbf{K}\Lambda/\Lambda)$, where $\mathbf{K}$ is the completion of $\mathbf{k}$ at ∞; note that $\mathbf{K} \cdot \Lambda/\Lambda$, and thus its image under $e(z)$, is compact. As S is infinite, the result follows.

Step 3: If $g(x) \in x^2 L[x]$ and $g(S) \subseteq S$, then we want to show that $g(x) \equiv 0$.

Suppose g is nonzero. The above hypothesis implies that if δ is small enough, then

$$0 < |x|_\infty < \delta \Rightarrow 0 < |g(x)|_\infty < |x|_\infty.$$

As S is a group which is *not* discrete, there exists $s \in S$ with $0 < |s|_\infty < \delta$. Consequently

$$|s|_\infty > |g(s)|_\infty > |g(g(s))|_\infty > \cdots;$$

thus there exists infinitely many torsion points defined over the finite extension $L(s)$ of L, and thus of $\mathbf{k}$.

Now let F be *any* finite extension of $\mathbf{k}$ with a Drinfeld module ϕ. Let $\wp_1$ and $\wp_2$ be two direct primes of good reduction for ψ. By reducing modulo these two primes, one sees that the torsion submodule of F, viewed as $\mathbf{A}$-module via ψ, must be finite. We thus immediately obtain a contradiction in the field $L(s)$ which completes Step 3.

Step 4: If $g(x) \in xL[x]$ and $g(S) \subseteq S$, then we want to show that $g \in \mathrm{End}_L(\phi)$.

But if $g(s) \subseteq S$, then for any $a \in \mathbf{A}$, the polynomial

$$h(x) := \phi_a(g(x)) - g(\phi_a(x))$$

maps S into itself also. However, one sees that

$$h(x) \in x^2 L[x]$$

and so is identically 0 by Step 3. This immediately shows that $g(x)$ commutes with ϕ_a for all $a \in \mathbf{A}$. Thus g takes $\phi[a]$ to itself for all $a \in \mathbf{A}$. This step is thus finished by Part 1 above.

Step 4 is due to A. Tamagawa and it completes the proof of the theorem.
□

It remains, finally, to discuss the analog of Theorem 4.13.9 for morphisms of Drinfeld modules. Thus let L be our field of generic characteristic and ψ another Drinfeld module over L. The reader may now easily reformulate Theorem 4.13.9 in terms of morphisms from ϕ to ψ. However, one now *must* also assume that $\mathrm{rank}(\phi) = \mathrm{rank}(\psi)$ as the following beautiful example points out.

Example 4.13.12. Let L be a perfect **A**-field and let ψ be a rank d Drinfeld module over L. Let $n \geq 1$. We define a new Drinfeld module ϕ over L by

$$\phi_a = \tau^{-n} \cdot \psi_{a^{r^n}} \cdot \tau^n.$$

One checks directly that the rank of ϕ is $r^n d$. On the other hand, one has the equation

$$\psi_{a^{r^n}} \tau^n = \tau^n \phi_a;$$

thus τ^n maps the a-torsion of ϕ to the a^{r^n}-torsion of ψ for all a. However τ^n is *not* a morphism of Drinfeld modules as there are no non-trivial morphisms between Drinfeld modules of different ranks.

The morphism version of Theorem 4.13.9 for Drinfeld modules of rank d is established in [Po1] modulo certain reasonable conjectures; we refer the reader there for details.

4.14. The Adjoint of a General Drinfeld Module

We present, in this subsection, the general construction of adjoints of Drinfeld modules. Although the *definition* of the adjoint is obvious from Subsection 3.7, where we discussed the adjoint of the Carlitz module, the theory of the adjoint is quite involved and deep. In fact, as of this writing, the theory is *still* evolving. Besides the relevant results presented in previous subsections, the main source for the results here is the paper [Po2] by B. Poonen. We refer the reader there for more details.

The adjoint of a Drinfeld module will give us yet another module structure arising from a given Drinfeld module. However, before discussing adjoints, it seems reasonable to review (and occasionally define) the various module structures arising *directly* (i.e., *without* the adjoint construction) from a Drinfeld module. If the reader chooses, this material may be safely skipped on a first reading and the reader may pass directly to the construction of adjoints which begins just before Definition 4.14.4.

Thus let L be an **A**-field where **A**, etc., are defined as in Subsection 4.1. Let $\imath\colon \mathbf{A} \to L$ be the structure map and let $\wp = \operatorname{char}(L)$ be the kernel of $\imath$. Let $S = \mathbf{A} - \wp$; clearly S is a multiplicative set. We put $\mathbf{A}_{(\wp)} = S^{-1}\mathbf{A} =$ the localization of **A** at $\wp$; so $\mathbf{A}_{(\wp)}$ is a local (but not complete) ring. Note that $\imath$ extends to a mapping from $\mathbf{A}_{(\wp)}$ to L.

Let $\psi\colon \mathbf{A} \to L\{\tau\}$ be a Drinfeld module over L of rank $d > 0$. Thus if $\alpha \in S$ then
$$\psi_\alpha = \imath(\alpha)\tau^0 + \{\text{higher terms}\}, \quad \imath(\alpha) \neq 0.$$

Let $L\{\{\tau\}\}$ be, as before, the "formal power series algebra in τ." In $L\{\{\tau\}\}$ we can formally invert ψ_α; we thus obtain a homomorphism
$$\psi\colon \mathbf{A}_{(\wp)} \to L\{\{\tau\}\}, \quad \psi_\alpha = \imath(\alpha)\tau^0 + \{\text{higher terms}\},$$
extending the action of ψ on **A**. For instance, if $\wp = (0)$ we obtain again the formal k-module of Subsection 4.6.

Suppose now that v is a non-trivial discrete valuation on L for which L is complete. We make the assumption that if $\alpha \in \mathbf{A}$, then $v(\imath(\alpha)) \geq 0$, and if $\alpha \in \wp$, then $v(\imath(\alpha)) > 0$. There are two basic types of examples:

Type 1: L is a finite extension of the completion of **k** at a finite prime $\mathcal{B}$. Here $\wp = (0)$ and $v = v_\mathcal{B}$.

Type 2: Let $L_0 = \mathbf{A}/\wp$, $0 \neq \wp \in \operatorname{Spec}(\mathbf{A})$. Let L be any field over L_0 which is complete with respect to a non-trivial discrete valuation v.

Let $R = \{\alpha \in L \mid v(\alpha) \geq 0\}$ be the valuation ring of v; our assumption on v implies that $\imath(\mathbf{A}) \subset R$. We set $M = \{\alpha \in L \mid v(\alpha) > 0\}$ and put $P = \imath^{-1}(M) \in \operatorname{Spec}(\mathbf{A})$. In the Type 1 example above we see that $P = \mathcal{B}$; in Type 2 we see that $P = \wp = \operatorname{char}(L)$.

Let $\psi\colon \mathbf{A} \to L\{\tau\}$ be our Drinfeld module which we now further assume has image in $R\{\tau\}$, i.e.,
$$\psi(\mathbf{A}) \subset R\{\tau\}.$$
Let $\mathcal{M}$ be the left ideal of $R\{\tau\}$ generated by $M\tau^0$ and τ; one sees that $\mathcal{M}$ is actually two-sided. As $\psi_\alpha \in \mathcal{M}$ for $\alpha \in \wp$, one sees directly that ψ "completes" to a formal $\mathbf{A}_P$-module
$$\psi\colon \mathbf{A}_P \to R\{\{\tau\}\}, \quad \psi_\alpha = \imath(\alpha)\tau^0 + \{\text{higher terms}\}.$$

Moreover, M itself becomes an $\mathbf{A}_P$-module via ψ.

Examples 4.14.1. 1. Let $\mathbf{A} = \mathbb{F}_r[T]$ and $\mathbf{k} = \mathbb{F}_r(T)$. Let $\psi = C$ be the Carlitz-module, $C_T = T\tau^0 + \tau \in \mathbf{k}\{\tau\}$. Let $\mathcal{B} \in \operatorname{Spec}(\mathbf{A})$ be a maximal ideal, and put $L = \mathbf{k}_\mathcal{B}$, $v = v_\mathcal{B}$. In this case $R = \mathbf{A}_\mathcal{B}$ and $P = \mathcal{B}$. We obtain the completion
$$C\colon \mathbf{A}_\mathcal{B} \to \mathbf{A}_\mathcal{B}\{\{\tau\}\} \subset L\{\{\tau\}\}.$$

2. Let $0 \neq \wp \in \operatorname{Spec}(\mathbf{A})$. Put $L = (\mathbf{A}/\wp)((x))$ for some indeterminate x. In this case, $v = v_x$ and $P = \wp$. Moreover $R = (\mathbf{A}/\wp)[[x]]$ and C gives rise to a map
$$C \colon \mathbf{A}_\wp \to R\{\{\tau\}\}.$$
When $\wp = (T)$ this example is particularly simple. Indeed, $\mathbf{A}/\wp \simeq \mathbb{F}_r$ and $C_T = \tau \in L\{\tau\}$. Thus C give rise to the obvious isomorphisms (as $\tau(x) = x^r$)
$$\mathbb{F}_r[T] \xrightarrow{\sim} \mathbb{F}_r\{\tau\} \subset R\{\tau\}$$
and
$$\mathbb{F}_r[[T]] \simeq \mathbb{F}_r\{\{\tau\}\} \subset R\{\{\tau\}\}$$
given by $T \mapsto \tau$!

Suppose now that $L \subset \mathbf{C}_\infty$ is a finite extension of $\mathbf{K}$ where $\mathbf{K}$ is the completion of $\mathbf{k}$ at ∞ and $\mathbf{C}_\infty$ is the completion of its algebraic closure. Let $v = v_\infty$ be the valuation on $\mathbf{K}$ extended to L in the usual way. Let $\psi \colon \mathbf{A} \to L\{\tau\}$ be a Drinfeld module over L of rank d. Let $e(z)$ be the exponential of ψ and $\log(z)$ its inverse as in Subsection 4.6. Let Λ be the lattice of ψ and set $X = \{v_\infty(\lambda) \mid 0 \neq \lambda \in \Lambda\} \subset \mathbb{R}$. As Λ is discrete, X has a largest element β.

Proposition 4.14.2. *The power series for* $\log(z)$ *converges exactly in* $U = \{z \in \mathbf{C}_\infty \mid v_\infty(z) > \beta\}$. *Moreover* $\log(z)$ *takes* U *to itself with inverse* $e(z)$.

Proof. Let $z \in U$. One then easily checks that
$$v_\infty(e(z)) = v_\infty(z).$$
Thus $e(z)$ takes U onto itself. Moreover $e(z)$ is injective on U as $U \cap \Lambda = \{0\}$. Finally $e'(z) \equiv 1$, so $e(z)$ is an étale mapping on U. In complex analysis we would now be done, and, in fact, rigid analysis [Go1] also allows us to conclude the result here also.

However, as was pointed out by Poonen, one can proceed in the following elementary fashion: First of all, one may assume that $\beta = 0$. Indeed, if not let λ_0 be an element with $v_0(\lambda_0) = \beta$ and consider $e_1(z) = e(\lambda_0 z)/\lambda_0$. Next let us write
$$e(\tau) = \tau^0 - g(\tau),$$
where $g(\tau) \in \tau \mathbf{C}_\infty\{\{\tau\}\}$. We then find that
$$\log(\tau) = \tau^0 + g(\tau) + g(\tau)^2 + \cdots;$$
thus one sees, with a little thought, that $\log(\tau)$ converges in the open unit ball.

Let $R = \{\alpha \in \mathbf{C}_\infty \mid v_\infty(\alpha) \geq 0\}$ and $M = \{\alpha \in \mathbf{C}_\infty \mid v_\infty(\alpha) > 0\}$. From the product expansion for $e(z)$, one sees that its coefficients are in R and that the reduction of $e(\tau)$ modulo $M \cdot R\{\{\tau\}\}$ is a nonzero, *nonconstant,*

4.14. The Adjoint of a General Drinfeld Module

polynomial. On the other hand, the reduction of $\log(\tau)$ is *still* an inverse for the reduction of $e(\tau)$. Thus infinitely many coefficients of the reduction of $\log(\tau)$ must be nonzero implying that the open unit ball is the *exact* domain of convergence of $\log(z)$. □

As we saw above, ψ extends to a formal k-module $\psi: \mathbf{k} \to L\{\{\tau\}\}$. Through the use of $e(z)$ and $\log(z)$, this can easily be extended to a formal **K**-module as follows: Let $\alpha \in \mathbf{K}$ and set

$$\psi_\alpha = e(\alpha \log(\tau)) \in L\{\{\tau\}\}; \quad \psi_\alpha = \alpha\tau^0 + \{\text{higher terms}\}.$$

Let $\mathbf{A}_\infty \subset \mathbf{K}$ be the ring of integers, i.e.,

$$\mathbf{A}_\infty = \{\alpha \in \mathbf{K} \mid v_\infty(\alpha) \geq 0\}.$$

Clearly the set U of Proposition 4.14.2 is a module over $\mathbf{A}_\infty$ under the usual multiplication of elements. Call this module $U^{(1)}$. The space U is also a module over $\mathbf{A}_\infty$ via ψ. Indeed let $\alpha \in \mathbf{A}_\infty$ and $u \in U$. Set

$$\alpha * u = \psi_\alpha(u) = e(\alpha \log(u)).$$

Call this module $U^{(2)}$. The impact of 4.14.2 is precisely that $U^{(1)}$ and $U^{(2)}$ are isomorphic as $\mathbf{A}_\infty$-modules.

Finally, let L be an **A**-field which we assume is perfect. Let $\imath: \mathbf{A} \to L$ be the structure map. As in Subsection 4.12, we can define the division ring $L((\tau))$ of "finite-tailed Laurent series in τ." Similarly, we have the division ring $L((\tau^{-1}))$. Let $\psi: \mathbf{A} \to L\{\tau\}$ be a Drinfeld module. As $L\{\tau\} \subset L((\tau^{-1}))$, we can extend ψ to an injection of k into $L((\tau^{-1}))$. Note, however, that the constant term of ψ_α, $\alpha \in \mathbf{k}$, is now *not* necessarily $\imath(\alpha)$, *unless* $\alpha \in \mathbf{A}$.

Example 4.14.3. Let $\mathbf{A} = \mathbb{F}_r[T]$ and let $0 \neq \wp \in \text{Spec}(\mathbf{A})$. Set $L = \mathbf{A}/\wp$; so L is perfect. Let ψ be the reduction of the Carlitz module C at $\wp$. Thus if $\imath: \mathbf{A} \to L$ is the structure map,

$$\begin{aligned}\psi_T &= \imath(T)\tau^0 + \tau \in L\{\tau\} \\ &= \tau(\imath(T)^{1/r}\tau^{-1} + \tau^0) \\ &= \tau(\tau^{-1} \cdot \imath(T) + \tau^0).\end{aligned}$$

Thus, in $L((\tau^{-1}))$ we see that

$$\begin{aligned}\psi_{T^{-1}} = (\psi_T)^{-1} &= (\tau^0 + \tau^{-1} \cdot \imath(T))^{-1}\tau^{-1} \\ &= (\tau^0 - \tau^{-1} \cdot \imath(T) + (\tau^{-1} \cdot \imath(T))^2 - \cdots)\tau^{-1}.\end{aligned}$$

Let $\mathbf{A}_\infty$ be as above and let $M_\infty \subset \mathbf{A}_\infty$ be its maximal ideal. As in the example one sees that

$$\psi(\mathbf{k} \cap \mathcal{M}_\infty) \subset \tau^{-1} L\{\{\tau^{-1}\}\}.$$

Thus ψ completes to an injection $\mathbf{K} \to L((\tau^{-1}))$ with $\psi(\mathbf{A}_\infty) \subseteq L\{\{\tau^{-1}\}\}$, [Dr2].

We turn next to a quick review of adjoints as presented in Subsection 1.7. Let k be a field of characteristic $p > 0$ with $\mathbb{F}_r \subseteq k$, $r = p^{m_0}$. Let $\bar{k}$ be a fixed algebraic closure of k and let $k^{\text{perf}} \subseteq \bar{k}$ be its perfection. As usual, $\tau: k^{\text{perf}} \to k^{\text{perf}}$ is the rth power mapping. We let $k^{\text{perf}}\{\tau\}$ be the composition ring of $\mathbb{F}_r$-linear polynomials in τ and $k^{\text{perf}}\{\tau^{-1}\}$ the composition ring of $\mathbb{F}_r$-linear "polynomials" in τ^{-1}. Finally, we let $k^{\text{perf}}\{\tau, \tau^{-1}\}$ be the ring of $\mathbb{F}_r$-linear polynomials in both τ and τ^{-1}.

Our next definition is just that of 1.7.1 reviewed for convenience.

Definition 4.14.4. Let $f(\tau) = \sum_{i=0}^{n} \alpha_i \tau^i$, $\alpha_n \neq 0$, $n > 0$, be an element of $k\{\tau\}$.

1. We set $f^*(\tau) = \sum_{i=0}^{n} \alpha_i^{1/r^i} \tau^{-i} \in k^{\text{perf}}\{\tau^{-1}\}$. This is the "$\tau^{-1}$-form" of the adjoint.

2. We set $f^{\text{ad}}(\tau) = \tau^n f^*(\tau) = \sum_{i=0}^{n} \alpha_i^{r^{n-i}} \tau^{n-i} \in k\{\tau\}$. This is the "$\tau$-form" of the adjoint.

As shown in Lemma 1.7.3, and easy to see,

$$(f(\tau) g(\tau))^* = g^*(\tau) f^*(\tau).$$

Thus one sees that "$*$" gives an isomorphism of $k^{\text{perf}}\{\tau\}$ with the opposite ring of $k^{\text{perf}}\{\tau^{-1}\}$, etc. It also gives an involution of $k^{\text{perf}}\{\tau, \tau^{-1}\}$. Moreover, it is trivial to see that $f^*(\tau)$ and $f^{\text{ad}}(\tau)$ have the same set of zeros.

Suppose now that $f(\tau)$ is separable (i.e., $\alpha_0 \neq 0$). Let $k_1 \subseteq \bar{k}$ be the splitting field of $f(x)$ and $k_2 \subseteq \bar{k}$ be the splitting field of $f^{\text{ad}}(x)$. As was shown in Theorem 1.7.11, we have the equality $k_1 = k_2$.

Our goal now is to reexamine the equality $k_1 = k_2$ from the point of view of certain Galois equivariant pairings of the roots of $f(\tau)$ and $f^*(\tau)$. This pairing is due to B. Poonen, and, independently, N. Elkies [E1]. Let W be the $\mathbb{F}_r$-vector space of roots of $f(\tau)$ and W^* the $\mathbb{F}_r$-vector space of roots of $f^*(\tau) = $ roots of $f^{\text{ad}}(\tau)$. Clearly

$$\dim_{\mathbb{F}_r}(W) = \dim_{\mathbb{F}_r}(W^*) = n.$$

Definition 4.14.5. Let $\alpha \in W$ and $\beta \in W^*$. Thus $f(\tau) \cdot \alpha \tau^0$ vanishes at $1 \in k$. By the results of Section 1, we may write

$$f(\tau) \cdot \alpha \tau^0 = g_\alpha(\tau)(\tau^0 - \tau),$$

for a unique $g_\alpha(\tau) \in \bar{k}\{\tau\}$. We define the symbol $\langle \alpha, \beta \rangle_f$ by $\langle \alpha, \beta \rangle_f := g_\alpha^*(\beta)$.

4.14. The Adjoint of a General Drinfeld Module 127

Proposition 4.14.6. *The symbol $\langle \alpha, \beta \rangle_f$ has values in $\mathbb{F}_r$ and gives rise to a bilinear pairing of $\mathbb{F}_r$-vector spaces $W \times W^* \to \mathbb{F}_r$.*

Proof. By definition we have
$$f(\tau) \cdot \alpha \tau^0 = g_\alpha(\tau)(\tau^0 - \tau).$$
By taking adjoints we obtain
$$\alpha f^*(\tau) = (\tau^0 - \tau^{-1})g_\alpha^*(\tau).$$
By evaluating at β, we obtain
$$\alpha f^*(\beta) = 0 = (\tau^0 - \tau^{-1})g_\alpha^*(\beta).$$
Thus $g_\alpha^*(\beta) = \langle \alpha, \beta \rangle_f \in \mathbb{F}_r$.

We leave to the reader the easy task of establishing the bilinearity of $\langle \alpha, \beta \rangle_f$. □

Lemma 4.14.7. *The pairing of Proposition 4.14.6 is non-degenerate.*

Proof. Suppose that we have α with $g_\alpha^*(\beta) = \langle \alpha, \beta \rangle_f = 0$ for all $\beta \in W^*$. By looking at degrees, we see that $g_\alpha(x) \equiv 0$ and this implies that $\alpha = 0$. As W and W^* are finite dimensional, the result is now easily deduced. □

Lemma 4.14.8. *The pairing of Proposition 4.14.6 is Galois-equivariant, i.e.,*
$$\langle \sigma(\alpha), \sigma(\beta) \rangle_f = \langle \alpha, \beta \rangle_f$$
for all $\sigma \in \mathrm{Gal}(\overline{k}/k)$.

Proof. Clear. □

The reader should note that Lemma 4.14.8 gives another proof of Theorem 1.7.11.

Our next goal is to present a different definition of the pairing $\langle \alpha, \beta \rangle_f$. As $k^{\mathrm{perf}}\{\tau\}$ has both left and right division algorithms, if $\alpha \in W$ and $\beta \in W^*$ we can write
$$\beta f(\tau) = (\tau^0 - \tau)h_\beta(\tau) + \delta$$
for $h_\beta(\tau) \in k^{\mathrm{perf}}\{\tau\}$ and $\delta \in k^{\mathrm{perf}}$. By taking adjoints and evaluating at 1, we see that $\delta = 0$. As $f(\alpha) = 0$, we see that
$$0 = \beta f(\alpha) = (\tau^0 - \tau)h_\beta(\alpha)$$
which implies that $h_\beta(\alpha) \in \mathbb{F}_r$. We set
$$\langle \alpha, \beta \rangle'_f = h_\beta(\alpha).$$

128   4. Drinfeld Modules

Lemma 4.14.9. *Let* $f(\tau), g(\tau) \in k^{\text{perf}}\{\tau\}$ *with*
$$f(\tau) \cdot (\tau^0 - \tau) = (\tau^0 - \tau)g(\tau).$$

Then
$$f^*(1) = g(1).$$

Proof. Note first of all that
$$0 = f(\tau) \cdot (\tau^0 - \tau) \mid_{x=1} = (\tau^0 - \tau)g(1);$$

thus $g(1) \in \mathbb{F}_r$. Now write
$$f(\tau) = \sum_{i=0}^{m_0} \alpha_i \tau^i,$$

and
$$g(\tau) = \sum_{i=0}^{m_1} \beta_i \tau^i.$$

From $f(\tau)(\tau^0 - \tau) = (\tau^0 - \tau)g(\tau)$ we conclude that $m_0 = m_1 = n$. Moreover in $k^{\text{perf}}\{\{\tau\}\}$ we have the equality
$$(\tau^0 + \tau + \tau^2 + \cdots)f(\tau) = g(\tau)(\tau^0 + \tau^1 + \tau^2 + \cdots).$$

Equating the coefficients of τ^n gives
$$\alpha_n + \alpha_{n-1}^r + \cdots + \alpha_0^{r^n} = \beta_n + \cdots + \beta_0,$$

or
$$f^*(1)^{r^n} = g(1).$$

As $g(1) \in \mathbb{F}_r$, the result follows. $\square$

Proposition 4.14.10. *Let* $\alpha \in W$ *and* $\beta \in W^*$. *Then*
$$\langle \alpha, \beta \rangle_f = \langle \alpha, \beta \rangle_f'.$$

Proof. We have
$$f(\tau) \cdot \alpha \tau^0 = g_\alpha(\tau) \cdot (\tau^0 - \tau)$$

and
$$\beta f(\tau) = (\tau^0 - \tau)h_\beta(\tau).$$

If we multiply the top equation by $\beta\tau^0$ on the left and the bottom equation by $\alpha\tau^0$ on the right, we obtain
$$\beta\tau^0 \cdot g_\alpha(\tau) \cdot (\tau^0 - \tau) = (\tau^0 - \tau) \cdot h_\beta(\tau) \cdot \alpha\tau^0.$$

From 4.14.9 we conclude that

4.14. The Adjoint of a General Drinfeld Module   129

$$(\beta \tau^0 \cdot g_\alpha(\tau))^* \mid_{x=1} = (h_\beta(\tau) \cdot \alpha \tau^0) \mid_{x=1}$$
$$g_\alpha^*(\beta) = h_\beta(\alpha),$$

or

$$\langle \alpha, \beta \rangle_f = \langle \alpha, \beta \rangle'_f. \qquad \square$$

Finally, let $r_0 = r^t$ for some t, and let $\mathbb{F}_0 = \mathbb{F}_{r_0}$ be the field of r_0-elements. Suppose that $\mathbb{F}_0 \subseteq k$ and that $f(\tau) \in k\{\tau\}$ is actually in $k\{\tau_0\}$, $\tau_0 = \tau^t$. Note that the adjoint of $f(\tau)$ is independent of which representation is used, and let W, W' be as before. Let $\langle ?, ? \rangle_f$ be the pairing of W, W' into $\mathbb{F}_r$ as above, and let $\langle ?, ? \rangle^0_f$ be the pairing into $\mathbb{F}_0$ given by viewing f as $\mathbb{F}_0$-linear. We then have the following result of Elkies [E1].

Proposition 4.14.11. *Let $\alpha \in W$ and $\beta \in W'$. Then*

$$\langle \alpha, \beta \rangle_f = \mathrm{Tr}^{\mathbb{F}_0}_{\mathbb{F}_r}(\langle \alpha, \beta \rangle^0_f).$$

Proof. Write
$$f \cdot \alpha \tau^0 = G_\alpha(1 - \tau^t)$$
$$= G_\alpha(1 + \tau + \cdots + \tau^{t-1})(1 - \tau).$$

So, by definition of the pairing,
$$\langle \alpha, \beta \rangle_f = (G_\alpha(1 + \tau + \cdots + \tau^{t-1}))^*(\beta)$$
$$= (1 + \tau^{-1} + \cdots + \tau^{1-t})(G_\alpha^*(\beta))$$
$$= (1 + \tau^{-1} + \cdots + \tau^{1-t})(\langle \alpha, \beta \rangle^0_f)$$
$$= \mathrm{Tr}^{\mathbb{F}_0}_{\mathbb{F}_r}(\langle \alpha, \beta \rangle^0_f). \qquad \square$$

Before passing to the theory for Drinfeld modules, we discuss two more results which indicate, yet again, the connection between τ and differentiation. The first result is due to Elkies (op. cit.) and shows that the above pairing may be obtained in *exactly* the same way as its differential analog. Let F be a complex differential operator with formal adjoint F^*; one then [I1,§5.3] has an identity $uF^*(v) - vF(u) = \mathcal{B}'$ where $\mathcal{B}$ is a bilinear form (the "bilinear concomitant") in u, v and their derivatives of order $\leq \deg F - 1$. Let δ be the operator of Subsection 1.9 and let f be an $\mathbb{F}_r$-linear polynomial with adjoint f^*. Then Elkies shows that $uf^*(v) - vf(u) = \delta B$ where B is a bilinear form. If u is a zero for f and v a zero for f^* then $\delta B(u,v) = 0$ and thus $B(u,v) \in \mathbb{F}_r$. The pairing B is then shown to equal the scalar pairing defined above! Secondly, Lemma 4.14.7 suggests an analogous result in differential Galois theory. This result is indeed true and can be seen, for example, from a result due to O. Gabber [K2, Lemma 1.5.3]. The reader now may also enjoy the paper [Tag9].

4. Drinfeld Modules

We will now define the adjoints of general Drinfeld modules; a definition which is almost obvious from the above and the case of the Carlitz module, Subsection 3.7.

Thus let $L = k$ be an **A**-field which we assume is perfect. Let $\mathbb{F}_r \subseteq L$ and let $\psi : \mathbf{A} \to L\{\tau\}$ be a Drinfeld module of rank d over L. Let $\alpha, \beta \in \mathbf{A}$. Note that
$$\psi^*_{\alpha\beta} = (\psi_\alpha \psi_\beta)^* = \psi^*_\beta \psi^*_\alpha = \psi^*_{\beta\alpha} = (\psi_\beta \psi_\alpha)^* = \psi^*_\alpha \psi^*_\beta ;$$
thus we obtain an $\mathbb{F}_r$-linear homomorphism $\mathbf{A} \to L\{\tau^{-1}\}$ given by $\alpha \mapsto \psi^*_\alpha$.

Definition 4.14.12. We define the *adjoint* of ψ to be the injection $\psi^* : \mathbf{A} \mapsto L\{\tau^{-1}\}$ given by
$$\psi^*_\alpha = (\psi_\alpha)^*$$
for $\alpha \in \mathbf{A}$.

Morphisms of adjoint Drinfeld modules ψ^*, ϕ^* are defined in the obvious fashion as the adjoint of morphisms of ψ, ϕ, etc. In this fashion, the functor $\psi \mapsto \psi^*$ is a contravariant equivalence of categories.

Let $\wp$ be the characteristic of L as an **A**-field, and let v be a prime of $\mathbf{A} \neq \wp$. As in Definition 4.10.9, we define the v-adic Tate module of ψ as
$$T_v(\psi) := \mathrm{Hom}_\mathbf{A}(\mathbf{k}_v / \mathbf{A}_v, \psi(\overline{L}))$$
where $\overline{L}$ is an algebraic closure of L. Similarly, we set
$$T_v(\psi^*) := \mathrm{Hom}_\mathbf{A}(\mathbf{k}_v / \mathbf{A}_v, \psi^*(\overline{L})),$$
where $\psi^*(\overline{L})$ is the group $\overline{L}$ viewed as **A**-module via ψ^*, etc. One knows that $T_v(\psi)$ and $T_v(\psi^*)$ are $\mathbf{A}_v$-module isomorphic to $\mathbf{A}_v^d$ (by the same proof). Moreover, both $T_v(\psi)$ and $T_v(\psi^*)$ are $\mathrm{Gal}(\overline{L}/L)$-modules (recall that L is perfect so any finite extension is therefore separable). Our goal now is to use the pairing $\langle ?, ? \rangle_f$ to establish a duality between $T_v(\psi)$ and $T_v(\psi^*)$ as Galois modules. This will be established via the next collection of results.

Let ϕ be another Drinfeld module over L and let $f(\tau)$ be a separable isogeny from ϕ to ψ. Let W be the zero set of $f(\tau)$ and W' the zero set of $f^*(\tau)$. Let $\langle ?, ? \rangle_f : W \times W' \to \mathbb{F}_r$ be defined as above and let $a \in \mathbf{A}$, $\alpha \in W$ and $\beta \in W'$.

Proposition 4.14.13. *We have*
$$\langle \phi_a(\alpha), \beta \rangle_f = \langle \alpha, \psi^*_a(\beta) \rangle_f .$$

Proof. We define $g_\alpha(\tau)$ as before, so
$$f(\tau) \cdot \alpha \tau^0 = g_\alpha(\tau) \cdot (\tau^0 - \tau)$$
and

4.14. The Adjoint of a General Drinfeld Module   131

$$f(\tau) \cdot \phi_a(\alpha)\tau^0 = g_{\phi_a(\alpha)}(\tau) \cdot (\tau^0 - \tau).$$

Multiply the first of the above equations by ψ_a on the left and then subtract the second. We obtain

$$\psi_a(\tau) \cdot f(\tau) \cdot \alpha\tau^0 - f(\tau) \cdot \phi_a(\alpha)\tau^0 = (\psi_a(\tau) \cdot g_\alpha - g_{\phi_a(\alpha)}(\tau)) \cdot (\tau^0 - \tau).$$

As f is an isogeny, we have

$$f(\tau) \cdot (\phi_a(\tau) \cdot \alpha\tau^0 - \phi_a(\alpha)\tau^0) = (\psi_a(\tau) \cdot g_\alpha(\tau) - g_{\phi_a(\alpha)}(\tau)) \cdot (\tau^0 - \tau).$$

As 1 is a root of $\phi_a(\tau) \cdot \alpha\tau^0 - \phi_a(\alpha)\tau^0$, we have

$$\phi_a(\tau) \cdot \alpha\tau^0 - \phi_a(\alpha)\tau^0 = h(\tau) \cdot (\tau^0 - \tau)$$

for some $h(\tau) \in \overline{L}\{\tau\}$. Upon cancelling $\tau^0 - \tau$ on the right, we obtain

$$f(\tau) \cdot h(\tau) = \psi_a(\tau) \cdot g_\alpha(\tau) - g_{\phi_a(\alpha)}(\tau)$$

or

$$h^*(\tau) \cdot f^*(\tau) = g_\alpha^*(\tau) \cdot \psi_a^*(\tau) - g_{\phi_a(\alpha)}^*(\tau).$$

If we apply both sides to $\beta \in \text{Ker}(f^*(\tau))$, we obtain

$$0 = g_\alpha^*(\psi_a^*(\beta)) - g_{\phi_a(\alpha)}^*(\beta)$$
$$= \langle \alpha, \psi_a^*(\beta) \rangle_f - \langle \phi_a(\alpha), \beta \rangle_f,$$

and the result is established. □

Let us set $\widetilde{\mathbf{A}} = \text{Hom}_{\mathbb{F}_r}(\mathbf{A}, \mathbb{F}_r)$. The $\mathbb{F}_r$-vector space $\widetilde{\mathbf{A}}$ becomes an **A**-module via the obvious action of **A** on itself. Let $\{\alpha, \beta, f\}$ be as in 4.4.13. We set

$$[?,?]_f: W \times W' \to \widetilde{\mathbf{A}}$$

by

$$[\alpha, \beta]_f = (a \mapsto \langle \phi_a(\alpha), \beta \rangle_f).$$

Let I be a nonzero ideal of **A** which annihilates W. Then the above pairing takes values in $\text{Hom}_{\mathbb{F}_r}(\mathbf{A}/I, \mathbb{F}_r) \subset \text{Hom}_{\mathbb{F}_r}(\mathbf{A}, \mathbb{F}_r)$ in the obvious fashion. It is easy to check that $\text{Hom}_{\mathbb{F}_r}(\mathbf{A}/I, \mathbb{F}_r)$ is non-canonically isomorphic to $\mathbf{A}/I$ itself.

Proposition 4.14.14. *The pairing $[?,?]_f$, as defined above, from $W \times W' \to \text{Hom}_{\mathbb{F}_r}(\mathbf{A}/I, \mathbb{F}_r) \subset \text{Hom}_{\mathbb{F}_r}(\mathbf{A}, \mathbb{F}_r) = \widetilde{\mathbf{A}}$ is a Galois-equivariant and non-degenerate pairing of **A**-modules (where $\widetilde{\mathbf{A}}$, etc., has trivial Galois action).*

Proof. The non-degeneracy and Galois-equivariance follow from properties of $\langle ?,? \rangle_f$. We need only check that $[?,?]_f$ is a pairing of **A**-modules. If $b \in \mathbf{A}$, $\alpha \in W$ and $\beta \in W'$, then for all $x \in \mathbf{A}$,

132   4. Drinfeld Modules

$$[\phi_b(\alpha), \beta]_f(x) = \langle \phi_x(\phi_b(\alpha)), \beta \rangle_f$$
$$= \langle \phi_{bx}(\alpha), \beta \rangle_f.$$

Thus by definition of $\widetilde{\mathbf{A}}$,

$$[\phi_b(\alpha), \beta]_f = b[\alpha, \beta]_f.$$

On the other hand,

$$[\alpha, \psi_b^*(\beta)]_f(x) = \langle \phi_x(\alpha), \psi_b^*(\beta) \rangle_f$$
$$= \langle \phi_b(\phi_x(\alpha)), \beta \rangle_f$$
$$= \langle \phi_{bx}(\alpha), \beta \rangle_f$$

by Proposition 4.14.13. Thus

$$[\alpha, \psi_b^*(\beta)]_f = b[\alpha, \beta]_f,$$

giving the result. □

Proposition 4.14.15. *Let $f(\tau)$ and $j(\tau) \in L\{\tau\}$ be separable elements. let $\alpha \in \mathrm{Ker}(j(\tau)f(\tau))$ and $\beta \in \mathrm{Ker}(j^*(\tau)) \subseteq \mathrm{Ker}((j(\tau) \cdot f(\tau))^*)$. Then we have*

$$\langle \alpha, \beta \rangle_{jf} = \langle f(\alpha), \beta \rangle_j.$$

Proof. Write

$$j(\tau) \cdot f(\tau) \cdot \alpha \tau^0 = g_\alpha(\tau) \cdot (\tau^0 - \tau)$$

and

$$j(\tau) \cdot f(\alpha) \tau^0 = g_{f(\alpha)}(\tau) \cdot (\tau^0 - \tau).$$

Subtract the second equation from the first to obtain

$$j(\tau) \cdot (f(\tau) \cdot \alpha \tau^0 - f(\alpha) \tau^0) = (g_\alpha(\tau) - g_{f(\alpha)}(\tau)) \cdot (\tau^0 - \tau).$$

As $f(\tau) \cdot \alpha \tau^0 - f(\alpha) \tau^0$ kills 1, we find

$$f(\tau) \cdot \alpha \tau^0 - f(\alpha) \tau^0 = h(\tau) \cdot (\tau^0 - \tau)$$

for some $h(\tau) \in \overline{L}\{\tau\}$. After using this equality in the above equation and cancelling by $(\tau^0 - \tau)$, we obtain

$$j(\tau) \cdot h(\tau) = g_\alpha(\tau) - g_{f(\alpha)}(\tau)$$

and

$$h^*(\tau) \cdot j^*(\tau) = g_\alpha^*(\tau) - g_{f(\alpha)}^*(\tau).$$

If we apply both sides to $\beta \in \mathrm{Ker}(j^*(\tau))$, we get

$$0 = g_\alpha^*(\beta) - g_{f(\alpha)}^*(\beta)$$
$$= \langle \alpha, \beta \rangle_{jf} - \langle f(\alpha), \beta \rangle_j$$

giving the result. □

4.14. The Adjoint of a General Drinfeld Module

Corollary 4.14.16. *Let $\{\phi, \psi, \rho\}$ be three Drinfeld modules over L. Let $f: \phi \to \psi$ and $j: \psi \to \rho$ be separable isogenies. then*
$$[\alpha, \beta]_{jf} = [f(\alpha), \beta]_j$$
where $\{\alpha, \beta\}$ are as in Proposition 4.14.15.

Proof. Let $x \in \mathbf{A}$. Then
$$\begin{aligned}[\alpha, \beta]_{jf}(x) &= \langle \phi_x(\alpha), \beta \rangle_{jf} \\ &= \langle f(\phi_x(\alpha)), \beta \rangle_j\end{aligned}$$
by what was just established. In turn, by the definition of isogeny,
$$\begin{aligned}\langle f(\phi_x(\alpha)), \beta \rangle_j &= \langle \psi_x(f(\alpha)), \beta \rangle_j \\ &= [f(\alpha), \beta]_j(x)\end{aligned}$$
and the result is established. $\square$

Corollary 4.14.17. *Let ϕ be a Drinfeld module over L and let $a \in \mathbf{A} - \wp$. For each integer $n \geq 0$ we have the pairing (constructed in the obvious fashion)*
$$[?, ?]_n : \phi[a^n] \times \phi^*[a^n] \to \mathrm{Hom}_{\mathbb{F}_r}(\mathbf{A}/(a^n), \mathbb{F}_r).$$
Let $\alpha \in \phi[a^{n+1}]$ and $\beta \in \phi^*[a^{n+1}]$. Then
$$[\phi_a(\alpha), \phi_a^*(\beta)]_n = a[\alpha, \beta]_{n+1}.$$

Proof. As our pairings are $\mathbf{A}$-module pairings, we find
$$[\phi_a(\alpha), \phi_a^*(\beta)]_n = a[\phi_a(\alpha), \beta]_n.$$
By Corollary 4.14.16 we have
$$a[\phi_a(\alpha), \beta]_n = a[\alpha, \beta]_{n+1},$$
and the result is finished. $\square$

We can now use these techniques to study the Tate modules $T_v(\psi)$ and $T_v(\psi^*)$ for ψ a Drinfeld module over L of rank d and v a prime of $\mathbf{A} \neq \wp$. Let h be the class number of $\mathbf{A}$; so $v^h = (a)$ for some $a \in \mathbf{A}$ prime to $\wp$. One then sees easily that
$$T_v(\psi) \simeq \varprojlim \psi[a^n]$$
where the limit is taken with respect to the map
$$\psi_a: \psi[a^{n+1}] \to \psi[a^n].$$
There is a similar description of $T_v(\psi^*)$.

The pairings of Corollary 4.14.17 piece together to give a pairing
$$[?,?]_v \colon T_v(\psi) \times T_v(\psi^*) \to \varprojlim \mathrm{Hom}_{\mathbb{F}_r}(\mathbf{A}/(a^n), \mathbb{F}_r)$$
where the inverse limit is taken with respect to the multiplication by a maps
$$\mathrm{Hom}_{\mathbb{F}_r}(\mathbf{A}/(a^{n+1}), \mathbb{F}_r) \to \mathrm{Hom}_{\mathbb{F}_r}(\mathbf{A}/(a^n), \mathbb{F}_r).$$
As $\mathbf{A}/(a^n)$ is isomorphic to $a^{-n}\mathbf{A}/\mathbf{A}$, the inverse limit becomes
$$\varprojlim \mathrm{Hom}_{\mathbb{F}_r}(a^{-n}\mathbf{A}/\mathbf{A}, \mathbb{F}_r)$$
with respect to inclusion maps. Finally, we have
$$\varprojlim \mathrm{Hom}_{\mathbb{F}_r}(a^{-n}\mathbf{A}/\mathbf{A}, \mathbb{F}_r) = \mathrm{Hom}_{\mathbb{F}_r}(\mathbf{k}_v/\mathbf{A}_v, \mathbb{F}_r);$$
thus we have established the following result of Poonen.

Proposition 4.14.18. *There is a non-degenerate, Galois-equivariant pairing of $\mathbf{A}_v$-modules*
$$[?,?]_v \colon T_v(\psi) \times T_v(\psi^*) \to \mathrm{Hom}_{\mathbb{F}_r}(\mathbf{k}_v/\mathbf{A}_v, \mathbb{F}_r). \qquad \square$$

It therefore remains to describe
$$\mathrm{Hom}_{\mathbb{F}_r}(\mathbf{k}_v/\mathbf{A}_v, \mathbb{F}_r)$$
as an $\mathbf{A}_v$-module.

Proposition 4.14.19. *The $\mathbf{A}_v$-module*
$$\mathrm{Hom}_{\mathbb{F}_r}(\mathbf{k}_v/\mathbf{A}_v, \mathbb{F}_r)$$
is free of rank 1.

Proof. Let $\pi_v \in \mathbf{k}$ be a uniformizing parameter at v, and let $\mathbb{F}_v$ be the finite field $\mathbf{A}/v$. Thus $\mathbf{k}_v \simeq \mathbb{F}_v[[\pi_v]]$. Now one has the *residue pairing*
$$\mathbb{F}_v((\pi_v)) \times \mathbb{F}_r((\pi_v))d\pi_v \to \mathbb{F}_r,$$
(i.e., $\mathbf{k}_v \times \mathbf{k}_v d\pi_v \to \mathbb{F}_r$), which takes $a \in \mathbb{F}_r((\pi_v))$ and $\omega \in \mathbb{F}_v((\pi_v))d\pi_v$ to
$$\mathrm{Res}_v(a\omega).$$

By definition, the set of ω which pair to 0 for all $a \in \mathbb{F}_v[[\pi_v]]$ is exactly $\mathbb{F}_v[[\pi_v]]d\pi_v$. Thus the dual of $\mathbb{F}_v((\pi_v))/\mathbb{F}_v[[v]]$ is directly seen to be
$$\mathbb{F}_r[[\pi_v]]d\pi_v$$
giving the result. $\qquad \square$

4.14. The Adjoint of a General Drinfeld Module

Summarizing these results we have the following important result.

Theorem 4.14.20. *There is a (non-canonical) perfect pairing of Galois-modules*
$$[?,?]_v : T_v(\psi) \times T_v(\psi^*) \to \mathbf{A}_v.$$ □

We therefore see that
$$T_v(\psi^*) \simeq \operatorname{Hom}_{\mathbf{A}_v}(T_v(\psi), \mathbf{A}_v)$$
as Galois-modules. We therefore have another justification for calling ψ^* the "adjoint" of ψ.

Example 4.14.21. Let $\mathbf{A} = \mathbb{F}_r[T]$ and let $\mathbf{k} = \mathbb{F}_r((T))$. Set L to be the perfection of $\mathbf{k}$ and let $\psi = C =$ the Carlitz module. Let v be a prime of $\mathbf{A}$. Let $\rho_1 : \operatorname{Gal}(\overline{L}/L) \to \mathbf{A}_v^*$ be the character giving the action of Galois on the Tate module of C and let ρ_2 be the character given by the action of Galois on $\psi^* = C^*$. By 4.14.20 we have
$$\rho_1 \rho_2 = 1,$$
which is also easy to see directly. In fact, this observation led the present author to suspect the existence of the general duality later established by Poonen.

Finally let $L = \mathbf{C}_\infty =$ completion of the algebraic closure of $\mathbf{k}_\infty$; so $\mathbf{C}_\infty$ is also algebraically closed. Let ψ be a Drinfeld module over L with lattice M and exponential function $e_M(z)$. By definition
$$e_M(az) = \psi_a(e_M(z))$$
for $a \in \mathbf{A}$. Taking adjoints we obtain (at least formally)
$$a\tau^0 e_M^*(\tau) = e_M^*(\tau)\psi_a^*(\tau).$$

One checks that $e_M^*(\tau)$ also converges for all $z \in \mathbf{C}_\infty$. Thus any a-division point of ψ^* is *automatically* a "zero" of $e_M^*(\tau)$. Continuity thus implies that the closure of the set of *all* torsion points of ψ^* is also among the zeros of $e_M^*(\tau)$. One can show that this closure is *compact* and *equals* the zero set M^* of $e_M^*(\tau)$ (the "dual lattice"). The Poonen pairing then extends to a compact/discrete duality (i.e., Pontryagin duality) between M and M^*. For details, we refer the reader to [Po2].

5. T-modules

In this section we will present an important generalization of Drinfeld modules called "T-modules." This theory is due to G. Anderson [A1]. Very roughly, if Drinfeld modules are analogous to elliptic curves, then T-modules are analogous to abelian varieties.

T-modules are realized as operators on n-dimensional space for arbitrary $n > 0$, in contrast to the 1-dimensional Drinfeld modules. Thus our first few subsections will give some background on n-dimensional notions; in particular, we will continue to stress the analogy between differentiation and the pth power mapping.

In this section the reader will also encounter, for the first time, the concept of the "two T's" where the the element T will play two distinct roles. The first, "T as operator," arises simply from the T-action of a Drinfeld module (or T-module). The second, "T as variable" occurs because our objects are defined over $\mathbb{F}_r[T]$-fields; e.g., the Carlitz module is defined by specifying the action *of* T as a polynomial *with* $\mathbb{F}_r[T]$-coefficients. The two different roles played by T will, in turn, be reflected in the theories of the Γ and L-functions presented in later sections.

5.1. Vector Bundles

Let L be a field (of arbitrary characteristic) and let X be a variety over L. In this subsection, we will review the basic dictionary between *locally-free* $\mathcal{O}_X$-*modules* of finite rank and *vector bundles* over X. For more, we refer the reader to [Har1] as well as for any elided details.

Let $U \subseteq X$ be a Zariski open subset. We set

$$\mathbb{A}_U^n := U \times \mathbb{A}_L^n,$$

where $\mathbb{A}_L^n$ is n-dimensional affine space over L.

Definition 5.1.1. 1. A *geometric vector bundle* of rank n over X is a scheme $f\colon Z \to X$ together with an open cover $\{U_i\}$ of X and isomorphisms $\psi_i\colon f^{-1}(U_i) \to \mathbb{A}_{U_i}^n$ such that, for any i, j and any open affine

$V = \text{Spec}(R) \subseteq U_i \cap U_j$, the automorphism $\psi = \psi_j \circ \psi_i^{-1}$ of $\mathbb{A}_V^n$ is *linear* (i.e., in $GL_n(R)$).

2. A *locally free sheaf* $\mathcal{E}$ of rank n is an $\mathcal{O}_X$-module $\mathcal{E}$ together with a covering $\{U_i\}$ of X such that there exist $\mathcal{O}_{U_i}$-module isomorphisms

$$\rho_i \colon \mathcal{E}|_{U_i} \simeq \mathcal{O}_{U_i}^n.$$

The concepts in parts 1 and 2 of the above definition are essentially the same as we now explain: First of all, let $\mathcal{E}$ be a locally free sheaf of rank n on X. Let $S(\mathcal{E})$ be the symmetric algebra on $\mathcal{E}$; i.e., $S(\mathcal{E})|_{U_i}$ isomorphic to the symmetric algebra of $\mathcal{E}|_{U_i}$ over $\mathcal{O}_{U_i}$ for affine U_i. Set

$$Z = \mathbf{Spec}(S(\mathcal{E}));$$

i.e., Z is the spectrum of the *sheaf* $S(\mathcal{E})$ of $\mathcal{O}_X$-algebras. The scheme Z comes equipped with the projection morphism $f \colon Z \to X$. One checks directly that $f \colon Z \to X$ is a geometric vector bundle of rank n over X; it is called the *geometric vector bundle associated to* $\mathcal{E}$ and denoted "$V(\mathcal{E})$."

Now let Z be a geometric vector bundle of rank n and let $U \subseteq X$ be an open set. A *section* of Z over U is a morphism $s \colon U \to Z$ such that $f \circ s = \text{id}|_U$. One then checks that the operation

$$U \mapsto \{\text{set of sections over } U\}$$

forms a locally free sheaf of rank n. This sheaf is denoted "$\mathcal{E}(Z)$."

Let $\mathcal{E}$ be a locally free sheaf of rank n and let $\check{\mathcal{E}} := \text{Hom}_{\mathcal{O}_X}(\mathcal{E}, \mathcal{O}_X) =$ the sheaf dual to $\mathcal{E}$. It is natural to ask about the relationship between $\mathcal{E}' := \mathcal{E}(V(\mathcal{E}))$ and $\mathcal{E}$ itself. Here a minor subtlety arises: $\mathcal{E}'$ is actually isomorphic to $\check{\mathcal{E}}$. Indeed let V be an open set and $s \in \Gamma(V, \check{\mathcal{E}})$. By viewing s as a homomorphism, it is easy to see that s gives rise to an $\mathcal{O}_V$-homomorphism $S(\mathcal{E}|_V) \to \mathcal{O}_V$, or a morphism

$$\mathbf{Spec}(\mathcal{O}_V) \to \mathbf{Spec}(S(\mathcal{E}|_V)).$$

This morphism is nothing but a section of $V(\mathcal{E})$, and leads to the isomorphism

$$\check{\mathcal{E}} \simeq \mathcal{E}(V(\mathcal{E})).$$

We finish this subsection by remarking that the above constructions may also be given in the theory of analytic spaces over $\mathbb{C}$ (or, for that matter, in the theory of rigid spaces over any complete non-Archimedean field).

5.2. Sheaves and Differential Equations

In this subsection, we will quickly review the sheaf-theoretic approach to differential equations. In the next subsection we will discuss an analog in characteristic p. Our basic reference is [De2]. As this review will not, in any sense, be complete, we refer the reader to [De2] for more information.

Suppose that we have an nth order complex differential equation

$$P(D)y := y^{(n)} + \sum_{i=0}^{n-1} a_i(x)y^{(i)} = 0$$

where $D = \frac{d}{dx}$ and we assume, for simplicity, that $\{a_i(x)\} \subset \mathbb{C}[x]$. To this equation we associate the equivalent system of first order equations

$$\begin{cases} Dy_i = y_{i+1} & (1 \leq i < n) \\ Dy_n = -\sum_{i=0}^{n-1} a_i(x)y_{i+1} \end{cases} \quad (*)$$

This is summarized by the $n \times n$ matrix

$$\partial = \begin{pmatrix} 0 & & & -a_0(x) \\ 1 & \ddots & & \vdots \\ & \ddots & \ddots & \vdots \\ & & 0 & \vdots \\ & & 1 & -a_{n-1}(x) \end{pmatrix}.$$

Set $X = \text{Spec}(\mathbb{C}[x])$ and let $\mathcal{E} = \mathcal{O}_X^n$. Let $U \subseteq X$ be an open set and $\varepsilon \in \Gamma(U, \mathcal{E})$ be a section of $\mathcal{E}$ over U. We represent ε by a column matrix $\begin{pmatrix} f_1(x) \\ \vdots \\ f_n(x) \end{pmatrix}$, $\{f_i(x)\} \subset \Gamma(U, \mathcal{O}_X)$. Thus, $\varepsilon = \sum f_i(x)e_i$, where $\{e_i\}$ is the canonical basis. We then set

$$D\varepsilon := \sum_{i=1}^{n}(Df_i(x)e_i - f_i(x)\partial^t e_i)$$

$$= \sum_{i=1}^{n}(Df_i(x)e_i + f_i(x)(-\partial^t)e_i) \in \mathcal{E}.$$

One checks that the solutions to $*$ are precisely the solutions to the first order equation

$$D\varepsilon = 0.$$

If we set $\Omega^1 := \Omega_X^1 = \mathcal{O}_X \, dx$, then the operator D may be given the equivalent form

$$\nabla \varepsilon := \sum (df_i(x)e_i - f_i(x)dx\eth^t e_i)$$
$$\in \Omega^1_U(\mathcal{E}) = \Omega^1_U \otimes_{\mathcal{O}_U} \mathcal{E}|_U.$$

Thus ∇ gives rise to a $\mathbb{C}$-linear map

$$\nabla : \mathcal{E} \to \Omega^1_X(\mathcal{E}),$$

such that for local sections f of $\mathcal{O}_X$ and ξ of $\mathcal{E}$ we have

$$\nabla(f\xi) = f \cdot \nabla(\xi) + df \cdot \xi.$$

This is the Leibniz identity.

Let X be an arbitrary smooth variety (or smooth complex analytic space, etc.). The above properties of ∇ translate immediately into the notion of a *connection* on a locally free sheaf $\mathcal{E}$ on X. Let $\mathcal{E}_1$ and $\mathcal{E}_2$ be two such locally free sheaves with connections ∇_1 and ∇_2.

5.2.1. We equip $\mathcal{E}_1 \oplus \mathcal{E}_2$ with the connection $\nabla = \nabla_1 \oplus \nabla_2$.

5.2.2. We equip $\mathcal{E}_1 \otimes \mathcal{E}_2$ with the connection ∇ such that for local sections $\varepsilon_1, \varepsilon_2$,

$$\nabla(\varepsilon_1 \otimes \varepsilon_2) = \nabla_1(\varepsilon_1) \otimes \varepsilon_2 + \varepsilon_1 \otimes \nabla_2(\varepsilon_2),,$$

(and where the obvious identifications of tensor products are understood).

Note that 5.2.2 is again the Leibniz identity.

Finally, let ∇_1 and ∇_2 be two connections on $\mathcal{E}$. Then $\nabla_1 - \nabla_2$ is easily seen to be an $\mathcal{O}_X$-linear map from $\mathcal{E}$ to $\Omega^1_X(\mathcal{E})$. Thus the set of connections forms a principal homogeneous space over

$$Hom(\mathcal{E}, \Omega^1_X(\mathcal{E})) \simeq \Omega^1_X(End(\mathcal{E})).$$

5.3. φ-sheaves

We turn now to a finite characteristic analog of a connection. This is based on the analogy between differentiation and the Frobenius map. Thus let $\mathbb{F}_r$, $r = p^{m_0}$, be our fixed finite field. Let L be an extension of $\mathbb{F}_r$ with algebraic closure $\overline{L}$. As usual, $\tau : \overline{L} \to \overline{L}$ is the rth power map.

Let $p(\tau) \in L\{\tau\}$ be an $\mathbb{F}_r$-linear polynomial with $\deg p = n$. We assume that $p(\tau)$ is separable. Thus, upon multiplication by a nonzero constant, we may assume that

$$p(\tau) = \tau^0 + \sum_{j=1}^{n} a_i \tau^i, \quad a_n \neq 0.$$

Associated to the equation $p(x) = 0$, we have the equivalent system of "first order equations:"

5.3. φ-sheaves

$$\begin{cases} x_{i+1} = \tau x_i \ (= x_i^r) \quad (1 \le i < n) \\ x_1 = -\sum_{i=1}^n a_i \tau x_i \end{cases} \quad (*)$$

This is summarized by the $n \times n$ matrix

$$\mathfrak{d} = \begin{pmatrix} -a_1 & \cdots & \cdots & \cdots & -a_n \\ 1 & 0 & & & \\ 0 & 1 & \ddots & & \\ & & \ddots & \ddots & \\ & & & \ddots & 0 \\ & & & 1 & 0 \end{pmatrix}.$$

Set $X = \mathrm{Spec}(L[x])$ and let $\mathcal{E} = \mathcal{O}_X^n$. As in Subsection 5.2, we let $U \subseteq X$ be an open set, and we represent $\varepsilon \in \Gamma(U, \mathcal{E})$ as

$$\varepsilon = \sum f_i(x) e_i = \begin{pmatrix} f_1(x) \\ \vdots \\ f_n(x) \end{pmatrix}.$$

We then set

$$\varphi \varepsilon := \sum f_i^r(x) \mathfrak{d} e_i$$

$$= \mathfrak{d} \begin{pmatrix} f_1^r(x) \\ \vdots \\ f_n^r(x) \end{pmatrix} \in \mathcal{E}.$$

One checks readily that the solutions to $*$ in $\overline{L}$ are precisely the solutions to the first order equation

$$\varphi \varepsilon = \varepsilon$$

or

$$\varphi \varepsilon - \varepsilon = 0,$$

on $\mathcal{E}$. The analogy with the previous subsection is clear. Also clear is how the above theory is a direct generalization of the operator δ of Subsection 1.9 whose zero set is $\mathbb{F}_r$.

Of course we could just have easily worked with $\mathrm{Spec}(L)$ itself. We chose to use $L[x]$ for purposes of comparison.

One sees readily that $\varphi \colon \mathcal{E} \to \mathcal{E}$ is a "Frobenius-linear" map. That is, it is additive and if f and $\mathcal{E}$ are local sections of $\mathcal{O}_X$ and $\mathcal{E}$ respectively, then

$$\varphi(f\mathcal{E}) = f^r \varphi(\mathcal{E}).$$

This leads to our next definition. Let X be a variety over $\mathbb{F}_r$.

Definition 5.3.1. [Dr3] A φ-*sheaf* on X is a finite dimensional locally-free sheaf $\mathcal{E}$ equipped with a Frobenius linear map $\varphi \colon \mathcal{E} \to \mathcal{E}$.

The rth power mapping on the structure sheaf $\mathcal{O}_X$ of X gives rise to a morphism $F\colon X \to X$. Let $F^*\mathcal{E}$ be the locally free sheaf (of the same rank) obtained by pulling back $\mathcal{E}$ via F. The mapping $\mathcal{E}$ to $F^*\mathcal{E}$ is Frobenius linear. Thus any $\mathcal{O}_X$-linear mapping $F^*\mathcal{E}$ gives rise to a φ-module. Conversely a Frobenius linear mapping must factor through $F^*\mathcal{E}$. Thus the φ module structures on $\mathcal{E}$ are in one to one correspondence with

$$\operatorname{Hom}_{\mathcal{O}_X}(F^*\mathcal{E}, \mathcal{E}).$$

Now let $\mathcal{E}_1$ and $\mathcal{E}_2$ be two φ-sheaves, with φ-structures φ_1, φ_2.

5.3.2. We equip $\mathcal{E}_1 \oplus \mathcal{E}_2$ with the φ-structure

$$\varphi = \varphi_1 \oplus \varphi_2.$$

5.3.3. We equip $\mathcal{E}_1 \otimes \mathcal{E}_2$ with the φ-structure

$$\varphi = \varphi_1 \otimes \varphi_2.$$

Finally, if $\mathcal{E}$ is a φ-sheaf then the solutions to the equation $\varphi\varepsilon = \varepsilon$ clearly form a vector space over $\mathbb{F}_r$.

5.4. Basic Concepts of *T*-modules

With the Frobenius formalism of Subsection 5.3 in mind, we now pass to related structures. These structures involve *both* the Frobenius action *and* the **A**-action of a Drinfeld module. The particular variant discussed in this section, "*T*-modules," is conceptually the closest to Drinfeld modules.

Our exposition here is based on Anderson's original paper [A1]. Also useful is [Go5].

We use the notation of Section 4. Let L be an **A**-field with $\imath\colon \mathbf{A} \to L$. We assume that L is *perfect*. Let $\psi\colon \mathbf{A} \to L\{\tau\}$ be a Drinfeld module over L of rank d and set

$$M = L\{\tau\}.$$

As an introduction to *T*-motives, we will firstly explain how to make M a module over a certain *non-commutative* ring of operators. This module is called the "**A**-motive" of ψ.

1. Let $\alpha \in L$ and $m = m(\tau) \in L\{\tau\}$. We then define

$$\alpha m := \alpha m(\tau) \in L\{\tau\}.$$

2. Let τ be the rth power mapping and m as in 1. We then define

$$\tau m := \tau m(\tau) \in L\{\tau\}.$$

3. Let $a \in \mathbf{A}$ and m as above. We then set

$$am := m(\psi_a(\tau)) \in L\{\tau\}.$$

Putting these actions together, we see that M is a left module over the ring

$$L \otimes_{\mathbb{F}_r} \mathbf{A}\{\tau\}$$

which is defined as follows: it is clear that we need only focus on how τ commutes with elements of $L \otimes_{\mathbb{F}_r} \mathbf{A}$. Thus let $\alpha = \sum l_i \otimes a_i \in L \otimes_{\mathbb{F}_r} \mathbf{A}$. Then we define

$$\tau \cdot \alpha = \tau \left(\sum l_i \otimes a_i \right) := \left(\sum l_i^r \otimes a_i \right) \tau.$$

One checks that $L \otimes_{\mathbb{F}_r} \mathbf{A}\{\tau\}$ indeed forms a ring which is non-commutative in general.

To study the properties of M as an $L \otimes_{\mathbb{F}_r} \mathbf{A}\{\tau\}$-module, we now fix a *non-constant* $T \in \mathbf{A}$. Thus $\mathbf{A}$ is a finite $\mathbb{F}_r[T]$-module. Let d_T =degree of (T) at ∞. By Lemma 4.5.1, ψ restricts to a Drinfeld module of rank $d_T d$ on $\mathbb{F}_r[T] \subseteq \mathbf{A}$.

Recall that we set d_∞ = the degree of ∞ over $\mathbb{F}_r$.

The module theoretic properties of M are now given by the following proposition.

Lemma 5.4.1. 1. *The module M is free of rank 1 over $L\{\tau\} \subset L \otimes_{\mathbb{F}_r} \mathbf{A}\{\tau\}$.*
2. *The module M is projective of rank d over $L \otimes_{\mathbb{F}_r} \mathbf{A} \subset L \otimes_{\mathbb{F}_r} \mathbf{A}\{\tau\}$.*
3. *τM is an $L \otimes_{\mathbb{F}_r} \mathbf{A}\{\tau\}$ submodule of M. The action of $\sum l_i \otimes a_i$ on $M/\tau M$ is left scalar multiplication by $\sum l_i \cdot \imath(a_i) \in L$.*

Proof. Part 1 is easy, using the fact that L is perfect.

To see Part 2, note that $\{\tau^i \mid 0 \leq i < dd_T\}$ generates M freely as $L[T] = L \otimes_{\mathbb{F}_r} \mathbb{F}_r[T]$ module. Thus M is finitely generated over $L \otimes_{\mathbb{F}_r} \mathbf{A}$. If M has torsion as an $L \otimes_{\mathbb{F}_r} \mathbf{A}$-module, then it must *also* as an $L[T]$-module; as it does not, we conclude that M is finitely generated and projective over $L \otimes_{\mathbb{F}_r} \mathbf{A}$.

Let δ_M be the rank of M over $L \otimes_{\mathbb{F}_r} \mathbf{A}$. Let $\{e_1, \ldots, e_{\delta_M}\}$ be linearly independent elements of M. Let $\{a_1, \ldots, a_{d_T}\}$ be a basis for $L \otimes_{\mathbb{F}_r} \mathbf{A}$ over $L[T]$. Then $\{a_j e_i\}$ generates a submodule of M of maximal rank over $L[T]$. We conclude that

$$\delta_M d_T = d d_T$$

giving the result.

Part 3 follows easily. $\square$

Looking at M, as opposed to ψ, has certain advantages. The most important is that it allows us to introduce a tensor product into the theory. This will be explained below. However, before proceeding, we will now introduce an important simplification. We set $\mathbf{A} := \mathbb{F}_r[T]$ = the simplest possible $\mathbb{F}_r$-algebra coming from a curve. We lose little from this and gain a good deal of ease of notation. In fact, the case of general $\mathbf{A}$ is viewed as being an

"$\mathbb{F}_r[T]$-object" ($T \in \mathbf{A}$ nonconstant) with "complex multiplication by $\mathbf{A}$." We even occasionally view these objects as being defined over $\mathbb{F}_p[T]$ with complex multiplication by $\mathbf{A}$. For more along these lines we refer the reader to [A1], [Go5].

The ring $L \otimes_{\mathbb{F}_r} \mathbf{A}\{\tau\}$ is denoted "$L[T,\tau]$" for $\mathbf{A} = \mathbb{F}_r[T]$. By definition

$$\tau T = T\tau, \quad T\alpha = \alpha T \ (\alpha \in L), \quad \tau\alpha = \alpha^r \tau \ (\alpha \in L).$$

Note that $L[\tau] \subset L[T,\tau]$ is canonically isomorphic to $L\{\tau\}$; we use both notations interchangeably.

Clearly $L[T,\tau] = L[\tau][T]$ = the ring of commuting polynomials in one variable over $L[\tau]$. As $L[\tau]$ has both left and right division algorithms, one sees easily (in the classical manner of Hilbert's Basis Theorem) that $L[T,\tau]$ is both left and right noetherian.

Set $\theta = \imath(T) \in L$. Abstracting from our discussion above of "$\mathbf{A}$-motives," we now have the following basic definition.

Definition 5.4.2. A *T-motive* M is a left $L[T,\tau]$-module which is free and finitely generated as an $L[\tau]$-module and such that

$$(T - \theta)^n M/\tau M = \{0\}$$

for $n \gg 0$. A morphism of two such T-motives is an $L[T,\tau]$-linear homomorphism of left $L[T,\tau]$-modules.

Remarks. 5.4.3. 1. It is very important for the reader to keep θ ($=$ "$T \in L$ as scalar") separate from $T \in L[T,\tau]$ ($=$ "T as operator"). This basic process of "two T's" (one scalar, one operator) is basic for the theory.

2. In the case of the motive of a Drinfeld module given above, we see that T acts as θ on $M/\tau M$. Thus $n = 1$ here; in general n may be strictly greater than 1. If $n > 1$, then scalar multiplication by θ and multiplication by T differ by a nilpotent matrix N;

$$T = \theta + N = \imath(T) + N.$$

3. The reason $n > 1$ is allowed lies with tensor products and is discussed below.

We now present the "geometric objects" which correspond to the motives. Let E be an algebraic group over L which is isomorphic to $\mathbb{G}_a^e$ for some $e > 0$, and fix one such isomorphism.

5.4. Basic Concepts of T-modules

Lemma 5.4.4. *Let* $x = \begin{pmatrix} x_1 \\ \vdots \\ x_e \end{pmatrix} \in E$ *and set*

$$\tau^i(x) = \begin{pmatrix} x_1^r \\ \vdots \\ x_e^r \end{pmatrix}.$$

Then any $f \in \mathrm{End}_{\mathbb{F}_r}(E) = \{composition\ ring\ of\ \mathbb{F}_r\text{-}linear\ endomorphisms\ of\ E\}$ *can be written uniquely as*

$$\sum g_i \tau^i$$

where $g_i \in M_e(L) = \{e \times e\ matrices\ over\ L\}$.

Proof. This is an elementary exercise as in Section 1, (see [Hul, §20.3] for details). □

Note also that $\mathrm{End}_{\mathbb{F}_r}(E)$ can be written as $M_e(L\{\tau\}) = \{e \times e$ matrices with coefficients in $L\{\tau\}\}$.

Let $\mathrm{Lie}(E)$ be the Lie algebra of E in the usual sense.

Definition 5.4.5. A *T-module* E/L is an algebraic group isomorphic to $\mathbb{G}_a^e$ equipped with an $\mathbb{F}_r$-linear endomorphism "T" over L such that

$$(T - \theta)^n \mathrm{Lie}(E) = 0$$

for $n \gg 0$ (and $\theta = \imath(T)$ as above). A *morphism* of T-modules is a T-equivariant morphism of algebraic groups over L. The *dimension* of E is e.

Remarks. 5.4.6. 1. A Drinfeld module for $\mathbb{F}_r[T]$ gives a 1-dimensional T-module in the obvious fashion. Clearly, $n = 1$ in this case. In fact, the converse is *almost* true; any 1-dimensional T-module with *non-scalar* action comes from a unique Drinfeld module.
2. The algebraic group $\mathbb{G}_a$ with the usual scalar action is a T-module. It is denoted "E_{linear}."
3. One does not need L to be perfect in the definition of T-module. It works over any **A**-field.

The categories of T-modules and T-motives over perfect fields are mirror images of each other as follows: Let E be a T-module and set

$$M := M(E) := \mathrm{Hom}_L^r(E, \mathbb{G}_a),$$

where $\mathrm{Hom}_L^r(?, ?)$ is the group of $\mathbb{F}_r$-linear morphisms of algebraic groups over L. We make M into an $L[T, \tau]$-module in the fashion given above:

$$(\alpha, m) \mapsto \alpha m(x) \colon E \to \mathbb{G}_a; \quad m \in M, \quad \alpha \in L;$$
$$(\tau, m) \mapsto m(x)^r \colon E \to \mathbb{G}_a; \quad m \in M;$$
$$(T, M) \mapsto m(Tx) \colon E \to \mathbb{G}_a; \quad m \in M.$$

It is an easy task to see that M is a T-motive and that $E \mapsto M(E)$ is a contravariant functor.

Lemma 5.4.7. *The groups* $\mathrm{Lie}(E)$ *and*

$$\mathrm{Hom}_L(\mathrm{Hom}_L^r(E, \mathbb{G}_a)/\tau\mathrm{Hom}_L^r(E, \mathbb{G}_a), L) = \mathrm{Hom}_L(M/\tau M, L)$$

are isomorphic under the natural map

$$x \mapsto (f \mapsto \partial_x f).$$

Proof. This is a direct consequence of the definitions. □

Proposition 5.4.8. *Let W be a $m \times n$ matrix with coefficients in $L\{\tau\}$. Then there exist matrices $U \in GL_m(L\{\tau\})$ and $V \in GL_n(L\{\tau\})$ such that all off-diagonal entries of UWV vanish.*

Proof. We may, of course, assume that $W \neq 0$. Moreover, upon exchanging some rows and columns, we may assume that $W_{11} \neq 0$. After multiplying W on the left and right by suitable invertible square matrices with entries in $L\{\tau\}$, we may also assume that for all $U \in GL_m(L\{\tau\})$ and $V \in GL_n(L\{\tau\})$ we have

$$(UWV)_{11} \neq 0 \Rightarrow \deg((UWV)_{11}) \geq \deg(W_{11}). \tag{$*$}$$

As L is perfect, we know from Section 1 that $L\{\tau\}$ has both left and right division algorithms. Thus, in order that $*$ not be violated, one sees that every element of the first row of W must belong to $W_{11} \cdot L\{\tau\}$, and every element of the first column to $L\{\tau\} \cdot W_{11}$. Thus replacing W with a matrix obtained by elementary row and column operations on W, we may assume that W_{11} is the unique nonzero entry in the first row and first column. The result now follows by induction. □

Proposition 5.4.9. *Every finitely generated left $L\{\tau\}$-module is of the form*

$$\bigoplus_{i=1}^m L\{\tau\}/L\{\tau\}f_i(\tau)$$

for $m \geq 0$ and $\{f_i(\tau)\} \subset L\{\tau\}$.

Proof. From the division algorithms, we know that every left ideal of $L\{\tau\}$ is principal. Thus every finitely generated left $L\{\tau\}$-module M is noetherian.

5.4. Basic Concepts of T-modules

Thus M admits a presentation as the cokernel of a map of free $L\{\tau\}$-modules of finite rank. Proposition 5.4.8 now finishes the proof. □

Lemma 5.4.10. *Let M be a left $L[T,\tau]$-module which is finitely generated as an $L[T]$-module and as an $L[\tau]$-module. Then M is free over $L[T]$ of finite rank if and only if it is free over $L[\tau]$ of finite rank.*

Proof. As T and τ commute, one checks readily the following equality:
$$\{m \in M \mid \dim_L(L[T]m) < \infty\} = \{m \in M \mid \dim_L(L[\tau]m) < \infty\}.$$
The vanishing of the left hand side is necessary and sufficient for M to be free of finite rank over $L[T]$. On the other hand, the vanishing of the right hand side is necessary and sufficient for M to be free as a module over $L[\tau]$ by 5.4.9. The result now follows. □

Recall that to the T-module E, we have associated its T-motive $M(E)$.

Theorem 5.4.11. *The functor $E \mapsto M(E)$ gives rise to an anti-equivalence between the the categories of T-modules and T-motives.*

Proof. This follows readily from the definitions and the previous lemmas (esp. Lemma 5.4.7). □

Let E be a T-module with motive $M = M(E)$.

Definition 5.4.12. We say that M is *abelian*, and E is an *abelian T-module*, if and only if M is finitely generated over $L[T]$.

Remarks. 5.4.13. 1. By 5.4.10, we see that M is abelian if and only if it is free of finite rank; this rank is the *rank* of E (and M).
2. We set $r(M) = r(E) := \operatorname{rank}_{L[T]} M$. We set $\rho(M) = \rho(E) := \operatorname{rank}_{L[\tau]} M =$ the dimension of E.
3. If $\dim E = 1$, then E is abelian if and only if it comes from a Drinfeld module.
4. The most general example of an e-dimensional T-module can be given as follows: Put $E = \mathbb{G}_a^e$ and represent $x \in \mathbb{G}_a^e$, etc., as in Lemma 5.4.4. Let N be a fixed nilpotent matrix. We then set
$$T \mapsto (Tx := (\theta I + N)(x) + g_1\tau(x) + \cdots + g_n\tau^n(x))$$
where $\{g_i\} \subseteq M_e(L)$.

5.5. Pure T-modules

In this subsection we discuss the notion of "purity" in the theory of T-modules. Before doing so, some introductory remarks are warranted. We shall be very brief, but perhaps the reader may gain some insight into the construction presented below. Let k be a number field and let X be a smooth projective variety over k. Let "$H^i(X)$" be the i-th cohomology group of X in some unspecified theory. So one always has

$$H^*(X) = \bigoplus_i H^i(X),$$

where $H^*(X)$ is the "total" cohomology of X.

In classical theory, one views $H^i(X)$ as the "pure" piece of the cohomology of weight i. This "purity" of $H^i(X)$ is reflected in many ways with, perhaps, the simplest coming from the action of Frobenius. Namely let $\wp$ be a finite place of k over which X has good reduction, and let $X_\wp$ be that reduction. Let $\mathbb{F}_\wp$ be the finite field of residues at $\wp$ with norm $N(\wp)$. Let $\overline{\mathbb{F}}_\wp$ be an algebraic closure of $\mathbb{F}_\wp$ and let $\overline{X}_\wp := X_\wp \times_{\mathbb{F}_\wp} \overline{\mathbb{F}}_\wp$ equipped with the Frobenius morphism $F := F_\wp \colon \overline{X}_\wp \to \overline{X}_\wp$. Via functionality, we have an action of F on $H^i(X)$ (e.g., on $H^i(\overline{X}_\wp, \mathbb{Q}_\ell) = \ell$-adic cohomology with $\ell \notin \wp$). In this situation, as is well-known, we have **Deligne's Theorem** (= the Weil Conjectures) which asserts the following: let $\{\alpha_{ij}\}$ be the eigenvalues of F on $H^i(X)$ and let L be the finitely generated extension of $\mathbb{Q}$ that they generate. Let $\sigma \colon L \to \mathbb{C}$ be *any* embedding. Then Deligne shows:

1. Each α_{ij} is an algebraic integer.
2. $|\sigma \alpha_{ij}| = N\wp^{i/2}$ for all σ.

In particular, the Frobenius action allows us to find $H^i(X)$ inside $H^*(X)$ (at least in ℓ-adic cohomology).

Suppose now that one wanted to introduce a tensor product formalism into $H^*(X)$. The natural place to start would be via the Künneth Formula which asserts very generally that

$$H^n(X \times Y) = \bigoplus_{p+q=n} H^p(X) \otimes H^q(Y).$$

Thus the tensor product is built out of the tensor product of pure pieces.

The functor $X \mapsto H^*(X)$ is the basic contravariant formalism of classical theory, just as $E \mapsto M(E)$ is (as will become evident) the basic contravariant formalism of Drinfeld modules and T-modules. So we see that to attach tensor products to T-modules we should first find a good notion of "purity." This is given by our next set of definitions.

5.5. Pure T-modules

Definition 5.5.1. Let M be a left $L[T, \tau]$-module. Put

$$M\left(\left(\frac{1}{T}\right)\right) := M \otimes_{L[T]} L\left(\left(\frac{1}{T}\right)\right).$$

The group M is made into an $L[T, \tau]$-module by using the obvious action of $L[T]$ and putting

$$\tau \left(m \otimes \sum_{j \gg -\infty} a_j T^{-j} \right) := \tau m \otimes \sum_{j \gg -\infty} a_j^r T^{-j}.$$

A *lattice* in $M\left(\left(\frac{1}{T}\right)\right)$ is a free $L\left[\left[\frac{1}{T}\right]\right]$-submodule giving $M\left(\left(\frac{1}{T}\right)\right)$ upon tensoring with $L\left(\left(\frac{1}{T}\right)\right)$.

Definition 5.5.2. 1. The module M is *pure* provided both of the following hold:
a. M is free and finitely generated over $L[T]$.
b. There exists an $L\left[\left[\frac{1}{T}\right]\right]$ lattice $W \subset M\left(\left(\frac{1}{T}\right)\right)$, and positive integers q and s, such that
$$\tau^s W = T^q W.$$

2. A T-module E is *pure* if and only if $M(E)$ is pure.

Note that a pure T-motive is also abelian.

Let M be the motive of a Drinfeld module ψ of rank d over L; so $M = L\{\tau\}$ with $L[T, \tau]$ action as before. Recall that we have defined the division ring $L((\tau^{-1}))$ of finite tailed Laurent series in τ^{-1}. Note that $L\{\tau\} \subset L((\tau^{-1}))$. We make $L((\tau^{-1}))$ into an $L[T, \tau]$-module by

$$(T, f(\tau)) \mapsto f(\psi_T);$$

thus the injection $L\{\tau\} \hookrightarrow L((\tau^{-1}))$ is one of $L[T, \tau]$-modules.

Proposition 5.5.3. $M\left(\left(\frac{1}{T}\right)\right) \simeq L((\tau^{-1}))$ as $L[T, \tau]$-modules.

Proof. In $L((\tau^{-1}))$ write

$$\psi_T = \imath(T)\tau^0 + \cdots + g_d \tau^d, \qquad g_d \neq 0$$

as

$$(\theta \tau^{-d} + \cdots + g_d) \tau^d.$$

Thus ψ_T^{-1} is constructed via the geometric series. Moreover, one sees that any finite tailed Laurent series in $1/T$ can be expressed in $L((\tau^{-1}))$ using $\psi_{T^{-1}} = \psi_T^{-1}$.

Now M is free over $L[T]$ on $\{\tau^i m_0 \mid 0 \le i \le d-1\}$ with m_0 the identity mapping (as in the proof of Lemma 5.4.1). Thus

$$M\left(\left(\tfrac{1}{T}\right)\right) = \bigoplus_{i=0}^{d-1} \tau^i m_0 \otimes L\left(\left(\tfrac{1}{T}\right)\right).$$

We map

$$\sum \tau^i m_0 \otimes f_i(T) \mapsto \sum \tau^i m_0 \circ f_i(\psi_T) \in L((\tau^{-1})).$$

One now checks that this gives an isomorphism $M\left(\left(\tfrac{1}{T}\right)\right) \simeq L((\tau^{-1}))$. □

Corollary 5.5.4. *The motive M of ψ is pure.*

Proof. We use the description of the proposition. Let $W = L\{\{\tau^{-1}\}\} \subset M\left(\left(\tfrac{1}{T}\right)\right)$. It is clear that W is an $L\left[\left[\tfrac{1}{T}\right]\right]$-lattice. One now sees directly that

$$TW = \tau^d W$$

by writing ψ_T as in the proof of 5.5.3. □

Remark. 5.5.5. Let E be a T-module with, for some $q \ge 1$,

$$T^q x = (\theta^q + N_1)x + \cdots + g_s \tau^s(x), \qquad s > 0$$

with N_1 nilpotent and $\det(g_s) \ne 0$. Then, reasoning as above, one can show that E is pure.

Proposition 5.5.6. *Let M be a pure left $L[T,\tau]$-module. Then M is free and finitely generated over $L[\tau]$. Moreover, let $M \ne \{0\}$ and q and s as in 5.5.2. Then*

$$q/s = \frac{\mathrm{rank}_{L[\tau]} M}{\mathrm{rank}_{L[T]} M}.$$

Proof. Let W be the lattice associated to M from the definition of pure module. Let f be a positive integer chosen large enough so that

$$M + T^f W = M\left(\left(\tfrac{1}{T}\right)\right).$$

Set $M_j = M \cap (T^{(j+f)q} W)$ for $j \ge 0$.

We then have isomorphisms

$$\begin{aligned} M_{j+1}/M_j &\simeq T^{(j+f+1)q} W / T^{(j+f)q} W \\ &= \tau^{(j+f+1)s} W / \tau^{(j+f)s} W. \end{aligned}$$

Thus $T^q M_j + M_j = M_{j+1} = \tau^s M_j + M_j$ for all $j \geq 0$. Therefore, M is finitely generated over $L\{\tau\}$. Thus the first part follows from Lemma 5.4.10.

The second part follows in a similar fashion. □

Corollary 5.5.7. *Let M be a pure left $L[T, \tau]$-module. If there exists $j \gg 0$ with*
$$(T - \theta)^j M / \tau M = 0,$$
then M is an abelian T-motive. □

Definition 5.5.8. Let M be a non-trivial pure T-motive. Then we define the *rank of M*, $r(M)$, by
$$r(M) := \operatorname{rank}_{L[T]} M,$$
and the *weight of M*, $w(M)$, by
$$w(M) := \operatorname{rank}_{L[\tau]} M / \operatorname{rank}_{L[T]} M.$$

We carry these definitions over to T-modules in the obvious fashion.

Thus $\dim E = \operatorname{rank}_{L[\tau]} M(E) = w(E) r(E)$ for a T-module E.

In our next subsection, we discuss the relationship between purity and the Frobenius. In the following subsection, we discuss the application of purity to tensor products.

5.6. Torsion Points

In this subsection, except near the end, we will let L be an algebraically closed **A**-field; thus L is automatically perfect. Let $f \in \mathbf{A} = \mathbb{F}_r[T] \subset L[T, \tau]$ be prime to the characteristic of L. Note that f is *central* in $L[T, \tau]$.

Let E be an abelian T-module over L and let $E(L)$ be the $L[T]$-module of L-valued points of E.

Definition 5.6.1. We set
$$E[f] := \{ e \in E(L) \mid f \cdot e = 0 \}.$$

The space $E[f]$ is the *module of f-division points*.

Of course Definition 5.6.1 is the obvious generalization of the definition of division points for Drinfeld modules. In particular, $E[f]$ is clearly an **A**-module.

Let $M := M(E)$ be the motive of E as defined in Subsection 5.4. As f is central, M/fM is an $L[T, \tau]$-module.

Definition 5.6.2. We set
$$(M/fM)^\tau := \{m \in M/fM \mid \tau m = m\}.$$

Thus $(M/fM)^\tau$ is an $\mathbb{F}_r$-vector space. The reader should note the formal similarity with the construction of fixed points of φ-sheaves (Subsection 5.3).

Proposition 5.6.3. *As $\mathbb{F}_r$-vector spaces,*
$$E[f] \simeq \mathrm{Hom}_{\mathbb{F}_r}((M/fM)^\tau, \mathbb{F}_r).$$

Proof. We begin by setting $\widetilde{L} = L$ and equipping it with the $L[\tau]$-module structure obtained by evaluation. One then readily sees that
$$E[L] \xrightarrow{\sim} \mathrm{Hom}_{L[\tau]}(M, \widetilde{L}).$$

As f is prime to the characteristic of L and M is finitely generated over $L[\tau]$, M/fM is finite dimensional over L with dimension $m := \deg(f)r(E)$. The map τ is Frobenius linear and injective on M/fM. Thus Lang's Theorem on GL_m assures us that
$$(M/fM)^\tau \otimes_{\mathbb{F}_r} L \simeq M/fM.$$

(Alternatively, use the structure theorem on Frobenius linear maps on finite dimensional vector spaces [Mu2, p. 143].)

Thus
$$\begin{aligned} E[f] &= \mathrm{Hom}_{L[\tau]}(M/fM, \widetilde{L}) \\ &= \mathrm{Hom}_{\mathbb{F}_r}((M/fM)^\tau, \widetilde{L}^\tau) \\ &= \mathrm{Hom}_{\mathbb{F}_r}((M/\tau M)^\tau, \mathbb{F}_r). \end{aligned}$$
$\square$

Note that the duality between $E[f]$ and M/fM is just the one induced by evaluation of an element $m \in M$ at $e \in E[f]$. If M is defined over a subfield $L_1 \subseteq L$, then $\mathrm{Aut}(L/L_1)$ acts on M (and M/fM) by $(\sigma, m) \mapsto \sigma m \sigma^{-1}$. The above isomorphism is then an isomorphism of $\mathrm{Aut}(L/L_1)$-modules.

Corollary 5.6.4. $E[f]$ *is a* $\mathbf{A}/(f)$*-module of rank* $= \mathrm{rank}_{L[T]}M$.

Proof. We know that M/fM is free of rank $r(E)$ over $L[T]/(f)$. Thus $(M/fM)^\tau$ is an $\mathbb{F}_r[T] = \mathbf{A}$-module. But by Lang's Theorem
$$(M/fM)^\tau \otimes L = M/fM$$
as vector spaces *and* $L[T]$-modules. Thus $(M/fM)^\tau \simeq \mathbf{A}/(f)^{r(E)}$, and the result follows by duality. $\square$

As in [A1], we now want to refine the above results to a duality of **A**-modules. Let $\Omega := \mathbf{A}\, dT = \mathbb{F}_r[T]\, dT$ be the module of Kähler differentials of $\mathbb{F}_r[T]$. In the standard fashion, we have the residue map

$$\mathrm{Res}_\infty \colon \mathbb{F}_r(T) \otimes_{\mathbb{F}_r[T]} \Omega \to \mathbb{F}_r$$

given by taking the residue at ∞.

Set $V := \mathrm{Hom}_{\mathbb{F}_r}(\mathbb{F}_r[T]/(f), L)$. We give V a left $L[T,\tau]$-module structure by

$$(a, v) \mapsto av(x), \quad a \in L, \ v \in V;$$
$$(T, v) \mapsto v(Tx), \quad v \in V;$$
$$(\tau, v) \mapsto v(x)^\tau.$$

With this definition, there is a map

$$E[f] \to \mathrm{Hom}_{L[T,\tau]}(M/fM, V)$$
$$e \mapsto (m \mapsto (a \mapsto m(ae))).$$

Lemma 5.6.5. *The above map is an isomorphism of **A**-modules.*

Proof. Trace through the definitions. □

Theorem 5.6.6. *As an **A**-module, $E[f]$ is canonically isomorphic to*

$$\mathrm{Hom}_{\mathbf{A}}((M/fM)^\tau, f^{-1}\Omega/\Omega).$$

Proof. As before (see the proof of 5.6.3), we have

$$\mathrm{Hom}_{L[T,\tau]}(M/fM, \mathrm{Hom}_{\mathbb{F}_r}(\mathbb{F}_r[T]/(f), L))$$

is isomorphic to

$$\mathrm{Hom}_{\mathbb{F}_r[T]}((M/fM)^\tau, \mathrm{Hom}_{\mathbb{F}_r}(\mathbb{F}_r[T]/(f), \mathbb{F}_r)).$$

The result now follows by noting that $\mathbb{F}_r[T]/(f)$ and $f^{-1}\Omega/\Omega$ are perfectly paired by

$$(a, \omega) \mapsto \mathrm{Res}_\infty(a\omega). \qquad \square$$

Remarks. 5.6.7. 1. The module $f^{-1}\Omega/\Omega$ is (non-canonically) isomorphic to $\mathbf{A}/(f)$. Indeed, such an isomorphism only arises after we choose a basis for Ω.
2. The reader should compare the results presented here with those of Subsection 4.14.
3. By using the map Res_∞ on $f^{-1}\Omega/\Omega$, one sees that Proposition 5.6.3 can be derived from Theorem 5.6.6.

154   5. *T-modules*

Suppose now that $(f) = v$ is a prime of **A** distinct from the characteristic of L. Then, exactly as for Drinfeld modules (or, elliptic curves, etc.) we can form the v-adic Tate module of E. We denote it by "$T_v(E)$."

Set
$$H_v^1(M) := \varprojlim (M/v^n M)^\tau$$

and put $\Omega_v = \mathbf{A}_v \otimes \Omega$. Note that Ω_v is *non-canonically* isomorphic to $\mathbf{A}_v$. We can then summarize the above results in the next theorem.

Theorem 5.6.8. 1. *As $\mathbf{A}_v$-modules, $T_v(E)$ and $H_v^1(M)$ are free of rank $r(E)$.*
2. *The association $E \mapsto T_v(E)$ is a covariant functor.*
3. *There is a duality of $\mathrm{Gal}(L/L_1)$-modules*

$$T_v(E) = \mathrm{Hom}_{\mathbf{A}_v}(H_v^1(M), \Omega_v)$$

whenever E (and M) is defined over a perfect subfield L_1 of L. □

Suppose now that L is the algebraic closure of a finite **A**-field L_1. Thus $L_1 \simeq \mathbb{F}_{r^t}$, for some integer t. Suppose further that E is defined over L_1. Thus the r^t-Frobenius, $\tau_1 := \tau^t$, is an endomorphism of E and so acts on $T_v(E)$ and $H_v^1(M)$. Let $Q(u)$ be the characteristic polynomial of τ_1 on $T_v(E)$.

Proposition 5.6.9. *$Q(u)$ has coefficients in **A** which are independent of the choice of v.*

Proof. By 5.6.8.3 we may compute $Q(u)$ on $H_v^1(M)$. Now, as previously mentioned, by Lang's Theorem, we have

$$(M/v^n M)^\tau \otimes_{\mathbb{F}_r} L = M/v^n M.$$

Thus to compute $Q(u)$ on $H_v^1(M)$ we need only compute it on $\varprojlim M/v^n M$ as a module over $L[T]/v^n L[T]$. But τ_1 is an endomorphism of the free $L[T]$-module M itself. Thus the coefficients of $Q(u)$ are in $L[T]$ and independent of v. From $(M/v^n M)^\tau \otimes L = M/v^n M$, we see that the coefficients of $Q(u)$ must be in $\mathbb{F}_r[T]$. □

We now finally assume that M is pure with weight $w(M)$. Let $\mathbf{k} = \mathbb{F}_r(T)$ and let $\mathbf{k}_1$ be the splitting field of $Q(u)$ in some algebraic closure of $\mathbf{k}$. Put $r_1 := r^t$ and let σ be any embedding of $\mathbf{k}_1$ into an algebraic closure of $\mathbf{K} = \mathbb{F}_r((\frac{1}{T}))$ equipped with the canonical extension of the standard normalized absolute value $|?|$.

Theorem 5.6.10. *The eigenvalues of τ_1 are pure of weight $w(M)$; i.e., $|\sigma(\alpha)| = r_1^{w(M)}$ for all σ.*

Proof. By definition, we have a lattice $W \subset M((\frac{1}{T}))$ with the property that

$$\tau^s W = T^q W$$

and $w(M) = q/s$. Thus

$$(\tau^t)^s W = T^{qt} W,$$

and the result follows easily. □

Theorem 5.6.10 thus relates the motivic definition of purity, Definition 5.5.2, with purity of eigenvalues of Frobenius over finite fields. Note also that the T-motive theory provides very compact proofs of some of the results of Subsection 4.12.

Let L now only be perfect, not necessarily algebraically closed. Let $\overline{L}$ be a fixed algebraic closure and E a T-module over L. Let $H \subset E(\overline{L})$ be a finite T-stable submodule which is stable under $\text{Gal}(\overline{L}/L)$. We now want to construct $E_1 := E/H$ as a T-module.

Note first that E_1 exists as a commutative affine algebraic group over L [Mu1] [Sh1]. We need to see that E_1 can be realized as $\mathbb{G}_a^d$, $d = \dim E/H = \dim E$. Let M be the motive of E over L and let M_1 be the group of $\mathbb{F}_r$-linear algebraic group morphisms from E_1 to $\mathbb{G}_a$ over L. Both M and M_1 are $L\{\tau\}$-modules in the standard fashion.

One can see that M_1 is "big" in the following sense. Let $m \in M$ and let $H_1 = m(H)$ be the image of H under m. Then $H_1 \subset \mathbb{G}_a$ is a finite $\text{Gal}(\overline{L}/L)$-stable $\mathbb{F}_r$-subspace. Let $P(\tau)$ be the monic separable polynomial with roots H_1 as in Subsection 1.8. Then $P(\tau) \in L\{\tau\}$ and $P(\tau) \circ m \in M_1$.

Let $\pi: E \to E_1$ be the projection; thus $\pi^*: M_1 \to M$ is an injection of $L\{\tau\}$-modules. The elementary divisor theory of 5.4.8 and 5.4.9 now implies that M_1 is free over $L\{\tau\}$ on elements $\{m_1, \ldots, m_d\}$. One then checks that the mapping

$$h: x \mapsto \begin{pmatrix} m_1(x) \\ \vdots \\ m_d(x) \end{pmatrix}$$

from E_1 to $\mathbb{G}_a^d$ is an isomorphism.

The mapping $\pi: E \to \mathbb{G}_a^d$ obtained by projecting to E_1 and then applying h, must be étale as H is finite. As π is $\mathbb{F}_r$-linear we see its differential $\pi_*: \text{Lie}(E) \to \text{Lie}(\mathbb{G}_a^d)$ is constant and invertible. By composing with a nonsingular matrix, we may assume that the differential is the identity. We now use π to transfer the T-structure on E/H to $\mathbb{G}_a^d$ obtaining a quotient T-module.

Similarly one could work with finite, T-stable, group schemes.

Remarks. 5.6.11. 1. Unlike the 1-dimensional case, π is *not* unique in general. In fact, one can always compose π with the map $P_1 = \text{id} + \alpha\tau$ where

$$\alpha = \begin{pmatrix} 0 & \cdots & 0 \\ \vdots & & \vdots \\ 1 & \cdots & 0 \end{pmatrix}.$$

156   5. T-modules

Indeed, P_1 is an automorphism of $\mathbb{G}_a^d$ with inverse $P_2 = \mathrm{id} - \alpha\tau$.

2. One needs L to be perfect as following elegant example of G. Faltings makes clear. Let $\mathcal{F}$ be a non-perfect field of characteristic p. Let $\alpha \notin \mathcal{F}^p$ and put
$$E := \ker((x,y) \mapsto x + x^p - \alpha y^p \colon \mathbb{G}_a^2 \to \mathbb{G}_a).$$
In $\mathcal{F}(\alpha^{1/p})$, E is isomorphic to the kernel of $x + y^p = 0$; thus E is easily seen to be parameterized by $x = -t^p$, $y = t$ and is isomorphic to $\mathbb{G}_a$. Suppose $(f,g)\colon \mathbb{G}_a \to E$ is any nonconstant map over $\mathcal{F}$. Then
$$f + f^p = \alpha g^p.$$
Upon examining coefficients of highest degree we see that this forces $\alpha \in \mathcal{F}^p$ which is a contradiction. So E is not isomorphic to $\mathbb{G}_a$ over $\mathcal{F}$. In this fashion, one obtains "potentially-additive" groups. For more, see [DeG1, IV, §3, no. 6].

5.7. Tensor Products

Having discussed in detail the concept of purity of T-motives, we can now pass to tensor products. Let M_1 and M_2 be two left $L[T,\tau]$-modules.

Definition 5.7.1. We define $M_1 \otimes M_2$ to be $M_1 \otimes_{L[T]} M_2$ equipped with diagonal τ-action; i.e.,
$$\tau(a \otimes b) = \tau(a) \otimes \tau(b).$$

Proposition 5.7.2. Let M_1 and M_2 be two pure T-motives. Then
1. $M_1 \otimes M_2$ is also a pure T-motive.
2. $r(M_1 \otimes M_2) = r(M_1)r(M_2)$
3. $w(M_1 \otimes M_2) = w(M_1) + w(M_2)$.

Proof. Let W_1 be the lattice associated to M_1 and W_2 the lattice associated to M_2. then $W_1 \otimes W_2 \subset (M_1 \otimes M_2)\left(\left(\frac{1}{T}\right)\right)$ is a lattice. Thus $M_1 \otimes M_2$ is pure and Part 1 follows from Corollary 5.5.7. Part 2 follows directly. To see Part 3, suppose
$$\tau^{s_1} W_1 = T^{q_1} W_1$$
and
$$\tau^{s_2} W_2 = T^{q_2} W_2.$$
Thus
$$\tau^{s_1 s_2} W_1 \otimes W_2 = T^{q_1 s_2 + q_2 s_1} W_1 \otimes W_2,$$
and the result follows. □

Let E_1, E_2 be two pure T-modules. Then $E_1 \otimes E_2$ is defined to be the T-module associated to $M(E_1) \otimes M(E_2)$.

More generally, one can define the tensor product of those *non-pure T-modules* which can be written as successive extensions of pure T-modules.

Proposition 5.7.3. *Let M_1, M_2 be two pure T-motives. Then*

$$H_v^1(M_1 \otimes M_2) = H_v^1(M_1) \otimes H_v^1(M_2).$$

Proof. This follows from the definitions and linear algebra. □

Remark. 5.7.4. Let E be a T-module with associated motive M. Proposition 5.7.3 implies that the functor

$$E \mapsto H_v^1(M) = \mathrm{Hom}_{\mathbf{A}_v}(T_v(E), \Omega_v)$$

commutes with tensor product. At the same time, we see that $E \mapsto T_v(E)$ does *not* (quite!) commute with tensor product.

5.8. The Tensor Powers of the Carlitz Module

We now pause to present the most basic example of a general T-motive. Let $\mathbf{A} = \mathbb{F}_r[T]$, as before, and put $\mathbf{k} = \mathbb{F}_r(T)$. Let $\overline{\mathbf{k}}$ be a fixed algebraic closure of $\mathbf{k}$ and let $L \subset \overline{\mathbf{k}}$ be the perfection of $\mathbf{k}$. Recall that, as in Section 3, the Carlitz module is the rank one Drinfeld module $C \colon \mathbf{A} \to \mathbf{k}\{\tau\}$ defined by

$$C_T(\tau) = T\tau^0 + \tau, \quad (\tau = r^{\mathrm{th}} \text{ power map}).$$

In this subsection we discuss the nth tensor power of C which we denote $C^{\otimes n}$.

The fields $\mathbf{k}$, L, etc., are obviously $\mathbf{A}$-fields via the canonical injection $\imath \colon \mathbf{A} \to \mathbf{k}$, etc. Set

$$\theta := \imath(T) \in \mathbf{k}$$

as before. Thus, as mentioned before, θ is "T considered as a scalar in L." The use of θ avoids confusion in $L[T]$; for instance without θ, $T - \theta$ would be "$T - T$."

As in the next subsection, we set $k = \mathbb{F}_r(\theta)$. Of course, in our situation, $k = \mathbf{k}$, but the use of non-bold symbols is useful in keeping track of when we are using a field of *operators* or *scalars*.

Set $M := M(C) \simeq L[\tau]$ ($\simeq L\{\tau\}$). Via C, M is a rank one $L[T]$-module with generator τ^0. Note that

$$(T - \theta)\tau^0 = \tau.$$

The $L[T]$-module $M^{\otimes n}$ is thus free of rank one on the element

$$\alpha = \tau^0 \otimes \cdots \otimes \tau^0.$$

By definition, on $M^{\otimes n}$,

$$\tau\alpha = \tau \otimes \cdots \otimes \tau = (T-\theta)^n(\tau^0 \otimes \cdots \otimes \tau^0).$$

Thus one see that $M^{\otimes n}$ is free of rank n over $L\{\tau\}$ with basis

$$\{\alpha, (T-\theta)\alpha, \ldots, (T-\theta)^{n-1}\alpha\}.$$

By playing the actions of T, τ off each other, one now easily finds the equations of $C^{\otimes n}$ which we present in our next definition.

Definition 5.8.1. $C^{\otimes n}$ is the abelian T-module with underlying group $\simeq \mathbb{G}_a^n$ and the following action of T: Let N_n be the $n \times n$ matrix

$$\begin{pmatrix} 0 & 1 & & & 0 \\ & \ddots & \ddots & & \\ & & \ddots & \ddots & \\ & & & \ddots & 1 \\ 0 & & & & 0 \end{pmatrix},$$

and V_n the $n \times n$ matrix

$$\begin{pmatrix} 0 & \cdots & 0 \\ \vdots & & \vdots \\ 1 & \cdots & 0 \end{pmatrix}.$$

Then if $x = \begin{pmatrix} x_1 \\ \vdots \\ x_n \end{pmatrix}$ we have

$$Tx := (\theta\tau^0 + N_n)x + V_n\tau x.$$

(Note that $\theta\tau^0 + N_n$ is a Jordan block.)

The reader should note that, while $C^{\otimes n}$ is a priori only defined over L, it actually makes sense over $\mathbf{k}$ itself, or even any $\mathbf{A}$-field $\mathcal{F}$. Let $\mathcal{F}$ be one such $\mathbf{A}$-field with algebraic closure $\overline{\mathcal{F}}$ and separable closure $\mathcal{F}^{\text{sep}} \subseteq \overline{\mathcal{F}}$. Let $\mathcal{F}^{\text{perf}} \subseteq \overline{\mathcal{F}}$ be the perfection of $\mathcal{F}$. We then have canonically

$$\text{Gal}(\mathcal{F}^{\text{sep}}/\mathcal{F}) \simeq \text{Gal}(\overline{\mathcal{F}}/\mathcal{F}^{\text{perf}}).$$

Let $f \in \mathbf{A}$ be prime to the characteristic of $\mathcal{F}$. For each $n \geq 1$, let $\chi_n \colon \text{Gal}(\mathcal{F}^{\text{sep}}/\mathcal{F}) \to \mathbf{A}/f^*$ be the character induced by the action of Galois on

$$C^{\otimes n}[f] \simeq \mathbf{A}/f.$$

As in Proposition 5.7.3, we deduce the basic fact

$$\chi_n = \chi^n.$$

5.9. Uniformization

As discussed at length in Section 4, every Drinfeld module ψ over $\mathbf{C}_\infty$ can be analytically uniformized. That is, ψ can be deduced from a lattice and its exponential function. This is, of course, analogous to the well-known theory for complex elliptic curves. However, just as the classical theory of lattices in $\mathbb{C}^n$, $n > 1$, is more complicated than for $n = 1$, so is the uniformization theory of T-modules more complicated in higher dimensions.

We let θ be an indeterminate and put $k = \mathbb{F}_r(\theta)$ and $K = \mathbb{F}_r\left(\left(\frac{1}{\theta}\right)\right)$. We will reserve the use of boldface for the analogous fields created out of T; i.e, $\mathbf{k} = \mathbb{F}_r(T)$, $\mathbf{K} = \mathbb{F}_r\left(\left(\frac{1}{T}\right)\right)$, etc. We make k and K into **A**-fields by setting $\theta = \imath(T)$.

Let $|?|$ be the normalized absolute value on K with

$$|1/\theta| = 1/r$$

and let $\overline{K}$ be a fixed algebraic closure of K equipped with the canonical extension of $|?|$.

Let X be any matrix with entries in $\overline{K}$. We set

$$|X| := \sup_{i,j}\{|X_{ij}|\}.$$

We also define

$$X^{(j)} := (X_{ij}^{r^j}),$$

for any $j \geq 0$.

Definition 5.9.1. A function $e\colon \overline{K}^m \to \overline{K}^n$ (where $\overline{K}^j$ is viewed as column vectors) is called *entire* $\mathbb{F}_r$-*linear* if and only if there exists a sequence $\{e_i\}$ of $n \times m$ matrices in $\overline{K}$ such that the following holds:
1. The coefficients of all $\{e_i\}$ are contained in a finite extension of K in $\overline{K}$;
2. $\lim\limits_{i \to \infty} r^{-i} \log|e_i| = -\infty$;
3. $e(x) = \sum\limits_{j=0}^{\infty} e_j x^{(j)}$.

Note that the second condition of 5.9.1 guarantees convergence for all $X \in \overline{K}^m$ and the first condition guarantees convergence in $\overline{K}^n$.

Proposition 5.9.2. *Let m, n and s be positive integers. Let $H \in M_m(\overline{K})$, $F \in M_{n,m}(\overline{K})$ and $G_0, G_1, \ldots, G_S \in M_n(\overline{K})$ be given such that*

$$FH = G_0 F, \qquad (H - \theta 1_m)^m = 0,$$

and

$$(G_0 - \theta 1_n)^n = 0.$$

Then there exists a unique entire $\mathbb{F}_r$-linear map $e\colon \overline{K}^m \to \overline{K}^n$ of the form

160   5. T-modules

$$e(x) = Fx + \{\text{higher order terms}\}$$

with the functional equation

$$e(Hx) = \sum_{j=0}^{s} G_j e(x)^{(j)}.$$

Proof. 1. Without loss of generality, we may assume that both G_0 and H are in Jordan canonical form. In particular, we have

$$G_0 \in \theta 1_n + M_n(\mathbb{F}_p)$$

and

$$H \in \theta 1_m + M_m(\mathbb{F}_p).$$

2. We now establish a lemma.

Lemma 5.9.3. *For all positive integers j and $Y \in M_{n,m}(\overline{K})$ there exists a unique $X \in M_{n,m}(\overline{K})$ such that*

$$Y = XH^{(j)} - G_0 X.$$

Moreover, for this X one has the estimate

$$|X| \leq |\theta^{-r^j} Y|,$$

and the extension field of K in $\overline{K}$ containing the coefficients of Y also contains the coefficients of X.

Proof of 5.9.3. We define a linear operator $P_j(Z): M_{n,m}(\overline{K}) \to M_{n,m}(\overline{K})$ by the rule

$$P_j(Z) := -ZH^{(j)} + G_0 Z + \theta^{r^j} Z - \theta Z.$$

One checks that P_j is *nilpotent*, and, by our assumptions on G_0, H,

$$|P_j(Z)| \leq |Z|$$

for all $Z \in M_{n,m}(\overline{K})$. Put $c := (\theta^{r^j} - \theta)^{-1}$. Then

$$X = cY + \sum_{i=1}^{\infty} c^{i+1} P_j^i(Y).$$

The uniqueness and desired properties of X follow from this and completes the proof of 5.9.3.

As a consequence of the lemma, one can find a unique sequence

$$\{e_0, e_1, \ldots\} \subset M_{n,m}(\overline{K})$$

such that $e_0 = F$ and

$$e_j H^{(j)} - G_0 e_j = \sum_{i=1}^{\inf(j,s)} G_i e_{j-i}^{(i)} \quad (*)$$

for all positive integers j. One checks readily that *all* the matrices $\{e_j\}$ have coefficients in a finite extension of K.

Put
$$b_j := r^{-j} \log |e_j|$$
and
$$c_i := \log |G_i|.$$

By Lemma 5.9.3 and $*$, one sees that

$$bj \leq \left(\sup_{i=1}^{\inf(j,s)} (b_{j-i} + r^{-j} c_i) \right) - \log |\theta|.$$

Thus $\lim_{j \to \infty} b_j = -\infty$. Therefore, if we put

$$e(x) := \sum_{i=0}^{\infty} e_i x^{(i)},$$

we see that $e(x)$ is the unique entire $\mathbb{F}_r$-linear map with the desired properties. □

Let E be an abelian T-module of dimension e over $\overline{K}$.

Definition 5.9.4. A *coordinate system* for E is an isomorphism of algebraic groups
$$\rho \colon E \xrightarrow{\sim} \mathbb{G}_a^e,$$

where we write $\rho = \begin{pmatrix} \rho_1 \\ \vdots \\ \rho_e \end{pmatrix}$. Associated to ρ we have the isomorphism

$$\rho_* \colon \mathrm{Lie}(E) \xrightarrow{\sim} \overline{K}^e$$

such that for $\varepsilon \in \mathrm{Lie}(E)$

$$\rho_*(\varepsilon) = \begin{pmatrix} \partial_\varepsilon \rho_1 \\ \vdots \\ \partial_\varepsilon \rho_e \end{pmatrix}.$$

Let E_1 and E_2 be two T-modules and $g \colon E_1 \to E_2$ a morphism. We let

$$g_* \colon \mathrm{Lie}(E_1) \to \mathrm{Lie}(E_2)$$

be the standard induced morphism.

Our next definitions involve the fundamental ideas associated with the *exponential function* of a T-module. The reader will see that it is the higher dimensional analog of the exponential function of a Drinfeld module.

Definition 5.9.5. Let E_1 and E_2 be two abelian T-modules of dimensions e_1 and e_2 respectively. Let $f\colon E_1 \to E_2$ be a morphism. The *exponential of f* is a map $\exp_f\colon \mathrm{Lie}(E_1) \to E_2(\overline{K})$ such that:

1. Let $\varepsilon \in \mathrm{Lie}(E_1)$. Then
$$\exp_f(T\varepsilon) = T\exp_f(\varepsilon)$$
(where $T\varepsilon$ is the induced Lie action of T).

2. Let $\exp_{f*}$ be the induced Lie morphism from $\mathrm{Lie}(E_1) \to \mathrm{Lie}(E_2)$. Then
$$\exp_{f*} = f_*\,.$$

3. Let ρ^1 and ρ^2 be coordinate systems for E_1 and E_2 respectively. Then there exists an entire $\mathbb{F}_r$-linear map $e\colon \overline{K}^{e_1} \to \overline{K}^{e_2}$ such that if $\varepsilon \in \mathrm{Lie}(E_1)$. Then
$$e(\rho^1_*\varepsilon) = \rho^2(\exp_f(\varepsilon))\,.$$

Theorem 5.9.6. *The exponential of f exists and is unique.*

Proof. This follows immediately from Proposition 5.9.2. □

Definition 5.9.7. Let E be an abelian T-module and let $\mathrm{id}\colon E \to E$ be the identity. The *exponential of E*, $\exp_E$, is defined by $\exp_E := \exp_{\mathrm{id}}$.

Let ρ be a coordinate system for E_1 and let $e(\tau) = \sum_{i=0}^{\infty} Q_i \tau^i$ be the entire $\mathbb{F}_r$-linear function associated to $\exp_E$, with respect to ρ. Then $Q_i \in M_e(\overline{K})$, $e = \dim(E)$ and $Q_0 = I_e$.

Let E_1 and E_2 be as before and let $f\colon E_1 \to E_2$ be a morphism of T-modules. One checks that
$$\exp_{E_2} \circ f_* = \exp_f = f \circ \exp_{E_1}\,.$$
Thus $\exp_E$ is *functorial* in E.

Remark. 5.9.8. Suppose that E_1 and E_2 correspond to Drinfeld modules ψ_1, ψ_2 with exponentials e_1 and e_2. Let f be a morphism from E_1 to E_2. In this situation, one obviously has canonical coordinate systems and
$$\exp_f = f \circ e_1\,.$$
The situation with general T-modules is similar once coordinate systems are chosen.

5.9. Uniformization   163

The exponential of a Drinfeld module is always surjective. Unfortunately, this is no longer true in higher dimensions as is seen in the following elegant example of G. Anderson and R. Coleman.

Example 5.9.9. Let $c \in \overline{K}$ be determined by $|c| < 1$ and $\theta = c^{-1} + c$. Put $E = \mathbb{G}_a^2$ over $\overline{K}$ with

$$Tx := \theta x + \begin{pmatrix} 0 & 1 - c^{r+1} \\ 1 - c^r & 0 \end{pmatrix} x^{(1)} + \begin{pmatrix} c^{1+r+r^2} & 0 \\ 0 & c^r \end{pmatrix} x^{(2)}.$$

As the coefficient of $x^{(2)}$ is invertible, one can see that E is abelian of rank 4. Therefore, the torsion submodule V of E is isomorphic to $(\mathbf{k}/\mathbf{A})^4 = (\mathbb{F}_r(T)/\mathbb{F}_r(T))^4$.

By definition E, comes equipped with a coordinate system and we let $e(x)$ be the entire function corresponding to $\exp_E$ in this system. We will show that $e(x)$ *cannot* be surjective by actually showing that it is *injective*. Indeed one then sees easily that the image of $e(x)$ must have null intersection with V. To establish injectivity, let

$$\lambda x := cx + \begin{pmatrix} 0 & -c^{r+1} \\ -c^{r+1} & 0 \end{pmatrix} x^{(1)} + \begin{pmatrix} c^{1+r+r^2} & 0 \\ 0 & 0 \end{pmatrix} x^{(2)}.$$

One checks that $\lambda_1 = T - \lambda$ is inverse to λ; so λ is an automorphism of $\mathbb{G}_a^2$. As

$$Tx = (\lambda + \lambda^{-1})x,$$

one sees $T\lambda = \lambda T$; thus λ is an automorphism of E. By functoriality, $e(cx) = \lambda e(x)$. Thus the kernel of $e(x)$ is *stable* under multiplication by c. But the inverse function theorem implies that this kernel is discrete. As $|c| < 1$ we conclude the kernel $= \{0\}$ and the example is complete!

We turn now to presenting conditions for the surjectivity of $\exp_E$ due to G. Anderson.

Definition 5.9.10. We let $\overline{K}\{T\}$ be the ring of power series $\sum_{i=0}^{\infty} a_i T^i$ such that
1. $\lim |a_i| = 0$;
2. $\{a_i\}$ is contained in a finite extension of $K \subset \overline{K}$.

The reader should not confuse the commutative ring $\overline{K}\{T\}$ with the non-commutative ring $\overline{K}\{\tau\}$.

Let $x \in \overline{K}$ with $|x| \leq 1$ and let $f(T) \in \overline{K}\{T\}$. Then $f(x)$ clearly converges to an element of $\overline{K}$.

Let M be a T-motive over $\overline{K}$.

Definition 5.9.11. 1. We put
$$M\{T\} := M \otimes_{\overline{K}[T]} \overline{K}\{T\}.$$
The module $M\{T\}$ has the obvious $\overline{K}[T]$-action; we let τ act by
$$\tau(m \otimes \sum a_i T^i) = \tau m \otimes \sum a_i^r T^i.$$
2. We set
$$H^1(M) := M\{T\}^\tau.$$
3. Given an abelian T-module E, i.e. set
$$H_1(E) := \text{kernel of } \exp_E \subset \text{Lie}(E).$$

Thus, both $H^1(M)$ and $H_1(E)$ are $\mathbf{A} = \mathbb{F}_r[T]$-modules in the obvious fashion.

Lemma 5.9.12. *Let E be as above. Then $H_1(E)$ is discrete in $\text{Lie}(E)$. It is also free of rank $\leq r(E)$.*

Proof. Choose a system ρ of coordinates of E; thus
$$\rho_* : \text{Lie}(E) \xrightarrow{\sim} \overline{K}^e$$
where e is the dimension of E (= rank over $L[\tau]$ of $M(E)$). We can, therefore, equip $\text{Lie}(E)$ with the *sup norm* $\|?\|$ via ρ_*. As the Jacobian of $e(x)$ is everywhere invertible, $e(x)$ is a local homeomorphism. Thus $H_1(E)$ is discrete.

By our assumption on T and θ, T acts invertibly on $\text{Lie}(E)$. Moreover, one checks that
$$\lim_{i \to \infty} \|T^{-i}\xi\| = 0$$
for all $\xi \in \text{Lie}(E)$. Thus $\text{Lie}(E)$ becomes a topological $\mathbf{K} = \mathbb{F}_r\left(\left(\frac{1}{T}\right)\right)$-vector space by the action
$$\left(\sum a_i T^i, v\right) \mapsto \sum a_i(T^i v).$$
Thus $H_1(E)$ is torsion free as an $\mathbf{A}$-module. The proof that $H_1(E)$ is free of rank $\leq r(E)$ now follows as in Subsection 4.6. □

Let E be an abelian T-module over $\overline{K}$.

Definition 5.9.13. We say that $M = M(E)$ is *rigid analytically trivial* if and only if the natural map
$$H^1(M) \otimes_\mathbf{A} \overline{K}\{T\} = M\{T\}^\tau \otimes_\mathbf{A} \overline{K}\{T\} \to M\{T\}$$
is an isomorphism.

The main result of this subsection can now be stated.

Theorem 5.9.14 (Anderson). *The following properties of an abelian T-module E are equivalent:*
1. $\operatorname{rank}_{\mathbf{A}} H_1(E) = r(E)$.
2. $\exp_E$ *is surjective.*
3. $M := M(E)$ *is rigid analytically trivial.*

The proof of Theorem 5.9.14 is quite involved. We will present it in stages interspersed with needed lemmas, etc. The proof will proceed "$2 \Rightarrow 1, 3 \Rightarrow 2, 1 \Rightarrow 3$."

Proof 5.9.14.2 $\Rightarrow$ 5.9.14.1. As $\exp_E$ is surjective, we have an isomorphism over $\overline{K}$
$$\operatorname{Lie}(E)/H_1(E) \widetilde{\to} E$$
compatible with T-modules. The result now follows directly. □

Definition 5.9.15. 1. Let B be a commutative topological group. We say that B is a *topological* **A**-*module* if and only if it is an **A**-module and the map $b \mapsto fb$ is continuous for all $f \in \mathbf{A}$ and $b \in B$.
2. Let B_0 and B_1 be two topological **A**-modules. We let $\operatorname{Hom}_{\mathbf{A}}^c(B_0, B_1)$ be the **A**-module of continuous **A**-morphisms from B_0 to B_1.

There is an obvious exact sequence
$$0 \to \mathbf{A} \to \mathbf{K} \to \mathbf{K}/\mathbf{A} \to 0 \qquad (5.9.16)$$
$$\left(0 \to \mathbb{F}_r[T] \to \mathbb{F}_r\left(\left(\frac{1}{T}\right)\right) \to \mathbb{F}_r\left(\left(\frac{1}{T}\right)\right)/\mathbb{F}_r[T] \to 0\right).$$

If we apply $\operatorname{Hom}_{\mathbf{A}}^c(?, E(\overline{K}))$ to 5.9.16, we obtain

$$0 \to \operatorname{Hom}_{\mathbf{A}}^c(\mathbf{K}/\mathbf{A}, E(\overline{K})) \to \operatorname{Hom}_{\mathbf{A}}^c(\mathbf{K}, E(\overline{K})) \to \operatorname{Hom}_{\mathbf{A}}^c(\mathbf{A}, E(\overline{K})). \quad (5.9.17)$$

Let $\lambda: \operatorname{Lie}(E) \to \operatorname{Hom}_{\mathbf{A}}^c(\mathbf{K}, E(\overline{K}))$ be defined by $x \mapsto (y \mapsto \exp_E(xy))$.

Lemma 5.9.18. *The map λ is an isomorphism.*

Proof. This follows as the exponential map is an isomorphism on an open neighborhood of the identity in $\operatorname{Lie}(E)$. □

Corollary 5.9.19. *The sequence 5.9.17 is isomorphic to the sequence*
$$0 \to H_1(E) \to \operatorname{Lie}(E) \xrightarrow{\exp_E} E(\overline{K}).$$
□

Remark. 5.9.20. The reader may find Corollary 5.9.19 more believable by examining the analogous statement in the theory of elliptic curves involving the sequence
$$0 \to \mathbb{Z} \to \mathbb{R} \to \mathbb{R}/\mathbb{Z} \to 0.$$

Definition 5.9.21. Let B be a topological **A**-module. We set
$$H := \text{Hom}^c(B, \overline{K}) = \text{the group of continuous maps } f \text{ from } B \text{ to } \overline{K}$$
such that $[K(f(B)):K] < \infty$. We make H into a left $\overline{K}[T, \tau]$-module by
$$(T, f) \mapsto (x \mapsto f(Tx))$$
and
$$(\tau, f) \mapsto f(x)^r.$$

In the category of topological groups, the exact sequence 5.9.16 is easily seen to split. Thus, if we apply $\text{Hom}^c(?, \overline{K})$ to 5.9.16, we obtain a new exact sequence of left $\overline{K}[T, \tau]$-modules
$$0 \to Z_1 \to Z_2 \to Z_3 \to 0 \tag{5.9.22}$$
where
$$Z_1 := \text{Hom}^c(\mathbf{K/A}, \overline{K})$$
$$Z_2 := \text{Hom}^c(\mathbf{K}, \overline{K})$$
and
$$Z_3 := \text{Hom}^c(\mathbf{A}, \overline{K}).$$

Let $B \in \{\mathbf{A}, \mathbf{K}, \mathbf{K/A}\}$ and let $f \in \text{Hom}_\mathbf{A}^c(B, E(\overline{K}))$.

Proposition 5.9.23. *Let $m: E \to \mathbb{G}_a$ be a morphism of $\overline{K}$-algebraic groups. Then $[K(m(f(B))):K] < \infty$. Thus the map $b \mapsto m(f(b)): B \to \overline{K}$ belongs to $\text{Hom}^c(B, \overline{K})$.*

Proof. For $B = \mathbf{A}$ the result is obvious. The rest follows, for instance, from Lemma 5.9.18. □

We now use Proposition 5.9.23 to reinterpret the exact sequence 5.9.17 and, therefore, the isomorphic exact sequence of 5.9.19. The idea is as follows: let B be as above; then by 5.9.23 there is a map
$$\text{Hom}_\mathbf{A}^c(B, E(\overline{K})) \to \text{Hom}_{\overline{K}[T,\tau]}(M(E), \text{Hom}^c(B, \overline{K}))$$
$$f \mapsto (m \mapsto (b \mapsto m(f(b)))).$$

Lemma 5.9.24. *The above map is an isomorphism.*

5.9. Uniformization

Proof. We leave this as an exercise for the reader (similar to the techniques used in the study of torsion in T-modules). □

Corollary 5.9.25. *The exact sequence 5.9.17 is functorially isomorphic to*

$$0 \to \mathrm{Hom}_{\overline{K}[T,\tau]}(M, Z_1) \to \mathrm{Hom}_{\overline{K}[T,\tau]}(M, Z_2) \to \mathrm{Hom}_{\overline{K}[T,\tau]}(M, Z_3). \quad \square$$

Remark. 5.9.26. Although somewhat convoluted, 5.9.24 plays a key role in the proof of the main result. Indeed, it allows us to translate the problem of surjectivity of $\exp_E$ into the homological question of whether

$$\mathrm{Ext}^1_{\overline{K}[T,\tau]}(M(E), Z_1)$$

vanishes.

We now pass to an over-ring of $\overline{K}[T,\tau]$. To do this, let $f(T) = \sum a_i T^i \in \overline{K}\{T\}$ and let n be a positive integer. We set

$$f^{(n)}(T) := \sum a_i^{r^n} T^i \in \overline{K}\{T\}.$$

We let $\overline{K}[T,\tau]_0$ be the set of finite sums $\sum_{j\geq 0} f_j \tau^j$, $f_j \in \overline{K}\{T\}$. We add these in the obvious fashion and multiply them by

$$\left(\sum f_i \tau^i\right)\left(\sum g_j \tau^j\right) := \sum\sum f_i g_j^{(i)} \tau^{i+j}.$$

Proposition 5.9.27. 1. $\overline{K}[T,\tau]_0$ *forms a ring containing* $\overline{K}[T,\tau]$ *as a subring.*
2. $\overline{K}\{T\}$ *becomes a* $\overline{K}[T,\tau]_0$-*module by* $(\tau^n, f) \mapsto f^{(n)}$.

Proof. These are simple exercises. □

Remark. 5.9.28. Via the definition of 5.9.27.2, $\overline{K}[T]$ is a $\overline{K}[T,\tau]$-submodule of $\overline{K}\{T\}$. Let M be any $\overline{K}[T,\tau]$-module. Then

$$M \otimes_{\overline{K}[T]} \overline{K}[T] \simeq M$$

as $\overline{K}[T,\tau]$-modules and where τ acts diagonally on the tensor product. Clearly, this generalizes to the case of any **A**-field L. Thus, $L[T]$, with τ-action as above, acts as the identity with respect to tensor products.

Note that $\mathbf{K}/\mathbf{A}$ is compact; thus any continuous function on it has bounded image. This allows us to make $Z_1 = \mathrm{Hom}^c(\mathbf{K}/\mathbf{A}, \overline{K})$ into a $\overline{K}[T,\tau]_0$-module by

$$(f, h) \mapsto \sum a_i h(T^i b)$$

for $f(T) = \sum a_i T^i$. Let γ be the function that takes $b \in \mathbf{K}/\mathbf{A}$ to $\mathrm{Res}_\infty(bdT) \in \mathbb{F}_r$; one sees that γ is τ-invariant and belongs to Z_1. One checks further that Z_1 is free of rank one over $\overline{K}\{T\}$ with basis γ. Thus Z_1 is isomorphic to $\overline{K}\{T\}$ as $\overline{K}[T,\tau]_0$-modules (where $\overline{K}\{T\}$ is given the action of 5.9.27.2).

Lemma 5.9.29. $\mathrm{Ext}^1_{\overline{K}[T,\tau]_0}(Z_1, Z_1) = \{0\}$.

Proof. Let $g_1: \overline{K}[T,\tau]_0 \to \overline{K}[T,\tau]_0$ be the map $q \mapsto q(1-\tau)$, and let $g_2: \overline{K}[T,\tau]_0 \to Z_1$ be the map $q \mapsto q\gamma$. One checks that

$$0 \to \overline{K}[T,\tau]_0 \xrightarrow{g_1} \overline{K}[T,\tau]_0 \xrightarrow{g_2} Z_1 \to 0$$

is a free resolution of Z_1. Thus, in the standard manner,

$$\mathrm{Ext}^1_{\overline{K}[T,\tau]_0}(Z_1, Z_1) = \overline{K}[T,\tau]_0/N$$

where $N = (1-\tau)\overline{K}[T,\tau]_0 + \overline{K}[T,\tau]_0(1-\tau)$. As

$$\overline{K}[T,\tau]_0/(\overline{K}[T,\tau]_0(1-\tau)) \simeq \overline{K}\{T\},$$

to see that Ext^1 vanishes, we are reduced to the following: Let $f \in \overline{K}\{T\}$; then we need to find $g \in \overline{K}\{T\}$ such that

$$g - g^{(1)} = f.$$

If $f \in \overline{K}[T]$, then it is easy to find such a g. If $f = \sum a_i T^i$ with $|a_i| < 1$ for all i, then we can set

$$g = \sum_{j=0}^{\infty} f^{(j)}.$$

In general, one can express f as $f_0 + f_1$ where $f_0 \in \overline{K}[T]$ and f_1 has coefficients of absolute value < 1. The result follows. □

Let M be a left $\overline{K}[T,\tau]$-module. Clearly $M\{T\}$ has a natural structure of $\overline{K}[T,\tau]_0$-module extending the $\overline{K}[T,\tau]$ action. Note also that

$$Z_1^\tau \simeq \mathbb{F}_r[T];$$

thus

$$Z_1^\tau \otimes_\mathbf{A} \overline{K}\{T\} \simeq Z_1.$$

The next lemma follows directly.

Lemma 5.9.30. *An abelian T-motive M is rigid-analytically trivial if and only if $M\{T\}$ is isomorphic to a direct sum of finitely many copies of Z_1.* □

We can now return to our proof of the main result.

Proof 5.9.14.3 $\Rightarrow$ 5.9.14.1. As mentioned in Remark 5.9.26, it is enough to show that
$$\operatorname{Ext}^1_{\overline{K}[T,\tau]}(M, Z_1) = \{0\}.$$

By definition, M is a finitely generated left $\overline{K}[T,\tau]$ module. As $\overline{K}[T,\tau]$ is left noetherian, there exists a resolution
$$P^\bullet \to M$$
by finitely generated free left $\overline{K}[T,\tau]$-modules. If we note that the functor
$$N \mapsto N \otimes_{\overline{K}[T]} \overline{K}\{T\}$$
is exact, we see that we obtain a resolution
$$P^\bullet \otimes \overline{K}\{T\} \to M\{T\}$$
of finitely generated free left $\overline{K}[T,\tau]_0$-modules. Further,
$$\operatorname{Hom}_{\overline{K}[T,\tau]}(P^\bullet, Z_1) = \operatorname{Hom}_{\overline{K}[T,\tau]_0}(P^\bullet \otimes \overline{K}\{T\}, Z_1);$$
thus
$$\operatorname{Ext}^1_{\overline{K}[T,\tau]}(M, Z_1) = \operatorname{Ext}^1_{\overline{K}[T,\tau]_0}(M\{T\}, Z_1).$$
But $\operatorname{Ext}^1_{\overline{K}[T,\tau]_0}(M\{T\}, Z_1)$ vanishes by Lemma 5.9.29. $\square$

Lemma 5.9.31. *$\overline{K}\{T\}$ is a principal ideal domain.*

Proof. This follows from the theory of Section 2. $\square$

A left $\overline{K}[T,\tau]_0$-module is *trivial* if and only if it is isomorphic to a finite direct sum of copies of Z_1.

Lemma 5.9.32. *Let M be a left $\overline{K}[T,\tau]_0$-module which is free of rank one over $\overline{K}\{T\}$ and has $\tau M = M$. Then M is trivial.*

Proof. Let m generate M over $\overline{K}\{T\}$ and let $f \in \overline{K}\{T\}$ satisfy
$$\tau m = fm;$$
thus f is a unit in $\overline{K}\{T\}$. To show that M is trivial we need an invertible $g \in \overline{K}\{T\}$ such that
$$\tau(gm) = gm.$$
But $\tau(gm) = g^{(1)}\tau m = g^{(1)}fm$. Thus we deduce that $g/g^{(1)} = f$. As f is invertible, the Newton Polygon tells us that we may assume f is of the form
$$1 + \sum a_i T^i, \quad |a_i| < 1 \text{ all } i, \quad a_i \to 0.$$

We, therefore, set $g := \prod_{j=0}^{\infty} f^{(j)}$. □

Lemma 5.9.33. *Every $\overline{K}[T,\tau]_0$-submodule M_1 of a trivial left $\overline{K}[T,\tau]_0$-module M is trivial.*

Proof. We proceed by induction on n. If $n = 1$, then the result is Lemma 5.9.32. Let us, therefore, assume that $n > 1$. As M is trivial, by hypothesis there exists an exact sequence

$$0 \to M_2 \xrightarrow{i} M \xrightarrow{\pi} Z_1 \to 0.$$

Thus there is an induced sequence

$$0 \to i^{-1}(M_1) \xrightarrow{i} M_1 \xrightarrow{\pi} \pi(M_1) \to 0.$$

By induction both ends are trivial. As

$$\mathrm{Ext}^1_{\overline{K}[T,\tau]_0}(Z_1, Z_1) = 0,$$

the result follows. □

Our next result is a sufficient condition for the triviality of a $\overline{K}[T,\tau]_0$-module.

Lemma 5.9.34. *Let M be a left $\overline{K}[T,\tau]_0$-module of rank n over $\overline{K}\{T\}$. Suppose that there exists $f_1, \ldots, f_n \in \mathrm{Hom}_{\overline{K}[T,\tau]_0}(M, Z_1)$ which are linearly independent over $\mathbf{A}$. Then M is trivial.*

Proof. Let

$$f = \begin{pmatrix} f_1 \\ \vdots \\ f_n \end{pmatrix} : M \to Z_1^n.$$

Then $f(M)$ is trivial by Lemma 5.9.33; thus

$$f(M) \simeq Z_1^m$$

for some $m \leq n$.

Let $e_1, \ldots, e_n$ be the coordinate projections, $Z_1^n \to Z_1$, and let e'_i be their restrictions to $f(M)$. By hypothesis the set $\{e'_i\}$ is linearly independent over $\mathbf{A}$. Now one can readily see that $\mathrm{End}_{\overline{K}[T,\tau]_0}(Z_1) = \mathbf{A}$; thus we deduce that $m = n$. As $\overline{K}\{T\}$ is a principal ideal domain, we deduce that f is *injective*. The result now follows from the previous lemma. □

We can now conclude our proof of the main result.

Proof of 5.9.14.1 ⇒ 5.9.14.3. We have seen that

$$H_1(E) = \mathrm{Hom}_{\overline{K}[T,\tau]}(M(E), Z_1) = \mathrm{Hom}_{\overline{K}[T,\tau]_0}(M\{T\}, Z_1).$$

Now by hypothesis $\mathrm{rank}_A H_1(E) = r(E)$. Thus Lemma 5.9.34 implies that $M\{T\}$ is trivial as a $\overline{K}[T,\tau]_0$-module, which, in turn, is readily seen to imply that M is rigid analytically trivial. □

Anderson's result has a number of important corollaries.

Corollary 5.9.35. *Let E be a uniformizable T-module. Then $H_1(E)$ is canonically isomorphic to*

$$\mathrm{Hom}_{\mathbf{A}}(H^1(M(E)), \mathbf{A}\, dT) = \mathrm{Hom}_{\mathbf{A}}(H^1(M(E)), \Omega).$$

Proof. We know that

$$H_1(E) = \mathrm{Hom}_{\overline{K}[T,\tau]_0}(M(E)\{T\}, Z_1).$$

As $M(E)$ is rigid analytically trivial,

$$H_1(E) = \mathrm{Hom}_{\mathbf{A}}(H^1(M(E)), Z_1^\tau).$$

But Z_1^τ is the group of continuous morphisms $\mathbf{K}$ to $\mathbb{F}_r$. In turn, the group of continuous morphisms $\mathbf{K} \to \mathbb{F}_r$ is isomorphic to Ω under the pairing

$$(x, \omega) \mapsto \mathrm{Res}_{T=\infty}(x\omega) \colon \mathbf{K} \times \Omega \to \mathbb{F}_r.$$
□

Example 5.9.36. Let E be the Carlitz module over k; so $T \mapsto \theta\tau^0 + \tau$ as an endomorphism of $\mathbb{G}_a$. Let M be the motive of E over $\overline{K}$. So $M \simeq \overline{K}\{\tau\}$ and is generated by $m_0 = \mathrm{id}$ with

$$\tau m_0 = (T - \theta)m_0.$$

By definition $H^1(M) = M\{T\}^\tau$, and, by the corollary $H^1(M)$ has rank 1 over $\mathbf{A}$ as the Carlitz module is a rank one T-module. We will show here how this fact may also be seen directly. So we need to solve the *Frobenius equation*

$$\tau m = m.$$

If we let $m = m_0 \otimes \sum a_i T^i$, then

$$\tau m = (T - \theta)m_0 \otimes \sum a_i^r T^i.$$

We are, therefore, reduced to solving the equation

$$(T - \theta)\sum a_i^r T^i = \sum a_i T^i. \qquad (*)$$

Using the theory of Section 2, one may find a *particular* solution Ξ to $*$ with no zeros in the disc $|T| < |\theta|$. Clearly $f\Xi$ is also a solution for $f \in \mathbf{A}$. Conversely, let $h(T)$ be any solution to $*$; then $g(T) = \frac{h(T)}{\Xi}$ is analytic on the disc $|T| \leq 1$, because $|\theta| > 1$, and satisfies

$$g^{(1)}(T) = g(T).$$

Thus $g(T)$ has coefficients in $\mathbb{F}_r$; but these coefficients also tend to 0 forcing $g(T)$ to be in $\mathbf{A}$.

Corollary 5.9.37. *The functor $E \mapsto H_1(E)$ of uniformizable abelian T-modules is faithful.*

Proof. Let M be a rigid analytically trivial T-motive. Then the functor $M \mapsto H^1(M)$ is easily seen to be faithful. Thus the result follows from the previous corollary. □

Corollary 5.9.38 *Let E_1 and E_2 be two uniformizable abelian T-modules. Then $E_1 \otimes E_2$ is also a uniformizable abelian T-module.*

Proof. Use Part 3 of Theorem 5.9.14. □

Let L be an $\mathbf{A}$-field containing k and let E be a abelian T-module over L. Let $\sigma \colon L \to \overline{K}$ be a k-embedding. For each such σ it makes sense to inquire as to whether $\sigma(E)$ is uniformizable. The actual dependence of uniformizability on σ is an important problem whose solution would be very interesting.

5.10. The Tensor Powers of the Carlitz Module redux

Recall that in Subsection 5.8, we discussed the nth tensor power $C^{\otimes n}$ of the Carlitz module. By Anderson's result (Theorem 5.9.14) and its corollaries, we now know that $C^{\otimes n}$ is uniformizable. In this subsection, we discuss the exponential and logarithm of $C^{\otimes n}$ as in [AT1]. Note also that [AT1] contains an elementary approach to the existence of these functions and their properties. The reader should look there for additional details and properties.

Recall further that in Definition 3.1.4, we introduced the elements L_i, D_i of $\mathbf{k}$. However, as we have seen, the analytic theory of T-modules is naturally defined over the fields $k = \mathbb{F}_r(\theta)$, $K = \mathbb{F}_r\left(\left(\frac{1}{\theta}\right)\right)$ and $\overline{K}$. Thus we will review Definition 3.1.4 here, but present the elements as lying in $A = \mathbb{F}_r[\theta]$, etc., by applying the homomorphism $\imath$.

Definition 5.10.1. 1. Let $i > 0$. We set $[i] := \theta^{r^i} - \theta \in \mathbb{F}_r[\theta]$.
2. We set $D_0 := 1 \in \mathbb{F}_r[\theta]$ and for $i > 0$,

5.10. The Tensor Powers of the Carlitz Module redux

$$D_i := [i][i-1]^r \cdots [1]^{r^{i-1}}$$
$$= (\theta^{r^i} - \theta)(\theta^{\sigma^i} - \theta^r) \cdots (\theta^{r^i} - \theta^{r^{i-1}}) \in A.$$

3. We set $L_0 := 1$, and for $i > 0$

$$L_i := [i][i-1] \cdots [1]$$
$$= (\theta^{r^i} - \theta)(\theta^{r^{i-1}} - \theta) \cdots (\theta^r - \theta) \in A.$$

Thus D_i is the product of all monics in A of degree i and L_i is the least common multiple of all monics of A of degree i.

The exponential, $\exp_n$, of $C^{\otimes n}$ is then found in the following fashion: one writes formally

$$\exp_n(\tau) = \sum_{i=0}^{\infty} Q_i \tau^i,$$

where $Q_i \in M_n(\overline{K})$ all i and $Q_0 = \mathrm{id}$. The matrices Q_i are then uniquely determined by

$$\exp_n(\tau) \circ (\theta + N_n)\tau^0 = ((\theta + N_n)\tau^0 + V_n \tau) \circ \exp_n(\tau);$$

one finds that all Q_i have coefficients in $k = \mathbb{F}_r(\theta)$. By general theory, of course, one knows that $\exp_n$ is entire and, when $n = 1$, we simply recover the exponential of the Carlitz module.

The logarithm, $\log_n(\tau)$, is defined to be the formal inverse of $\exp_n(\tau)$. It can be determined by writing

$$\log_n(\tau) = \sum_{i=0}^{\infty} P_i \tau^i,$$

where $P_i \in M_n(\overline{K})$ and $P_0 = \mathrm{id}$ and then using

$$(\theta + N_n)\tau^0 \circ \log_n(\tau) = \log_n(\tau) \circ ((\theta + N_n)\tau^0 + V_n \tau).$$

In [AT1,2.4.3] it is shown that $\log_n(x)$ converges for all $x = \begin{pmatrix} x_1 \\ \vdots \\ x_n \end{pmatrix}$ with

$$|x_i| < |\theta|^{i-n+(nr/(r-1))}, \text{ all } i.$$

Thus in this range $\log_n(x)$ converges to the inverse to $\exp_n(x)$, i.e.,

$$\exp_n(\log_n(x)) = x.$$

We pause, for a moment, to recall some facts from classical complex analysis. Through the use of the geometric series, one deduces

$$-\log(1-x) = \sum_{k=0}^{\infty} \frac{x^k}{k}.$$

The n-fold Hadamard product of the above power series,

$$\sum \frac{x^k}{k^n},$$

is the nth *multilogarithm function*.

Returning to characteristic p, we know that

$$\sum_{j=0}^{\infty} \frac{(-1)^j x^{r^j}}{L_j}$$

is the logarithm of the Carlitz module. It's n-fold Hadamard product is the nth *Carlitz multilogarithm* and equals

$$\sum_{j=0}^{\infty} \frac{(-1)^{jn} x^{r^j}}{L_j^n}.$$

The connection with $\log_n(x)$ is given by the formula [AT1]

$$\log_n \begin{pmatrix} 0 \\ \vdots \\ 0 \\ x_n \end{pmatrix} = \begin{pmatrix} \vdots \\ \vdots \\ \sum_{j=0}^{\infty} \frac{(-1)^{jn} x_n^{r^j}}{L_j^n} \end{pmatrix}.$$

Similarly, one can define the nth *Carlitz multiexponential* as the nth Hadamard convolution of the Carlitz exponential. Anderson and Thakur [AT1] then find

$$\exp_n \begin{pmatrix} x_1 \\ 0 \\ \vdots \\ 0 \end{pmatrix} = \begin{pmatrix} \sum_{i=0}^{\infty} \frac{x^{r^i}}{D_i^n} \\ \vdots \\ \end{pmatrix}.$$

There are further, rather remarkable, connections with Thakur's new theory of *hypergeometric functions* in function fields, [Th2].

Finally, we know that the rank of $C^{\otimes n}$ is 1. Thus the lattice $H_1(C^{\otimes n})$ is 1-dimensional over $\mathbf{A}$. It is shown in [AT1] that $H_1(C^{\otimes n})$ is generated by a vector Ξ with last coordinate equal to ξ^n, where ξ is the period of the Carlitz module.

5.11. Scattering Matrices

There are a few more basic results on T-modules and T-motives that we will mention in this last subsection. These results involve the technique of *scattering matrices*. As these techniques are similar to those of Subsection 5.9, we shall be very brief and refer the reader to [A1, §3] for details.

We retain the notation of Subsection 5.10.

Definition 5.11.1. Let $f(T) = \sum_{j=-\infty}^{\infty} a_j T^j$ be a formal Laurent series with coefficients in $\overline{K}$. As before, we set

$$f^{(n)}(T) := \sum_{j=-\infty}^{\infty} a_j^{r^n} T^j.$$

If $N = (f_{ij})$ is a matrix of such functions, we set

$$N^{(n)} := (f_{ij}^{(n)}).$$

Let α, β be two rational numbers with $\alpha \leq \beta$.

Definition 5.11.2. Let L be a subfield of $\overline{K}$.
1. We let $L\{T/\theta^\beta, \theta^\alpha/T\}$ be the ring of Laurent series $f(T) = \sum_{i=-\infty}^{\infty} a_i T^i$ with $\{a_i\} \subset L_1 \subseteq \overline{K}$ such that $[L_1: K] < \infty$ and such that $f(T)$ is convergent for $|\theta|^\alpha \leq |T| \leq |\theta|^\beta$.
2. We let $L\{T/\theta^\beta\}$ denote the subring of $L\{T/\theta^\beta, \theta^\alpha/T\}$ consisting of those $f(T)$ (as above) with no negative terms. Similarly $L\{\theta^\alpha/T\}$ will be those $f(T)$ with no positive terms.
3. Put $\overline{K}\{T\}^+ := \bigcup_{\beta>0} \overline{K}\{T/\theta^\beta\}$.

Thus $\overline{K}\{T\}^+$ is the ring of all power series $f(T)$ in T with coefficients in $\overline{K}$ such that the coefficients of $f(T)$ generate a finite extension of K and such that $f(T)$ has non-trivial radius of convergence.

Definition 5.11.3. Let n, N be positive integers. A *scattering matrix* Ψ of rank n and index N is an element of $GL_n(\overline{K}\{T\}^+)$ with the following properties:
1. $\Psi^{(s)}\Psi^{-1} \in M_n(\overline{K}[T])$ for all positive integers s.
2. $\det(\Psi^{(1)}\Psi^{-1}) = \lambda(T-\theta)^N$ for some $0 \neq \lambda \in \overline{K}$.
3. There exists $U \in GL_n(\overline{K}((1/T)))$ and positive integers r and s such that

$$U^{(s)}(\Psi^{(s)}\Psi^{-1})T^{-r}U^{-1} \in GL_n(\overline{K}[[1/T]]).$$

For such a Ψ we set $r(\Psi) := n$, $\rho(\Psi) := N$ and $w(\Psi) := N/n$.

The set of scattering matrices is made into a category by defining

$$\mathrm{Hom}(\Psi_1, \Psi_2) := M_{n_2, n_1}(\mathbb{F}_r[T]) \cap \Psi_2^{-1} M_{n_2, n_1}(\overline{K}[T]) \Psi_1 ,$$

where $n_1 = r(\Psi_1)$ and $n_2 = r(\Psi_2)$. Composition of morphisms is then given by matrix multiplication.

The importance of scattering matrices lies in the fact that they give a parameterization of the category (suitably modified) of pure, uniformizable abelian T-modules. This we now describe.

Let R be any ring and set $R^n = \{n \times 1 \text{ column vectors}\}$ as usual. Thus $(R^n)^t$ is the set of $1 \times n$ row vectors over R.

Definition 5.11.4. An *abelian T-module with σ-structure* is a pure, nonzero, uniformizable abelian T-module E together with:
1. A $\overline{K}[T]$-linear isomorphism $\sigma_E^1 \colon M(E) \xrightarrow{\sim} (\overline{K}[T]^d)^t$ where $d = r(E)$.
2. A $\mathbb{F}_r[T]$-linear isomorphism $\sigma_E^2 \colon H_1(E) \xrightarrow{\sim} \mathbb{F}_r[T]^d$, $d = r(E)$.

A *morphism of abelian T-modules with σ-structure* is just a morphism of the underlying T-modules.

The connection between T-modules and scattering matrices arises from the following construction. Let E be an abelian T-module with σ-structure and let $m \in M(E)$ and $v \in H_1(E)$. Set

$$g(m, v, T) := - \sum_{j=0}^{\infty} m(\exp_E(T^{-(j+1)} v)) T^j .$$

One checks that $g(m, v, T)$ is convergent for all $|T| < |\theta|$, and is $\mathbb{F}_r[T]$-linear in v and $\overline{K}[T]$-linear in m. Thus there exists a unique matrix $\Psi(E)$ with

$$g(m, v, T) = \sigma_E^1(m) \Psi(E) \sigma_E^2(v) .$$

Lemma 5.11.5. $\Psi(E)$ *is a scattering matrix of rank $r(E)$ and index $\rho(E)$.* □

We then have the basic result.

Theorem 5.11.6. *The functor $E \mapsto \Psi(E)$ is an equivalence of the categories of abelian T-modules with σ-structure and scattering matrices.* □

Using the theory of scattering matrices, Anderson shows the following remarkable result [A1, Cor. 3.3.6].

Theorem 5.11.7. *Let E be a pure, uniformizable abelian T-module. Then $H_1(E)$ generates $\mathrm{Lie}(E)$ as a $\overline{K}[T]$-module.* □

Remark. 5.11.8. Let E be a pure, uniformizable T-module over $\overline{K}$ of the form

$$T \mapsto (T\tau^0 + N)\tau^0 + \{\text{higher terms}\}.$$

What is so remarkable about the above result is the connection it establishes between $\text{rank}_{\mathbf{A}} H_1(E)$ and the nilpotent matrix N. For instance, if $\text{rank}_{\mathbf{A}} H_1(E) = 1$, then $(TI + N)$ must now be equivalent to a Jordan block (i.e., the minimal polynomial of N is u^d, $d = \text{dimension } E$). As an example, one can look at the nth tensor power of the Carlitz module.

Another important technique involving scattering matrices is contained in the next result. Let

$$\overline{K}\langle T\rangle := \left(\bigcup_{\alpha > 0} \overline{K}\{\theta^\alpha/T\}\right)[T].$$

Theorem 5.11.9. *Let Ψ be a scattering matrix of rank d and*

$$W \in GL_d\left(\mathbb{F}_r\left(\left(\frac{1}{T}\right)\right)\right).$$

Then there exists $W_1 \in GL_d(\overline{K}\langle T\rangle)$ such that $W_1^{-1}\Psi W$ is again a scattering matrix. □

Theorem 5.11.9 then has the following important corollary.

Corollary 5.11.10. *Let E be a pure uniformizable T-module and let $\Gamma \subset \text{Lie}(E)$ be a cocompact discrete $\mathbf{A}$-submodule of the $\mathbb{F}_r\left(\left(\frac{1}{T}\right)\right)$-span of $H_1(E)$ in $\text{Lie}(E)$. Then there exists a pure uniformizable abelian T-module E_1 and a $\overline{K}[T]$-linear isomorphism $f\colon \text{Lie}(E_1) \xrightarrow{\sim} \text{Lie}(E)$ such that*

$$f(H_1(E_1)) = \Gamma.$$ □

6. Shtukas

In this section we will introduce "shtukas" which are also called "F-sheaves" or "FH-sheaves." Let $\mathbf{A}$ and $\mathbf{k}$ be defined as in Subsection 4.1; so $\mathbf{k}$ is a global field over the finite field $\mathbb{F}_r$ and $\mathbf{A}$ is the subring of functions regular away from a fixed place ∞. As in Sections 4 and 5, we have seen that Drinfeld modules and T-modules correspond to representing $\mathbf{A}$ as a ring of operators on $\mathbb{G}_a^d$ for some d. The notion of a shtuka, then corresponds to a proper model of this action, i.e., the shtukas will be certain locally free sheaves on the *complete* curve X corresponding to $\mathbf{k}$ (or X base changed to an overfield of $\mathbb{F}_r$). One can then study shtukas through powerful projective methods.

The shtukas are due to Drinfeld [Dr3], [Dr4]. Our exposition will follow that of Mumford [Mu2]. However, we will begin by reviewing some relevant algebraic geometry as in [Har1]. For any concepts not covered in 6.1, the reader should refer there.

6.1. Review of Some Algebraic Geometry

Definition 6.1.1. 1. A *graded ring* is a commutative ring R together with a decomposition $R = \bigoplus_{d=0}^{\infty} R_d$ where R_d is an abelian group for all d and such that for $d, e \geq 0$
$$R_d \cdot R_e \subseteq R_{d+e}.$$
The elements of R_d are *homogeneous of degree d*.
2. An ideal $I \subseteq R$ is *homogeneous* if and only if
$$I = \bigoplus_{d \geq 0}(I \cap R_d).$$

It is easy to see that an ideal is homogeneous if and only if it can be generated by homogeneous elements. Note also that $R_0 \subseteq R$ is a subring.

Let R be graded as above and let R_+ be the ideal $\bigoplus_{d>0} R_d$. As is standard, one defines $\mathrm{Proj}(R)$ to be the set of all homogeneous prime ideals of R *not* containing all of R_+. The set $\mathrm{Proj}(R)$ is given the structure of a locally ringed space as a scheme. This ringed structure may be visualized as follows. Let $f \in R_+$ be a homogeneous element. One sets

$$D_+(f) := \{\wp \in \mathrm{Proj}(R) \mid f \notin \wp\};$$

the subset $D_+(f)$ is then affine open and isomorphic to $\mathrm{Spec}(R_{(f)})$ where $R_{(f)}$ is the set of elements of degree 0 in the localization R_f. Moreover, $\mathrm{Proj}(R)$ is covered by the collection of affines $D_+(f)$ for $f \in R_+$.

Example 6.1.2. Let A be a fixed commutative ring (with unit) and set $R = A[x_0, \ldots, x_n]$ with the standard gradation of homogeneous polynomials of degree d. Then $\mathrm{Proj}(R)$ is just projective n-space over A.

Let L be a field. There is another way to obtain projective n-space over L from $L[x_1, \ldots, x_n]$ (n-variables) which is relevant to shtukas, and which we now discuss. Let $f \in L[x_1, \ldots, x_n]$ be a polynomial of *total degree* d. Recall that this means that f can be written $f = f_d + \sum_{i=0}^{d-1} f_i$ where $f_d \neq 0$ is homogeneous of degree d and the elements f_i are homogeneous (and possibly trivial) of degree $i < d$. We give the constant polynomial 0 total degree $-\infty$. It is not true that two elements of total degree d need add up to an element of total degree d, as is trivially seen. However, the elements of degree $\leq d$ do indeed form an abelian group. Thus let

$$R := \bigoplus_{d=0}^{\infty} R_d$$

where

$$R_d := \{f \in L[x_1, \ldots, x_n] \mid \text{total degree } f \leq d\}.$$

It is easy to see that R is a graded ring.

Example 6.1.3. Let $n = 1$. Then $R_0 \simeq L$, $R_1 = \{L + Lx_1\}$, and $R_d = \{\text{polynomials } P(x_1) \mid \deg P \leq d\}$. Let $e \in R_1$ correspond to 1. Multiplication by e is the operation of taking a polynomial of degree $\leq d$ and considering it having degree $\leq d + 1$. Let x correspond to $x_1 \in R_1$. Then

$$R = L[e, x] \simeq L[x_0, x_1]$$

with the usual gradation.

As in the example, for arbitrary n, one sees $R \simeq L[x_0, \ldots, x_n]$. Thus $\mathrm{Proj}(R)$ is isomorphic to $\mathbb{P}_L^n$.

In the above examples, R is always generated by R_1 over R_0. This need not be true in general. But, as we shall see presently, the general case is not far from our examples. Let v be a positive integer and set

$$R^{(v)} := \bigoplus_{d=0}^{\infty} R_{dv};$$

note that $R^{(v)}$ is a graded subring of R. In [Bo1] the following result is established.

Proposition 6.1.4. *Let R be as above and assume that R is finitely generated over R_0. Then there exists $v \geq 1$ such that $R^{(mv)}$ is generated by R_{mv} over R_0 for all positive integers m.* □

It is easy to see that for any v, $\mathrm{Proj}(R^{(v)}) \simeq \mathrm{Proj}(R)$. Thus, in many situations, Proposition 6.1.4 allows us to reduce to looking at graded rings generated by R_1 over R_0.

In the applications of Proposition 6.1.4 of interest to us, R_0 will be a field and $\dim_{R_0} R_d \leq \infty$ for all d. Thus if v is chosen as in 6.1.4, we deduce that $R^{(v)}$ is finitely generated over R_0.

Finally, we need to present some basic results of Serre on coherent $\mathcal{O}_X$-modules with $X = \mathrm{Proj}(R)$. From the above discussion, we can, and will, assume that R is finitely generated by R_1 over R_0.

Definition 6.1.5. 1. A *graded R-module* is an R-module M together with a decomposition

$$M = \bigoplus_{d \in \mathbb{Z}} M_d$$

such that $R_e \cdot M_d \subseteq M_{d+e}$.
2. Let $n \in \mathbb{Z}$. We then define the *twisted module* $M(n)$ by $M(n)_d := M_{d+n}$ for all d.

Given a graded module M, one defines a quasi-coherent $\mathcal{O}_X$-module $\widetilde{M}$ on $X = \mathrm{Proj}(R)$. This can be described as follows. Let $f \in R_+$ be a homogeneous element and let $D_+(f)$ be the affine subscheme of X as before. Over $D_+(f)$, $\widetilde{M}$ corresponds to the $R_{(f)}$-module $M_{(f)}$ consisting of elements of degree 0 in the localization M_f of M.

Definition 6.1.6. We define $\mathcal{O}_X(n)$ to be the coherent $\mathcal{O}_X$-module corresponding to $R(n)$.

Thus $\mathcal{O}_X(1)$ is the familiar twisting invertible sheaf of Serre (remember R is generated by R_1 over R_0!) with the usual properties such as

$$\mathcal{O}_X(m) \otimes \mathcal{O}_X(n) \simeq \mathcal{O}_X(m+n).$$

See [Har1, §II.5] for more.

It turns out that every quasi-coherent sheaf of $\mathcal{O}_X$-modules may be obtained from a graded R-module in the above fashion (and still assuming that $R_0[R_1] = R$). This is established as follows.

Definition 6.1.7. 1. Let $\mathcal{F}$ be a quasi-coherent sheaf of $\mathcal{O}_X$-modules. We set
$$\mathcal{F}(n) := \mathcal{F} \otimes \mathcal{O}_X(n).$$

2. We set
$$\Gamma_*(\mathcal{F}) := \bigoplus_{n \in \mathbb{Z}} H^0(X, \mathcal{F}(n)),$$
where $H^0(X, \mathcal{F}(n))$ is the space of global sections of $\mathcal{F}(n)$.

We can make $\Gamma_*(\mathcal{F})$ into a graded R-module in the following fashion. An element $r \in R_d$ gives rise to a global section of $\mathcal{O}_X(d)$. Let $\gamma \in H^0(X, \mathcal{F}(n))$. Then "$r \cdot \gamma$" is that element in $H^0(X, \mathcal{F}(n+d))$ coming from $r \otimes \gamma$ under the natural isomorphism
$$\mathcal{F}(n) \otimes \mathcal{O}_X(d) \simeq \mathcal{F}(n+d).$$

Finally, under the assumption that R is finitely generated by R_1 over R_0 one concludes [Har1, Prop. 5.15] that
$$\widetilde{\Gamma_*(\mathcal{F})} \xrightarrow{\sim} \mathcal{F}.$$

6.2. The Shtuka Correspondence

In this subsection we present the basic and elegant dictionary of shtukas giving an equivalence between two very distinct types of data. The first type of data concerns certain subrings of $L\{\tau\}$ and the second concerns certain types of coherent sheaves on complete curves.

As there is little added difficulty in doing so, we will actually present the dictionary in a more general setting. We let L be any field (of arbitrary characteristic) and let $\sigma \in \mathrm{Aut}(L)$ be an automorphism of infinite order with L_0 being the fixed field. (So, of course, the basic example will be where L is a perfect infinite extension of $\mathbb{F}_r = L_0$ and $\sigma = \tau =$ the rth power map.)

We then have two types of data.

Data A. 1. Let X be a reduced and irreducible complete curve over L_0, and set
$$\overline{X} := L \otimes_{L_0} X.$$
We assume that $\overline{X}$ is also irreducible.

2. Let $P_0 \in X$ be a regular closed point and set
$$P := L \otimes_{L_0} P_0 \subset \overline{X}.$$

3. A torsion-free coherent $\mathcal{O}_{\overline{X}}$-module $\mathcal{F}$ on $\overline{X}$ such that
$$h^0(\mathcal{F}) = h^1(\mathcal{F}) = 0$$
(where $h^i(\mathcal{F}) := \dim_L H^i(\overline{X}, \mathcal{F})$).

4. A maximal flag of coherent sub-$\mathcal{O}_{\overline{X}}$-modules

$$\mathcal{F} = \mathcal{F}_0 \supset \mathcal{F}_{-1} \supset \cdots \supset \mathcal{F}_{-t} = \mathcal{F}(-P)$$

with length $(\mathcal{F}_{k+1}/\mathcal{F}_k) = 1$.

5. An $\mathcal{O}_{\overline{X}}$-module homomorphism

$$\alpha \colon \tilde{\sigma}^*\mathcal{F}/\overline{X} - P \to \mathcal{F}/\overline{X} - P$$

where $\tilde{\sigma}$ is *fixed* to be the morphism

$$\operatorname{Spec}(\sigma) \times 1_{\overline{X}} \colon \overline{X} \to \overline{X}.$$

Moreover, α should not be surjective on $\overline{X} - P$ and on $\overline{X}$, α takes $\tilde{\sigma}^*\mathcal{F}_k$ to $\mathcal{F}_{k+1}$.

Data B. A commutative subring $R \subset L\{\sigma\}$ (with the obvious definition) such that
1. L_0 is strictly contained in R as $L_0 \cdot \sigma^0$.
2. $R \cap L = L_0 \cdot \sigma^0$.

Two such rings R_1 and R_2 are said to be *equivalent* if and only if there exists $\alpha \in L^*$ with

$$R_1 = \alpha R_2 \alpha^{-1}.$$

Remarks. 6.2.1. 1. In the case where $\sigma = \tau$ and L contains $\mathbb{F}_r$, we have seen that examples of Data B are given by the Drinfeld module realization of the ring **A** as a ring of operators. In the present case, the rings R may *not* be regular and so more general affine rings may occur. Also, as will be apparent, the point P_0 will correspond to ∞ in the Drinfeld module case.
2. Let $x \in \overline{X}(L)$ be an L-point. Then $\tilde{\sigma}(x)$ has the effect of applying σ^{-1} to the coordinates of x.
3. Note the similarity between Data A.5 and the notion of φ-sheaf as a Subsection 5.3.
4. In our next section we will use the notation "$\mathcal{F}^{(1)}$" for "$\tilde{\sigma}^*\mathcal{F}$" etc.

Proposition 6.2.2. *Let $\overline{X}$ be as in Data A.1. Then $\overline{X}$ is automatically an integral scheme.*

Proof. Integral is equivalent to reduced and irreducible. Thus we need to see that $\overline{X}$ is reduced. This will follow once we know that L is separable over L_0. But suppose that there is a sequence $\alpha_1, \ldots, \alpha_n \in L_0^{1/p}$ which are linearly independent over L_0 but dependent over L with n minimal. Without loss, we can write

$$\alpha_1 + a_2\alpha_2 + \cdots + a_n\alpha_n = 0 \qquad \{a_i\} \subset L,$$

and such that not all $a_i = 0$. We deduce that

$$\alpha_1 + \sigma(a_2)\alpha_2 + \cdots + \sigma(a_n)\alpha_n = 0$$

also, and that

$$(a_2 - \sigma(a_2))\alpha_2 + \cdots + (a_n - \sigma(a_n))\alpha_n = 0.$$

As n is minimal, we conclude that $a_i = \sigma(a_i)$ for all i. Thus $a_i \in L_0$ for all i which gives a contradiction. □

We now begin establishing the equivalence of Data A and Data B by proving a basic observation of Drinfeld on the sheaf $\mathcal{F}$ of Data A.

Proposition 6.2.3. *Let $\mathcal{F}$ be the sheaf of Data A except that we assume instead (the a-priori weaker statement) that*

$$\chi(\mathcal{F}) = h^0(\mathcal{F}) - h^1(\mathcal{F}) = 0.$$

Then we conclude that $h^0(\mathcal{F}) = h^1(\mathcal{F}) = 0$ (so that $\mathcal{F}$ perfectly fits Data A).

Proof. We begin by extending the definition of $\mathcal{F}_j$, $j = 0, \ldots, -t$, to all j by requiring

$$\mathcal{F}_{j+t} := \mathcal{F}_j(P).$$

From general properties of coherent cohomology, one sees that

$$\chi(\mathcal{F}_n) = n.$$

As $\mathcal{F}_n \subset \mathcal{F}_{n+1}$, one finds that

$$h^0(\mathcal{F}_n) \leq h^0(\mathcal{F}_{n+1})$$

and

$$h^1(\mathcal{F}_n) \geq h^1(\mathcal{F}_{n+1}).$$

Let n_0 be the smallest n such that $h^0(\mathcal{F}_n) \neq 0$, and let $0 \neq s_0 \in H^0(\overline{X}, \mathcal{F}_{n_0})$. One must have $n_0 \leq 1$ since

$$h^0(\mathcal{F}_1) \geq h^0(\mathcal{F}_1) - h^1(\mathcal{F}_1) = \chi(\mathcal{F}_1) = 1.$$

Now consider the maps

$$\alpha: \tilde{\sigma}^* \mathcal{F}_n \to \mathcal{F}_{n+1},$$

and define inductively sections $s_n \in H^0(\overline{X}, \mathcal{F}_{n+n_0})$ by

$$s_n := \alpha(\tilde{\sigma}^* s_{n-1}).$$

Since $\chi(\mathcal{F}_{n+1}) = \chi(\tilde{\sigma}^* \mathcal{F}_n) + 1$, we see that the length of the cokernel of α must be 1. As α is assumed *not* to be surjective on $\overline{X} - P$, we conclude that it *must* be surjective at P. From this, one sees readily (by thinking locally) that

$$\mathcal{F}_{n+1} = \mathcal{F}_n + \alpha(\tilde{\sigma}^* \mathcal{F}_n).$$

Moreover, it also follows that in $\mathcal{F}_{n+1}$,
$$\alpha(\tilde{\sigma}^*\mathcal{F}_{n-1}) = \mathcal{F}_n \cap \alpha(\tilde{\sigma}^*\mathcal{F}_n).$$
Thus the sequence of sections s_n satisfies
$$s_n \in H^0(\overline{X}, \mathcal{F}_{n+n_0-1}) \Rightarrow \alpha(\tilde{\sigma}^*s_{n-1}) \in H^0(\overline{X}, \mathcal{F}_{n+n_0-1} \cap \alpha\tilde{\sigma}^*(\mathcal{F}_{n+n_0-1}))$$
$$\Rightarrow s_{n-1} \in H^0(\overline{X}, \mathcal{F}_{n_0+n-2}).$$

But, as $s_0 \notin H^0(\overline{X}, \mathcal{F}_{n_0-1})$ we conclude that for all n, $s_n \notin H^0(\overline{X}, \mathcal{F}_{n_0+n-1})$ and so the sections $\{s_0, \ldots, s_n\}$ are linearly independent. Thus
$$n + 1 \leq h^0(\mathcal{F}_{n+n_0})$$
$$= \chi(\mathcal{F}_{n+n_0}) \quad \text{if } n \gg 0$$
$$= n + n_0.$$

Thus $n_0 \geq 1$ and so $n_0 = 1$. By definition, we see that $h^0(\mathcal{F}_0) = 0 = h^1(\mathcal{F}_0)$, and the result is established. $\square$

Data A $\Rightarrow$ Data B. We can now explain how to proceed from Data A to Data B using the proof of Proposition 6.2.3. Indeed, as in that proof, we have elements $s_n \in H^0(\overline{X}, \mathcal{F}_{n+1})$ (remember $n_0 = 1$) for all n. One sees that $h^1(\mathcal{F}_n) = 0$ for $n \geq 0$ as
$$h^1(\mathcal{F}_n) \leq h^1(\mathcal{F}_0) = 0.$$
Therefore, we conclude that $\{s_n\}_{n \geq 0}$ is a basis for
$$H^0(\overline{X} - P, \mathcal{F}).$$
Now let R be the affine ring of $X - P_0$; clearly R acts on $H^0(\overline{X} - P, \mathcal{F})$ by multiplication. Thus if $f \in R$, then
$$f \cdot s_0 = \sum_{n=0}^{N} a_n(f) \cdot s_n$$
for suitable elements $\{a_n(f)\} \subset L$. Let $\psi \colon R \to L\{\sigma\}$ by
$$f \mapsto \sum_{n=0}^{N} a_n(f)\sigma^n.$$
It is easy to see that ψ is injective and L_0-linear. One then uses the definitions to see that it is multiplicative also. Thus we realize R as a subring of $L\{\sigma\}$. Different choices of s_0 give rise to conjugate embeddings. This completes the construction.

186   6. Shtukas

As we have seen with Drinfeld modules, once one realizes R as a ring of operators, the natural object of study is the division points. Our next result explains how these division points can be interpreted in terms of Data A.

Let $R \subset L\{\sigma\}$ come from $(X, P_0, \{\mathcal{F}_i\}, \alpha)$ via ψ as above. Let S be an extension field of L equipped with an extension of σ; we will also denote this extension by "σ." Finally, let $x_0 \in X - P_0$ be a closed point corresponding to the maximal ideal $M := M_{x_0}$.

By construction as above, we see that

$$H^0(\overline{X} - P, \mathcal{F}) = \bigoplus_{n=0}^{\infty} L \cdot s_n.$$

Moreover, this module is free of rank 1 over $L\{\sigma\}$ with basis s_0. As the mapping $\widetilde{\sigma}^*$ on functions is R-linear, we see that the action of R on $H^0(\overline{X} - P, \mathcal{F})$ is realized via ψ as composition on $L\{\sigma\}$. That is, if $f \in R$, and $\gamma \in H^0(\overline{X} - P, \mathcal{F})$ corresponds to $P(\sigma) \in L\{\sigma\}$, then $f \cdot \gamma$ corresponds to $P(\psi(f)) \in L\{\sigma\}$. Thus $\mathcal{F}/M\mathcal{F}$ inherits a left $L\{\sigma\}$-module structure as left multiplication by σ and multiplication by f commute.

Proposition 6.2.4. *Let S be equipped with the left $L\{\sigma\}$-module structure given by evaluation. Then there is an isomorphism between the following spaces.*
1. *The L_0-vector space of all $s \in S$ with*

$$\psi(m)(s) = 0, \qquad \forall m \in M.$$

2. $\mathrm{Hom}_{L\{\sigma\}}(\mathcal{F}/M\mathcal{F}, S)$.

Proof. From the discussion just above, the proof is now reduced to tracing through definitions. □

Remarks. 6.2.5. 1. The reader should note the exact similarity between Proposition 6.2.4 and the description of torsion given in the proof of Proposition 5.6.3.
2. The main application of the above construction of interest to us occurs when $\mathbb{F}_r \subset L$ and $\sigma(x) = \tau(x) = x^r$. Strictly speaking, we must pass to a perfection of L in order for τ to be an automorphism. However, the reader will see that the construction of Data B from Data A may be performed over L itself.

Data B $\Rightarrow$ Data A. We now want to explain how to pass from Data B to Data A. The reader will note the similarity with the construction, given in the previous section, of T-motives; in fact both constructions begin with the same basic objects. Historically, however, the construction of shtukas preceded that of T-motives.

So let R be an instance of Data B. We now proceed much like in Example 6.1.3. Note first of all that there is a degree function, deg, on R given by

$$\deg\left(\sum_{i=0}^{n} \alpha_i \sigma^i, \quad \alpha_n \neq 0\right) = n.$$

Definition 6.2.6. We define the graded ring $\mathcal{R} := \bigoplus_{d=0}^{\infty} R_d$, by setting

$$R_d := \{r \in R \mid \deg(r) \leq d\}.$$

We set
$$X := \text{Proj}(\mathcal{R}).$$

Note that $R_0 \simeq L_0$.

Proposition 6.2.7. 1. R and $\mathcal{R}$ are integral domains.
2. R and $\mathcal{R}$ are finitely generated L_0-algebras.

Proof. As $L\{\sigma\}$ has no zero divisors, Part 1 is immediate.

For Part 2, let $t =$ g.c.d. of $\{\deg(r) \mid r \in R\}$. Let A_1 and A_2 have degree mt, and write
$$A_i = a_i \sigma^{mt} + \{\text{lower terms}\}.$$
Let $L((\sigma^{-1}))$ be the division ring of finite-tailed Laurent series in σ^{-1}. In $L((\sigma^{-1}))$, one can find an element α (in the quotient field of R) of degree t. Using α and the commutativity of R, we deduce that

$$\sigma^t(a_1/a_2) = a_1/a_2.$$

If we let $L_2 \subset L$ be the fixed field of σ^t, then L_2/L_0 is Galois with group $\simeq \mathbb{Z}/t\mathbb{Z}$. Let $L_1 \subseteq L_2$ be the subfield generated by the ratios a_1/a_2 as above; then there exist positive integers d, w such that

$$t = d \cdot w$$

and L_1/L_0 is Galois of degree w. Thus for all n

$$\dim_{L_0}(R_{(n+1)t}/R_{nt}) \leq w$$

with equality for $n \gg 0$.

Next, let $e \in R_1$ represent 1 and let A be any element of R with

$$\deg(A) = mt, \quad m \neq 0;$$

so, in particular, A is non-constant. One then sees that $\mathcal{R}$ is finite over $L_0[e, A]$. The result follows. $\square$

Let $\{e, A\}$ be as in the proof, so $\mathcal{R}$ is finite over $L_0[e, A]$.

Corollary 6.2.8. *X is an integral, proper curve over L_0.* □

It is not necessarily true that $\mathcal{R}$ is generated over $R_0 \simeq L_0$ by R_1. Let v be a positive integer and set, as before,

$$\mathcal{R}^{(v)} = \bigoplus_{n=0}^{\infty} R_{nv}.$$

General theory (Proposition 6.1.4) assures us that for some v, $\mathcal{R}^{(v)}$ is generated over R_0 by R_v. In our particular case, however, this can also be seen directly using 6.2.7.

Our two chosen elements give rise to subsets of X.

1. $D_+(e)$ ($=$ "$e \neq 0$") is affine and open in X.
It is isomorphic to
$$\mathrm{Spec}(\mathcal{R}_{(e)}) \simeq \mathrm{Spec}(R).$$

2. $D_+(A)$ ($=$ "$A \neq 0$") is affine and open.
It is isomorphic to
$$\mathrm{Spec}(\mathcal{R}_{(A)}) = \mathrm{Spec}(R_1)$$
where R_1 is the ring of fractions
$$\{C/A^k \mid \deg C \leq k \deg A\}.$$

3. $V(e)$ ($=$ "Cartier divisor $e = 0$ on X") is a closed subscheme of X. It is isomorphic to
$$\mathrm{Proj}(\mathcal{R}/e\mathcal{R}) = \mathrm{Proj}(G)$$
where $G = \bigoplus_{d=0}^{\infty} R_d/R_{d-1}$. Let $B = L\{\sigma\}$ and form B_d in the obvious fashion.
Thus
$$G \subseteq \bigoplus_{d=0}^{\infty} B_d/B_{d-1} \simeq B.$$

Now let L_1 be the subfield defined in the proof of 6.2.7. One then sees that

$$G \simeq \{\text{subring of finite codimension in } L_1\{\sigma^t\} \simeq L_1[u]\}.$$

Thus $V(e) \simeq \mathrm{Spec}(L_1)$. As L_1/L_0 is Galois, we see that $V(e)$ consists of one regular closed point with residue field L_1.

Finally, it is easy to see that L_0 is algebraically closed in R; thus $\overline{X} := X \otimes L$ is irreducible. By Proposition 6.2.2 it is also integral.

Remark. 6.2.9. There is a corollary of the above construction of importance for Drinfeld modules. So, in the notation of Section 4, let L be an **A**-field and let ψ be a Drinfeld module over L. Let ∞ have degree w and form the subfield

6.2. The Shtuka Correspondence 189

$L_1 \subset L$ of quotients as above. Then L_1 is an extension of $\mathbb{F}_r$ of degree w. In other words, we have discovered an a-priori restriction on the coefficients of a Drinfeld module. For another, slightly more elementary, approach see [Ha2, §6], or Subsection 7.2.

Having found the proper curves X and $\overline{X}$, we must now equip them with sheaves as in Data A. To find these, we set $M = L\{\sigma\}$ and view it as a module over $L \otimes_{L_0} R$ *exactly* as in Subsection 5.4.

Definition 6.2.10. 1. Let $n \geq 0$. We set

$$M_n := \{f \in M \mid \deg f \leq n\},$$

and $\mathcal{M} := \bigoplus_{n=0}^{\infty} M_n = \bigoplus_{n=0}^{\infty} \mathcal{M}_n$.

2. Let $j \in \mathbb{Z}$. We set $\mathcal{M}[j] := \bigoplus_{n=0}^{\infty} \mathcal{M}[j]_n$ where $\mathcal{M}[j]_n = M_{j+n}$.

The reader should note that we use the notation "[?]" instead of "(?)" *precisely* because $\mathcal{R}$ may not be generated by R_1 over R_0. On the other hand, one sees directly that all $\mathcal{M}[n]$ are finitely generated as $L \otimes \mathcal{R}$-modules.

Definition 6.2.11. We define the coherent $\mathcal{O}_{\overline{X}}$-modules on $\overline{X}$ by $\mathcal{F}_{n+1} := \widetilde{\mathcal{M}[n]}$.

Proposition 6.2.12. 1. *There is an injection $\mathcal{F}_{n+1} \hookrightarrow \mathcal{F}_{n+2}$ which is the identity on $\overline{X} - P$.*
2. *$\mathcal{F}_{n+2}/\mathcal{F}_{n+1}$ has length 1.*
3. *Let t be the g.c.d. of $\{\deg(x) \mid x \in R\}$. Then*

$$\mathcal{F}_{n-t} = \mathcal{F}_n(-P).$$

4. *$h^0(\mathcal{F}_0) = h^1(\mathcal{F}_0) = 0$. Thus $\mathcal{F}_0$ is torsion-free.*
5. *The sheaves $\mathcal{F}_n$ are locally free of rank t/w in a neighborhood of P (where P has w distinct points).*
6. *The map α is derived from the degree preserving map $\sigma: \mathcal{M}[n] \to \mathcal{M}[n+1]$.*

Proof. 1. The element $e \in \mathcal{R}$ gives a degree preserving injection

$$\mathcal{M}[n] \hookrightarrow \mathcal{M}[n+1];$$

thus an injection $\mathcal{F}_{n+1} \hookrightarrow \mathcal{F}_{n+2}$. On $\overline{X} - P = \{e \neq 0\}$, this reduces to the identity.

2. The cokernel of $e: \mathcal{M}[n] \to \mathcal{M}[n+1]$ is a graded module with graded pieces isomorphic to L. Thus $\mathcal{F}_{n+2}/\mathcal{F}_{n+1}$ is isomorphic to L (as a sheaf) and so has length 1.

3. We now choose v large enough so that $\mathcal{R}^{(v)}$ is generated by R_v. By possibly replacing v with vt, we may assume that v is divisible by t. In this situation, we then have $\mathcal{F}_n(-P)$ corresponds to

$$[(\text{graded ideal of } P) \cdot (\text{graded module of } \mathcal{F}_n)]^\sim,$$

which equals

$$\left[\left(\bigoplus_{m=1}^{\infty} \operatorname{Im}(R_{mv-1} \to R_{mv})\right) \cdot \left(\bigoplus_{m=0}^{\infty} M_{n+mv}\right)\right]^\sim.$$

Note that as $t \mid v$, we have $R_{mv-t} = R_{mv-1}$ for any $m \geq 0$. We can also choose v large enough (still $\equiv 0\ (t)$) so that $R_{v-t} M_k = M_{v-t+k}$ for all $k \gg 0$. Thus

$$\mathcal{F}_n(-P) = \left[\left(\bigoplus_{m=1}^{\infty} \operatorname{Im}(R_{mv-t} \to R_{mv})\right) \cdot \left(\bigoplus_{m=0}^{\infty} M_{n+mv}\right)\right]^\sim$$

$$= \left(\bigoplus_{m=0}^{\infty} R_{v-t} \cdot M_{n+(m-1)v}\right)^\sim$$

$$= \left(\bigoplus_{m=0}^{\infty} M_{n+mv-t}\right)^\sim = \mathcal{F}_{n-t}.$$

4. Using the dictionary sketched in the previous subsection, and standard results on coherent modules over proper varieties, we see that

$$\left.\begin{array}{l} M_n \simeq H^0(\overline{X}, \mathcal{F}_{n+1}) \\ H^1(\overline{X}, \mathcal{F}_{n+1}) = (0) \end{array}\right\} n \gg 0. \qquad (*)$$

The long exact sequence of cohomology gives a commutative diagram

$$\begin{array}{ccccccccc} 0 & \to & H^0(\mathcal{F}_n) & \to & H^0(\mathcal{F}_{n+1}) & \to & H^0(\mathcal{F}_{n+1}/\mathcal{F}_n) & \to & H^1(\mathcal{F}_n) & \to 0 \\ & & h_0 \uparrow & & \uparrow \approx & & \uparrow h_1 & & \\ 0 & \to & M_{n-1} & \to & M_n & \to & L \cdot \sigma^n & \to & 0 \end{array}$$

we have dropped the reference to $\overline{X}$ for brevity. By definition it follows that h_1 is an isomorphism, thus h_0 must be an isomorphism also and $H^1(\overline{X}, \mathcal{F}_n) = 0$. Thus $(*)$ holds at n also. In particular,

$$H^0(\overline{X}, \mathcal{F}_0) = M_{-1} = (0)$$

and

$$H^1(\overline{X}, \mathcal{F}_0) = 0;$$

thus giving Part 4.

5. From Part 4 we see that $\mathcal{F}_0$ is torsion free. Thus as P_0 is a smooth point, $\mathcal{F}_0$ is locally-free in a neighborhood of P_0. Part 5 then follows from a local calculation.

6. Multiplication by σ gives a map $\mathcal{M}[n] \to \mathcal{M}[n+1]$. This is linear with respect to R, but σ-linear with respect to L. Thus (as with the φ-sheaves of Subsection 5.3) we obtain the map α. $\square$

The "characteristic" of L as a field over R (in the obvious fashion) can be found as follows. Over $\operatorname{Spec}(R)$, all the sheaves $\mathcal{F}_n$ reduce to M and α to multiplication by σ. Thus the cokernel of α is $\widetilde{(M/\sigma M)} \simeq L \cdot \sigma^0$. It sits over over the L-valued point x of $\operatorname{Spec}(L \otimes R)$ given by

$$x((a_0 + a_1\sigma + \cdots) \otimes b) := a_0 b.$$

The characteristic of L is then the point of $\operatorname{Spec}(R)$ lying under x.

We will now work in the Drinfeld module case. So $\mathbb{F}_r \subset \mathbf{A}$ is the field of constants, $\sigma = \tau = \tau_r$, L is an $\mathbf{A}$-field, $\operatorname{Spec}(\mathbf{A}) = X - \infty$ and we set $\deg_{\mathbb{F}_r} \infty = w = d_\infty$.

Example 6.2.13. Let $\phi \colon \mathbf{A} \to L\{\tau\}$ be a Drinfeld module of rank d. By Lemma 4.5.1, we see that dw, is the greatest common divisor of $\deg_\tau \phi_a$, $a \in \mathbf{A}$. Thus ϕ corresponds to the tower of locally free sheaves $\{\mathcal{F}_n\}$ as in the construction and each $\mathcal{F}_n$ has rank d. Moreover, as $h^0(\mathcal{F}_0) = h^1(\mathcal{F}_0) = 0$, we have

$$\chi(\mathcal{F}_0) = 0.$$

When $d = 1$, all $\mathcal{F}_n$ are invertible sheaves. By the Riemann-Roch Theorem, we must have $\deg(\mathcal{F}_0) = g - 1$.

Suppose that we have the instance $(X, P_0, \{\mathcal{F}_n\}, \alpha)$ of Data A as above. Then as in the proof of Proposition 6.2.3, we conclude that *all* information actually comes from $\mathcal{F}_0$ *and* the diagram:

$$\begin{array}{ccc} \widetilde{\sigma}^* \mathcal{F}_0 & & \\ & \searrow \alpha & \\ & & \mathcal{F}_1 \\ & \beta \nearrow & \\ \mathcal{F}_0 & & \end{array} \qquad (*)$$

where β is just the inclusion. The diagram $(*)$ is the *shtuka*. More generally, any such diagram $(*)$ is called a shtuka if both α and β have cokernels of length 1. (Our shtukas are *right F-sheaves*. As in [Dr3], *left F-sheaves* are obtained by "reversing the arrows.") From the Drinfeld module examples, we call the support of the cokernel of α the *pole* of the shtuka, and the support of the cokernel of β is called its *zero*.

Remarks. 6.2.14. 1. As in [Mu2] one checks that the shtukas coming from Drinfeld modules are characterized by the following.

a. Let $P = \{$the pole and all conjugates over $\mathbb{F}_r\}$. Then the zero is disjoint from P.

b. Restricted to P, $\alpha^{-1} \cdot \beta$ defines a τ^{-1}-semi-linear endomorphism of the $\mathbb{F}_r$-vector space

$$\mathcal{F}_0 / M_P \mathcal{F}_0.$$

This map should be nilpotent. (The reader may wonder where the "$\chi(\mathcal{F}_0) = 0$" condition has gone. The point being that whatever the Euler-Poincaré characteristic of $\mathcal{F}_0$ is, the construction eventually produces some sheaf $\mathcal{F}_j$ with $\chi(\mathcal{F}_j) = 0$.)

2. Let rank $\mathcal{F}_0 = 1$ and let P_0 be a $\mathbb{F}_r$-rational point. Then 1.b is automatically satisfied and, in this case, rank 1 shtukas and rank 1 Drinfeld modules are almost equivalent constructions!

3. Let P be the pole of the shtuka and Z its zero (as L-rational points). Then a line bundle $\mathcal{L}$ has the structure as a shtuka if and only if

$$\mathcal{L}^{-1} \otimes \mathcal{L}^\sigma \simeq \mathcal{O}_{\overline{X}}(P - Z),$$

where $\mathcal{L}^\sigma := \tilde{\sigma}^* \mathcal{L}$. For more along these lines see Subsection 7.11.

4. The above theory also encompasses the adjoint of a Drinfeld module ψ. Indeed, instead of using $\sigma = \tau$, we use $\sigma = \tau^{-1}$; then everything else goes through. Let ψ correspond to the tower $\{\mathcal{F}_n\}$ and let ψ^* correspond to $\{\mathcal{F}_n^*\}$. Is it true that

$$\mathcal{F}_0^* \simeq Hom(\mathcal{F}_0, \Omega),$$

where Ω is the coherent sheaf of 1-forms on $\overline{X}$? It is very natural to expect that this is so.

Finally, we have stressed the similarity between τ and differentiation (over fields of characteristic 0). There is indeed a differential version of the shtuka correspondence (at least for rank 1). This is due, in modern times, to I. Krichever and actually was the *motivation* for Drinfeld's construction. We refer the reader to [Mu2] for the details.

7. Sign Normalized Rank 1 Drinfeld Modules

In this section we will present the construction of "sign-normalized" rank one Drinfeld modules. These play the role of the Carlitz module for general **A**. This basic construction is due to David Hayes [Ha3] [Ha2]. Thus we will also call them "Hayes-modules." We will use Hayes-modules to construct a "cyclotomic theory" of function fields. The reader will find much that is familiar in these extensions from classical theory.

We will be heavily guided by Hayes' elegant exposition [Ha2].

Subsection 7.2 will begin our discussion of Hayes-modules. Subsection 7.1 will present a review of Drinfeld's moduli-theoretic construction of class fields. We will then, on occasion, use this construction in later subsections (while [Ha2] is totally self-contained and elementary).

We will use the notation of Section 4 in this section also. Recall that d_∞ is the degree of ∞ over $\mathbb{F}_r$. Set $W := r^{d_\infty} - 1$ and $w := r - 1$. Thus W is the number of elements in $\mathbb{F}_\infty^*$, etc.

7.1. Class-fields as Moduli

Let L be an **A**-field with structure map $\imath$ and fixed algebraic closure $\overline{L}$. Let $\psi \colon \mathbf{A} \to L\{\tau\}$ be a rank one Drinfeld module and let $f \in \mathbf{A}$ be prime to the characteristic of L. Let $L_1 \subseteq \overline{L}$ be the splitting field of $\psi_f(x)$ which is separable over L.

Lemma 7.1.1. *L_1/L is an abelian extension.*

Proof. $\psi[f] \subset \overline{L}$ is **A**-module isomorphic to $\mathbf{A}/(f)$. Let α be any generator of $\psi[f]$ as **A**-module and let $\sigma \in \mathrm{Gal}(L_1/L)$. As ψ_f and σ commute, $\sigma(\alpha) = \psi_g(\alpha)$ for some $g \in \mathbf{A}/(f)^*$. The map $\sigma \mapsto g$ thus gives an injection of $\mathrm{Gal}(L_1/L)$ into $\mathbf{A}/(f)^*$. □

The above lemma, which is a prototype for much that follows, establishes that rank one Drinfeld modules provide a good means of constructing abelian extensions.

Drinfeld's construction of class-fields is via moduli spaces which we now briefly describe.

Let d be a fixed integer ≥ 1 and let ψ be a Drinfeld module over L of rank d. Let $0 \neq I \subseteq \mathbf{A}$ be an ideal. Recall that the monic polynomial $\psi_I(\tau) \in L\{\tau\}$, $\tau = r^{\text{th}}$ power map, was defined in 4.4.4.

Definition 7.1.2. A *level I structure on ψ* is a homomorphism $\rho\colon (I^{-1}/\mathbf{A})^d \to \psi(L)$ of $\mathbf{A}$-modules such that

$$\psi_I(x) = \prod_{i \in (I^{-1}/A)^d} (x - \rho(i)).$$

Let $\wp = $ the characteristic of L.

Remarks. 7.1.3. 1. If I is prime to $\wp$, then a level I structure is just an isomorphism $(I^{-1}/\mathbf{A})^d \simeq \psi[I]$.
2. Let $\mathrm{Aut}(\psi, L)$ be the automorphism group of ψ over L. Thus

$$\mathrm{Aut}(\psi, L) = \{\alpha \in L^* \mid \alpha \psi_a \alpha^{-1} = \psi_a,\ \forall a \in \mathbf{A}\}.$$

Clearly $\mathrm{Aut}(\psi, L)$ is finite and consists of roots of unity. Suppose that I is divisible by a prime $\neq \wp$. Then one sees that the only automorphism of ψ fixing a level structure ρ must be the identity.

The following result is found in [Dr1].

Theorem 7.1.4. 1. *Let I be an ideal of $\mathbf{A}$ such that $V(I) \subset \mathrm{Spec}(\mathbf{A})$ has at least two points. Then there is a fine moduli scheme M_I^d for Drinfeld modules with level I structure and rank d.*
2. *M_I^d is affine and of finite type over $\mathrm{Spec}(\mathbf{A})$. It is d-dimensional and regular.*
3. *The structure map $M_I^d \to \mathrm{Spec}(\mathbf{A})$ is flat with $d-1$ dimensional fibres. It is smooth outside of $V(I)$. If $I \subseteq J$, then the morphism $M_I^d \to M_J^d$ is finite and flat.* □

The construction of M_I^d is elementary in that no geometric invariant theory is used. Let $\{\alpha_1, \ldots, \alpha_j\}$ be a generating set for $\mathbf{A}$ over $\mathbb{F}_r$. The idea is that a non-trivial point of level I gives a canonical basis for $\mathbf{G}_a$. One then uses this basis to write down the equations for the coefficients of $\{\phi_{\alpha_1}, \ldots, \phi_{\alpha_j}\}$ so that these come from a Drinfeld module ϕ (e.g., $\phi_{\alpha_i}\phi_{\alpha_j} = \phi_{\alpha_j}\phi_{\alpha_i}$, etc.). Then one writes down the equations necessary for a level I structure to exist on ϕ.

Definition 7.1.5. We set $M^d := \varprojlim M_I^d$.

The scheme M^d has an action of the adeles on it as follows. Let A be the adele ring of $\mathbf{k}$ and let $A_f \subset A$ be the subring of finite adeles. Let $\hat{\mathbf{A}}$ be the completion of **A** with respect to the topology given by *all* non-trivial ideals; so $\widehat{\mathbf{A}} \subset A_f$. Let $g \in GL_d(A_f)$ have coefficients in $\widehat{\mathbf{A}}$; thus g can be viewed as an endomorphism of $(\mathbf{k}/\mathbf{A})^d$ with finite kernel H.

Now a point of M^d corresponds to a Drinfeld module ψ together with a homomorphism $\rho \colon (\mathbf{k}/\mathbf{A})^d \to \psi(L)$ such that the restriction to each $(I^{-1}/\mathbf{A})^d$ is a level I structure for all I. The quotient $\psi_1 := \psi/\rho(H)$ is a Drinfeld module equipped with a morphism $\widehat{\rho} \colon (\mathbf{k}/\mathbf{A})^d \to \psi_1(L)$. One shows that $\widehat{\rho}$ restricted to $(I^{-1}/\mathbf{A})^d$ is a level I structure for all I. Thus we obtain a map g_* on the geometric points of M^d. With a little work, one sees that g_* actually arises from a morphism $M^d \to M^d$ which we also denote "g_*."

It is now easy to see that if g corresponds to the scalar matrix

$$a \cdot 1_d$$

for $0 \neq a \in \mathbf{A}$, then g_* is trivial. Thus we obtain an action of $GL_d(A_f)/\mathbf{k}^*$ on M^d.

We then have the following basic result [Dr1].

Theorem 7.1.6 (Drinfeld). *The scheme M^1 is isomorphic to the spectrum of the ring of integers in the maximal abelian extension of $\mathbf{k}$ which is totally split at ∞. The action of $A_f^*/\mathbf{k}^*$ is that of class field theory.* □

Definition 7.1.7. We let $\mathbf{H} \subset \mathbf{C}_\infty$ be the Hilbert class field of $\mathbf{k}$ which is totally split at ∞.

Thus $\mathrm{Gal}(\mathbf{H}/\mathbf{k}) \simeq \mathrm{Pic}(\mathbf{A})$ via Artin Reciprocity. Let $U \subset A_f^*$ be the group $\prod_{v \in \mathrm{Spec}(\mathbf{A})} U_v$, where U_v is the group of local units, and let M_1 be the quotient of M^1 by U. Then M_1 is the *coarse* moduli space of Drinfeld modules of rank one and its function field is isomorphic to $\mathbf{H}$ by Theorem 7.1.6.

Remarks. 7.1.8. 1. The field $\mathbf{H}$ is basic to the theory of Drinfeld modules. Let L be an **A**-field of generic characteristic which has a Drinfeld module ψ of rank one over it. Then L *must* contain an isomorphic copy of $\mathbf{H}$ by Theorem 7.1.6. Thus $\mathbf{H}$ is the smallest *possible* field of definition of a rank one Drinfeld module ψ defined in "characteristic (0)." We shall see later on that ψ is indeed isomorphic, over an algebraic closure of L, to a rank one Drinfeld module defined over $\mathbf{H}$.

2. The result of Part 1 actually holds for all Drinfeld modules of arbitrary rank. Indeed let ψ be a Drinfeld module of arbitrary rank defined over an extension L of $\mathbf{k}$. Now view ψ as being defined over the perfection L^{perf} of L. By using the highest exterior power, either for T-modules or shtukas, one can produce a rank one Drinfeld module out of ψ. (See the introduction to [A1] where such a construction is sketched over an algebraically closed field; however there is no difficulty working over the perfection.) Thus Part 1 applies, and we deduce an injection of $\mathbf{H}$ into L^{perf}. As $\mathbf{H}$ is separable over

k; this immediately implies that L itself must contain the image of **H**. (It would be nice to have an argument that does not use the perfection of L...)
3. Let d_∞ be the degree of ∞ over $\mathbb{F}_r$ as usual. Then it is easy to see by class field theory that **H** contains a subfield $\simeq \mathbb{F}_{r^{d_\infty}} \simeq \mathbb{F}_\infty$. This field is also the full field of constants of **H**.

As ∞ splits completely in M^1, it is clear that M^1 is not the spectrum of the *full* ring of **A**-integers in the maximal abelian extension of **k**. In [Dr2] a covering $\widetilde{M}^1$ of M^1 is constructed via, loosely, a "level ∞ structure" on the universal Drinfeld module over M^1. In the spirit of Subsection 1.9, this level ∞ structure is based on the theory of the *Baker function* for the non-linear Korteweg-deVries differential equation (defined in Subsection 10.2), [Mu2], [SegW1]. The reader should also note that in the analogy $\tau \leftrightarrow D$ as described in Subsection 1.9, pseudo-differential operators in D correspond to finite tailed Laurent series in τ^{-1} (as defined in Subsection 4.14). Drinfeld then establishes that $\widetilde{M}^1$ is indeed the spectrum of the *full* ring of abelian **A**-integers. We refer the reader to [Dr2] for more details as we will not use this construction further.

Remark. 7.1.9. Let L be any finite extension of **k**. It is actually impossible to obtain the *full* abelian closure of L inside the division fields associated to *any* Drinfeld module ψ (of *any* rank) defined over L. The point being that such division fields can only contain at most a *finite dimensional* constant field extension. Indeed, consider ψ as being defined over some finite $\mathbf{K}_1$ of **K** (via an embedding of L into $\mathbf{C}_\infty$, etc.). Let $\mathbf{K}_2$ be a finite extension of $\mathbf{K}_1$ which contains the lattice associated to ψ. Then the analytic construction of Drinfeld modules implies that $\mathbf{K}_2$ actually contains *all* the division points of ψ. But $[\mathbf{K}_2 : \mathbf{K}] < \infty$, and so $\mathbf{K}_2$ can contain at most a finite constant field extension.

7.2. Sign Normalization

It is, of course, very well known that all abelian extensions of $\mathbb{Q}$ are cyclotomic (the Kronecker-Weber Theorem). Thus the ring of integers in the maximal abelian extension of $\mathbb{Q}$ is

$$R := \bigcup_n \mathbb{Z}[\zeta_n],$$

ζ_n a primitive n^{th} root of unity. The ring of integers in the maximal totally-real (="totally split at ∞") subfield is then precisely

$$R^+ := \bigcup_n \mathbb{Z}[\zeta_n + \overline{\zeta}_n].$$

7.2. Sign Normalization

From the discussion of the previous subsection one sees the close analogy of the spectrum of R^+ and Drinfeld's pro-scheme M^1.

Therefore a good "cyclotomic" theory of abelian extensions of **A** would proceed as follows: One should find *canonical* rank one Drinfeld modules such that their division fields give an explicit class field theory. While one can never hope to obtain *all* class fields of **k** this way (as in Remark 7.1.9), those fields that *are* produced would be the cyclotomic ones.

The canonical rank one Drinfeld modules are precisely the sign-normalized ones. To define them, we need to introduce a function field version of the classical notion of "sign of a number."

Recall that **K** is the completion of the base field **k** at the prime ∞ with $\mathbb{F}_\infty \subset \mathbf{K}$ the field of constants.

Definition 7.2.1. 1. A *sign function* on $\mathbf{K}^*$ is a homomorphism $\mathrm{sgn} \colon \mathbf{K}^* \to \mathbb{F}_\infty^*$ which is the identity on $\mathbb{F}_\infty^*$. We also set $\mathrm{sgn}(0) = 0$.
2. Let $\sigma \in \mathrm{Gal}(\mathbb{F}_\infty/\mathbb{F}_r)$. The composite $\sigma \circ \mathrm{sgn}$ is a *twisted sign-function*.

Let $\mathbf{A}_\infty \subset \mathbf{K}$ be the ring of integers and let M be its maximal ideal. Let $U_1 \subset \mathbf{K}^*$ be the group of 1-units, i.e., those elements $u \in \mathbf{A}_\infty$ with $u \equiv 1 \ (M)$. It is well-known that U_1 is a pro-p-group. Thus, as $\mathbb{F}_\infty^*$ has order prime to p, we see that any sign function must be *trivial* on U_1.

Example 7.2.2. Sign functions may be easily constructed in the following fashion. Let $v_\infty \colon \mathbf{K} \to \mathbb{Z}$ be the valuation at ∞ and let $\pi \in \mathbf{K}$ be an element with $v_\infty(\pi) = 1$. Let $x \in \mathbf{K}^*$; then x may be written

$$x = \pi^a \cdot \zeta \cdot u$$

where $a \in \mathbb{Z}$, $\zeta \in \mathbb{F}_\infty^*$, and u is a 1-unit. Set

$$\mathrm{sgn}(x) := \zeta.$$

It is now trivial to see that sgn is a sign function.

Proposition 7.2.3. *Let* sgn_1 *and* sgn_2 *be two sign functions on* **K**. *Then there is an element* $\zeta \in \mathbb{F}_\infty^*$ *such that*

$$\mathrm{sgn}_1(x) = \mathrm{sgn}_2(x)\zeta^{\deg(x)/d}$$

where $d = d_\infty = [\mathbb{F}_\infty : \mathbb{F}_r]$.

Proof. The map $x \mapsto \mathrm{sgn}_1(x)/\mathrm{sgn}_2(x)$ is trivial on the *full* group of units in $\mathbf{K}^*$. Thus it factors through v_∞. The proof is complete upon recalling that

$$\deg(x) = -dv_\infty(x).$$

□

Recall that we set $W = r^{d_\infty} - 1$.

Corollary 7.2.4. *There are exactly W sign-functions on* **K**. □

We now fix a distinguished sign-function "sgn." We view the pair (**A**, sgn) as being the analog of ℤ equipped with the canonical notion of sign. In view of this, we have the following definition.

Definition 7.2.5. *Let $x \in \mathbf{K}^*$. The element x is positive (or monic) if and only if* $\mathrm{sgn}(x) = 1$.

Note that the positive elements form a subgroup of $\mathbf{K}^*$.
Now let ψ be a Drinfeld module of rank d over an **A**-field L.

Definition 7.2.6. *Let $x \in \mathbf{A}$. We set $\mu_\psi(x) \in L$ to be the leading coefficient (i.e., coefficient of highest degree) in ψ_x.*

The following lemma is then easy to see.

Lemma 7.2.7. $\mu_\psi(xy) = \mu_\psi(x)\mu_\psi(y)^{N(x)^d}$ *where* $N(x) := r^{\deg(x)}$. □

Definition 7.2.8. *We define the graded ring of* **K** *to be*

$$\mathrm{gr}(\mathbf{K}) := \bigoplus_{m \in \mathbb{Z}} M^m / M^{m+1}.$$

Following Deligne, we can extend μ_ψ to homogeneous elements of $\mathrm{gr}(\mathbf{K})$ as follows. Let $n \gg 0$. Then the Riemann-Roch Theorem implies that every element a of M^{-n}/M^{-n+1} is the image $\bar{x}$ of an element

$$x \in M^{-n} \cap \mathbf{A}.$$

We set

$$\mu_\psi(a) := \mu_\psi(\bar{x}) := \mu_\psi(x).$$

Lemma 7.2.9. *The above definition is independent of the choice of x.*

Proof. If $\deg x = \deg y$ and $\bar{x} = \bar{y}$, then

$$\deg(x - y) < \deg(x).$$

Thus

$$\deg(\psi_{x-y}) = \deg(\psi_x - \psi_y) < \deg \psi_x,$$

and the result follows. □

$\mu_\psi(x)$ is thus defined on homogeneous elements of $\mathrm{gr}(\mathbf{K})$ of sufficiently small degree. Note that on these elements $\mu_\psi(x)$ satisfies

7.2. Sign Normalization   199

$$\deg(a) = \deg(b) \Rightarrow \mu_\psi(a+b) = \mu_\psi(a) + \mu_\psi(b) \qquad (7.2.10)$$

$$\mu_\psi(ab) = \mu_\psi(a)\mu_\psi(b)^{N(a)^d} \qquad (7.2.11)$$

where $N(a)$ is as in Lemma 7.2.7.

To finish the definition of μ_ψ on $\mathrm{gr}(\mathbf{K})$, let $a \in \mathrm{gr}(\mathbf{K})$ be arbitrary homogeneous. We set

$$\mu_\psi(a) := \mu_\psi(ab)\mu_\psi(b)^{-N(a)^d}$$

where b is of sufficiently negative grade.

Proposition 7.2.12. 1. $\mu_\psi(a)$ *is independent of the choice of* b.
2. *On homogeneous elements of* $\mathrm{gr}(\mathbf{K})$, *7.2.10 and 7.2.11 are still valid.*

Proof. Part 1 follows from 7.2.11. Part 2 can be checked in a straightforward fashion. □

Remark. 7.2.13. By restricting $\mu_\psi(x)$ to $M^0/M \simeq \mathbb{F}_\infty$, we obtain an injection $\imath_\infty$ of $\mathbb{F}_\infty$ into L. This gives us a *third* proof of this fact; the first was via shtukas (Remark 6.2.9) and the second arises via Remark 7.1.8.3 and the fact that M_1 is the coarse moduli space of rank one Drinfeld modules.

Although μ_ψ is defined on $\mathrm{gr}(\mathbf{K})$, it can be easily lifted to $\mathbf{K}$ itself. Indeed, set $\mu_\psi(0) = 0$, and let $x \neq 0 \in \mathbf{K}$. Suppose $v_\infty(x) = j$. Then we set

$$\mu_\psi(x) := \mu_\psi(\overline{x})$$

where $\overline{x} = x + M^{j+1} \in M^j/M^{j+1}$.

We now restrict the rank, d, of the Drinfeld module to be 1 and L to be a subfield of $\mathbf{C}_\infty$.

Definition 7.2.14. We say that ψ is sgn-*normalized*, for our fixed sign function sgn, if and only if the map

$$x \mapsto \mu_\psi(x)$$

is a twisting of sgn. We also say that ψ is a *Hayes-module for* sgn.

In general, without referring to sgn, we say that ψ is *sign-normalized*. We can now establish the main result of this subsection.

Theorem 7.2.15. ψ *is isomorphic over* $\mathbf{C}_\infty$ *to a* sgn-*normalized rank* 1 *Drinfeld module.*

Proof. Let $\pi \in \mathbf{K}$ be a uniformizer with $\mathrm{sgn}(\pi) = 1$. Choose $\alpha \in \mathbf{C}_\infty$ so that $\alpha^W = \mu_\psi(\pi^{-1})$ where $W = \#\mathbb{F}_\infty^*$. It is then easy to see that if $\psi_1 = \alpha\psi\alpha^{-1}$, then

$$\mu_{\psi_1}(\pi^{-1}) = 1.$$

Let $x \in \mathbf{A}$ be written as $x = \zeta \pi^j u$ where $\zeta \in \mathbb{F}_\infty^*$ and u is a 1-unit, we see that
$$\mu_{\psi_1}(x) = \imath_\infty(\zeta) = \imath_\infty(\mathrm{sgn}(x))$$
with $\imath_\infty$ as in 7.2.13. This finishes the proof. □

Example 7.2.16. Let $\mathbf{A} = \mathbb{F}_r[T]$ and let sgn correspond to the usual notion of monic. Then the Carlitz module is sgn-normalized.

Over $\mathbf{C}_\infty$ every rank one Drinfeld module corresponds to a lattice M of rank one. Thus M is isomorphic to $I\varepsilon$ where $I \subseteq \mathbf{A}$ is an ideal and $\varepsilon \in \mathbf{C}_\infty^*$. Rank one lattices M_1 and M_2 are in the same isomorphism class over $\mathbf{C}_\infty$ if and only if there is a non-zero $\alpha \in \mathbf{C}_\infty$ with $\alpha M_1 = M_2$. Thus our next result follows immediately.

Proposition 7.2.17. *Let $h(\mathbf{A})$ be the class number of $\mathbf{A}$. Then there are exactly $h(\mathbf{A})$ isomorphism classes of rank one Drinfeld modules over $\mathbf{C}_\infty$.* □

Lemma 7.2.18. *Let ψ be a Hayes module for sgn and let $t \in \mathbf{C}_\infty^*$. If $\psi_1 := t\psi t^{-1}$ is also a Hayes module for sgn, then $t \in \mathbb{F}_\infty^*$.*

Proof. Let π be a positive uniformizer for sgn. It is elementary to see that
$$\mu_{\psi_1}(\pi^{-1}) = t^{1-r^{d_\infty}} \mu_\psi(\pi^{-1}) = t^{-W} \mu_\psi(\pi^{-1})$$
with $d_\infty = [\mathbb{F}_\infty : \mathbb{F}_r]$. The result now follows. □

Corollary 7.2.19. *There are exactly $h(\mathbf{A}) \cdot \frac{r^{d_\infty}-1}{r-1} = h(\mathbf{A})\frac{W}{w}$ sgn-normalized rank one Drinfeld modules.* □

Recall that in Subsection 4.9 we introduced the action of ideals on Drinfeld modules. If $I \subseteq \mathbf{A}$ is an ideal and ψ is a Drinfeld module, then we defined the Drinfeld module $I * \psi$.

Let $\mathcal{I}$ be the group of fractional ideals of $\mathbf{A}$ and $\mathcal{P}$ the subgroup of principal ideals. Inside $\mathcal{P}$, we let $\mathcal{P}^+$ be the group of principal fractional ideals generated by positive elements. The group $\mathcal{P}/\mathcal{P}^+$ is isomorphic to $\mathbb{F}_\infty^*/\mathbb{F}_r^*$. We call the finite group $\mathcal{I}/\mathcal{P}^+$ the *narrow class group of $\mathbf{A}$ relative to* sgn.

Let Isom be the isomorphism classes of rank one Drinfeld modules over $\mathbf{C}_\infty$ and let Isom$^+$ be the isomorphism classes of sgn-normalized rank one modules.

Proposition 7.2.20. *The action of ideals on Drinfeld modules makes* Isom *a principal homogeneous space for $\mathcal{I}/\mathcal{P}$ and* Isom$^+$ *a principal homogeneous space under $\mathcal{I}/\mathcal{P}^+$.*

7.3. Fields of Definition of Drinfeld Modules

Proof. It is easy to see that if ψ is sgn-normalized, then so is $I * \psi$. The result now follows from Corollary 4.9.5. □

7.3. Fields of Definition of Drinfeld Modules

Before proceeding further with our discussion of sgn-normalized rank one Drinfeld modules (equipped with our fixed choice of sgn) we pause to present some needed general results. In this subsection, we let ψ be a Drinfeld module of arbitrary rank d defined over $\mathbf{C}_\infty$.

Definition 7.3.1. A subfield $\mathcal{F} \subseteq \mathbf{C}_\infty$ is a *field of definition* for ψ if and only if ψ is isomorphic over $\mathbf{C}_\infty$ to a Drinfeld module ψ_1 defined over $\mathcal{F}$.

Our proof of the next result is taken from [Ha2].

Theorem 7.3.2. *There exists a field of definition, $\mathcal{F}_\psi$, of ψ which is contained in every field of definition for ψ. Moreover, $\mathcal{F}_\psi$ is finitely generated over $\mathbf{k}$.*

Proof. Let $\alpha \in \mathbf{A}$ and set

$$\psi_\alpha = \alpha \tau^0 + \sum_{i=1}^{d \deg \alpha} c_i(\psi, \alpha) \tau^i,$$

where τ is the r-th power mapping. Let $\lambda \in \mathbf{C}_\infty^*$. It is then easy to see that

$$c_i(\lambda \psi \lambda^{-1}, \alpha) = \lambda^{1-r^i} c_i(\psi, \alpha). \tag{7.3.3}$$

Let us now fix an $\alpha \in \mathbf{A}$ which is non-constant.

Let $\{i_1, \ldots, i_s\}$ be the indices for which $c_i(\psi, \alpha)$ are non-zero. Let g be the g.c.d. of the integers

$$\{r^{i_j} - 1 \mid j = 1, \ldots, s\};$$

so we have

$$g = \sum_{j=1}^{s} e_j (r^{i_j} - 1).$$

Let

$$I_{i_j}(\psi, \alpha) := c_{i_j}(\psi, \alpha) \left(\prod_{m=1}^{s} c_{i_m}(\psi, \alpha)^{e_m} \right)^{\frac{1 - r^{i_j}}{g}} \in \mathbf{C}_\infty$$

for $j = 1, \ldots, s$. From 7.3.3 it follows that $I_{i_j}(\psi, \alpha)$ depends *only* on the isomorphism class of ψ. Thus if we set

$$\mathcal{F}_\psi := \mathbf{k}(\{I_{i_j}(\psi,\alpha)\}),$$

we see that $\mathcal{F}_\psi$ is contained in *every* field of definition of ψ. We thus need only show that $\mathcal{F}_\psi$ is a field of definition of ψ.

Choose $\lambda \in \mathbf{C}_\infty$ so that

$$\lambda^g = \prod_{j=1}^{s} c_{i_j}(\psi,\alpha)^{e_j}.$$

Thus, for $j = 1, \ldots, s$, we have

$$I_{i_j}(\psi,\alpha) = \lambda^{1-r^{i_j}} c_{i_j}(\psi,\alpha).$$

Consequently $\lambda\psi_\alpha\lambda^{-1}$ has coefficients in $\mathcal{F}_\psi$. The result now follows from the theory of Subsection 4.6, especially 4.6.7. □

Remark. 7.3.4. From Drinfeld's moduli-theoretic viewpoint, the functions

$$I_{i_j}(\psi,\alpha)$$

are rational *modular functions* on the coarse moduli space of Drinfeld modules of rank d.

7.4. The Normalizing Field

Let ψ be a fixed Hayes-module over $\mathbf{C}_\infty$ for our chosen sign function, sgn. Let $\alpha \in \mathbf{A}$ be any non-constant element.

Definition 7.4.1. We let $\mathbf{H}^+ \subset \mathbf{C}_\infty$ be the subfield generated by $\mathbf{k}$ and the coefficients of ψ_α.

Proposition 7.4.2. *The field $\mathbf{H}^+$ is independent of the choice of α.*

Proof. This follows, as before, from the theory of Subsection 4.6, especially 4.6.7. □

Thus we see $\mathbf{H} \subseteq \mathbf{H}^+$. Moreover, one can see that $\mathbf{H}^+$ depends only on the choice of the sign function.

Proposition 7.4.3. *The extension $\mathbf{H}^+$ is finite and normal over $\mathbf{k}$.*

Proof. Let σ be an automorphism of $\mathbf{C}_\infty$ over $\mathbf{k}$. Let $\sigma\psi$ be the rank one Drinfeld module obtained by applying σ to the coefficients of ψ. It is easy to see that $\sigma\psi$ is sgn-normalized. But there are only finitely many sgn-normalized Hayes-modules. The result follows. □

7.4. The Normalizing Field

Proposition 7.4.4. *The extension* $\mathbf{H}^+/\mathbf{k}$ *is Galois with* $\mathrm{Gal}(\mathbf{H}^+/\mathbf{k})$ *isomorphic to a subgroup of* $\mathcal{I}/\mathcal{P}^+$, *where* $\mathcal{I}$ *and* $\mathcal{P}^+$ *are as in Subsection 7.2.*

Proof. We know that $\mathbf{H}^+$ must contain the field of definition $\mathcal{F}_\psi$ of ψ. On the other hand, as ψ is rank one, the analytic theory implies that $\mathbf{K} = \mathbf{k}_\infty$ is a field of definition of ψ. Now $\mathbf{K}/\mathbf{k}$ is well-known to be a separable extension. Thus we conclude that $\mathcal{F}_\psi$ is separable over $\mathbf{k}$.

Let λ be chosen so that $\lambda\psi\lambda^{-1} = \psi_1$ is defined over $\mathcal{F}_\psi$. Let $\alpha \in \mathbf{A}$ be positive. By looking at coefficients of highest degree, we conclude that

$$\lambda^{1-r^{\deg(\alpha)}} \in \mathcal{F}_\psi.$$

Thus $\mathbf{H}^+ \subseteq \mathcal{F}_\psi(\lambda)$ and is thus also separable over $\mathbf{k}$; thus it is Galois.

One sees easily that the action of the ideals commutes with the Galois action. Thus we obtain an injection of $\mathrm{Gal}(\mathbf{H}^+/\mathbf{k})$ into $\mathcal{I}/\mathcal{P}^+$ finishing the result. $\square$

Let $\mathcal{O}^+ \subset \mathbf{H}^+$ be the ring of $\mathbf{A}$-integers. We say a Drinfeld module ϕ is *defined over* $\mathcal{O}^+$ if and only if all the coefficients of ϕ_α, $\alpha \in \mathbf{A}$, are in $\mathcal{O}^+$ and the highest coefficient is a unit. (One says that ϕ is a *Drinfeld module over* $\mathrm{Spec}(\mathcal{O}^+)$, etc.).

Lemma 7.4.5. *Let ψ be a Hayes-module for* sgn. *Then ψ is defined over* $\mathcal{O}^+$.

Proof. From 4.10.4.2 we know that ψ has potential good reduction at every prime of $\mathcal{O}^+$. Thus, given a prime $\wp$ of $\mathcal{O}^+$, there exists an extension (R, v) of $(\mathcal{O}^+, \wp)$ and an element λ such that $\lambda\psi\lambda^{-1}$ has good reduction at v. But, as ψ is sgn-normalized, we conclude that λ is a *unit* at v. The result follows. $\square$

Lemma 7.4.6. *Let ψ and ϕ be two Hayes-modules for* sgn. *Let $\wp \in \mathrm{Spec}(\mathcal{O}^+)$. If ψ and ϕ reduce to the same module over $\mathcal{O}^+/\wp$, then $\psi = \phi$.*

Proof. By 7.2.20 we may assume

$$\psi = I * \phi$$

for some ideal I. Moreover, by combining 7.2.20 and weak approximation, we can assume that I is prime to $\wp$.

By definition

$$\phi_I \phi_\alpha = \psi_\alpha \phi_I.$$

By reducing this modulo $\wp$ we obtain

$$\phi_I \phi_\alpha \equiv \phi_\alpha \phi_I \pmod{\wp}.$$

Thus ϕ_I is an endomorphism of ϕ mod $\wp$. But the endomorphism ring of a rank one module is $\mathbf{A}$ itself. So $\phi_I \equiv \phi_a$ (mod $\wp$) for some a. As the leading coefficient of ϕ_I is 1, we conclude that a is positive.

We want to show that $I = (a)$. Let $J = I + (a)$. As $\phi_I \equiv \phi_a$, we see that (a), I, and J have the *same* torsion module in an algebraic closure of $\mathcal{O}^+/\wp$. From this we conclude $I = (a)$ and the result is complete. □

Corollary 7.4.7. *The extension $\mathbf{H}^+/\mathbf{k}$ is unramified at every finite place of $\mathbf{A}$.*

Proof. Let $\wp \in \text{Spec}(\mathbf{A})$ and let $I_\wp \subseteq \text{Gal}(\mathbf{H}^+/\mathbf{k})$ be the inertia group. By definition if $\sigma \in I_\wp \Rightarrow \sigma\psi \equiv \psi$ (mod $\wp$). By 7.4.6 we find that $\sigma\psi = \psi$. Thus σ fixes the coefficients of ψ_a for all $a \in \mathbf{A}$. As these generate $\mathbf{H}^+$ over $\mathbf{k}$, the result follows. □

Theorem 7.4.8. *Let ψ be our fixed Hayes-module for sgn. Let $I \subseteq \mathbf{A}$ be a non-zero ideal and $\sigma_I \in \text{Gal}(\mathbf{H}^+/\mathbf{k})$ the Artin symbol of I. Then*

$$\sigma_I \psi = I * \psi.$$

Proof. Suppose first that $I = \wp$ is prime. Then $\psi_\wp(\tau)$ is congruent to $\tau^{\deg(\wp)}$ modulo $\wp$; indeed, it is clearly an inseparable polynomial modulo $\wp$, and thus counting as in 4.5.3 implies that $\psi_\wp$ must be *purely* inseparable modulo $\wp$. Thus by 7.4.6, $\sigma_\wp \psi$ equals $\wp * \psi$. One knows that

$$I * (J * \psi) = (IJ * \psi);$$

thus the proof follows by induction. □

Corollary 7.4.9. $\text{Gal}(\mathbf{H}^+/\mathbf{k}) \simeq \mathcal{I}/\mathcal{P}^+$, *in line with class field theory.*

Proof. This follows from the equation

$$\sigma_I \psi = I * \psi$$

just established. □

Let $\alpha \in \mathbf{k}^*$. We then let $\sigma_\alpha \in \text{Gal}(\mathbf{H}^+/\mathbf{k})$ be the Artin symbol of the principal fractional ideal $\mathbf{A}\alpha$. By 7.4.9 one sees that $\sigma_\alpha \psi$ depends *only* on sgn α and one finds

$$\sigma_\alpha \psi = \mu_\psi(\alpha)^{-1} \psi \mu_\psi(\alpha).$$

Moreover, via the map $\alpha \mapsto \sigma_\alpha$, we see that $\text{Gal}(\mathbf{H}^+/\mathbf{k})$ contains a subgroup isomorphic to $\mathbb{F}_\infty^*/\mathbb{F}_r^*$. Class field theory then implies that this subgroup is *both* the inertia and decomposition group at ∞. As

$$\text{Gal}(\mathbf{H}/\mathbf{k}) \simeq \mathcal{I}/\mathcal{P}$$

via the Artin map, we conclude the following result.

Proposition 7.4.10. $\mathrm{Gal}(\mathbf{H}^+/\mathbf{H}) \simeq \mathbb{F}_\infty^*/\mathbb{F}_r^*$. *This extension is totally and tamely ramified at all infinite primes of* $\mathbf{H}$. □

Finally, let ψ be our Hayes module for sgn with field of definition $\mathcal{F}_\psi$. From Remark 7.3.4 we conclude that $\mathcal{F}_\psi \subseteq \mathbf{H}$. Thus we deduce that $\mathcal{F}_\psi = \mathbf{H}$ or that ψ is actually defined over $\mathbf{H}$. Later on, we will see that one can choose a model for ψ which is defined over the $\mathbf{A}$-integers $\mathcal{O}$ of $\mathbf{H}$ in that its highest coefficient is a unit (so we have a Drinfeld module over $\mathrm{Spec}(\mathcal{O})$).

7.5. Division Fields

We continue with our fixed sgn-normalized rank one Drinfeld module ψ. In this subsection we will adjoin division points of ψ to $\mathbf{H}^+$ to obtain *cyclotomic abelian extensions* of $\mathbf{k}$.

Let $I \subseteq \mathbf{A}$ be an ideal.

Definition 7.5.1. 1. We let $\mathcal{I}(I) \subseteq \mathcal{I}$ be the subgroup generated by primes $\wp$ of $\mathbf{A}$ *not* dividing I. Let
$$\mathcal{P}(I) \subseteq \mathcal{I}(I)$$
be the subgroup of principal ideals.
2. Let $\mathcal{P}_I \subseteq \mathcal{P}(I)$ be the subgroup of principal fractional ideals generated by elements $\alpha \in \mathbf{k}$ such that
$$\alpha \equiv 1 \pmod{I}.$$
3. Let $\mathcal{P}_I^+ \subseteq \mathcal{P}_I$ be the subgroup generated by *positive* $\alpha \equiv 1 \pmod{I}$.,

The group $\mathcal{I}(I)/\mathcal{P}_I^+$ is the *narrow ray class group modulo I associated to* sgn. It is a standard exercise to see that it is finite.

If ϕ is any sgn-normalized rank one Drinfeld module in $\mathbf{C}_\infty$, then, as before, we set $\phi[I] \subset \phi(\mathbf{C}_\infty)$ to be its $\mathbf{A}$-module of I-division points. As ϕ has rank one, one knows that $\phi[I]$ is an $\mathbf{A}$-module isomorphic to $\mathbf{A}/I$. If we define
$$\Phi(I) := \#(\mathbf{A}/I)^*,$$
then $\phi[I]$ has $\Phi(I)$ generators as an $\mathbf{A}$-module.

Let $\mathrm{Isom}(I)$ be the isomorphism classes of pairs (ϕ, λ) where ϕ is as above and λ is a generator of $\phi[I]$. Let $J \subseteq \mathbf{A}$ be an ideal prime to I. We then set
$$J * (\phi, \lambda) := (J * \phi, \phi_J(\lambda)).$$

Our next result is then easy to see.

Theorem 7.5.2. *The above action of ideals on* $\mathrm{Isom}(I)$ *factors through* $\mathcal{I}(I)/\mathcal{P}_I^+$. *Further,* $\mathrm{Isom}(I)$ *is a principal homogeneous space for* $\mathcal{I}(I)/\mathcal{P}_I^+$. □

We can now proceed to the construction of cyclotomic extensions of function fields.

Definition 7.5.3. We set $\mathbf{k}(I) := \mathbf{H}^+(\psi[I]) =$ the smallest extension of $\mathbf{H}^+$ containing all $\lambda \in \psi[I]$. We let $\mathcal{O}(I) \subset \mathbf{k}(I)$ be its ring of $\mathbf{A}$-integers.

Via the action of ideals on $\mathrm{Isom}(I)$, one sees readily that $\mathbf{k}(I)$ is independent of the choice of the sgn-normalized Drinfeld module used.

Proposition 7.5.4. 1. The extension $\mathbf{k}(I)/\mathbf{k}$ is Galois and abelian.
2. The extension $\mathbf{k}(I)/\mathbf{k}$ is unramified at the primes $\wp$ of $\mathbf{A}$ not dividing I.
3. Let $\lambda \in \psi[I]$ and $\mathcal{B}$ an ideal of $\mathbf{A}$ prime to I with Artin symbol $\sigma_\mathcal{B} \in \mathrm{Gal}(\mathbf{k}(I)/\mathbf{k})$. Then
$$\sigma_\mathcal{B}(\lambda) = \psi_\mathcal{B}(\lambda).$$

Proof. Let $h = h(\mathbf{A})$ be the class number of $\mathbf{A}$; then $I^h = (i)$ is principal. If $\wp \nmid I$, then $\wp \nmid i$. It is now easy to see that $\mathbf{k}(i)/\mathbf{k}$ is separable and unramified at $\wp$. Thus so is $\mathbf{k}(I)/\mathbf{k}$. The action of the ideals on $\mathrm{Isom}(I)$ now allows us to conclude that $\mathbf{k}(I)/\mathbf{k}$ is Galois with an injection of $\mathrm{Gal}(\mathbf{k}(I)/\mathbf{k})$ into $\mathcal{I}(I)/\mathcal{P}_I^+$. Part 3 now follows as in Theorem 7.4.8. $\square$

Corollary 7.5.5. $\mathrm{Gal}(\mathbf{k}(I)/\mathbf{k}) \simeq \mathcal{I}(I)/\mathcal{P}_I^+$.

Proof. This follows from 7.5.4.3 and 7.5.2. $\square$

Corollary 7.5.6. $\mathrm{Gal}(\mathbf{k}(I)/\mathbf{H}^+) \simeq (\mathbf{A}/I)^*$.

Proof. Every class in $\mathbf{A}/I$ can be represented by a positive element. Thus the corollary follows from 7.5.5. $\square$

Corollary 7.5.7. Let $\alpha \in \mathbf{k}^*$ with $\alpha \equiv 1 \pmod{I}$. Then for $\lambda \in \psi[I]$ we have
$$\sigma_\alpha(\lambda) := \sigma_{(\alpha)}(\lambda) = \mu_\psi(\alpha)^{-1}\lambda,$$
with $\mu_\psi(\alpha)$ the highest coefficient in ψ_α.

Proof. This follows from 7.5.4.3 as $\psi_{(\alpha)}(x) = \mu_\psi^{-1}\psi_\alpha(x)$. $\square$

From Corollary 7.5.7 we see that $\mathrm{Gal}(\mathbf{k}(I)/\mathbf{k})$ contains a subgroup G_∞ isomorphic to $\mathbb{F}_\infty^* \subset \mathbf{K}^*$.

Proposition 7.5.8. The subgroup G_∞ is both the decomposition and inertia group at ∞ in $\mathrm{Gal}(\mathbf{k}(I)/\mathbf{k})$.

Proof. This follows from class field theory. $\square$

7.5. Division Fields

Corollary 7.5.9. *The extension* $k(I)/k$ *is tamely ramified at* ∞. □

Definition 7.5.10. Let $k(I)^+$ be the subfield of $k(I)$ fixed by G_∞.

Lemma 7.5.11. $k(I)^+/k$ *is totally split at* ∞. □

Remarks. 7.5.12. 1. The reader should be aware that our notation, which is standard, may contain a small confusion. Indeed, the field **H** *is* totally split at ∞, while the field $\mathbf{H}^+$ is *not*.
2. The field $k(I)^+$ is k-isomorphic to the function field of the moduli scheme M_I^1.
3. Let ∞_1 be another place of **k** not equal to ∞. Let k_∞^+ be the compositum of all the fields $k(I)^+$ defined above, with $k_{\infty_1}^+$ defined similarly. Then, as in [Ha4], one can see that the compositum $k_{\infty_1}^+ \cdot k_\infty^+$ is the *maximal* abelian extension of **k**. Along these lines, it would be very useful to have a modification of Drinfeld's construction in [Dr2] which can be applied *directly* to ψ to give the *full* abelian closure.

For any finite extension L of **k**, we let $\mathcal{O}(L)/\mathbf{A}$ be the ring of **A**-integers. It is standard fact that $\mathcal{O}(L)$ is always *finite* over **A** regardless of whether $L/\mathbf{k}$ is separable or not, (see e.g., the discussion in Subsection 8.6).

Suppose now that L contains $\mathbf{H}^+$ so that

$$\mathcal{O}(L) \supseteq \mathcal{O}^+ = \mathcal{O}(\mathbf{H}^+).$$

Class field theory then implies that the ideal norm down to **A** of *any* non-zero ideal $\mathcal{B}$ of $\mathcal{O}(L)$ is principal *and* positively generated.

Definition 7.5.13. We define $n\mathcal{B}$ to be the positive generator of the ideal norm (down to **A**) of $\mathcal{B}$.

From Corollary 7.5.6 we see that $\text{Gal}(k(I)/\mathbf{H}^+) \simeq (\mathbf{A}/I)^*$. Let $L(I)$ be the compositum of $k(I)$ and L. Then the formalism of class field theory immediately gives the following result.

Proposition 7.5.14. *Let* $\mathcal{B} \subseteq \mathcal{O}(L)$ *be prime to* I *and let* $\sigma_\mathcal{B} \in \text{Gal}(L(I)/L)$ *be its Artin symbol. Then* $\sigma_\mathcal{B}$ *corresponds to* $n\mathcal{B}+I$ *via the canonical injection of* $\text{Gal}(L(I)/L)$ *into* $(\mathbf{A}/I)^*$. □

Let $\lambda \in \psi[I] \in \mathbf{C}_\infty$ be any fixed generator of $\psi[I]$ as an **A**-module. As $k(I)/k$ is abelian, so is $k(\lambda)/k$. The proof of the following pleasing result was shown to us by David Hayes.

Theorem 7.5.15. *We have* $k(I) = k(\lambda)$.

Proof. By Galois theory we will be done if we show that the only Galois automorphism of $\mathbf{k}(I)/\mathbf{k}$ fixing λ is the identity. In turn, by 7.5.4.3, this is equivalent to showing that for $\mathcal{B}$ an ideal of $\mathbf{A}$ prime to I we have

$$\psi_{\mathcal{B}}(\lambda) = \lambda \Rightarrow \mathcal{B} = (\beta)$$

with $\beta \equiv 1 \pmod{I}$ and β positive.

Put $\widehat{\psi} = \mathcal{B} * \psi$. As $\mathcal{B}$ is prime to I, one sees readily that ψ and $\widehat{\psi}$ must have the same I-division points. Thus

$$\psi_I = \widehat{\psi}_I.$$

Let e be chosen so that $I^e = (\alpha)$ for α positive. Clearly, any α-division point maps to an I-division point upon dividing ψ by $\psi[I^a]$ for some a. If one notes that $\psi_I(\tau)$ has coefficients in $\mathcal{O}(\mathbf{H}^+)$, one see that

$$\psi_\alpha(\tau) = \sigma_I^{e-1}\psi_I(\tau) \cdot \sigma_I^{e-2}\psi_I(\tau) \cdots \psi_I(\tau).$$

We thus conclude that $\psi_\alpha = \widehat{\psi}_\alpha$.

However, over $\mathbf{C}_\infty$, knowledge of the Drinfeld action at *any* non-constant is enough to determine the exponential function. Thus we deduce $\psi = \widehat{\psi}$; in turn this forces $\mathcal{B}$ to be principal and positive by 7.2.20. The result now follows easily. $\square$

The above proof shows that $\psi_I(\tau)$ *always* determines ψ as sgn-normalized rank one module. There are examples, due to L. Shu, which show this to be false without the sgn-normalization hypothesis.

Proposition 7.5.16. *Let $\Lambda = \xi J$ be the lattice associated to ψ where $J \subseteq \mathbf{A}$ is an ideal. Then, inside $\mathbf{C}_\infty$, we have*

$$\mathbf{K}(\lambda) = \mathbf{K}(\xi),$$

for any torsion $\lambda \neq 0$.

Proof. Let $e(z) = e_\Lambda(z) = z \prod_{\substack{\alpha \in \Lambda \\ \alpha \neq 0}} (1 - z/\alpha)$ be the exponential function of ψ. By taking logarithmic derivatives we find

$$\frac{1}{e(z)} = \frac{e'(z)}{e(z)} = \sum_{\alpha \in \Lambda} \frac{1}{z + \alpha}.$$

We know $\lambda = e(t\xi)$ for some $t \in \mathbf{k}$. Then we see

$$\lambda^{-1} = \sum_{\beta \in J}(t\xi + \beta\xi)^{-1} = \xi^{-1}\sum_{\beta \in J}(t + \beta)^{-1}.$$

As $\sum_{\beta \in J}(t + \beta)^{-1}$ converges in $\mathbf{K}$, the result follows. $\square$

Remark. 7.5.17. By 7.5.7 and 7.5.8, we know that
$$\lambda^W = \lambda^{r^{d_\infty}-1} \in \mathbf{K}^*.$$
The proof of 7.5.16 thus implies that
$$\xi^W \in \mathbf{K}^*$$
also.

Finally, we examine the ramification properties of the extensions
$$\mathbf{k}(\wp^m)/\mathbf{H}^+$$
for $\wp \in \mathrm{Spec}(\mathbf{A})$. From above we need only discuss the primes of $\mathbf{H}^+$ lying above $\wp$.

Proposition 7.5.18. *The extension $\mathbf{k}(\wp^m)/\mathbf{H}^+$ is totally ramified at all primes of $\mathbf{H}^+$ above $\wp$.*

Proof. This follows directly from the formalism of class field theory. □

Another, more "cyclotomic," proof of 7.5.18 can be given as follows. Let $\lambda_1, \lambda_2 \in \psi[\wp^m]$ be two generators as $\mathbf{A}$-module. Thus $\lambda_1 = \psi_a(\lambda_2)$ and $\lambda_2 = \psi_b(\lambda_1)$ for $a, b \in \mathbf{A}$. Then the following result is easily seen.

Lemma 7.5.19. λ_1/λ_2 *is a unit of the ring of $\mathbf{A}$-integers of $\mathbf{k}(\wp^m)$.* □

Corollary 7.5.20. λ_1, λ_2 *have the same valuation at all finite primes of $\mathbf{k}(\wp^m)$.* □

To motivate a construction that will be presented shortly, we pause for a moment to discuss some classical theory. Recall that if j and h are prime to p, then
$$\frac{1-\zeta_{p^m}^j}{1-\zeta_{p^m}^h}$$
is a unit in $\mathbb{Q}(\zeta_{p^m})$ where ζ_{p^m} is a primitive p^m-th root of one. From this one then readily establishes that $\mathbb{Q}(\zeta_{p^m})/\mathbb{Q}$ is totally ramified above (p) and that $(1-\zeta_{p^m})$ is the unique prime above (p).

To finish with our cyclotomic proof for $\mathbf{k}$, we will need the following result which will be shown in Theorem 7.6.2:
$$\wp^m \mathcal{O}^+ = \left(\prod_{0 \neq \lambda \in \psi[\wp^m]} \lambda \right).$$
Thus
$$\wp \mathcal{O}^+ = \left(\prod \lambda_* \right), \tag{7.5.21}$$

210   7. Sign Normalized Rank 1 Drinfeld Modules

where λ_* runs over the *generators* of $\psi[\wp^m]$. Note that the number of such generators is $\Phi(\wp^m) = \#(\mathbf{A}/\wp^m)^*$.

Recall that $\mathcal{O}(\wp^m)$ is the ring of **A**-integers in the abelian extension $\mathbf{k}(\wp^m)$ of **k**.

Proposition 7.5.22. *Let $\lambda \in \mathcal{O}(\wp^m)$ be any generator of $\psi[\wp^m]$.*
1. *We have $\lambda \mathcal{O}(\wp^m) = (\lambda) = \prod \mathcal{B}$ where $\mathcal{B}$ runs over the primes of $\mathcal{O}(\wp^m)$ above $\wp$.*
2. *Each such $\mathcal{B}$ is totally ramified over $\mathcal{O}^+$.*

Proof. Note that $\mathcal{O}^+/\mathbf{A}$ is étale and that $\mathrm{Gal}(\mathbf{k}(\wp^m)/\mathbf{H}^+) \simeq (\mathbf{A}/\wp^m)^*$. Thus the result follows by 7.5.20 and 7.5.21, and an easy calculation. □

Of course, 7.5.22.2 is 7.5.18 and our "cyclotomic proof" is at hand.

Let ψ and ρ be *any* two sgn-normalized Drinfeld modules. Let λ_ψ and λ_ρ be *any* two generators of $\psi[\wp^m]$ and $\rho[\wp^m]$ respectively.

Corollary 7.5.23. *$\lambda_\psi/\lambda_\rho$ is a unit in $\mathcal{O}(\wp^m)$.*

Proof. This follows directly from 7.5.22.1. □

The group of units spanned by *all* such $\lambda_\psi/\lambda_\rho$ is *the group of cyclotomic units of level $\wp^m$*. The behavior of units in the completions at finite primes is discussed in [Ki1].

It remains to examine the ramification properties of $\mathbf{k}(\wp^m)^+/\mathbf{H}$. As above, this extension turns out to be totally ramified at the primes of **H** above $\wp$ by class field theory. A more cyclotomic proof can also be given in the same spirit as above. Indeed, by Takahashi's Theorem (Theorem 7.6.4 in our next subsection) there is a Drinfeld module of rank one defined over the **A**-integers $\mathcal{O}$ of **H** (so all coefficients are in **H** with the highest being a unit – *not necessarily constant* – in $\mathcal{O}^+$). Then, as above, the extension of **H** obtained by adjoining $\wp^m$-division points is totally-ramified at the primes above $\wp$. (N.B.: this extension may well *not* equal $\mathbf{k}(\wp^m)$.) As $\mathbf{k}(\wp^m)^+$ is contained in this extension, we are finished.

Another proof, due to D. Hayes, goes as follows: as $\mathbf{H}^+/\mathbf{H}$ is totally ramified at the primes above ∞, and $\mathbf{k}(\wp^m)^+$ is totally-split, the fields $\mathbf{H}^+$ and $\mathbf{k}(\wp^m)^+$ are linearly disjoint over **H**. Set

$$G := \mathrm{Gal}(\mathbf{k}(\wp^m)/\mathbf{H}^+).$$

As $\mathbf{H}^+$ and $\mathbf{k}(\wp^m)^+$ are linearly disjoint, G surjects onto $\mathrm{Gal}(\mathbf{k}(\wp^m)^+/\mathbf{H})$. But G is the ramification group of $\mathbf{k}(\wp^m)/\mathbf{k}$ and the result follows.

As a corollary, since $\mathbf{H}^+/\mathbf{H}$ is étale above $\wp$, we see that $\mathbf{k}(\wp^m)/\mathbf{k}(\wp^m)^+$ has ramification index $r-1$ at the primes above $\wp$.

7.6. Principal Ideal Theorems

Let sgn continue to be our fixed sign function and let ψ be our chosen Hayes-module for sgn. Let $I \subseteq \mathbf{A}$ be an ideal and let

$$\psi_I(x) = \prod_{\alpha \in \psi[I]} (x+\alpha)$$

as before. Thus $\psi_I(x)$ is $\mathbb{F}_r$-linear with coefficients in the $\mathbf{A}$-integers $\mathcal{O}^+$ of $\mathbf{H}^+$. In particular, the derivative $D\psi_I := \frac{d}{dx}\psi_I(x)$ is a *constant* in $\mathcal{O}^+$.

Let $J \subseteq \mathbf{A}$ be another ideal. Using Lemma 4.9.2, the reader may easily establish the following result (which actually works for *arbitrary ψ*).

Lemma 7.6.1. $D\psi_{IJ} = D(J*\psi)_I D\psi_J$. □

We can now state and prove our first Principal Ideal Theorem.

Theorem 7.6.2. $I\mathcal{O}^+ = (D\psi_I)$.

Proof. By 7.6.1 we are reduced to having $I = \wp$ prime.

Let $\mathfrak{B}$ be a prime of $\mathcal{O}^+$ dividing $\wp$. As $\mathcal{O}^+/\mathbf{A}$ is unramified at $\wp$, we need to show, first of all, that the valuation $v_\mathfrak{B}(D\psi_\wp)$ equals 1. But reducing ψ modulo $\mathfrak{B}$ gives a rank one module of characteristic $\wp$; thus it has *no* non-trivial $\wp$-division points. This implies $v_\mathfrak{B}(D\psi_\wp) \geq 1$.

On the other hand, let $a \in \mathbf{A}$ have a zero of order 1 at $\wp$ and put

$$(a) = \wp\wp'$$

where $(\wp, \wp') = 1$. By 7.6.1 we obtain

$$\psi_a = \mu_\psi(a)(\wp*\psi)_{\wp'}\psi_\wp,$$

where, as above, $\mu_\psi(a)$ is the leading coefficient. Thus we obtain

$$1 = v_\wp(a) = v_\mathfrak{B}(a) = v_\mathfrak{B}(D(\wp*\psi)_{\wp'}) + v_\mathfrak{B}(D\psi_\wp).$$

So $1 = v_\mathfrak{B}(D\psi_\wp)$.

Suppose now that $\mathfrak{B}$ is a prime of $\mathcal{O}^+$ not dividing $\wp$. Then we want to show that

$$\mathfrak{B} \nmid D\psi_\wp.$$

Let e be chosen so that $\wp^e = (\beta)$ is principal. The result now follows via the same formalism as in the proof of Theorem 7.5.15. □

Corollary 7.6.3. *Let ψ be as above. Let $\sigma \in \mathrm{Gal}(\mathbf{H}^+/k)$. Then for any ideal I of $\mathbf{A}$, $(D\psi_I)^{1-\sigma}$ is a unit in $\mathcal{O}^+$.* □

Recall that **H** ⊆ **H**⁺ is the Hilbert class field which is split totally at ∞ with **A**-integers $\mathcal{O} = \mathcal{O}(\mathbf{H})$. In order to prove a result like 7.6.2 for $\mathcal{O}$, we first present Hayes' proof of the following result of Takahashi.

Theorem 7.6.4. *Every rank one Drinfeld module defined over* $\mathbf{C}_\infty$ *is isomorphic to one defined over* $\mathcal{O}$.

Recall further that for a Drinfeld module to be defined over $\mathcal{O}$ means that all its coefficients are in $\mathcal{O}$ with the highest coefficient being a unit.

The proof of 7.6.4 requires some lemmas. We begin by recalling a result from elementary number theory.

Lemma 7.6.5. *Let* $i_1, \ldots, i_j$ *be positive integers with greatest common divisor* t. *Then the greatest common divisor of* $x^{i_1} - 1, \ldots, x^{i_j} - 1, \ldots, x^{i_j} - 1$ *is* $x^t - 1$. □

Lemma 7.6.6. *Let* ψ *be as above and let* I *be a non-trivial ideal of* **A**. *Then there exists an element* $\gamma_I \in \mathbf{k}(I)$ *such that*

$$\gamma_I^{r-1} = (-1)^{\frac{r \cdot \deg(I) - 1}{r-1}} \cdot D\psi_I.$$

Moreover, let $x \in \mathbf{k}^*$ *with* $x \equiv 1 \pmod{I}$. *Then*

$$\gamma_I^{\sigma_x} = \sigma_x(\gamma_I) = \mu_\psi(x)^{\frac{1-r^{\deg I}}{r-1}} \cdot \gamma_I,$$

where σ_x *is defined in 7.5.7 and* $\mu_\psi(x)$ *is defined in 7.2.6.*

Proof. It is clear that

$$D\psi_I = \prod_{0 \neq \lambda \in \psi[I]} \lambda.$$

Note that $\psi[I]$ is an $\mathbb{F}_r$-vector space of dimension equal to $\deg I$. Let $\{\lambda_i\} \subseteq \psi[I]$ be a set of orbit representatives of $\psi[I] - \{0\}$ under $\mathbb{F}_r^*$, and set

$$\gamma_I := \prod \lambda_i.$$

By following the action of the elements of $\mathbb{F}_r^*$, one computes

$$D\psi_I = (-1)^{\frac{r^{\deg I} - 1}{r-1}} \gamma_I^{r-1}.$$

If x is as above, then from our explicit knowledge of the action of σ_x (7.5.7), we see that

$$\gamma_I^{\sigma_x} = \prod \mu_\psi(x)^{-1} \lambda_i = \mu_\psi(x)^{\frac{1-r^{\deg I}}{r-1}} \gamma_I,$$

where, as always, $\mu_\psi(x)$ is the highest coefficient. This completes the proof. □

7.6. Principal Ideal Theorems   213

Proof of Theorem 7.6.4. It is well-known that **k** contains a divisor of degree 1. Thus by 7.6.5 we can find ideals $I_j \subseteq \mathbf{A}$ and integers n_j, $j = 1, \ldots, e$, such that

$$\sum_{j=1}^{e} n_j(r^{\deg I_j} - 1) = r - 1. \tag{7.6.7}$$

For each j, let γ_{I_j} be as in 7.6.6 and set

$$\gamma_0 = \prod_{j=1}^{e} \gamma_{I_j}^{n_j}.$$

By 7.6.7 and Lemma 7.6.6, we have

$$\gamma_0^{r-1} = -\prod_{j=1}^{e} D\psi_{I_j}^{n_j}. \tag{7.6.8}$$

Set $I = \prod I_j$ and

$$\widetilde{P}_I = \{x \in \mathbf{k}^* \mid x \equiv 1 \pmod{I}\}.$$

Let $x \in \widetilde{P}_I$. Then one computes that

$$\gamma_0^{\sigma_x} = \mu_\psi(x)^{-1}\gamma_0, \tag{7.6.9}$$

with σ_x the Artin symbol.

Let J be an ideal of **A** with Artin symbol σ_J. Set $\xi_0 = \gamma_0^{1-\sigma_J}$; so by 7.6.6 and 7.6.8 we know that ξ_0^{r-1} is a unit of $\mathcal{O}^+$. We shall show next that $\xi_0 \in \mathcal{O}^+$; so it is also a unit.

Thus, let $\delta \in \mathrm{Gal}(\mathbf{H}^+(\gamma_0)/\mathbf{H}^+)$; so

$$\gamma_0^\delta = \delta\gamma_0 = \zeta\gamma_0$$

for some $\zeta \in \mathbb{F}_r^*$ by 7.6.8. Moreover, $\mathbf{H}^+(\gamma_0)/\mathbf{k}$ is abelian and so $\xi_0 = \gamma_0^{1-\sigma_J}$ also belongs to $\mathbf{H}^+(\gamma_0)$. Therefore for $\delta \in \mathrm{Gal}(\mathbf{H}^+(\gamma_0)/\mathbf{H}^+)$ we see that

$$\xi_0^\delta = \gamma_0^{(1-\sigma_J)\delta} = \gamma_0^{\delta(1-\sigma_J)} = \xi_0.$$

So $\xi_0 \in \mathbf{H}^+$ as claimed.

We can now finish the proof. Let ϕ be any rank one Drinfeld module defined over $\mathbf{C}_\infty$. By Theorem 7.2.15 we may assume that $\phi = \psi$ is sgn-normalized. Let J now be chosen so that $\deg(J) \equiv 1 \pmod{d_\infty}$ and set

$$\rho := \xi\psi\xi^{-1}$$

where $\xi^{r-1} = \xi_0$.

From the fact that $\deg(J) \equiv 1 \pmod{d_\infty}$, 7.6.9 and the remark following Corollary 7.4.9, one computes that the coefficients of ρ lie in **H**. As the top coefficient of ρ must be a unit by construction, we see that ρ is actually defined over $\mathcal{O}$ and the theorem is proved. □

One now readily deduces the following Principal Ideal Theorem for $\mathcal{O}$.

Theorem 7.6.10. *Let ρ be the Drinfeld module defined over $\mathcal{O}$ constructed above. Let $I \subseteq \mathbf{A}$ be an ideal and let ρ_I, $D\rho_I$ be constructed in the obvious fashion (following the sgn-normalized case). Then*

$$I\mathcal{O} = (D\rho_I).$$
□

Let ϕ be *any* rank one Drinfeld module defined over $\mathbf{H}$ and $I \subseteq \mathbf{A}$ an ideal with Artin symbol $\sigma_I \in \text{Gal}(\mathbf{H}/\mathbf{k})$. then one checks that

$$\sigma_I \phi \simeq I * \phi$$

over the algebraic closure of $\mathbf{k}$ in $\mathbf{C}_\infty$.

Finally, the Principal Ideal Theorem holds in fact for a large family of "Hilbert class fields." See [Ros2] for an exposition of this theory.

7.7. A Rank One Version of Serre's Theorem

In the previous subsections we presented the basic theory of the abelian extension arising from division points of sign-normalized rank one Drinfeld modules. The reader may now be wondering exactly what may be said about the division fields of *arbitrary* rank one Drinfeld modules. In this subsection, we handle this situation by proving a *rank one* analog of Serre's famous theorem on Galois groups of elliptic curves *without* complex multiplication [Se3].

Let $\mathbf{k}$ be our base field, as usual, and L be a finite extension of $\mathbf{k}$ inside a fixed algebraic closure $\overline{\mathbf{k}}$.

Theorem 7.7.1. *Let ψ be a Drinfeld module of rank 1 over L. Let $\wp \in \text{Spec}(\mathbf{A})$ and let $\rho_\wp \colon \text{Gal}(L^{\text{sep}}/L) \to \mathbf{A}_\wp^*$ be the $\wp$-adic Tate character arising from the Galois action on $T_\wp(\psi)$. Then the following hold.*
1. *The image of $\rho_\wp$ is an open subgroup of $\mathbf{A}_\wp^*$ for all $\wp$.*
2. *For almost all $\wp$, the image of $\rho_\wp$ is $\mathbf{A}_\wp^*$.*

The proof of Theorem 7.7.1 will use the moduli theory of rank one Drinfeld modules as given in Subsection 7.1. In particular, we will use the description of the moduli space of Drinfeld modules of rank 1 together with level I structure as the spectrum M_I^1 of the ring of $\mathbf{A}$-integers in the maximal abelian extension of $\mathbf{k}$ which is totally split at ∞ and which has conductor I. We have denoted the field of functions of M_I^1 by $\mathbf{k}(I)^+$. The fields $\mathbf{k}(I)^+$ contain the Hilbert class field, $\mathbf{H}$, which is split totally at ∞ and one has

$$\mathrm{Gal}(\mathbf{k}(I)^+/\mathbf{H}) \simeq (\mathbf{A}/I)^*/\mathbb{F}_r^*.$$

Let $\wp$ be a prime of $\mathbf{A}$ and let $L(\wp^m)$ be the extension of L obtained by adjoining the group of $\wp^m$-division points of ψ as before. In this situation, one has both a rank 1 Drinfeld module and a level $\wp^m$-structure; thus we deduce an injection of $\mathbf{k}(\wp^m)^+$ into $L(\wp^m)$. (The reader should note that this fact can also be seen via Hayes' fields of definition of Drinfeld modules as in Subsection 7.3.) The idea of the proof of the theorem is, essentially, to use known facts about the Galois groups of the tower

$$\mathbf{H} \subseteq \mathbf{k}(\wp)^+ \subset \cdots \subset \mathbf{k}(\wp^m)^+$$

to show that the Galois groups in question are "big."

Lemma 7.7.2. *Let $m > 1$. Then the compositum of $L(\wp)$ and $\mathbf{k}(\wp^m)^+$ is $L(\wp^m)$.*

Proof. Indeed, one knows that the compositum is contained in $L(\wp^m)$ by the above remarks. On the other hand, let λ_0 be any non-trivial division point of order $\wp$ and let λ_1 be any division point of order $\wp^m$. Then one checks readily that $\frac{\lambda_1}{\lambda_0} \in \mathbf{k}(\wp^m)^+$. The result now follows easily. $\square$

Lemma 7.7.3. *Let K be a finite extension of L. In order to establish Theorem 7.7.1 for $\{L, \psi\}$, it is enough to establish it for $\{K, \psi\}$.*

Proof. Indeed, standard Galois theory shows that the image of the $\wp$-adic Tate character associated to $\{K, \psi\}$ is an open subgroup of finite index in the image of the $\wp$-adic Tate character associated to $\{L, \psi\}$. $\square$

Proof of Part 1 of Theorem 7.7.1. Let $\wp$ be a prime of $\mathbf{A}$ and $\mathbf{k}(\wp^\infty)^+$ be the union of all $\mathbf{k}(\wp^t)^+$ for $t \geq 0$.

Let $U_t \subset \mathbf{A}_\wp^*$ be the group of units $\equiv 1 \mod (\wp^t)$ for $t \geq 0$.

By Lemma 7.7.3 we can replace L by $L(\wp)$. By Lemma 7.7.2 we have

$$L \cdot \mathbf{k}(\wp^j) = L(\wp^j),$$

for all $j > 1$.

Let $L_1 \subset L$ be the intersection of L and $\mathbf{k}(\wp^\infty)^+$ with conductor $\wp^m$, for some $m \geq 1$. Galois theory now implies that for $j \geq m$

$$\mathrm{Gal}(L(\wp^j)/L(\wp^m)) \simeq \mathrm{Gal}(\mathbf{k}(\wp^j)^+/\mathbf{k}(\wp^m)^+) \simeq U_m/U_j.$$

A priori one knows that $\mathrm{Gal}(L(\wp^t)/L) \subseteq (\mathbf{A}/\wp^t)^*$ for all t. Thus the image of the $\wp$-adic Tate character $\rho_\wp$ must be a union of cosets mod $\wp^m$. This completes the proof. $\square$

We now turn towards the proof of Part 2 of Theorem 7.7.1. The reader will note some complications arising in this proof from possible inseparable elements in L. Before giving the proof we will need a few more results.

Let $L_2 \subseteq L$ be the maximal separable (over **k**) subfield, (the reader will note that the field L_1 defined above is a subfield of L_2). Let p^δ be the inseparable degree of L/L_2. Let $\mathfrak{P}$ be a prime of **A** which is unramified in L_2. Let $\mathfrak{P}_1$ be a fixed prime of L_2 above $\mathfrak{P}$; so if $\mathfrak{b} \in \mathfrak{P}$ is of order 1 at $\mathfrak{P}$, then it is also of order 1 at $\mathfrak{P}_1$. One knows that there is a unique prime $\mathfrak{P}_2$ of L above $\mathfrak{P}_1$ and we let $\mathfrak{b}_1$ be an element of order 1 at $\mathfrak{P}_2$. One knows from Dedekind theory that

$$\mathfrak{b} = \mathfrak{b}_1^e \cdot u$$

where u is a unit at $\mathfrak{b}_1$ and $e > 0$.

Lemma 7.7.4. *The index e is a power of p.*

Proof. One knows that $\mathfrak{b}_1^{p^\delta} \in L_2$; thus

$$\mathfrak{b}_1^{p^\delta} = \mathfrak{b}^v \cdot u_1$$

for u_1 a unit. Now substitute the above formula for $\mathfrak{b}$ into this formula. One sees that $p^\delta = ve$. □

Let L_3 be the intersection of L and the maximal abelian extension of **k** split totally at ∞ and let $\mathcal{T}$ be its conductor. Let $\wp$ now be a prime *not* dividing $\mathcal{T}$.

Lemma 7.7.5.
$$L(\wp) \cap \mathbf{k}(\wp^\infty)^+ = \mathbf{k}(\wp)^+.$$

Proof. Indeed, we know that $\mathbf{k}(\wp)^+$ is contained in the intersection. Let α be an element in the intersection which is not contained in $\mathbf{k}(\wp)^+$. We show that this is impossible. By definition $\alpha \in \mathbf{k}(\wp^j)^+$ for some $j > 1$. Moreover, α is also an abelian element over **H**; its group over **H** is a quotient of $(\mathbf{A}/\wp^j)^*/\mathbb{F}_r^*$. We would have a contradiction if we show that every element of $\mathrm{Gal}(\mathbf{k}(\wp^j)^+/\mathbf{k}(\wp)^+)$ fixes α.

We know that $\mathrm{Gal}(\mathbf{k}(\wp^j)^+/\mathbf{k}(\wp)^+)$ is a p-group. By assumption

$$\mathbf{H}(\alpha) \cap L = \mathbf{H}.$$

Thus
$$\mathrm{Gal}(\mathbf{H}(\alpha)/\mathbf{H}) \simeq \mathrm{Gal}(L(\alpha)/L).$$

But the latter Galois group is a subgroup of $(\mathbf{A}/\wp)^*$ and so has order prime to p which gives the result. □

7.7. A Rank One Version of Serre's Theorem

Proof of Part 2 *of* Theorem 7.7.1. Let $\wp$ now be a prime of $\mathbf{A}$ with the following properties:
1. ψ has good reduction at the primes of L above $\wp$;
2. $\wp$ is unramified in L/L_2;

It is clear that almost all primes of $\mathbf{A}$ satisfy these assumptions. Moreover, by our knowledge of the ramification properties of the extensions $\mathbf{k}(I)^+/\mathbf{k}$ we see that such a $\wp$ is also prime to the conductor $\mathcal{T}$ of L_3 defined above, (note also that $L_1 \subseteq L_3 \subseteq L_2 \subseteq L$). The proof of Part 1 of the theorem and Lemma 7.7.5 thus imply that in order to establish Part 2 for $\wp$, we need only establish that the injection

$$\mathrm{Gal}(L(\wp)/L) \hookrightarrow (\mathbf{A}/\wp)^*,$$

is actually an isomorphism.

Let λ_1 and λ_2 be two non-trivial $\wp$-division points. As we have seen, the quotient λ_1/λ_2 is a unit.

Now every rank one Drinfeld module is absolutely isomorphic to a rank one Drinfeld module defined over the $\mathbf{A}$-integers of $\mathbf{H}$ by Takahashi's Theorem. Let ϕ be one such Drinfeld module associated to ψ and let

$$\phi_\wp(x) = \prod(x - \tilde{\lambda}_j)$$

where $\tilde{\lambda}_j$ runs over the $\wp$-division points of ϕ. The Principal Ideal Theorem (7.6.10) tells us that

$$\tilde{\mathfrak{b}} = \prod_{0 \neq \tilde{\lambda}_j \in \phi[\wp]} \tilde{\lambda}_j$$

is an element in the integers of $\mathbf{H}$ which has order 1 at the primes of $\mathbf{H}$ over $\wp$. As ψ has good reduction at the primes over $\wp$ we *also* conclude that the element of L

$$\mathfrak{b} = \prod_{0 \neq \lambda_j \in \psi[\wp]} \lambda_j$$

has the same valuation as $\tilde{\mathfrak{b}}$ at the primes above $\wp$.

Set $\mathfrak{P} = \wp$ and let $\mathfrak{b}_1$ be as before; so $\mathfrak{b}_1$ is an element of order 1 at a prime of L over $\wp$. We conclude that for fixed non-zero $\lambda \in \psi[\wp]$

$$v_{\mathfrak{b}_1}(\lambda) = \frac{v_{\mathfrak{b}_1}(\mathfrak{b})}{r^{\deg \wp} - 1} = \frac{v_{\mathfrak{b}_1}(\tilde{\mathfrak{b}})}{r^{\deg \wp} - 1}$$

and by Lemma 7.7.4 we conclude that $v_{\mathfrak{b}_1}(\mathfrak{b})$ is a power of p. Thus

$$v_{\mathfrak{b}_1}(\lambda) = \frac{p^t}{r^{\deg \wp} - 1},$$

for some $t \geq 0$. The only way this can happen is that $L(\wp)/L$ is totally ramified of degree $r^{\deg \wp} - 1$ at $\mathfrak{b}_1$ which completes the proof. □

7.8. Classical Partial Zeta Functions

The results of this subsection, and the next, are background for the computation of lattice invariants given in Subsection 7.10 which is due to E.-U. Gekeler. In this subsection we discuss some "classical" partial zeta functions. By "classical" we shall always mean complex valued functions in the standard fashion of Artin and Weil. For more, see [Ge2, §III].

We begin by recalling the classical zeta function of X itself. Let X_0 be the set of closed points of X. For $x \in X_0$, let $\mathcal{O}_x$ be the associated local ring with maximal ideal M_x. Set

$$\mathbb{F}_x := \mathcal{O}_x/M_x$$

and

$$Nx := \#(\mathcal{O}_x/M_x).$$

Definition 7.8.1. Let $s \in \mathbb{C}$. We set

$$\zeta_X(s) := \zeta_\mathbf{k}(s) := \prod_{x \in X_0} (1 - Nx^{-s})^{-1}.$$

Let D be a divisor of X and set

$$ND := r^{\deg D}.$$

One then sees easily that

$$\zeta_\mathbf{k}(s) = \sum_D ND^{-s}$$

where D runs over the *positive* divisors of X. The reader should note that ∞ plays no special role in this definition.

Set $u := r^{-s}$. Then it is very well-known that

$$\zeta_\mathbf{k}(s) = Z_\mathbf{k}(u)$$

where

$$Z_\mathbf{k}(u) = \frac{P_\mathbf{k}(u)}{(1-u)(1-ru)}.$$

Moreover, if g = genus X = genus of $\mathbf{k}$, then $P_\mathbf{k}(u)$ is a polynomial of degree $2g$ in u with the following (*incomplete*) list of properties:

1. $P_\mathbf{k}(0) = 1$.
2. $P_\mathbf{k}(1) = h(\mathbf{k})$, where $h(\mathbf{k})$ is the class number of the field $\mathbf{k}$.
3. $P_\mathbf{k}(u)$ satisfies the functional equation

$$P_\mathbf{k}(u) = r^g u^{2g} P_\mathbf{k}(1/ru).$$

We set
$$\zeta_{\mathbf{A}}(s) := (1 - N\infty^{-s})\zeta_{\mathbf{k}}(s)$$
and
$$Z_{\mathbf{A}}(u) := (1 - u^{d_\infty})Z_{\mathbf{k}}(u).$$
It is thus easy to see that
$$Z_{\mathbf{A}}(u) = P_{\mathbf{A}}(u)/(1 - ru)$$
where $P_{\mathbf{A}}(u) := (1 + u + \cdots + u^{d_\infty - 1})P_{\mathbf{k}}(u)$. Thus
$$P_{\mathbf{A}}(1) = d_\infty h(\mathbf{k}) = h(\mathbf{A})$$
where $h(\mathbf{A})$ is the class number of $\mathbf{A}$ as a Dedekind domain.

We can now define the partial zeta functions of interest to us for describing periods. If $I \subseteq \mathbf{A}$ is an ideal and $j \geq 0$, then we set
$$I_j := \{i \in I \mid \deg(i) \leq j\}.$$
Clearly I_j is a finite dimensional $\mathbb{F}_r$-vector space. For $j \gg 0$, this dimension is given exactly by the Riemann-Roch Theorem.

Definition 7.8.2. 1. Let α be an ideal class of $\mathbf{A}$. We set
$$\zeta_\alpha(s) := Z_\alpha(u) := \sum_{\substack{I \in \alpha \\ I \subseteq \mathbf{A}}} NI^{-s}.$$

2. Let $I \subseteq \mathbf{A}$ be an ideal and $a \in \mathbf{A}$. We then set
$$\zeta_{a,I}(s) := Z_{a,I}(u) := \sum_{\substack{b \in \mathbf{A} \\ b \equiv a\,(I)}} Nb^{-s} = \sum u^{\deg b},$$
where $Nb := N(b\mathbf{A})$.

Note that if $b = 0$, then Nb^{-s} is considered to be identically 0 and so contributes nothing to the sum.

Proposition 7.8.3. 1. $Z_{a,I}(u) = Z_{b,I}(u)$ if $a \equiv b\,(I)$.
2. $\sum_{\substack{a \,(\mathrm{mod}\,IJ) \\ a \equiv b\,(I)}} Z_{a,IJ}(u) = Z_{b,I}(u)$.
3. $Z_{ba,bI}(u) = u^{\deg b} Z_{a,I}(u)$ for $0 \neq b \in \mathbf{A}$.
4. Let $\zeta \in \mathbb{F}_r^*$. Then
$$Z_{\zeta a, I}(u) = Z_{a,I}(u).$$

Proof. All easy exercises. □

For any fractional ideal I of $\mathbf{A}$ we let $\tilde{I}$ be its ideal class. One then sees readily that if $J \subseteq \mathbf{A}$ is an ideal, then

$$(r-1)Z_{\tilde{J}}(u) = u^{-\delta} Z_{0,J}(u),$$

$\delta = \deg J$.

We now summarize some of the calculations of [Ge2, III]. They follow by direct checking, the Riemann-Roch Theorem, or the theory of Weierstrass gaps.

First of all, let $a \in \mathbf{A}$ and set

$$t(a) := \inf\{\deg b \mid b \equiv a \,(\mathrm{mod}\ I)\}$$

and

$$w(a) := \dim_{\mathbb{F}_r} I_{t(a)}.$$

If $f(u) = \sum_{j=0}^{\infty} a_j u^j$ is a power series, then for $i \geq 0$ we set

$$Q_i f(u) := \sum_{j=i+1}^{\infty} a_j u^j.$$

Proposition 7.8.4. 1. $Z_{a,I}(u)$ is a rational function in u.
2. $Z_{a,I}(u) = Q_{t(a)} Z_{0,I}(u) + r^{w(a)} u^{t(a)}$. □

Let N be a positive integer $\equiv 0\ (d_\infty)$ and set

$$_N Z_{a,I}(u) := \sum_{\substack{b \equiv a\ (I) \\ \deg b \leq N}} u^{\deg b}.$$

Thus, if "$\prime$" denotes $\frac{d}{du}$, then

$$_N Z'_{a,I}(1) = \sum_{\substack{b \equiv a\ (I) \\ \deg b \leq N}} \deg(b).$$

As $N \to \infty$, $_N Z_{a,I}(u)$ approaches $Z_{a,I}(u)$, etc.

Proposition 7.8.5. Let d be a positive integer $\equiv 0\ (\mathrm{mod}\ d_\infty)$. Then for $N \gg 0$, N as above, we have

$$(r^d - 1)Z'_{a,I}(1) = r^d \,_N Z'_{a,I}(1) - \,_{N+d} Z_{a,I}(1) + dr^{1-g+N+d-\deg(I)},$$

where, as above, g is the genus of $\mathbf{k}$. □

7.9. Unit Calculations

Let sgn continue to be our fixed sign function. Let π be our fixed *positive* uniformizer in $\mathbf{K} \subset \mathbf{C}_\infty$. Let $x \in \mathbf{K}^*$, then x can be written *uniquely* as

$$x = \mathrm{sgn}(x)\pi^j u$$

where $j \in \mathbb{Z}$ and $u \in \mathbf{K}^*$ is a 1-unit.

Definition 7.9.1. We set $\langle x \rangle := \langle x \rangle_\infty := \langle x \rangle_\pi := u$.

Lemma 7.9.2. $\langle xy \rangle = \langle x \rangle \langle y \rangle$. □

Thus x is written uniquely as

$$x = \mathrm{sgn}(x)\pi^j \langle x \rangle,$$

where $j = v_\infty(x) = -\deg(x)/d_\infty$, $d_\infty = [\mathbb{F}_\infty : \mathbb{F}_r]$.

Remark. 7.9.3. There are somewhat analogous decompositions in classical Archimedean theory. In $\mathbb{R}^*$ we have

$$x = \pm|x|,$$

and in $\mathbb{C}^*$ we have

$$x = |x|e^{i\theta}$$

for $\theta = \arg(x)$. Note, of course, that $|e^{i\theta}| = 1$.

Let I be an ideal of $\mathbf{A}$ and $a \in \mathbf{A}$ with $d = \deg(a)$. Let C be a system of representatives of I/aI. It is easy to see that we may choose C to be an $\mathbb{F}_r$-vector space. Moreover, by [Ge2, Lemma III. 2.7 and p.30] we may choose the elements of C to have degree $\leq m$ where m is the *smallest* integer $\geq 2g - 1 + d + \deg I$ which is $\equiv 0 \ (d_\infty)$, and where g is the genus of $\mathbf{k}$.

Let N be a natural number $\equiv 0 \ (d_\infty)$ in what follows. Recall that for $j \in \mathbb{Z}$, we let $I_j = \{i \in I \mid \deg(i) \leq j\}$. There is a map

$$I_N \times I/aI \to I_{N+d}$$

given by

$$(b, c) \mapsto ba + c,$$

with $c \in C$. For $N \gg 0$ this map is bijective by the Riemann-Roch Theorem. In this case one sees readily that each $b \in I_N$ corresponds to r^d such elements $ba + c$. Moreover,

$$\mathrm{sgn}(ba + c) = \mathrm{sgn}(b)\mathrm{sgn}(a).$$

Lemma 7.9.4. *For large N, the product*
$$\varepsilon(I, N) := \prod_{\substack{b \in I \\ \deg b = N}} \mathrm{sgn}(b)$$
does not depend on N. The limit as $N \to \infty$ depends only on the ideal class of I.

Proof. From above, we see that the product is d periodic for $d = \deg(a)$. By Riemann-Roch we see that
$$d_\infty = \mathrm{g.c.d.}\{\deg a \mid a \in \mathbf{A}\}.$$
Thus the first assertion. The second follows in a similar fashion. $\square$

Definition 7.9.5. Let $\tilde{I}$ be the ideal class of I. We denote the limit of Lemma 7.9.4 by $\varepsilon_{\tilde{I}}$.

Let N be as above.

Definition 7.9.6. Set
$$U_N := \prod_{\substack{i \in I \\ \deg i = N}} \langle i \rangle.$$

Proposition 7.9.7. *We have $U_N \to 1$ as $N \to \infty$.*

Proof. Let $j \geq 0$ be given with $j \equiv 0 \ (d_\infty)$. We want to show that for $N \geq N_0$, $u_N \equiv 1 \ (\pi^j)$. We can, of course, assume that j is large enough so that there exists $a \in \mathbf{A}$ with $\deg(a) = j$. Recall now that if $N \gg 0$, with $N \equiv 0 \ (d_\infty)$, then every element $i \in I_{j+N}$ can be written
$$i = ab + c$$
with $b \in I_N$ and c of degree $\leq m$, where m is the smallest integer $\geq 2g - 1 + j + \deg I$, $g = $ genus $\mathbf{k}$, which is divisible by d_∞.

Now choose $N_0 \gg 0$ and so that
$$N_0 + j - m \geq N_0 + 1 - 2g - \deg I \ \text{is} \ \geq j.$$
One then sees that for $N \geq N_0$, and i as above,
$$\prod \langle i \rangle \equiv (\prod \langle a \rangle)^{r^j} \langle b \rangle^{r^t} \ (\mathrm{mod} \ \pi^j)$$
where $t \geq j$. But as we are in characteristic p, it is trivial to see that the product on the right is $\equiv 1 \ (\pi^j)$. $\square$

We shall use the notation
$$\text{``} \prod{}' \alpha_n \text{''}$$
to always mean that we take the product over the *nonzero* α_n, etc.

Definition 7.9.8. 1. We set
$$u_{0,I} = \lim_{N \to \infty} {\prod_{i \in I, \deg i \leq N}}' \langle i \rangle.$$

2. Let $b \in \mathbf{k} - I$. Then we set
$$u_{b,I} = \lim_{N \to \infty} \prod_{\substack{i \equiv b \ (I) \\ \deg i \leq N}} \langle i \rangle.$$

By 7.9.7 both products converge to 1-units in $\mathbf{K}^*$.

Remark. 7.9.9. Let π_1 be another positive uniformizer. Thus $v := \pi_1/\pi$ is a 1-unit. Note that $v^{p^t} \to 1$ as $t \to \infty$ as above. Thus one can use v, and its p-adic powers, to relate $u_{0,I}$ as defined through the use of either $\langle ? \rangle_\pi$ or $\langle ? \rangle_{\pi_1}$.

7.10. Period Computations

In this subsection we finally present the computations of the periods of sign normalized rank one Drinfeld modules. These are due to E.-U. Gekeler following ideas of D. Hayes. They generalize the formula of Subsection 3.2 (see 3.2.10.3, for instance) for the Carlitz module.

Let sgn continue to be our fixed sign function with π a fixed positive uniformizer in $\mathbf{K} \subset \mathbf{C}_\infty$.

Definition 7.10.1. Let $\Lambda \subset \mathbf{C}_\infty$ be a rank one $\mathbf{A}$-lattice. The lattice Λ is *special* if and only if the associated rank one Drinfeld module is sgn-normalized.

We know, by Theorem 7.2.15, that every rank one lattice is isomorphic over $\mathbf{C}_\infty$ to a special lattice.

Definition 7.10.2. We let $\xi(\Lambda) \in \mathbf{C}_\infty^*$ be defined by the requirement that $\xi(\Lambda) \cdot \Lambda$ is special.

Remark. 7.10.3. Let $\zeta \in \mathbb{F}_\infty^*$. As we have seen,
$$\zeta \xi(\Lambda) \Lambda$$
is *also* sign-normalized. Thus $\xi(\Lambda)$ is defined *only* up to an element of $\mathbb{F}_\infty^*$. Note also that if $\zeta \in \mathbb{F}_\infty^* - \mathbb{F}_r^*$, then $\zeta \xi(\Lambda) \Lambda$ is a *distinct* lattice from $\xi(\Lambda) \Lambda$.

Let $I \subseteq \mathbf{A}$ be an ideal. Let $a \in \mathbf{A}$ be a *positive* element with $d = \deg(a)$ and
$$N(a) := r^d = |a|_\infty$$

as before. Let $L = \xi(I) \cdot I$ and let $\psi := \psi_L$ be the sgn-normalized rank one Drinfeld module associated to L.

Lemma 7.10.4. *Let $z \in \mathbf{C}_\infty$. Then*
$$e_L(\xi(I)z) = \xi(I)e_I(z),$$
where $e_I(z) := z \prod_{0 \neq i \in I} (1 - z/i) = z {\prod}'_{i \in I} (1 - z/i).$

Proof. Obvious from the infinite product for $e_L(z)$ and $e_I(z)$. □

Note that
$$\psi_a(\tau) = a\tau^0 + \cdots + \tau^{\deg(a)}.$$
We thus deduce that
$$a = {\prod}'_{\alpha \in I/aI} e_L(\xi(I)\alpha/a) = {\prod}'_\alpha (\xi(I)e_I(\alpha/a)).$$
(Where we recall $\prod'$ is over *nonzero* elements.) Thus
$$a = \xi(I)^{N(a)-1} {\prod}'_\alpha e_I(\alpha/a). \tag{7.10.5}$$
As
$$e_I(z) = z \prod_{0 \neq i \in I} (1 - z/i),$$
we see
$$\xi(I)^{1-r^d} = \xi(I)^{1-N(a)} = a^{-N(a)} {\prod}'\alpha {\prod}'_{i \in I}(1 - \alpha/ai)$$
$$= a^{-N(a)} {\prod}'\alpha {\prod}'_{i \in I}\left(\frac{ai - \alpha}{ai}\right). \tag{7.10.6}$$

Recall that for $N \geq 0$, $I_N = \{i \in I \mid \deg(i) \leq N\}$.

Lemma 7.10.7. *Let $N \gg 0$. Then $\xi(I)^{1-N(a)}$ is the limit as $N \to \infty$ of the finite expression*
$$a^{-N(a)} {\prod}'_{b \in I_{N+d}} b / {\prod}'_{i \in I_N} (ai)^{N(a)}.$$
(Note that without loss of generality, we can always assume $d_\infty \mid N$.)

Proof. As $\alpha \neq 0$ varies, for $N \gg 0$ we obtain all elements of I_{N+d} except for those in aI. But letting $\alpha = 0$ is harmless as we are dividing by the same elements. □

Lemma 7.10.8. *Let $\{x_i\} \subset \mathbf{C}_\infty^*$ be a finite collection of numbers with weight g_i assigned to x_i. We assume that all g_i are prime to the characteristic p of $\mathbf{C}_\infty^*$. Assume also that*

$$\prod x_i^{t_i} = 1 \text{ whenever } \sum t_i g_i = 0.$$

Then there exists a well defined number y of weight $g = \text{g.c.d.}\{g_i\}$ such that

$$x_i = y^{g_i/g}$$

and y lies in the multiplicative group generated by the x_i.

Proof. The proof will be by induction on the number N of elements x_i. Suppose, to being with, we have two elements x_1, x_2 with weights g_1, g_2. There are now two cases to consider depending on whether

$$(g_1, g_2) = 1 \text{ or is } > 1.$$

In the first case, we see

$$x_1^{g_2} = x_2^{g_1}.$$

Set $\widetilde{y} =$ equal to a fixed $(g_1 \cdot g_2)$-th root of $x_1^{g_2}$ which is also a $(g_1 \cdot g_2)$-th root of $x_2^{g_1}$. Notice that $\widetilde{y}^{g_1} = \zeta_1 x_1$ and $\widetilde{y}^{g_2} = \zeta_2 x_2$ where ζ_1 (resp. ζ_2) is a g_2-th root of 1 (resp. g_1-st root of 1).

Let μ_t be the group of t-th roots of 1 in $\mathbf{C}_\infty^*$. Note that the map

$$\mu_{g_1 g_2} \xrightarrow{(g_1, g_2)} \mu_{g_2} \times \mu_{g_1}$$

$$\zeta \mapsto (\zeta^{g_1}, \zeta^{g_2})$$

is an isomorphism. Thus we can find an element $\beta \in \mu_{g_1 g_2}$ such that $\beta^{g_1} = \zeta_1$ and $\beta^{g_2} = \zeta_2$. To finish, set $y = \widetilde{y}/\beta$. If $sg_1 + tg_2 = 1 \Rightarrow x_1^s x_2^t = y$.

If $k = (g_1, g_2) \neq 1$, then replace g_1 by g_1/k and g_2 with g_2/k.

If the number N is > 2, choose the first $N-1$ elements and use induction to get an element y of weight $\widetilde{g} = \text{g.c.d.}\{g_1, \ldots, g_{N-1}\}$. Then apply the above argument to $\{y, x_N\}$. $\square$

We now return to the formula of Lemma 7.10.7. We now decompose by splitting all the elements into their sgn, 1-unit and π^j parts. Thus the formula of Lemma 7.10.7 can be written as

$$\text{sgn}(a)^{-j} \cdot \varepsilon_N \cdot \langle a \rangle^{-j} \pi^k \prod {}'\langle b \rangle / \prod {}'\langle i \rangle^{r^d}, \tag{7.10.9}$$

where

7. Sign Normalized Rank 1 Drinfeld Modules

$$d = \deg(a),$$
$$j = r^d(\#I_N - 1) + r^d$$
$$= r^{1-g+N+d-\deg(I)}, \quad \text{(by Riemann - Roch, } g = \text{genus } \mathbf{k}\text{)}$$
$$\varepsilon_N = \prod{}' \operatorname{sgn}(b) / \prod{}' \operatorname{sgn}(i),$$

and
$$k = (r^d \sum{}' \deg(i) - \sum{}' \deg(b) + dj)/d_\infty,$$

where "$\sum{}'$" always means sum over nonzero elements.

We now let $N \to \infty$ with $d_\infty \mid N$. Note that, as $\langle a \rangle$ is a 1-unit, $\langle a \rangle^j \to 1$. Also as a is positive, $\operatorname{sgn}(a) = 1$. By examining the products in ε_N, we see that
$$\lim_{N \to \infty} \varepsilon_N = \varepsilon_{\widetilde{I}}^{d/d_\infty},$$
where $\varepsilon_{\widetilde{I}}$ is defined in Definition 7.9.5. By 7.8.5, $\lim_{N \to \infty} k = (r^d - 1) Z'_{0,I}(1)/d_\infty$.
Finally
$$\lim_{N \to \infty} \prod{}' \langle b \rangle / \prod{}' \langle i \rangle^{r^d} = u_{0,I}^{1-r^d},$$
where $u_{0,I}$ is as in Definition 7.9.8.

Recall that we set $W = r^{d_\infty} - 1$ and $w = r - 1$. By Riemann-Roch we see that
$$d_\infty = \text{g.c.d.}\{\deg(a) \mid a \in \mathbf{A} \text{ is positive}\}.$$
Thus, by Lemma 7.6.5,
$$r^{d_\infty} - 1 = \text{g.c.d.}\{r^{\deg(a)} - 1 \mid a \in \mathbf{A} \text{ is positive}\}.$$

Theorem 7.10.10. *We have*
$$\xi(I)^W = \varepsilon_{\widetilde{I}}^{-1} \pi^{-t} u_{0,I}^W \in \mathbf{K},$$
where $t = W \cdot Z'_{0,I}(1)/d_\infty$.

Proof. From above we have
$$\xi(I)^{1-N(a)} = \varepsilon_{\widetilde{I}}^{d/d_\infty} \pi^k u_{0,I}^{1-N(a)},$$
or
$$\xi(I)^{r^d-1} = \varepsilon_{\widetilde{I}}^{-d/d_\infty} \pi^{-k} u_{0,I}^{r^d-1}.$$
Note that $\varepsilon_{\widetilde{I}}$ is a W-th root of unity and also that for any $\zeta \in \mu_W$,
$$\zeta^{(r^d-1)/W} = \zeta^{d/d_\infty}.$$
The result now follows from 7.10.8. □

Therefore, we know $\xi(I)$ up to a W-th root of 1.

Corollary 7.10.11. $|\xi(I)|_\infty = r^{Z'_{0,I}(1)}$. □

Example 7.10.12. Let $\mathbf{A} = \mathbb{F}_r[T]$ and $I = \mathbf{A}$. Thus $Z_\mathbf{A}(u) = \frac{1}{1-ru}$ and $Z_{0,\mathbf{A}}(u) = \frac{r-1}{1-ru}$. Consequently

$$Z'_{0,\mathbf{A}}(1) = \frac{r}{r-1}$$

in agreement with the calculations of Subsection 3.2 for the period of the Carlitz module.

Theorem 7.10.10 is analogous to the classical formula

$$\pi = 2 \prod_{n=1}^{\infty} (1 - 1/4n^2)^{-1}$$
$$= \lim_{N \to \infty} \prod_{\substack{a \in \mathbb{Z} \\ (a) \leq N}}' (1 - 1/2a)^{-1}$$

obtained from $\sin z$.

7.11. The Connection with Shtukas and Examples

We establish here the connection with shtukas as originally presented in Section 6. Our main reference here is [Th1].

We begin by placing ourselves in the situation of "two T's" as explained in Remark 5.4.3.1. Let $\mathbf{A}$, $\mathbf{k}$, $\mathbf{H}$, $\mathbf{H}^+$, $\mathbf{K}$, $\mathbf{C}_\infty$ be as before, and let A, k, H, H^+, K, C_∞ be another copy of these rings. Let $\overline{K} \subset C_\infty$ be the algebraic closure with $\overline{K} \subset C_\infty$ its copy. We always have the canonical isomorphisms of these rings from "bold" to "non-bold;" we denote the maps by "θ_{con}."

The mapping

$$\mathbf{A} \overset{\theta_{\text{con}}}{\to} A \hookrightarrow H^+ \subset C_\infty$$

gives a C_∞-point of $\text{Spec}(\mathbf{A})$. We denote this point by "Ξ", (N.B.: in [Th1] this point is denoted by "ξ" but we have used that symbol for periods). If X is the curve associated to $\mathbf{k}$, then $\Xi \in X(C_\infty)$. Note, of course, that Ξ lies over the generic point of $\text{Spec}(\mathbf{A})$ and so of X as a scheme.

Let sgn be a fixed sign function for $\mathbf{K}$. Let ψ be a fixed Hayes module for sgn over $\mathbf{H}^+$. Via θ_{con} we view ψ as being defined over $H^+ \subset C_\infty$.

We now introduce some useful notation as in [Th1]. Let $\overline{X} = C_\infty \otimes_{\mathbb{F}_r} X$; so Ξ is a closed point of $\overline{X}$. Let $\tau^i: C_\infty \to C_\infty$ be the r^i-th power mapping, and $\text{Spec}(\tau^i)$ the associated map $\text{Spec}(C_\infty) \to \text{Spec}(C_\infty)$ as schemes over $\mathbb{F}_r$. If f is a meromorphic function on $\overline{X}$ we let $f^{(i)}$ be its pullback via

$$\text{Spec}(\tau^i) \times 1_X : \overline{X} \to \overline{X}.$$

We extend this notation to locally free sheaves in the obvious fashion. If $\mathcal{F}$ is locally free, then $\mathcal{F}^{(1)}$ is just $\tilde{\sigma}^*\mathcal{F}$ in the notation of Section 6. If $x \in X(C_\infty)$, then $x^{(i)} \in X(C_\infty)$ will denote the point obtained by raising the co-ordinates of X to the r^i-th power. If $\{x_1, \ldots, x_t\}$ is the set of zeros (or poles) of f, then $\{x_1^{(i)}, \ldots, x_t^{(i)}\}$ is the set of zeros (or poles) of $f^{(i)}$. We extend the definition of $x^{(i)}$ to divisors in the obvious linear fashion.

As before, we let $\pi \in \mathbf{K}$ be a fixed positive uniformizer. One knows that there are d_∞ points

$$\{\infty_1, \ldots, \infty_{d_\infty}\}$$

over ∞ in $X(C_\infty)$. Each ∞_i corresponds to a class of embeddings of H into C_∞ over k. Let $\overline{\infty} \in \{\infty_1, \ldots, \infty_{d_\infty}\}$ be one such point; the point $\overline{\infty}$ will be chosen precisely later on. The uniformizer $\pi \in \mathbf{K}$ pulls back to a uniformizer at $\overline{\infty}$. If f is a meromorphic function on $\overline{X}$, then f can be written

$$\sum_{j=-m}^{\infty} a_j \pi^j, \quad \{a_j\} \subset C_\infty$$

with $a_{-m} \neq 0$. We set

$$\widetilde{\mathrm{sgn}}\, f = a_{-m} \in C_\infty.$$

Thus $\widetilde{\mathrm{sgn}}$ extends sgn; if $f \in \mathbf{k}$, then $\widetilde{\mathrm{sgn}}\, f$ can be seen to be a twisting of $\mathrm{sgn}\, f$.

Let D be a line bundle on $\overline{X}$ of degree $g-1$ where $g = \mathrm{genus}\, \overline{X}$. By the Riemann-Roch Theorem

$$\chi(D) := h^0(D) - h^1(D) = 0.$$

From Section 6 we know that ψ corresponds to a line bundle $\mathcal{F}_0$ of degree $g-1$ on $\overline{X}$. By definition

$$\mathcal{F}_0^{(1)} \simeq \mathcal{F}_0(-\Xi + \infty_i)$$

for some closed point ∞_i of $\overline{X}$ over ∞. We now set

$$\overline{\infty} := \infty_i.$$

Set $\mathcal{L} := \mathcal{F}_1 := \mathcal{F}_0(\overline{\infty})$. Thus

$$\mathcal{L}^{(1)} = \mathcal{F}_0^{(1)}(\overline{\infty}^{(1)}) \simeq \mathcal{F}_0(-\Xi + \overline{\infty})(\overline{\infty}^{(1)})$$
$$\simeq \mathcal{F}_0(\overline{\infty})(-\Xi + \overline{\infty}^{(1)})$$
$$\simeq \mathcal{L}(-\Xi + \overline{\infty}^{(1)}).$$

Note that $\deg(\mathcal{L}) = g$. Thus, again by Riemann-Roch, $\chi(\mathcal{L}) = 1$. In particular, $h^0(\mathcal{L}) \geq 1$; so there is an *effective* divisor V which represents the class of $\mathcal{L}$. By Drinfeld's result, Proposition 6.2.3, $h^0(V - (\overline{\infty})) = h^1(V - (\overline{\infty})) = 0$. Thus we see

7.11. The Connection with Shtukas and Examples

$$h^0(\mathcal{L}) = h^0(V) = 1.$$

In particular, V is the *unique* effective divisor in its class.

Definition 7.11.1. The divisor V is the *Drinfeld divisor* associated to ψ.

By definition
$$V^{(1)} - V + (\Xi) - (\overline{\infty}^{(1)}) = (f)$$
for some function $f = f_V$ on $\overline{X}$. We normalize f so that $\widetilde{\mathrm{sgn}}(f) = 1$ in which case f is *unique*.

Definition 7.11.2. The function f_V is the *shtuka function* associated to ψ.

Given that $\Xi \neq \overline{\infty}^{(i)}$ for any i, once we have f, we can find V and thus $\mathcal{L}$ and ψ. In other words, the whole theory may be read off from f_V and, in particular, the general theory will turn out to be very similar to that of the Carlitz module! We shall now explain how this is so by calculating the exponential, $e(z)$, and logarithm, $\log(z)$, of ψ in terms of f, (see also the discussion after Theorem 9.8.9).

Lemma 7.11.3. *Let V be the Drinfeld divisor associated to ψ as above. Then $\overline{\infty}$ and Ξ cannot belong to the support of V.*

Proof. If $\overline{\infty}$ is in the support of V, then $V - (\overline{\infty})$ is still effective. Thus $h^0(V - (\overline{\infty})) \geq 1$ contradicting Proposition 6.2.3. Suppose now that Ξ is in the support of V. Set
$$W := V - (\Xi) + (\overline{\infty}^{(1)}).$$
Notice that W is an effective divisor which represents the class of $\mathcal{L}^{(1)}$. Thus
$$W^{(-1)} = V^{(-1)} - (\Xi^{(-1)}) + (\overline{\infty})$$
is an effective divisor representing the class of $\mathcal{L}$ itself with $(\overline{\infty})$ in its support. This is again a contradiction by the reasoning just given. $\square$

Proposition 7.11.4. *Let f be the shtuka function associated to ψ as above, and let $e(z)$ be the exponential of ψ. There is a unique $\frac{r^{d_\infty}-1}{r-1}$-st root of unity ζ such that if we put $\hat{f} = \zeta f$, then*
$$e(z) = z + \sum_{n=1}^{\infty} \frac{z^{r^n}}{(\hat{f}^{(0)} \cdots \hat{f}^{(n-1)})\,|_{\Xi^{(n)}}}.$$

Proof. Recall that $\mathcal{L} = \mathcal{F}_0(\overline{\infty})$ where $\mathcal{F}_0$ is the line bundle of degree $g-1$ associated to ψ. We now represent $\mathcal{L}$ as $\mathcal{O}_{\overline{X}}(V)$, where V is the Drinfeld divisor.

From the "Data A $\Rightarrow$ Data B" construction of Subsection 6.2, we need a basis for the space of sections of $\mathcal{O}_{\overline{X}}(V)$ over $\overline{X} - \infty$. For this basis we use

$$1, f^{(0)}, f^{(0)}f^{(1)}, f^{(0)}f^{(1)}f^{(2)}, \ldots .$$

Let $a \in \mathbf{A}$; we now express a as a function operating on $\mathcal{O}_{\overline{X}}(V)$ over $\overline{X}_\infty$ as

$$a = a \cdot 1 := \sum c_{a,j} f^{(0)} \cdots f^{(j-1)}, \quad \{c_{a,j}\} \subset C_\infty; \qquad (7.11.5)$$

from our choice of f and its normalization, we see that

$$a \mapsto f_a = \sum c_{a,j} \tau^j$$

is a sgn-normalized rank one Drinfeld module which is isomorphic over C_∞ to ψ.

Thus, for some $\zeta \neq 0$, if we put $\widehat{f} = \zeta f$ and

$$a = a \cdot 1 := \sum \psi_{a,j} \widehat{f}^{(0)} \cdots \widehat{f}^{(j-1)},$$

we find

$$\psi_a = \sum \psi_{a,j} \tau^j .$$

As ψ is also sgn-normalized, we find $\zeta^{\frac{r^{d_\infty}-1}{r-1}} = 1$.

Now put

$$\widehat{e}(z) := z + \sum_{n=1}^{\infty} \frac{z^{r^n}}{(\widehat{f}^{(0)} \cdots \widehat{f}^{(n-1)})\mid_{\Xi^{(n)}}} .$$

From 7.11.3, we deduce that the coefficients of $\widehat{e}(z)$ are never infinite. To see that $\widehat{e}(z)$ satisfies the correct functional equation, we divide both sides of 7.11.5 by $\widehat{f}^{(0)} \cdots \widehat{f}^{(n-1)}$ and then evaluate at $\Xi^{(n)}$. $\square$

If $d_\infty = 1$, then $f = \widehat{f}$.

Another way to find $\widehat{f}$, suggested by G. Anderson, is as follows. Consider $M := C_\infty[\tau]$ as a $C_\infty \otimes_{\mathbb{F}_r} \mathbf{A}$-module (as in Subsection 5.4) via ψ. As ψ has rank 1, M is projective of rank one over $C_\infty \otimes \mathbf{A}$. Thus there exists a unique $g \in C_\infty \otimes \mathbf{A}$ with

$$g \cdot 1 = \tau,$$

and $\widehat{f} = g$.

Let Ω be the $\mathcal{O}_{\overline{X}}$-module of 1-forms. By Riemann-Roch and Proposition 6.2.3, the space of global sections of

$$\Omega(-V + 2(\overline{\infty}))$$

is 1-dimensional. Let ω be a non-zero section. Let $\omega^{(i)}$ be defined through action on the coefficients as before. We normalize ω so that

$$1 = \operatorname{Res}_\Xi \frac{\omega^{(1)}}{\widehat{f}} .$$

7.11. The Connection with Shtukas and Examples 231

Let $\log z$ be the $\mathbb{F}_r$-linear inverse to $e(z)$; i.e., the logarithm of ψ.

Proposition 7.11.6 (Anderson). *We have*

$$\log z = \sum_{n=0}^{\infty} \left(\operatorname{Res}_{\Xi} \frac{\omega^{(n+1)}}{\widehat{f}(0) \cdots \widehat{f}(n)} \right) z^{r^n}.$$

Proof. Let $l(z)$ be the above expression. We want to show that

$$e(l(z)) = z.$$

Thus, we need to show that for $i > 0$, the coefficient of z^{r^i} in $e(l(z))$ vanishes. When one writes out this coefficient, and the smoke clears, one sees that it vanishes as it is *precisely* the sum of the residues of

$$\frac{\omega^{(i+1)}}{\widehat{f}(0) \cdots \widehat{f}(i)}.$$

□

Proposition 7.11.6 is remarkable in that it relates the logarithm of ψ to the geometry of differential forms on $\overline{X}$.

In [Th1] the following results are shown.

Proposition 7.11.7. 1. *The coefficients of $\log z$ are never zero.*
2. *The coefficients of $e(z)$ are never zero if X has a closed point of degree 1.* □

We can now present examples of the above results. The first case is obviously that of the Carlitz module C for $\mathbf{A} = \mathbb{F}_r[T]$. Once the $\mathbb{F}_r[T]$-case is understood, the general situation will be seen to follow a very similar pattern.

Example 7.11.8. Set $\mathbf{A} = \mathbb{F}_r[T]$, $\psi = C$ as above. Let $\theta := \overline{T} := \theta_{\operatorname{con}}(T) = T\vert_{\Xi} \in C_\infty$. One sees easily that $f = T - \theta$ is the shtuka function. Indeed,

$$f^{(0)} \cdots f^{(n-1)} \vert_{\Xi} = (T - \theta)(T - \theta^r) \cdots (T - \theta^{r^{(n-1)}}) \vert_{\Xi^{(n)}}$$
$$= (\theta^{r^n} - \theta) \cdots (\theta^{r^n} - \theta^{r^{n-1}})$$
$$= D_n$$

as given in Definition 5.10.1; thus 7.11.4 gives the Carlitz exponential. On the other hand, it is easy to see that

$$\omega = dT,$$

and that

$$\operatorname{Res}_\Xi \frac{\omega^{(n+1)}}{f^{(0)}\ldots f^{(n)}} = \frac{1}{(\theta - \theta^r)\cdots(\theta - \theta^{r^n})}$$
$$= \frac{(-1)^n}{L_n},$$

with L_n as in 5.10.1. So 7.11.6 gives the Carlitz logarithm.

Outside of $\mathbb{F}_r[T]$, there are exactly four **A**'s with $h(\mathbf{A}) = 1$ by [LMQ]. Note, in particular, that this implies that ∞ is rational. For these four **A**'s we will give ψ and f as in [Ha1] and [Th1]. In keeping with the notation given above, we view these modules as being defined over k. For $\alpha \in \mathbf{A}$ we set

$$\overline{\alpha} := \theta_{\operatorname{con}}(\alpha).$$

Examples 7.11.9. 1. Let $\mathbf{A} = \mathbb{F}_2[x,y]/(y^2 + y = x^3 + x + 1)$. We have
$$\psi_x = \overline{x}\tau^0 + (\overline{x}^2 + \overline{x})\tau + \tau^2$$
$$\psi_y = \overline{y}\tau^0 + (\overline{y}^2 + \overline{y})\tau + \overline{x}(\overline{y}^2 + \overline{y})\tau^2 + \tau^3,$$

and
$$f = \frac{\overline{x}(x + \overline{x}) + y + \overline{y}}{x + \overline{x} + 1}.$$

2. $\mathbf{A} = \mathbb{F}_4[x,y]/(y^2 + y = x^3 + \zeta)$, $\zeta^2 + \zeta + 1 = 0$. Then
$$\psi_x = \overline{x}\tau^0 + (\overline{x}^8 + \overline{x}^2)\tau + \tau^2$$
$$\psi_y = \overline{y}\tau^0 + (\overline{x}^{10} + \overline{x})\tau + (\overline{x}^{32} + \overline{x}^8 + \overline{x}^2)\tau^2 + \tau^3$$

where $\tau = \tau_4$. Moreover,
$$f = \frac{\overline{x}^2(x + \overline{x}) + y + \overline{y}}{x + \overline{x}}.$$

3. $\mathbf{A} = \mathbb{F}_3[x,y]/(y^2 = x^3 - x - 1)$. Then
$$\psi_x = \overline{x}\tau^0 + \overline{\alpha}\tau + \tau^2$$
$$\psi_y = \overline{y}\tau^0 + \overline{\beta}\tau + \overline{\gamma}\tau^2 + \tau^3$$

with $\alpha = y(x^3 - x)$, $\beta = y^4 - y^2$ and
$$\gamma = (y^3 - y)(y(x^3 - x + 1) - 1)(y(x^3 - x + 1) + 1).$$

Moreover,
$$f = \frac{-\overline{y}(x - \overline{x}) + y - \overline{y}}{x - \overline{x} - 1}.$$

4. $\mathbf{A} = \mathbb{F}_2[x,y]/(y^2 + y = x^5 + x^3 + 1)$. Then

7.11. The Connection with Shtukas and Examples

$$\psi_x = \overline{x}\tau^0 + (\overline{x}^2 + \overline{x})^2\tau + \tau^2$$
$$\psi_y = \overline{y}\tau^0 + \overline{\alpha}_1\tau + \overline{\alpha}_2\tau^2 + \overline{\alpha}_3\tau^3 + \overline{\alpha}_4\tau^4 + \tau^5,$$

with

$$\alpha_1 = (y^2 + y)(x^2 + x)$$
$$\alpha_2 = x^2(x+1)(y^2+y)(x^3+y)(x^3+y+1)$$
$$\alpha_3 = y(y+1)(x^5+x^3+x^2+x+1)((1+x^2+x^3)y+x^2+x^4+x^7)$$
$$\quad\times ((1+x^2+x^3)y+1+x^3+x^4+x^7)$$
$$\alpha_4 = (x(y^2+y)(x^5+x^2+1)(x+y)(x+1+y))^2.$$

Moreover,

$$f = \frac{(\overline{x}+x)(\overline{x}^4+\overline{x}^3+(1+x)\overline{x}^2)+y+\overline{y}}{\overline{x}^3+x\overline{x}^2+(1+x)\overline{x}+x^2+x}.$$

8. *L*-series

We retain the notation of Section 4; thus X is our base curve over $\mathbb{F}_r$ ($r = p^{m_0}$) with fixed closed point $\infty \in X$, function field $\mathbf{k}$ and where $\mathrm{Spec}(\mathbf{A}) = X - \infty$. We let $\mathbf{C}_\infty$ be the completion of an algebraic closure of the completion $\mathbf{K}$ of $\mathbf{k}$ at ∞. Let $\overline{\mathbf{k}} \subset \mathbf{C}_\infty$ be the algebraic closure and $\mathbf{k}^{\mathrm{sep}} \subset \overline{\mathbf{k}}$ the separable closure. Set
$$G := \mathrm{Gal}(\mathbf{k}^{\mathrm{sep}}/\mathbf{k}).$$

Let ℓ be a prime of $\mathbb{Z} \neq p$ and let $\rho \colon G \to GL(V)$ be a continuous finite dimensional ℓ-adic representation. For instance, let Z be an abelian variety over $\mathbf{k}$ with ℓ-adic Tate module $T_\ell := T_\ell(Z)$; set
$$V := \mathrm{Hom}_{\mathbb{Z}_\ell}(T_\ell, \mathbb{Q}_\ell) \simeq H^1_{\mathrm{\acute{e}t}}(Z, \mathbb{Q}_\ell)$$

with the canonical action of G on T_ℓ and V. Let $w \in X$ be a closed point and let $\overline{w}$ be an extension of w to $\overline{\mathbf{k}}$. Let $D_{\overline{w}} \subseteq G$ be the decomposition group with $I_{\overline{w}} \subseteq D_{\overline{w}}$ the inertia group. Let $V^{I_{\overline{w}}}$ be the fixed subspace under $I_{\overline{w}}$ and let $F_{\overline{w}} \in D_{\overline{w}}/I_{\overline{w}}$ be the geometric Frobenius (i.e., the inverse of the usual Frobenius). We then set
$$f_{\overline{w}}(u) := f_{V,\overline{w}}(u) := \det(1 - F_{\overline{w}} u \mid V^{I_{\overline{w}}}).$$

This polynomial in fact is easily seen to be *independent* of the choice of $\overline{w}$; so we put
$$f_w(u) := f_{\overline{w}}(u)$$
for *any* $\overline{w}$ over w. In general, it is expected that these polynomials have coefficients in $\mathbb{Z}$ which are independent of ℓ. This is certainly true in the abelian variety case and almost always (in terms of places w of $\mathbf{k}$) true, in any case, when V comes from geometry. Thus with this expectation we can form the classical *L*-series of V by
$$L(V, s) := \prod_w f_w(Nw^{-s})^{-1},$$

where Nw is the number of elements in the residue field $\mathbb{F}_w$ at w. *L*-series of this sort, for global fields, are expected to possess all sorts of important properties (see [Ta1] for instance).

Now let us return to the situation of Drinfeld modules. Let ϕ be a Drinfeld module over a finite $\mathbf{A}$-field L. Let w be a prime of $\mathbf{A}$ different from the characteristic of L. Then the Tate module $T_w(\phi)$ of ϕ at w is clearly an $\mathbf{A}_w$-module. In particular, the characteristic polynomial of the Frobenius morphism on $T_w(\phi)$ lies in $\mathbf{A}_w[u]$ (in fact, as we have seen in Subsection 4.12, it lies in $\mathbf{A}[u]$); thus it is a purely characteristic p object. It, therefore, seems only natural that an L-series associated to ϕ should *also* be a purely characteristic p object. Such functions are presented in this section.

From the viewpoint of the "two T's" (where we distinguish between T as scalar and T as operator; cf. Remark 5.4.3.1 and Subsection 7.11), we see that the L-series *precisely* arise on the "T as operator" side. As will be seen in our next section, the Γ-functions will actually arise on the "T as scalar" side.

Of course, to define $L(\phi, s)$ we first need to understand what "s" is. This will constitute our first few subsections. We then go on to present the state of the theory as of this writing. Thus we will let $\mathbf{A}$ be general in what follows. However, the reader *may* find it a very profitable exercise to first read this section with the example $\mathbf{A} = \mathbb{F}_r[T]$ and $L = \mathbf{k} = \mathbb{F}_r(T)$ in mind. Indeed, with these assumptions *many* technicalities can be dispensed with.

8.1. The "Complex Plane" S_∞

Recall that in Subsection 7.9 we obtained a decomposition of an element $\alpha \in \mathbf{K}^*$ as follows: let sgn be a fixed sign function (Definition 7.2.1), let $\pi \in \mathbf{K}^*$ be a fixed positive (i.e., $\text{sgn}(\pi) = 1$) uniformizer, then

$$\alpha = \text{sgn}(\alpha)\pi^j \langle \alpha \rangle \qquad (8.1.1)$$

where $j = v_\infty(\alpha)$ and $\langle \alpha \rangle$ is a 1-unit which depends on π. So if α is positive, then

$$\alpha = \pi^j \langle \alpha \rangle.$$

Next we recall some very basic facts from complex analysis. Let $s = x + iy$ be a complex number and n any positive number. Then, of course,

$$n^s = e^{(x+iy)\log n} = e^{x \log n} \cdot e^{iy \log n}$$

where

$$|n^s| = |e^{x \log n}|$$

and

$$|e^{iy \log n}| = 1.$$

In the function field case, we now use (8.1.1) to do something similar.

8.1. The "Complex Plane" S_∞

Definition 8.1.2. 1. We set
$$S_\infty := \mathbf{C}_\infty^* \times \mathbb{Z}_p.$$
We make S_∞ into a topological group in the obvious fashion with group action written additively.

2. If $\alpha \in \mathbf{K}^*$ is positive and $s = (x, y) \in S_\infty$, then we set
$$\alpha^s := x^{\deg(\alpha)} \langle\alpha\rangle^y = x^{-d_\infty v_\infty(\alpha)} \langle\alpha\rangle^y.$$

We note that $\langle\alpha\rangle^y = \sum_{j=0}^{\infty} \binom{y}{j}(\langle\alpha\rangle - 1)^j$ converges *precisely* because $\langle\alpha\rangle$ is a 1-unit.

The basic definition, 8.1.2.2, is motivated by the following simple observations: first of all, if α is positive, then 8.1.1 simply becomes
$$\alpha = \pi^j \langle\alpha\rangle.$$

Now the subgroup of $\mathbf{K}^*$ given by the powers of π is obviously infinite cyclic; thus any homomorphism of it into $\mathbf{C}_\infty^*$ is determined by what π maps to, and this we call "x." On the other hand, $\langle\alpha\rangle$ is a 1-unit in $\mathbf{K}^*$; as is well known the group of 1-units is isomorphic to the countable product of $\mathbb{Z}_p$ with itself. Thus, this group has a huge group of endomorphisms and an even larger group of homomorphisms into $\mathbf{C}_\infty^*$. The simplest and most natural of these endomorphisms is raising an element to the y-th power for $y \in \mathbb{Z}_p$, and these are the ones used in the definition (i.e., $\langle\alpha\rangle \mapsto \langle\alpha\rangle^y$). As of this writing, there has been no indication that the theory needs to use any larger group of endomorphisms for exponentiating.

Proposition 8.1.3. 1. Let α and β be positive elements and let $s = (x, y) \in S_\infty$. Then
$$(\alpha\beta)^s = \alpha^s \beta^s.$$

2. Let $s_0, s_1 \in S_\infty$ and α positive. Then
$$\alpha^{s_0 + s_1} = \alpha^{s_0} \alpha^{s_1}.$$

Proof. We have $\langle\alpha\beta\rangle = \langle\alpha\rangle\langle\beta\rangle$ and $\deg(\alpha\beta) = \deg\alpha + \deg\beta$. Thus Part 1 follows easily, and the second part follows in a similar fashion. □

Let $j \in \mathbb{Z}$ and $\alpha \in \mathbf{K}^*$ be positive. Let "α^j" be defined in the standard fashion of $\mathbb{Z}$-modules. We next show how $\alpha^j = \alpha^{s_j}$ for *all* α and fixed $s_j \in S_\infty$. Recall that $\deg(\alpha) = -d_\infty v_\infty(\alpha)$. Let $\pi_* \in \mathbf{C}_\infty^*$ be a fixed d_∞-th root of π. Set
$$s_j := (\pi_*^{-j}, j) \in S_\infty.$$
Note that s_j depends on the choices of π and π_*.

Proposition 8.1.4. *Let $\alpha \in \mathbf{K}$ be positive. Then*
$$\alpha^{s_j} = \alpha^j.$$

Proof. By definition
$$\begin{aligned}
\alpha^{s_j} &= (\pi_*^{-j})^{\deg(\alpha)} \langle \alpha \rangle^j \\
&= (\pi_*)^{jd_\infty v_\infty(\alpha)} (\pi^{-v_\infty(\alpha)} \alpha)^j \\
&= \pi^{jv_\infty(\alpha)} \pi^{-jv_\infty(\alpha)} \alpha^j \\
&= \alpha^j.
\end{aligned}$$
□

It is important to note that the morphism $f \colon \mathbb{Z} \to S_\infty$, $f(j) = s_j$, has *discrete* image. We identify j and $f(j)$ in the future, but will use "s_j" when, as in our next subsection, there may be confusion.

Before passing on to more general theory, just to show the reader that the functions defined in this section are truly easily understood, we point out that we are now *already* able to define the *zeta function of* $\mathbf{A} = \mathbb{F}_r[T]$. Indeed, this function is simply given as

$$\zeta_{\mathbf{A}}(s) := \sum_{a \in \mathbf{A} \text{ monic}} a^{-s},$$

where the convergence, etc., will be discussed in later subsections. This function is the prototype for *all* the L-series we will eventually define.

8.2. Exponentiation of Ideals

The ring $\mathbb{Z}$ is a principal ideal domain; so ideals may be identified with their positive generators. However, our base ring $\mathbf{A}$ will not be a principal ideal domain in general (note that as $\mathbf{A}$ is a Dedekind domain, it is a principal ideal domain if and only if it is a factorial domain). Thus exponentiating positive elements, as in Subsection 8.1, is *not* sufficient to define L-series. We will show here how easily the exponentiation of elements of 8.1 can be lifted to ideals.

Recall that we let $\mathcal{I}$ be the group of fractional ideals of $\mathbf{A}$. We let $\mathcal{P}$ be the subgroup of principal ideals and $\mathcal{P}^+ \subseteq \mathcal{P}$ the subgroup generated by positive elements. Let $U_1 \subset \mathbf{K}^*$ be the 1-units.

Definition 8.2.1. *We let $\widehat{U}_1 \subset \mathbf{C}_\infty^*$ be the group of 1-units in $\mathbf{C}_\infty$.*

Lemma 8.2.2. *The natural action of $\mathbb{Z}_p$ on $\widehat{U}_1$ may be extended uniquely to an action of $\mathbb{Q}_p$.*

Proof. Let $u = 1 + m$, $|m|_\infty < 1$. Let $y = \sum_{j > -\infty} c_j p^j$, $0 \le c_j < p$, be in $\mathbb{Q}_p$. We then set
$$u^y := \prod_{j > -\infty} (1 + m^{p^j})^{c_j}.$$
It is easy to see that this action has all the correct properties. □

Corollary 8.2.3. *The group $\widehat{U}_1$ is (uniquely) divisible.* □

As $\widehat{U}_1$ is divisible, it is *injective*. As such, if G is *any* abelian group with subgroup H, and $f: H \to \widehat{U}_1$ is a morphism, then f *automatically* extends to a map $\widehat{f}: G \to \widehat{U}_1$. Moreover, if G/H is finite, then, as $\widehat{U}_1$ is uniquely divisible, we see that this extension must be unique.

Corollary 8.2.4. *The map $\mathcal{P}^+ \to \widehat{U}_1$, $(\alpha) \mapsto \langle \alpha \rangle$ (α positive) uniquely extends to a map $\mathcal{I} \to \widehat{U}_1$ (which will also be denoted $\langle ? \rangle$).*

Proof. $\mathcal{I}/\mathcal{P}^+$ is finite. □

Let I be an ideal of $\mathbf{A}$. We call $\langle I \rangle$ the *1-unit part of I* (and it depends, of course, on π).

Definition 8.2.5. Let $I \subseteq \mathbf{A}$ be a fractional ideal and $s = (x, y) \in S_\infty$. We then set
$$I^s := x^{\deg I} \langle I \rangle^y,$$
where $\langle ? \rangle$ is the canonical extension to $\mathcal{I}$ of $\langle ? \rangle: \mathcal{P}^+ \to U_1$.

Clearly
$$(IJ)^s = I^s J^s$$
and
$$I^{s+t} = I^s I^t,$$
for $s, t \in S_\infty$. Moreover, if $I = (i)$ with i positive, then
$$I^s = i^s.$$
More generally, let $I = (i)$ with i arbitrary. Write
$$i = \operatorname{sgn}(i) \pi^j \langle i \rangle.$$

Proposition 8.2.6. *Let $s = (x, y) \in S_\infty$ and $I = (i)$ as above. Then*
$$I^s = x^{-jd_\infty} \langle i \rangle^y.$$

Proof. Set $t = r^{d_\infty} - 1$, so i^t is positive, and

$$i^t = \pi^{jt}\langle i\rangle^t = \pi^{jt}\langle i^t\rangle.$$

Thus

$$\begin{aligned}(I^t)^s &= x^{-jtd_\infty}\langle i^t\rangle^y \\ &= x^{-jtd_\infty}\langle i\rangle^{ty} \\ &= (I^s)^t.\end{aligned}$$

The result now follows easily. □

Recall that for $j \in \mathbb{Z}$ we defined in Subsection 8.1 elements $s_j \in S_\infty$ with $\alpha^{s_j} = \alpha^j$ for all positive α.

Corollary 8.2.7. *Let* $I = (i)$. *Then* $I^{s_1} = i/\mathrm{sgn}(i)$. □

In general, I^s, $s = (x,y)$, can be computed as follows: Let e be the order of I in $\mathcal{I}/\mathcal{P}^+$; so $I^e = (\lambda)$ with positive $\lambda \in \mathbf{k}^*$. Then

$$I^s = x^{\deg(I)}\langle\lambda\rangle^{y/e},$$

where $y/e \in \mathbb{Q}_p$.

Definition 8.2.8. Let $\mathbf{V} \subseteq \mathbf{C}_\infty$ be defined by

$$\mathbf{V} := \mathbf{k}(\{I^{s_1} \mid I \in \mathcal{I}\});$$

i.e., $\mathbf{V}$ is the smallest subfield of $\mathbf{C}_\infty$ containing $\mathbf{k}$ and the values, $I \mapsto I^{s_1}$, for all I. The field $\mathbf{V}$ is the *value field associated to* sgn *(and* π_**)*.

Proposition 8.2.9. *The field* $\mathbf{V}$ *is finite over* $\mathbf{k}$.

Proof. $\mathcal{I}/\mathcal{P}^+$ is finite. □

One can say more about $\mathbf{V}$. For instance suppose $d_\infty = 1$ and $\mathrm{Pic}(\mathbf{A})$ has order prime to p. Then $\mathbf{V} \subset \mathbf{K}$ (as we can take the appropriate roots in the 1-units of $\mathbf{K}$). However, if p divides the class number of $\mathbf{A}$, then $\mathbf{V}/\mathbf{k}$ will not be separable.

Note that I^{s_1} depends on the choice of π_* such that $\pi_*^{d_\infty} = \pi$; thus so does $\mathbf{V}$. This is made precise in Proposition 8.2.16 where the effect of changing parameters is seen to be quite minor. Note also that 8.2.7 implies that $\mathbb{F}_\infty^* \subset \mathbf{V}^*$.

Next we establish a simple but very useful version of the principal ideal theorem for $\mathbf{V}$. Let $\mathcal{O}_\mathbf{V} \subset \mathbf{V}$ be the ring of $\mathbf{A}$-integers.

Proposition 8.2.10. *Let* $I \subseteq \mathbf{A}$ *be an ideal. Then*

$$I\mathcal{O}_\mathbf{V} = (I^{s_1}).$$

8.2. Exponentiation of Ideals   241

Proof. Let e be chosen so that I^e is principal and positively generated; say

$$I^e = (i), \qquad \text{sgn}(i) = 1.$$

Let $\lambda = I^{s_1}$; we can then also conclude that $\lambda^e = i$. Thus (λ) and $I\mathcal{O}_\mathbf{V}$ are divisible by the same primes to the same order. □

Remarks. 8.2.11. 1. In [Go4] we defined exponentiation of ideals in a far more general fashion. Here we have based our exposition on the fact that the 1-units in $\mathbf{C}_\infty$ form a divisible abelian group. We believe that nothing essential is lost in this approach and, in fact, much is gained in the way of simplicity. The approach presented here arose out of correspondence with D. Thakur about his paper [Th4].
2. We have not used "I^j" for I^{s_j} in order to avoid confusion with ideal multiplication. However, when it is clear that we are talking about elements of $\mathbf{V}$, no such confusion can take place and we may use the "I^j" notation.

By decomposing $\mathcal{I}/\mathcal{P}^+$ into its primary components, it is easy to see that $[\mathbf{V}\colon \mathbf{k}] \leq h(\mathbf{A})^+$ where $h(\mathbf{A})^+$ is the order of the finite group $\mathcal{I}/\mathcal{P}^+$. We will now show that, in certain circumstances, the inequality is strict. We begin with a well-known lemma (and thank M. Rosen for showing us the simple proof given below). Let F be any *perfect* field in characteristic p. Let K be a finitely generated function field of transcendency one over K (which is exactly the type of function field of interest to us). Let K^p be the subfield of K obtained by raising all elements of K to the p-th power; note that, as the p-th power mapping is injective, K and K^p are isomorphic fields.

Proposition 8.2.12. *We have* $[K\colon K^p] = p$.

Proof. We can find an element $x \in K$ which is transcendental over F and such that K is a finite extension of $F(x)$. Note that as F is perfect we have $F(x^p) = F(x)^p$ and $[F(x)\colon F(x^p)] = p$. Next we compute $[K\colon F(x^p)]$ in *two* distinct ways. First, we have

$$[K\colon F(x^p)] = [K\colon F(x)][F(x)\colon F(x^p)] = p \cdot [K\colon F(x)].$$

Next, we have

$$[K\colon F(x^p)] = [K\colon K^p][K^p\colon F(x^p)].$$

Now, as we discussed above, K is isomorphic to K^p via the p-th power mapping. Thus

$$[K\colon F(x)] = [K^p\colon F(x^p)],$$

and so we deduce $[K\colon K^p] = p$ which is the proposition. □

Let $K^{1/p}$ be the extension of K obtained by adjoining the p-th root of *all* elements of K. By the proposition, and the action of the p-th power map, we know that $[K^{1/p}\colon K] = p$.

Corollary 8.2.13. *Let K be as above and let $\alpha \in K - K^p$ be any element. Then $K^{1/p} = K(\alpha^{1/p})$.* □

It is clear that the above results extend in the obvious fashion to arbitrary powers of p.

As above, we defined $h(\mathbf{A})^+$ to be the order of $\mathcal{I}/\mathcal{P}^+$. Let us write $h(\mathbf{A})^+ = p^{e_1} \cdot \widetilde{h(\mathbf{A})}^+$ where $\widetilde{h(\mathbf{A})}^+$ is prime to p. If $e_1 = 0$ set $e = 0$. If $e_1 > 0$ then we let p^e be the p-exponent of $\mathcal{I}/\mathcal{P}^+$, i.e., the smallest nontrivial power of p which kills *all* p-power torsion. Thus, of course, $e \leq e_1$. If this equality is strict, then the above corollary immediately implies that $h(\mathbf{A})^+ > [\mathbf{V}:\mathbf{k}]$, where $\mathbf{V}$ is defined in 8.2.8. Indeed, one can decompose $\mathcal{I}/\mathcal{P}^+$ into a product of cyclic abelian groups of prime power order. Let M be one such subgroup and let I be a generator. Thus $I^t \in \mathcal{P}^+$ for some prime power t. If t is a power of p, then I^{s_1} is obviously a totally inseparable element over $\mathbf{k}$ (raising it to the t-th power lands it in $\mathbf{k}$). If t is prime to p, then I^{s_1} is separable over $\mathbf{k}$ and adjoining it to $\mathbf{k}$ gives an extension of degree at most the order of M; the observation now follows by continuing this process over all such subgroups.

Question 8.2.14. Does $[\mathbf{V}:\mathbf{k}] = p^e \cdot \widetilde{h(\mathbf{A})}^+$?

Finally we need to discuss what happens when we change from one positive uniformizer $_1\pi \in \mathbf{K}$ to another positive uniformizer $_2\pi \in \mathbf{K}$. Set $u := {_1\pi}/{_2\pi}$; so $u \in \mathbf{K}$ is a 1-unit. Let I be an ideal of $\mathbf{A}$ and let $\langle I \rangle_i$, $i = 1, 2$, be the 1-unit parts defined with respect to $_i\pi$ as given by Corollary 8.2.4.

Proposition 8.2.15. *We have*
$$\langle I \rangle_1 = \left(u^{1/d_\infty}\right)^{\deg I} \cdot \langle I \rangle_2,$$
where we take the 1-unit root of u.

Proof. Suppose that $I^j = (i)$ where $i \in \mathbf{A}$ is positive. Thus
$$\langle I \rangle_1^j = {_1\pi}^{\deg i/d_\infty} \cdot i$$
$$= {_1\pi}^{(j \deg I)/d_\infty} \cdot i.$$

Similarly
$$\langle I \rangle_2^j = {_2\pi}^{(j \deg I)/d_\infty} \cdot i.$$

Therefore
$$(\langle I \rangle_1 / \langle I \rangle_2)^j = \left(({_1\pi}/{_2\pi})^{\deg I/d_\infty}\right)^j.$$

As we are working with 1-units, we find
$$\langle I \rangle_i / \langle I \rangle_2 = u^{\deg I/d_\infty},$$
which is the result. □

Therefore for $y \in \mathbb{Z}_p$ we find

$$\langle I \rangle_1^y = u^{y \deg I/d_\infty} \langle I \rangle_2^y,$$

and

$$x^{\deg I} \langle I \rangle_1^y = \left(xu^{y/d_\infty}\right)^{\deg I} \langle I \rangle_2^y,$$

etc.

Similarly we would like to see how I^{s_1} varies with different choices of positive uniformizer. Let ${}_i\pi_* \in \mathbf{C}_\infty$ be a fixed d_∞-th root of ${}_i\pi$, for $i = 1, 2$. Let $I_i^{s_1}$ be defined with respect to the parameter ${}_i\pi$ for $i = 1, 2$.

Proposition 8.2.16. *There exists a d_∞-th root of unity ζ such that*

$$I_1^{s_1} = \zeta^{\deg I} I_2^{s_1}.$$

Proof. Let u be as in 8.2.15. Thus by definition and Proposition 8.2.15 we have
$$\begin{aligned}
I_1^{s_1} &= {}_1\pi_*^{-\deg I} \langle I \rangle_1 \\
&= {}_1\pi_*^{-\deg I} u^{\deg I/d_\infty} \langle I \rangle_2 \\
&= {}_1\pi_*^{-\deg I} u^{\deg I/d_\infty} {}_2\pi_*^{\deg I} I_2^{s_1} \\
&= \zeta^{\deg I} I_2^{s_1},
\end{aligned}$$
where $\zeta = u^{1/d_\infty} {}_2\pi_*/{}_1\pi_*$. But clearly $\zeta^{d_\infty} = 1$. □

Thus if d_∞ is a power of p (including $d_\infty = 1$), then the value field $\mathbf{V}$ is independent of the choice of positive uniformizer.

8.3. v-adic Exponentiation of Ideals

In this subsection we explain how the ∞-adic theory of our previous two subsection can also be given a v-adic analog for $v \in \mathrm{Spec}(\mathbf{A})$.

Fix d_v to be the degree of v. Let $\mathbf{A}_v$ be the completion of $\mathbf{A}$ at v with units $\mathbf{A}_v^*$. It is completely standard that every element $\alpha \in \mathbf{A}_v^*$ has a canonical decomposition

$$\alpha = \omega_v(\alpha) \langle \alpha \rangle_v \tag{8.3.1}$$

where $\omega_v(\alpha) \in \mu_{r^{d_v}-1} = (r^{d_v} - 1)$-st roots of 1 in $\mathbf{A}_v^*$, and $\langle \alpha \rangle_v$ is a 1-unit at v.

Let $\mathbf{k}_v$ be the completion of $\mathbf{k}$ at v with its normalized absolute value $|?|_v$. Let $\overline{\mathbf{k}}_v$ be a fixed algebraic closure of $\mathbf{k}_v$ equipped with the canonical extension of $|?|_v$. Let $\sigma: \mathbf{V} \to \overline{\mathbf{k}}_v$ be a fixed embedding of the value field $\mathbf{V}$ into $\overline{\mathbf{k}}_v$ over $\mathbf{k}$. Of course, σ gives rise to a finite place of $\mathbf{V}$.

244   8. *L*-series

Definition 8.3.2. We set

$$\mathbf{k}_{\sigma,v} := \mathbf{k}_v(\sigma(\mathbf{V})).$$

By Proposition 8.2.9, the field $\mathbf{k}_{\sigma,v}$ is finite over $\mathbf{k}_v$. Let $\mathbf{A}_{\sigma,v} \subset \mathbf{k}_{\sigma,v}$ be its ring of integers. Let $f = f_\sigma$ be the residue degree of $\mathbf{k}_{\sigma,v}$ over $\mathbf{k}_v$ and let $\beta \in \mathbf{A}^*_{\sigma,v}$. Clearly β can be decomposed into

$$\beta = \omega_{\sigma,v}(\beta)\langle\beta\rangle_{\sigma,v} \qquad (8.3.3)$$

where $\omega_{\sigma,v}(\beta) \in \mu_{r^{d_v f}-1}$ = roots of unity in $\mathbf{A}_{\sigma,v}$ and $\langle\beta\rangle_{\sigma,v}$ is a 1-unit. Note that 8.3.3 generalizes 8.3.1.

Definition 8.3.4. We set

$$S_{\sigma,v} := \varprojlim_j \mathbb{Z}/(p^j(r^{d_v f} - 1)) \simeq \mathbb{Z}_p \times \mathbb{Z}/(r^{d_v f} - 1).$$

Note that $S_{\sigma,v}$ actually depends only on $f = f_\sigma$.

If $s_v = (s_{v,0}, s_{v,1}) \in S_{\sigma,v}$ and β is as in 8.3.3, then we set

$$\beta^{s_v} := \omega_{\sigma,v}(\beta)^{s_{v,1}} \langle\beta\rangle_{\sigma,v}^{s_{v,0}}.$$

In our previous subsection we defined the homomorphism $I \mapsto I^{s_1}$ from $\mathcal{I}$ to $\mathbf{V}^*$. Thus $\sigma(I^{s_1}) \in \mathbf{k}^*_{\sigma,v}$. If (α) is principal with α positive, then $\sigma((\alpha)^{s_1}) = \alpha$; thus if I is prime to v, then $I^{s_1} \in \mathbf{V}^*$ is prime to all the places of $\mathbf{V}$ above v.
  Let $\mathbf{C}_v$ be the completion of $\overline{\mathbf{k}}_v$.

Definition 8.3.5. Let I be a fractional ideal of $\mathbf{k}$ prime to v and let $e_v := (x_v, s_v) \in \mathbf{C}_v^* \times S_{\sigma,v}$ (viewed as an additive topological group). Then we set

$$I^{e_v} := I^{(x_v, s_v)} := x_v^{\deg I}(\sigma(I^{s_1}))^{s_v}.$$

Thus, for instance, if $J = (\alpha)$ with α positive and prime to v, then

$$J^{e_v} = x_v^{\deg \alpha} \cdot \alpha^{s_v}.$$

The mapping $I \mapsto I^{e_v}$ has the same formal properties as I^s, $s \in S_\infty$, except that if α is as above and $j \in \mathbb{Z}$, then

$$\alpha^j = \alpha^{(1,j,j)},$$

and the set $\{(1, j, j)\}$ is *dense* in $1 \times S_v \subset \mathbf{C}_v^* \times S_v$.
  Let $\mathbf{A} = \mathbb{F}_r[T]$ and let v be a prime of $\mathbf{A}$. As an example of how the above theory is used in practice, we now define the v-adic version of $\zeta_\mathbf{A}(s)$ (as defined in Subsection 8.1). This function will be denoted $\zeta_v(\mathbf{A}, e_v)$ (or

$\zeta_{\mathbf{A},v}(e_v))$ and is defined simply as $\sum\limits_{\substack{a \text{ monic} \\ v \nmid a}} a^{-e_v}$. As with the theory at infinity, the convergence properties will be presented in later subsections.

The reader may well wonder how the different avatars of $\zeta_{\mathbf{A}}(s)$ are related to each other. This is precisely what the concept of "essential algebraicity" does and will be discussed in Subsection 8.5.

8.4. Continuous Functions on $\mathbb{Z}_p$

Let L be any field of characteristic p which is complete with respect to a nontrivial absolute value $|?|$. Let $\mathcal{O} \subset L$ be the ring of integers with maximal ideal M. We shall give a description of the space of continuous functions from $\mathbb{Z}_p$ to L which is a variant of the classical theorem of Mahler. Our exposition will follow that of [L2, §4.1].

Definition 8.4.1. Let $C(\mathbb{Z}_p, \mathcal{O})$ be the space of continuous $\mathcal{O}$-valued functions on $\mathbb{Z}_p$. Let $C(\mathbb{Z}_p, L)$ be the space of continuous L-valued functions.

Note that every $f \in C(\mathbb{Z}_p, L)$ has compact image as $\mathbb{Z}_p$ is compact. Thus there exists $\alpha \in L^*$ with $\alpha f \in C(\mathbb{Z}_p, \mathcal{O})$. Note also that $C(\mathbb{Z}_p, L)$ is an L-vector space and $C(\mathbb{Z}_p, \mathcal{O})$ is an $\mathcal{O}$-module. Finally $C(\mathbb{Z}_p, L) = C(\mathbb{Z}_p, \mathcal{O}) \otimes_\mathcal{O} L$.

We can give these spaces norm as follows: let $f \in C(\mathbb{Z}_p, L)$, then we set

$$\|f\| := \max_{s \in \mathbb{Z}_p}\{|f(s)|\}.$$

Both $C(\mathbb{Z}_p, L)$ and $C(\mathbb{Z}_p, \mathcal{O})$ are complete with respect to $\|f\|$.

Next we present some basic functions from $\mathbb{Z}_p$ to L.

Definition 8.4.2. Set $\binom{x}{0} \equiv 1$. For k a positive integer set

$$\binom{x}{k} := \frac{x(x-1)\cdots(x-k+1)}{k!} \in \mathbb{Q}[x].$$

Thus $\binom{x}{k}$ gives rise to a continuous function from $\mathbb{Q}_p$ to itself. If x is a non-negative integer, then $\binom{x}{k} \in \mathbb{Z} \subset \mathbb{Z}_p$. As such integers are dense, we see that

$$\binom{x}{k}: \mathbb{Z}_p \to \mathbb{Z}_p.$$

By reducing these modulo (p) we can consider them as continuous functions with values in $\mathbb{F}_p \subset \mathcal{O} \subset L$; we will also denote these functions by "$\binom{x}{k}$".

Let $\{a_k\} \subset L$ be a sequence of elements in L with $a_k \to 0$ as $k \to \infty$. Set

$$\varphi(x) := \sum_{k=0}^{\infty} a_k \binom{x}{k}.$$

It is clear that $\varphi(x)$ is continuous from $\mathbb{Z}_p$ to L.

Definition 8.4.3. We set $\Delta^0 \varphi(x) := \varphi(x)$ and
$$\Delta \varphi(x) := \varphi(x+1) - \varphi(x).$$

Lemma 8.4.4. *We have for $j \geq 0$*
$$a_j = \Delta^j \varphi(0)$$
$$= \sum_{i=0}^{j} (-1)^{j-i} \binom{j}{i} \varphi(i).$$

Proof. One checks easily that
$$\Delta \binom{x}{k} = \binom{x}{k-1}.$$

Thus
$$\Delta^j \varphi(x) = \sum_{k=0}^{\infty} a_{k+j} \binom{x}{k}.$$

The result follows immediately. □

Lemma 8.4.5. *Let $0 \leq k < p^N$. Then the function $\binom{x}{k} \colon \mathbb{Z}_p \to \mathbb{F}_p$ is periodic modulo (p^N).*

Proof. We have to show that if $\alpha \in \mathbb{Z}_p$ is in (p^N), then
$$\binom{x+\alpha}{k} = \binom{x}{k}$$

as elements of $\mathbb{F}_p$ for all $x \in \mathbb{Z}_p$. Write α as $p^N u$ with $u \in \mathbb{Z}_p^*$. Let T be a variable and look at
$$(1+T)^{x+\alpha} = (1+T)^x (1+T)^\alpha = (1+T)^x (1+T)^{p^N u}$$

as elements of $\mathbb{F}_p[[T]]$. But $(1+T)^{p^N u} = (1+T^{p^N})^u$, and the result now follows. □

Theorem 8.4.6. *Every $\varphi(x) \in C(\mathbb{Z}_p, L)$ can be expanded uniquely as*
$$\varphi(x) = \sum a_k \binom{x}{k}$$

where $\{a_k\} \subset L$ and $a_k \to 0$ as $k \to \infty$. Moreover, the a_k are given as in Lemma 8.4.4.

Proof. The uniqueness is clear from Lemma 8.4.4. We need only show that all such $\varphi(x)$ can be expanded as above.

By perhaps multiplying φ by a constant, we may assume that φ has image in $\mathcal{O}$. From uniqueness, we need only show that for all $j \geq 0$, the function

$$\varphi^{(j)}(x) := \varphi(x) \pmod{M^j},$$

where $M \subset \mathcal{O}$ is the maximal ideal, can be expressed as a finite sum of $\{\binom{x}{k}\}$.

As $\mathcal{O}/M^j$ is discrete, the function $\varphi^{(j)}(x)$ *must* factor modulo (p^N) for some $N \geq 0$. Thus we need only show that every map (as discrete spaces) from $\mathbb{Z}/(p^N) \to \mathcal{O}/M^j$ can be written uniquely as a finite sum of $\{\binom{x}{k}\}$. For any ring R, let $\text{Map}(\mathbb{Z}/(p^N), R)$ be the R module of maps from $\mathbb{Z}/(p^N)$ to R. If $\mathbb{F}_p \subseteq R$, then one sees trivially that

$$\text{Map}(\mathbb{Z}/(p^N), R) = \text{Map}(\mathbb{Z}/(p^N), \mathbb{F}_p) \otimes_{\mathbb{F}_p} R.$$

We are thus reduced to establishing that *any* map from $\mathbb{Z}/(p^N)$ to $\mathbb{F}_p$ can be written as above.

Let S be the set of sequences $\{a_j \mid j = 0, 1, \ldots, p^N - 1\}$ of elements of $\mathbb{F}_p$. For each such j, by 8.4.5 we know that $\binom{x}{j}$ factors modulo (p^N). Thus to $s = (a_j) \in S$ we associate the function

$$\varphi_s(x) := \sum_{j=0}^{p^N-1} a_j \binom{x}{j}$$

from $\mathbb{Z}/(p^N)$ to $\mathbb{F}_p$. Since Lemma 8.4.4 is also valid in this context, we deduce that all such $\varphi_s(x)$ are *distinct*. Thus there are exactly p^N such φ_s which is exactly the number of maps from $\mathbb{Z}/(p^N)$ to $\mathbb{F}_p$. The result now follows. □

Corollary 8.4.7. *Let $\varphi(x) = \sum a_k \binom{x}{k}$ be a continuous function from $\mathbb{Z}_p$ to L. Then $\varphi(x) \in C(\mathbb{Z}_p, \mathcal{O})$ if and only if $\{a_k\} \subset \mathcal{O}$.* □

Let $f \in C(\mathbb{Z}_p, L)$ be written as

$$f(x) = \sum a_k \binom{x}{k}.$$

We set

$$\|f\|_1 := \max_k \{|a_k|\}.$$

Proposition 8.4.8. *We have*

$$\|f\| = \|f\|_1.$$

Proof. It is clear that $\|f\| \leq \|f\|_1$. We just need to show the inequality in the other direction. But by 8.4.4 each a_j can be expressed as an $\mathbb{F}_p$-linear sum of the values of f on non-negative integers. Thus $\|f\|_1$ must be $\leq \|f\|$. □

8.5. Entire Functions on S_∞

We recall that $S_\infty = \mathbf{C}_\infty^* \times \mathbb{Z}_p$. Let $s = (x,y) \in S_\infty$ and let $f(s) \colon S_\infty \to \mathbf{C}_\infty$ be a continuous function. We can view $f(s)$ as a continuous family of $\mathbf{C}_\infty$-valued functions on $\mathbf{C}_\infty^*$ or as a continuous family of continuous $\mathbf{C}_\infty$-valued functions on $\mathbb{Z}_p$. Both approaches have utility depending on the given situation.

The prototype of such $f(s)$, as above, is

$$f(s) = \alpha^s$$

for $\alpha \in \mathbf{K}^*$, α positive. Recall that

$$\alpha^s = x^{\deg(\alpha)} \langle \alpha \rangle^y \, ;$$

thus α^s is *analytic* in the $\mathbf{C}_\infty^*$-variable.

This suggests that we should restrict our attention to continuous functions on S_∞ which are analytic in the first variable. This is almost correct; we also want the roots of these analytic functions to "flow continuously" in the $\mathbb{Z}_p$-variable. It is well-known that this is accomplished by requiring uniform continuity on bounded subsets of $\mathbf{C}_\infty^*$. We are thus lead to the next definition.

Definition 8.5.1. We define an *entire function on* S_∞ to be a continuous family of $\mathbf{C}_\infty$-valued entire power series in x^{-1} which is parameterized by $\mathbb{Z}_p$. Moreover, this family is required to be uniformly convergent on bounded subsets of $\mathbf{C}_\infty$.

Thus, in down to earth terms, to give an entire function for S_∞ means that for each $y \in \mathbb{Z}_p$ we are given a power series $g_y(u)$ which converges for all u and

$$f(x,y) = g_y(1/x) \, .$$

Furthermore, let $B \subset \mathbf{C}_\infty$ be any bounded subset and $\varepsilon > 0$. Then we require the existence of $\delta := \delta(B) > 0$ so that if $y_0, y_1 \in \mathbb{Z}_p$ and $|y_0 - y_1|_p < \delta$ then $|g_{y_0}(u) - g_{y_1}(u)| < \varepsilon$ for all $u \in B$.

Remark. 8.5.2. Although Definition 8.5.1 will suffice for most applications, it can be easily generalized by allowing $g_y(u)$ to be a Laurent series convergent for all $0 \neq u$, and with appropriate modifications of our boundedness condition. We leave the easy details to the reader. If we need to distinguish the different cases, we will refer to the type of functions just mentioned as *entire Laurent functions*.

In practice, the boundedness condition of 8.5.1 is difficult to verify directly. Thus our next goal is to give a reformulation which is easier to handle.

Definition 8.5.3. Let $g(u) = \sum_{i=0}^{\infty} a_i u^i$ be a power series with $\{a_i\} \subset \mathbf{C}_\infty$. Let $r \in \mathbb{R}$. We define

$$\|g\|_r := \max_i \{|a_i| r^i\} \in \mathbb{R} \cup \{\infty\}.$$

For instance, $\|g\|_1 = \max_i \{|a_i|\}$. If $\|g\|_1 < \infty$, then $g(u)$ converges for all u with $|u| < 1$, etc. Note also that if $g(u)$ is entire, then $\|g\|_r < \infty$ for *all* r.

Let $g(u)$ now be a power series that converges on the closed unit ball $\{u \mid |u| \leq 1\}$. This is clearly equivalent to having $a_i \to 0$ as $i \to \infty$.

Claim 8.5.4. $\|g\|_1 = \max_{|u| \leq 1} \{|g(u)|\}$.

Proof. Clearly $\|g\|_1 \geq \max_{|u| \leq 1} \{|g(u)|\}$. Conversely, let $\|g\|_1 = r_0 \in \mathbb{R}$. The number r_0 must be in the value group of $\mathbf{C}_\infty$ since $a_i \to 0$ as $i \to \infty$. If $r_0 = 0$, then there is nothing to show; thus suppose $r_0 > 0$ and let $\alpha \in \mathbf{C}_\infty$ have $|\alpha| = r_0$. Upon replacing g with g/α we may assume that $\|g\|_1 = 1$. We simply need to establish that $\max_{|u| \leq 1} \{|g(u)|\} \geq 1$. But, by assumption, the reduction of g modulo the maximal ideal of $\mathbf{C}_\infty$ is a nonzero polynomial. Thus there are elements in the residue field at ∞ *not* annihilated by the reduction. If $\bar{u}$ is one such, with lift to $u \in \mathbf{C}_\infty$, then $|g(u)| = 1$, and the result is established. □

Thus $\|?\|_1$ forces uniform convergence on the closed unit disc. In a similar vein, if we let G be the value group of $\mathbf{C}_\infty$ and $r \in G$, then $\|?\|_r$ forces convergence on the closed disc of radius r. We summarize all this in our next result.

Proposition 8.5.5. *To give an entire function on S_∞ is equivalent to giving an entire power series $g_y(u)$ for $y \in \mathbb{Z}_p$ with the following property: let $r \in \mathbb{R}_+$ and $\varepsilon > 0$, then there exists $\delta = \delta(r) > 0$ so that if $y_0, y_1 \in \mathbb{Z}_p$, then $|y_0 - y_1| < \delta$ implies $\|g_{y_0}(u) - g_{y_1}(u)\|_r < \varepsilon$.* □

Let $f(s) \colon S_\infty \to \mathbf{C}_\infty$ be an entire function and let $y \in \mathbb{Z}_p$. By definition

$$f(s) = f(x, y) = \sum_{i=0}^{\infty} f_i(y) x^{-i}$$

where $f_i(y) \in \mathbf{C}_\infty$. By above we see that $f_i(y) \colon \mathbb{Z}_p \to \mathbf{C}_\infty$ is continuous.

Set
$$m_i := \{\max_{y \in \mathbb{Z}_p} |f_i(y)|\}.$$
By compactness, m_i is finite.

Theorem 8.5.6. *Let $r \in \mathbb{R}_+$. Then*
$$m_i r^i \to 0$$
as $i \to \infty$.

Proof. Let $\varepsilon > 0$. We want to show the existence of $N \in \mathbb{N}$ so that if $i > N$ then $m_i r^i < \varepsilon$. Cover $\mathbb{Z}_p$ by a collection of open balls $\{B_j\}$ such that if $y, y' \in B_j$, then
$$\|f(x^{-1}, y) - f(x^{-1}, y')\|_r < \varepsilon.$$
By compactness we can find a finite sub-cover $\{B_{i_1}, \ldots, B_{i_t}\}$. Let $y_j \in B_{i_j}$ for $j = 1, \ldots, t$. For each j let N_j be chosen so that if $i > N_j$, then $|f_i(y_j)| r^i < \varepsilon$. Let $N = \max_j \{N_j\}$.

Suppose now that $i > N$. Let $\widehat{y}_i \in \mathbb{Z}_p$ be chosen so that $|f_i(\widehat{y}_i)| = m_i$. There exists $j \in \{1, \ldots, t\}$ with $\widehat{y}_i \in B_{i_j}$. Thus
$$\begin{aligned}
m_i r^i &= |f_i(\widehat{y}_i)| r^i \\
&= |f_i(y_j) + f_i(\widehat{y}_i) - f_i(y_j)| r^i \\
&\leq \max\{|f_i(y_j)|, |f_i(\widehat{y}_i) - f_i(y_j)|\} r^i \\
&\leq \varepsilon.
\end{aligned}$$
□

Conversely, let $f_i(y) : \mathbb{Z}_p \to \mathbf{C}_\infty$, $i = 0, \ldots, \infty$, be a collection of continuous functions. Let
$$m_i := \max_y \{|f_i(y)|\},$$
and suppose $m_i r^i \to 0$ as $i \to \infty$ for all $r \in \mathbb{R}_+$. We then have the following converse to Theorem 8.5.6.

Theorem 8.5.7. *Set $f(x, y) := \sum_{j=0}^{\infty} f_j(y) x^{-j}$. Then $f(x, y)$ is an entire function on S_∞.*

Proof. It is clear that for fixed y, $f(x, y)$ is an entire power series in x^{-1}. We thus need to establish uniform continuity. Let $r \in \mathbb{R}_+$ and let $\varepsilon > 0$. Let $N \in \mathbb{N}$ be such that if $i > N$, then $m_i r^i < \varepsilon$.

Let j be a non-negative integer $\leq N$. Since $f_j(y)$ is continuous on $\mathbb{Z}_p$ it is uniformly continuous. Then we can find $\delta_j > 0$ so that if $|y_0 - y_1|_p < \delta_j$, then $|f_j(y_0) - f_j(y_1)| < \varepsilon/r^j$. Let $\delta := \min_j \{\delta_j\}$. Thus if $|y_0 - y_1| < \delta$, then

$$\|f(x^{-1},y_0) - f(x^{-1},y_1)\|_r$$
$$\leq \max\{\varepsilon, |f_0(y_0) - f_1(y_0)|r^0, \ldots, |f_N(y_0) - f_N(y_1)|r^N\}$$
$$\leq \varepsilon.$$

□

Thus Theorems 8.5.6 and 8.5.7 characterize entire functions on S_∞ as continuous families of power series. Let $f(x,y) = \sum_{j=0}^{\infty} f_j(y)x^{-j}$ be as in 8.5.7. By Theorem 8.4.6 for each j we can write

$$f_j(y) = \sum_{k=0}^{\infty} f_{j,k} \binom{y}{k}$$

for $\{f_{j,k}\} \subset \mathbf{C}_\infty$ and $f_{j,k} \to 0$ as $k \to \infty$. Thus

$$f(x,y) = \sum_{j=0}^{\infty} \left(\sum_{k=0}^{\infty} f_{j,k} \binom{y}{k} \right) x^{-j}$$
$$= \sum_{k=0}^{\infty} \left(\sum_{j=0}^{\infty} f_{j,k} x^{-j} \right) \binom{y}{k};$$

the change of order being justified by uniform convergence. Let us write this as

$$f(x,y) = \sum_{k=0}^{\infty} \widehat{f}_k(x^{-1}) \binom{y}{k}.$$

By Lemma 8.4.4,

$$\widehat{f}_j(x^{-1}) = \sum_{i=0}^{j} (-1)^{j-i} \binom{j}{i} f(x,i); \qquad (8.5.8)$$

in any case $\widehat{f}_j(u)$ is certainly entire in u.

Let $r \in \mathbb{R}_+$. Then the set

$$\{\|\widehat{f}_j\|_r\} \subset \mathbb{R}_+$$

is bounded (use Theorem 8.5.6). Indeed we can use this property to characterize entire functions as follows. Let $r \in \mathbb{R}_+$. We define two families of norms on $f(x,y)$ in our next definition.

Definition 8.5.9. Let $\{m_i\}$ be as in 8.5.6 and $r \in \mathbb{R}_+$. We then set

$$\|f(x,y)\|_r^{(1)} := \max_i \{m_i r^i\} < \infty.$$

Let $\widehat{f}_j(x^{-1})$ be as in 8.5.8. Then we set

$$\|f(x,y)\|_r^{(2)} := \max_j \{\|\widehat{f}_j(u)\|_r\} < \infty.$$

Lemma 8.5.10. Let $f(x,y)$ be entire on S_∞. Then
$$\|f(x,y)\|_r^{(1)} = \|f(x,y)\|_r^{(2)}$$
for all $r \in \mathbb{R}_+$.

Proof. This follows as in 8.4.8. □

We thus define $\|f(x,y)\|_r$ to be the common value of $\|f(x,y)\|_r^{(1)}$ and $\|f(x,y)\|_r^{(2)}$.

Theorem 8.5.11. Let $\widehat{f}_j$ be as in 8.5.8. Then for $r \in \mathbb{R}_+$, $\|\widehat{f}_j\|_r \to 0$ as $j \to \infty$. Conversely, let $\{\widehat{f}_j(u)\}$ be a collection of entire functions such that $\|\widehat{f}_j\|_r \to 0$ as $j \to \infty$ for $r \in \mathbb{R}_+$. Set
$$f(x,y) := \sum_{j=0}^{\infty} \widehat{f}_j(x^{-1}) \binom{y}{j}.$$
Then $f(x,y)$ is an entire function on S_∞.

Proof. This is left to the reader. □

Let v be a prime of $\mathbf{A}$ and $\mathbf{V}$ the value field as in Subsection 8.3. In Definition 8.3.4 we defined the complete ring
$$S_{\sigma,v} \simeq \mathbb{Z}_p \times \mathbb{Z}/(r^{d_v f} - 1)$$
where $\sigma \colon \mathbf{V} \to \overline{k}_v$ is a $\mathbf{k}$-embedding of $\mathbf{V}$ into a fixed algebraic closure of $\mathbf{k}_v$, d_v is the degree of v, and f is the residue class degree of $\mathbf{k}_v(\sigma(\mathbf{V}))$ over $\mathbf{k}_v$.

We leave it to the reader to give the obvious details of "entire functions" (with values in $\mathbf{C}_v$) on
$$\mathbf{C}_v^* \times S_{\sigma,v}.$$
The only thing which needs to be mentioned is that, in order to use the results of Subsection 8.4, we need to restrict ourselves to a fixed residue class modulo $r^{d_v f} - 1$ of $S_{\sigma,v}$.

The L-series that we will define will always turn out to be entire functions on S_∞. However, they will also possess a *very* strong algebraicity component. This algebraicity mirrors that of exponentiation itself, and so we begin again by examining the entire function
$$f(s) := a^{-s}$$
where $a \in \mathbf{A}$ is positive. Thus
$$f(s) = x^{-d} \langle a \rangle^{-y}$$

where $s = (x, y)$, $d = \deg(a)$ and $\langle a \rangle = \pi^{d/d_\infty} \cdot a$. Recall that we choose π_* to be a fixed d_∞-th root of π in $\mathbf{C}_\infty$. Suppose now that $y = -j$ for $j \geq 0$. Set

$$h_f(x, -j) := f(x\pi_*^j, -j) = x^{-d} \pi_*^{-dj} \langle a \rangle^j$$
$$= x^{-d} \pi_*^{-dj} \pi^{jd/d_\infty} \cdot a^j$$
$$= x^{-d} a^j.$$

Thus, by using the substitution, $x \mapsto x\pi_*^j$ at $y = -j$, we have succeeded in *removing* the 1-unit part of a. This leads to our next definition.

Definition 8.5.12. Let $f(s) = f(x, y)$ be an entire function on S_∞. We say that $f(s)$ is *essentially algebraic* if and only if

$$h_f(x, -j) := f(x\pi_*^j, -j)$$

is a *polynomial* in x^{-1} with algebraic coefficients for all $j \geq 0$. We further require that *all* such coefficients (for *all* j) generate a finite extension of $\mathbf{k}$.

It is easy to see that this notion has nothing to do with our choice of π_* as in the example.

In practice the coefficients of $h_f(x, -j)$ will be elements of $\mathbf{A}$. Or there will exist an entire function g with $h_g(x, -j)$ having $\mathbf{A}$ coefficients and f sits in a factorization of g (e.g., f is some sort of L-function). Thus in practice the following results will be sufficient for detecting essentially algebraic entire functions.

Proposition 8.5.13. *Let $f(s)$ be an entire function such that the power series*

$$h_f(x, -j) = f(x\pi_*^j, -j)$$

has $\mathbf{A}$-coefficients of all $j \geq 0$. Then f is essentially algebraic.

Proof. By assumption $h_f(x, -j)$ has $\mathbf{A}$-coefficients which are *discrete* in $\mathbf{C}_\infty$. On the other hand, $h_f(x, -j)$ is still an entire power series of x^{-1}. Thus the coefficient of x^{-t} in $h_f(x, -j)$ must go to 0 as $t \to \infty$. The only way this can happen is for $h_f(x, -j)$ to be a polynomial. □

Corollary 8.5.14. *Let $f(s)$ be as in 8.5.13 and suppose $f(s) = g(s)h(s)$ where both $g(s)$ and $h(s)$ are entire. Then $h_g(x, -j)$ and $h_h(x, -j)$, $j \geq 0$, (defined as in 8.5.13) are both polynomials.*

Proof. We know that

$$h_f(x, -j) = h_g(x, -j) h_h(x, -j)$$

where $h_f(x,-j)$ is polynomial by 8.5.13, and $h_g(x,-j)$, $h_h(x,-j)$ are entire functions in x^{-1}. As entire functions always factor, as in Theorem 2.14, the result follows immediately. □

Thus, in practice, essential algebraicity follows very easily from a function being entire. *Essential algebraicity is extremely important precisely because it is the "glue" holding together the theories at the different places of* **k**. Indeed, one is able to pass between the theories at the different primes *only* by using the *polynomials* of 8.5.12. These polynomials interpolate to the functions defined at *all* the places of **k**.

For instance in our prototypical case of the zeta function of $\mathbf{A} = \mathbb{F}_r[T]$, we will see (in Subsection 8.9) that $\zeta_\mathbf{A}(s)$ is entire on S_∞. We immediately deduce from 8.5.13 that it is essentially algebraic. Thus the power series

$$h_\zeta(x,-j) = \sum_{t=0}^\infty x^{-t} \left(\sum_{\substack{a \text{ monic} \\ \deg a = t}} a^j \right)$$

are, in fact, *polynomials*. The v-adic theory, for $v \in \mathrm{Spec}(\mathbf{A})$, (also given in Subsection 8.9) *and* continuity now imply that these polynomials *also* determine the v-adic functions $\zeta_v(\mathbf{A}, e_v)$. The general theory always works the same way!

Note that *no* such result as 8.5.14 can be established complex analytically. Indeed, using e^x, counterexamples are easily found.

8.6. *L*-series of Characteristic *p* Arithmetic

We can now present the basic definition of the L-series of interest to us. The simplest way to do this is via *compatible systems of v-adic representations*. This concept is, of course, modeled on the classical concept of compatible systems of ℓ-adic representations as in [Se4, §1.2]. We refer the reader there for more details.

Let L be a finite extension of **k** with ring of **A**-integers $\mathcal{O}_L$. One knows, of course, that $\mathcal{O}_L$ is finite over **A**. If L is separable, then this is due, essentially, to Dedekind. If L is not separable, let L' be its maximal separable subfield. We thus reduce, by Dedekind, to handling totally inseparable extensions. Without loss of generality, we need only handle the case $L/\mathbf{k}$ totally inseparable. Let p^e be the degree of $L/\mathbf{k}$ and let $\{\alpha_1, \ldots, \alpha_m\}$ be generators of **A** as $\mathbb{F}_r$-algebra. Then $\mathcal{O}_L \subseteq \mathbb{F}_r[\alpha_1^{p^{-e}}, \ldots, \alpha_m^{p^{-e}}]$ and is also finite.

For each prime w of L, let $\mathbb{F}_w$ be the corresponding finite field and let $\deg w = [\mathbb{F}_w : \mathbb{F}_r]$ be its degree over $\mathbb{F}_r$. Set

$$Nw := \#\mathbb{F}_w.$$

If w lies over a prime v of $\mathrm{Spec}(\mathbf{A})$, then we set $\mathbb{F}_v := \mathbf{A}/v$, and

$$nw := v^{\deg w / \deg v}$$
$$= v^{f_w},$$

where $f_w = [\mathbb{F}_w : \mathbb{F}_v]$. We view nw as an ideal of $\mathbf{A}$ and extend n and N to all fractional $\mathcal{O}_L$-ideals in the obvious multiplicative fashion.

Clearly,
$$Nw = Nnw,$$
so both ways of forming norms are compatible.

Let $\alpha \in L^*$ and let $N_{\mathbf{k}}^L(\alpha)$ be the usual multiplicative norm. Let (α) be the $\mathbf{A}$-fractional ideal generated by α. Then we have

$$(N_{\mathbf{k}}^L(\alpha)) = n(\alpha).$$

If $L/\mathbf{k}$ is separable, this is standard. We can reduce, as in the above case of integral finiteness, to $L/\mathbf{k}$ totally inseparable. Now use the fact that the local completions of $\mathbf{k}$ are separable extensions of $\mathbf{k}$ to reduce to the case of a totally inseparable extension of local fields. Here we need only take p^n-th roots of a uniformizing parameter. Compare Corollary 8.2.13.

Let $\bar{\mathbf{k}}$ be an algebraic closure of $\mathbf{k}$ which contains L and let $L^{\mathrm{sep}} \subset \bar{\mathbf{k}}$ be the separable closure of L. Set $G := \mathrm{Gal}(L^{\mathrm{sep}}/L)$. Let $v \in \mathrm{Spec}(\mathbf{A})$ and $\mathbf{k}_v$ the completion of $\mathbf{k}$ at v.

Definition 8.6.1. A *v-adic representation* of G is a continuous homomorphism
$$\rho \colon G \to \mathrm{Aut}(V)$$
where V is a finite dimensional $\mathbf{k}_v$-vector space and $\mathrm{Aut}(V)$ inherits its topology from $\mathrm{End}_{\mathbf{k}_v}(V)$.

Let w be any place of L and let $\bar{w}$ be a place of L^{sep} lying over w. Let $D_{\bar{w}}$ and $I_{\bar{w}}$ be the decomposition and inertia groups at $\bar{w}$. Thus $D_{\bar{w}}/I_{\bar{w}}$ is isomorphic to $\mathrm{Gal}(\bar{\mathbb{F}}_w/\mathbb{F}_w)$ where $\mathbb{F}_w$ is the finite field at w. Let $F_{\bar{w}} \in D_{\bar{w}}/I_{\bar{w}}$ be the *geometric Frobenius* at $\bar{w}$, that is $F_{\bar{w}}$ is the *inverse* of the automorphism

$$x \mapsto x^{Nw}.$$

Let $\hat{w}$ be another place of L^{sep} over w. Then $D_{\hat{w}}$, $I_{\hat{w}}$ and $F_{\hat{w}}$ are all conjugate to $D_{\bar{w}}$, $I_{\bar{w}}$, $F_{\bar{w}}$, respectively.

Definition 8.6.2. Let ρ be as in Definition 8.6.1. We say that ρ is *unramified* at w if and only if $\rho(I_{\bar{w}})$ is the identity.

Note that this notion only depends on w and not on any choice of place above it.

Let ρ be as in Definition 8.6.2 and unramified at w. Set
$$P_{\rho,w}(u) := \det(1 - \rho(F_{\overline{w}})u \mid V);$$
notice that this characteristic polynomial also *only* depends on w.

Definition 8.6.3. Let ρ be as in 8.6.2. We say that ρ is *rational* (resp. *integral*) if and only if there exists a finite set B of places of L such that the following hold.
1. ρ is unramified at all finite places of L *not* in B.
2. Let w be a finite place of L not in B. Then $P_{\rho,w}(u)$ has coefficients in **k** (resp. **A**).

Let L'/L be a finite separable extension. Clearly ρ restricts to a v-adic representation of $\mathrm{Gal}(L^{\mathrm{sep}}/L')$ which will be rational (resp. integral) if ρ is.

Let $v' \in \mathrm{Spec}(\mathbf{A})$ be a prime not equal to v. Let ρ' be a v'-adic representation of G.

Definition 8.6.4. Assume that both ρ and ρ' are rational. Then ρ and ρ' are said to be *compatible* if there is a finite subset B of places of L such that ρ and ρ' are unramified at the finite places outside of B and
$$P_{\rho,w}(u) = P_{\rho',w}(u)$$
for all such w not in B.

Finally, we have the following basic notion.

Definition 8.6.5. For each finite prime v of **A**, let ρ_v be a rational v-adic representation. The family (ρ_v) is said to be *compatible* if ρ_v, $\rho_{v'}$ are compatible for any two primes v, v'. The family (ρ_v) is said to be *strictly compatible* if there exists a finite subset B of places of L such that the following conditions hold:
1. Let B_v be the set of places of L above v. Then for every finite w not in $B \cup B_v$, the representation ρ_v is unramified at w and $P_{\rho_v,w}(u)$ has rational coefficients.
2. Let $w \notin B \cup B_v \cup B_{v'}$. Then
$$P_{\rho_v,w}(u) = P_{\rho_{v'},w}(u).$$

It is easy to see that there is a *smallest* set B of finite places satisfying 1 and 2 above. We call B *the set of bad places for the family*.

Examples 8.6.6. 1. Let $\rho_v = 1$ be the trivial character for all v. Then (ρ_v) is clearly a strictly compatible, integral, family.

2. Let ψ be a Drinfeld module of rank d over L. Let $v \in \mathrm{Spec}(\mathbf{A})$ and let $T_v := T_v(\psi)$ be the v-adic Tate module of ψ. Let $H_v^1(\psi, \mathbf{k}_v) := \mathrm{Hom}_{\mathbf{A}_v}(T_v, \mathbf{k}_v)$ with the natural dual action of G. Let $\wp$ be a prime of $\mathcal{O}_L$ not lying over v such that ψ has *good reduction* at $\wp$. Then $T_v(\psi)$ and $H_v^1(\psi, \mathbf{k}_v)$ are both unramified at $\wp$. Moreover the action of the geometric Frobenius on $H_v^1(\psi, \mathbf{k}_v)$ may be computed (Subsection 4.12) over the finite field $\mathbb{F}_\wp = \mathcal{O}_L/\wp$ via the action of the Frobenius endomorphism of of the reduction of ψ at $\wp$. Let ρ_v be the action of G on $H_v^1(\psi, \mathbf{k}_v)$. Then the discussion of Subsection 4.12 assures us that (ρ_v) is a strictly compatible integral family. We can define B to be the set of finite primes where ψ has bad reduction. This construction is basic to the theory of Drinfeld modules (and is totally analogous to its classical counterpart for elliptic curves and abelian varieties).

3. Let E be a T-module over L where $\mathbf{A}$ acts via "complex multiplication." Although we have not worked out explicitly any of the details, it is clear that E – which only involves finitely many equations – has "good" reduction at almost all primes of $\mathcal{O}_L$, etc. Thus one can also obtain strictly compatible integral families of representations here also. As an example, one has the tensor product of two Drinfeld modules. As we will not pursue this further, the details are left to the reader.

Let $\widehat{\rho} := (\rho_v)$ be a strictly compatible family of representations with B the set of bad places. Let w be a finite place of L not in B and let $P_{\widehat{\rho},w}(u)$ be the polynomial $P_{\rho_v,w}(u)$ for v not equal to w. By definition, this polynomial does not depend on the choice of v.

Let S_∞, etc., be as in Subsections 8.1 and 8.2.

Definition 8.6.7. Let $s \in S_\infty$. We set

$$L(\widehat{\rho}, s) := \prod_{\substack{w \text{ finite} \\ w \notin B}} P_{\widehat{\rho},w}(nw^{-s})^{-1}.$$

The above definition is not complete in that we would also like to define Euler factors for those primes in B. The "obvious" way to proceed is to follow classical theory and associate to a finite prime $w \in S$ the polynomial

$$P_{\rho_v,w}(u) := \det(1 - \rho_v(F_{\overline{w}})u \mid V_v^{I_{\overline{w}}}),$$

where v does not lie under w; this polynomial is independent of the choice of prime $\overline{w}$ over w. It *seems* reasonable to expect that this polynomial has coefficients in $\mathbf{A}$ and is independent of the choice of v when the representation arises from a Drinfeld module or T-module. However, very little is known about these polynomials and we will not pursue them further here.

It is also easy to give a v-adic version of 8.6.7. Let $\mathbf{V}$ be the value field as in Subsection 8.3. Let $\sigma: \mathbf{V} \to \overline{\mathbf{k}}_v$ be our fixed embedding and let $S_{\sigma,v}$ be as in Definition 8.3.4, etc.

Definition 8.6.8. Let $e_v = (x_v, s_v) \in \mathbf{C}_v^* \times S_{\sigma,v}$ (cf. Definition 8.3.5). We set
$$L_{\sigma,v}(\hat{\rho}, e_v) := \prod_{\substack{w \text{ finite} \\ w \notin B \\ w \nmid v}} P_{\hat{\rho},w}(nw^{-e_v})^{-1}.$$

Of course, we also need to define factors at the bad primes not dividing v, etc.

Proposition 8.6.9. 1. Let $\hat{\rho}$ be the family of trivial representations. Then $L(\hat{\rho}, s)$ converges for all $\{s = (x,y) \in S_\infty \mid |x|_\infty > 1\}$.
2. Let $\hat{\rho}$ arise from a Drinfeld module ψ of rank d. Then $L(\hat{\rho}, s)$ converges for all
$$\{s = (x,y) \in S_\infty \mid |x|_\infty > r^{1/d}\}.$$
3. Let $\hat{\rho}$ be as in 1 or 2. Then $L_{\sigma,v}(\hat{\rho}, e_v)$ converges for all $\{e_v = (x_v, s_v) \mid |x|_v > 1\}$.

Proof. Part 1 follows easily by definition. Part 2 follows because of the Riemann Hypothesis for Drinfeld modules over finite fields (see 4.12.8.5). Part 3 follows because the characteristic polynomials in question all have **A**-coefficients and so are v-adic integers. □

The above subspaces of S_∞ and $\mathbf{C}_v^* \times S_{\sigma,v}$ are the *half planes of convergence* of the Euler products for the L-functions. In order to obtain a half plane of convergence for the L-series of an arbitrary T-module we would need to add on purity conditions.

We use the notation "$L(\psi, s)$" if the family $\hat{\rho}$ arises from a Drinfeld module ψ over L, etc. We use the notation "$\zeta_{\mathcal{O}_L}(s)$" for $L(\hat{\rho}, s)$ if $\hat{\rho}$ is the trivial family and $\mathcal{O}_L$ is the ring of **A**-integers in L. This function is completely analogous the classical Dedekind zeta functions. Indeed by definition we have
$$\zeta_{\mathcal{O}_L}(s) := \prod_{\wp}(1 - n\wp^{-s})^{-1},$$
where $\wp$ runs over *all* primes of $\mathcal{O}_L$. Thus
$$\zeta_{\mathcal{O}_L}(s) = \sum_I nI^{-s},$$
where now I runs over *all* non-zero ideals of $\mathcal{O}_L$. The v-adic interpolation of this function is
$$\zeta_{\sigma,v}(\mathcal{O}_L, e_v) := \prod_{v \nmid \wp}(1 - n\wp^{-e_v})^{-1}$$
etc., as a function on $\mathbf{C}_v^* \times S_{\sigma,v}$. These definitions clearly encompass those presented earlier for $\mathbf{A} = \mathbb{F}_r[T]$ and $L = \mathbf{k}$.

Remark. 8.6.10. We can extend the notion of compatible families, etc., of v-adic representations as follows. Let $\mathcal{F}$ be a fixed finite extension of $\mathbf{k}$ with integers $\mathcal{O}_{\mathcal{F}}$. For each finite prime w of $\mathcal{O}_{\mathcal{F}}$ we have the complete field $\mathcal{F}_w$. We then define a *compatible family of w-adic representations rational over* $\mathcal{F}$ to be a family of continuous homomorphism $\rho_w \colon G \to \operatorname{Aut}_{\mathcal{F}_w}(W_w)$ where W_w is now a finite dimensional $\mathcal{F}_w$-vector space. Further, almost all primes are to be unramified, and at the unramified primes we want the characteristic polynomial (as before 8.6.3) to have coefficients in $\mathcal{F}$ with the obvious compatibility requirements, etc.

Example 8.6.11. Here is an example of how 8.6.10 works in practice. Let $L' \subset L^{\text{sep}}$ be finite and Galois over L with group G'. We have the obvious continuous surjection $s \colon G \to G'$. Let $r \colon G' \to \operatorname{Aut}_{\mathcal{F}}(W')$ be a homomorphism where W' is a finite dimensional $\mathcal{F}$-vector space. By injecting $\operatorname{Aut}_{\mathcal{F}}(W')$ into $\operatorname{Aut}_{\mathcal{F}_w}(W_w)$, $W_w := W' \otimes_{\mathcal{F}} \mathcal{F}_w$, we obtain a strictly compatible, integral family of w-adic representations. There are subtleties at the finite ramified primes of L' in terms of the definition of L-series, and the reader is advised to consult Section 8.10.

Finally, we finish this section by remarking that the adjoints of Drinfeld modules can be given L-series also. Indeed, let L be as before and let $L^{\text{perf}} \subset \mathbf{C}_\infty$ be its perfection. Let ψ be a Drinfeld module over L with adjoint ψ^* defined over L^{perf}. Let $\mathcal{O}^{\text{perf}} \subset L^{\text{perf}}$ be the ring of $\mathbf{A}$-integers and let $w \subset \mathcal{O}^{\text{perf}}$ be a prime. As L^{perf}/L is totally inseparable, we see that w is the *unique* prime over $w' = w \cap \mathcal{O}_L$. Moreover, $\mathcal{O}^{\text{perf}}/w \simeq \mathcal{O}_L/w'$. We thus define $nw := nw'$. With this simple observation, the definition of $L(\psi^*, s)$ is now obvious, and will be left to the reader. We also leave it to the reader to establish that $L(\psi^*, s)$, and its v-adic analogs, converge on half-planes, etc. The only subtlety here is to note that, because, we are dealing with adjoint representations, the function $L(\psi^*, s)$ should not be essentially algebraic. However it is easily checked from 4.12.19 that the translated function $L(\psi^*, s-1)$ will be essentially algebraic if $L(\psi^*, s)$ is entire.

8.7. Formal Dirichlet Series

Let $\mathbf{K}_1 \subset \mathbf{C}_\infty$ be a finite extension of $\mathbf{K}$. For each ideal I of $\mathbf{A}$ let $c(I) \in \mathbf{K}_1$. Let $s \in S_\infty$. We call a formal series of the form

$$L(s) := \sum_{I \subseteq \mathbf{A}} c(I) I^{-s}$$

a *Dirichlet series.*

For instance the functions of our previous subsection precisely expand out to such Dirichlet series. Clearly $L(s)$ converges on some (possibly empty) half-plane H of S_∞. Moreover, H contains those integral powers j such that $j \geq N$ for some N.

It is clear what is meant by an *Euler product* for such Dirichlet series.

Theorem 8.7.1. *Let $L(s)$ be a Dirichlet series with non-trivial half-plane of convergence H as above. Then $L(s)$ is uniquely determined by $\{L(j)\}$ such that $j \geq N$ (N as above) and*

$$j \equiv 0 \ (h(\mathbf{A})(r^{d_\infty} - 1)).$$

Proof. Let $j \geq N$ and write j as

$$j = j_1(h(\mathbf{A})(r^{d_\infty} - 1)).$$

Thus

$$I^j = (I^{h(\mathbf{A})(r^{d_\infty}-1)})^{j_1},$$

where $I^{h(\mathbf{A})(r^{d_\infty}-1)}$ is now principal and positively generated.

By a simple counting argument one can show that if I_0, I_1 are two ideals with

$$I_0^t = I_1^t \in \mathbf{V} \subseteq \overline{\mathbf{K}}$$

for some positive integer t, then $I_0 = I_1$. Thus, we are reduced to considering our Dirichlet series as being a sum over positively generated principal ideals.

Clearly it suffices to show that if $L(j) = 0$ for $j \gg 0$, then $L(s)$ is the trivial Dirichlet series. To see this, we will use a version of *Artin's independence of characters*. Thus let

$$L(s) = c(a_1)a_1^{-s} + \cdots + c(a_h)a_h^{-s} + \sum_{|a|>|a_h|} c(a)a^{-s},$$

where all a_h have the same absolute value.

As $a_1 \neq a_2$, for some $j_0 \geq N$ we have $a_1^{j_0} \neq a_2^{j_0}$. Thus, for $j \gg 0$ we find

$$0 = c(a_1)a_1^{-(j+j_0)} + \cdots + c(a_h)a_h^{-(j+j_0)} + \sum_{|a|>|a_h|} c(a)a^{(-j-j_0)}$$

$$= c(a_1)a_1^{-j}a_1^{-j_0} + \cdots + c(a_h)a_h^{-j}a_h^{-j_0} + \sum_{|a|>|a_h|} c(a)a^{-j}a^{-j_0}.$$

Now divide by $a_2^{-j_0}$ and subtract to find

$$0 = c(a_1)\left(\left(\frac{a_2}{a_1}\right)^{j_0} - 1\right)a_1^{-j} + 0 + \cdots + c(a_h)\left(\left(\frac{a_2}{a_h}\right)^{j_0} - 1\right)a_h^{-j} +$$

$$\sum_{|a|>|a_h|} c(a)\left(\left(\frac{a_2}{a}\right)^{j_0} - 1\right)a^{-j}.$$

Upon repeating the process, we arrive finally at an equation of the form, for $j \gg 0$,
$$0 = \bar{c}(a_1)a_1^{-j} + \sum_{|a|>|a_1|} \bar{c}(a)a^{-j},$$
where
$$\bar{c}(a_1) = 0 \iff c(a_1) = 0.$$
But, for $j \gg 0$,
$$|\bar{c}(a_1)| = \left| \sum_{|a|>|a_1|} \bar{c}(a) \left(\frac{a_1}{a}\right)^j \right|$$
and the right hand side approaches 0 as j tends to ∞. □

Corollary 8.7.2. *The elements $\{I^{-s}\}$ are linearly independent.* □

Corollary 8.7.3. *Suppose $L(s)$ has an Euler product over the primes of $\mathbf{A}$. Then this Euler product is unique.* □

8.8. Estimates

We now present the estimates necessary for proving that $L(\hat{\rho}, s)$ is entire for certain families of compatible representations.

Lemma 8.8.1. 1. *Let J, J_1 be two fields over $\mathbb{F}_r$. Let $W \subseteq J$ be a finite dimensional $\mathbb{F}_r$-vector space of dimension α and let $\{\mathcal{L}_1, \ldots, \mathcal{L}_t\}$ be $\mathbb{F}_r$-linear maps of J into J_1. Finally, let $x \in J$ and let $\{i_1, \ldots, i_t\}$ be non-negative integers so that*
$$\sum_{h=1}^{t} i_h < (r-1)\alpha.$$
Then
$$\sum_{w \in W} \left(\prod_{h=1}^{t} \mathcal{L}_h(x+w)^{i_h} \right) = 0.$$

2. *We assume now that J_1 has an additive discrete valuation v with $v(\mathcal{L}_h(w)) > 0$ for all h and w. Let $\{i_h\}$ now be an arbitrary collection of non-negative integers, and for $j > 0$ put*
$$W_j = \{w \in W \mid v(\mathcal{L}_h(w)) \geq j \text{ for all } h\}.$$
Then
$$v\left(\sum_{w \in W} \prod_{h=1}^{t} \mathcal{L}_h(w)^{i_h} \right) \geq (r-1)Q,$$

for
$$Q = \sum_j \dim_{\mathbb{F}_r}(W_j).$$

Proof. To see the first part, let $\{e_1, \ldots, e_\alpha\}$ be a basis for W over $\mathbb{F}_r$. Then our sum becomes
$$\sum_{c_1, \ldots, c_\alpha \in \mathbb{F}_r} \prod_{h=1}^t (\mathcal{L}_h(x) + c_1 \mathcal{L}_h(e_1) + \cdots + c_\alpha \mathcal{L}_h(e_\alpha))^{i_h}.$$

By the multinomial theorem we see that $(\mathcal{L}_h(x) + c_1 \mathcal{L}_h(e_1) + \cdots + c_\alpha \mathcal{L}_h(e_\alpha))^{i_h}$ equals
$$\sum_{j_1^h, \ldots, j_\alpha^h} \frac{i_h! \mathcal{L}_h(x)^{(i_h - j_1^h - \cdots - j_\alpha^h)}}{j_1^h! \cdots j_\alpha^h! (i_h - j_1^h - \cdots - j_\alpha^h)!} \mathcal{L}_h(e_1)^{j_1^h} \cdots \mathcal{L}_h(e_\alpha)^{j_\alpha^h} c_1^{j_1^h} \cdots c_\alpha^{j_\alpha^h}.$$

The product then contains elements of the form
$$(\text{junk}) c_1^{j_1^1 + \cdots + j_1^t} \cdots c_\alpha^{j_\alpha^1 + \cdots + j_\alpha^t}; \qquad (*)$$

where (junk) belongs to J_1 and the sum of the exponents is $\leq (i_1 + \cdots + i_t) < (r-1)\alpha$.

On the other hand, the sum over $\{c_1, \ldots, c_\alpha\}$ in $(*)$ will clearly be zero unless each exponent is divisible by $(r-1)$. The first part follows easily.

For the second part, choose j_0 so that $W_{j_0+1} = \{0\}$ but $W_{j_0} \neq \{0\}$. Pick a basis $\{e_1, \ldots, e_\alpha\}$ for W so that $\{e_1, \ldots, e_{\dim_{\mathbb{F}_r}(W_j)}\}$ is a basis for W_j for all j. This choice of basis allows us to write $w \in W$ as $w_1 + w_0$ with w_0 being the projection of w onto W_{j_0}.

Now expand out $\mathcal{L}_h(w)^{i_h} = (\mathcal{L}_h(w_1) + \mathcal{L}_h(w_0))^{i_h}$ by the binomial theorem and use the first part applied to W_{j_0}. We obtain by induction on W/W_{j_0} that
$$v \left(\sum_{w \in W} \prod_{h=1}^t \mathcal{L}_h(w)^{i_h} \right) \text{ is}$$
$$\geq (r-1) \sum_{j=1}^{j_0-1} (\dim_{\mathbb{F}_r}(W_j) - \dim_{\mathbb{F}_r}(W_{j_0})) + j_0(r-1) \dim_{\mathbb{F}_r}(W_{j_0}).$$

This gives the result. □

The reader should be aware that the estimate in 8.8.1.2 depends *only* on $(v, \mathcal{L}_h, W)$ and *not* on the choice of $\{i_1, \ldots, i_t\}$.

To illustrate the technique of 8.8.1.2 we present the following result. Of course it is a corollary of 8.8.1.2; however, we give a slightly more elementary derivation of it.

Let $W(d) \subset \mathbb{F}_r[u]$ be the subspace of polynomials of degree $\leq d$ with trivial constant term. Let v measure the vanishing order at $u = 0$, and let $j \geq 0$.

Proposition 8.8.2.

$$v\left(\sum_{f \in W(d)} f(u)^j\right) \geq (r-1)\frac{d(d+1)}{2}.$$

Proof. We again use induction on d. The result is trivial for $d = 1$. Let $f(u) \in W(d)$ be written as

$$f(u) = cu^d + h(u),$$

where $c \in \mathbb{F}_r$ and $\deg(h) \leq d - 1$. Then

$$f(u)^j = \sum_{t=0}^{j} \binom{j}{t}(cu^d)^t h(u)^{j-t}.$$

Summing over c kills the terms associated to $t \leq r - 2$. The result follows by summing over $h(u)$ and using induction as

$$d + \frac{d(d-1)}{2} = \frac{d(d+1)}{2}.$$

$\square$

As a basic example of how these estimates are used, we examine the x-adic behavior of the function

$$L(x, j) = \sum_{j=0}^{\infty} x^d \left(\sum_{f \in W(d)} f^j\right),$$

where $j \geq 0$. By 8.8.2 we see immediately that the u-adic valuation of the coefficient of x^d grows *quadratically* in d. Thus $L(x, j)$ is *entire* in x. Our method of handling arbitrary abelian L-series is simply a generalization of this.

8.9. L-series of Finite Characters

In this subsection we will prove that the L-series associated to finite characters are entire. Let $\mathbf{A}$, $\mathbf{k}$, $\mathbf{K}$, $\mathbf{C}_\infty$, etc., be as usual. Let $\bar{\mathbf{k}}$ be an algebraic closure of $\mathbf{k}$ and let $L \subset \bar{\mathbf{k}}$ be a finite extension which may have non-trivial inseparability degree. (In [Go4] we assumed that L was indeed separable. However, the remarks at the beginning of Subsection 8.6 assure us that this assumption is not necessary.) Let $\mathcal{O} := \mathcal{O}_L \subset L$ be its ring of $\mathbf{A}$-integers.

Let $G = \text{Gal}(L^{\text{sep}}/L)$, where $L^{\text{sep}} \subset \bar{\mathbf{k}}$ is the separable closure. Let $\chi \colon G \to \mathbf{C}_\infty^*$ be a homomorphism of Galois type i.e., there exists a finite abelian extension L_1/L such that $L_1 \subset L^{\text{sep}}$ and χ factors through $G_1 := \text{Gal}(L_1/L)$. Let $\mathcal{B}$ be the conductor of L_1 which we write as $\mathcal{B} = \mathcal{B}_\infty \mathcal{B}_f$; here $\mathcal{B}_f$ is made up of finite primes of L (and so can be considered as an ideal of $\mathcal{O}_L$) and $\mathcal{B}_\infty$ is made up of infinite primes.

Note that no assumption is being made that $\mathcal{B}$ is the exact conductor of χ; in fact it is important to proceed *without* such an assumption as we will see later on.

Let $\wp$ be a finite prime of L. If $\wp$ ramifies in L_1, then we set $\chi(\wp) := 0$. If $\wp$ is unramified, then we set $\chi(\wp) := \chi((\wp, L_1/L)^{-1})$ where $(\wp, L_1/L)$ is the Artin symbol at $\wp$. (The reader should note again that we are using the *inverse* of the Artin symbol in keeping with previous conventions.)

Definition 8.9.1. We set
$$L(\chi, s) := \prod_{\wp \text{ prime of } \mathcal{O}} (1 - \chi(\wp) n\wp^{-s})^{-1}.$$

As before $L(\chi, s)$ converges on a half-plane of S_∞ and can be written
$$\sum_{I \subseteq \mathcal{O}} \chi(I) nI^{-s},$$
where I is an ideal of $\mathcal{O}$ and $\chi(I)$ is defined multiplicatively. When $\chi \equiv 1$, then $L(\chi, s) = \zeta_\mathcal{O}(s)$, etc.

Our main result here is the following.

Theorem 8.9.2. $L(\chi, s)$ *is entire in the sense of* Subsection 8.5.

Before giving the proof we introduce some needed notation. Let $\mathbb{F} \subset L$ be the field of constants and set $t_\mathbb{F} := [\mathbb{F} : \mathbb{F}_r]$. For a divisor D of L we let $\deg_1(D)$ be its degree over $\mathbb{F}$ and $\deg(D)$ its degree over $\mathbb{F}_r$; so
$$\deg(D) = t_\mathbb{F} \deg_1(D).$$
Let $\{\infty_1, \ldots, \infty_\varepsilon\}$ be the infinite primes of L with degrees $\delta_1, \ldots, \delta_\varepsilon$ over $\mathbb{F}$. Finally, let p^λ be the inseparable degree of L over $\mathbf{k}$.

Let $L_0 \subseteq L$ be its maximal separable over **k** subfield. Thus $[L:L_0] = p^\lambda$ and is purely inseparable. Note that each ∞_i lies over a unique infinite prime of L_0. Note also that we can view ∞_i as a class of embeddings of L into $\mathbf{C}_\infty$ over **k**. Each such embedding arises uniquely from its restriction to L_0.

Proof of Theorem 8.9.2. Let $\{\hat{L}_1, \ldots, \hat{L}_\varepsilon\}$ be the completions of L at the places $\{\infty_1, \ldots, \infty_\varepsilon\}$. Let $K_i \subseteq \hat{L}_i^*$ be the local kernels of the reciprocity map on the ideles of L. In particular, K_i is certainly contained in the kernel of the local component, χ_i, of χ at ∞_i.

Let $\mathcal{I}_L$ be the group of $\mathcal{O}_L$-fractional ideals and let $\mathcal{I}_L(\mathcal{B}) \subseteq \mathcal{I}_L$ be the subgroup generated by the primes of $\mathcal{O}_L$ which are unramified in L_1. Let $\mathcal{P}_L(\mathcal{B}) \subseteq \mathcal{I}_L(\mathcal{B})$ be the subgroup generated by those $\alpha \in L^*$ such that for a prime w of $\mathcal{O}_L$

$$\alpha \equiv 1 \ (\mathcal{B}_f \mathcal{O}_w)$$

and such that for all infinite primes ∞_i,

$$\alpha \in K_i.$$

(N.B.: We have abused notation here as we are using the groups K_i in $\mathcal{P}_L(\mathcal{B})$ at ∞_i instead of a subgroup defined by congruence mod $\mathcal{B}_\infty$.)

It is a standard fact that the group $\mathcal{I}_L(\mathcal{B})/\mathcal{P}_L(\mathcal{B})$ is finite. (Indeed, it is easy to see that $\mathcal{P}_L(\mathcal{B})$ is a subgroup of finite index of the group of principal ideals (β) such that $\beta \equiv 1 \ (\mathcal{B}_f \mathcal{O}_w)$ for all primes w of $\mathcal{O}_L$.) Moreover, the abelian reciprocity law implies that there is a surjection from $\mathcal{I}_L(\mathcal{B})/\mathcal{P}_L(\mathcal{B})$ to G_1 given by the Artin symbol.

Thus we are reduced to viewing χ as a $\mathbf{C}_\infty^*$-valued character on the group $\mathcal{I}_L(\mathcal{B})/\mathcal{P}_L(\mathcal{B})$. The L-series of such a character χ is defined as

$$L(\chi, s) := \prod_{\substack{\wp \text{ prime} \\ \wp \in \mathcal{I}_L(\mathcal{B})}} (1 - \chi^{-1}(\wp) n\wp^{-s})^{-1},$$

and agrees, over those primes $\wp \in \mathcal{I}_L(\mathcal{B})$, with the L-series of χ as a morphism on G_1.

By expanding out the Euler product, we have

$$L(\chi, s) = \sum_{\substack{I \in \mathcal{I}_L(\mathcal{B}) \\ I \subseteq \mathcal{O}_L}} \chi^{-1}(I) nI^{-s}.$$

Let m be the number of element in $\mathcal{I}_L(\mathcal{B})/\mathcal{P}_L(\mathcal{B})$ and let $\{\alpha_1, \ldots, \alpha_m\}$ be fixed representatives of these classes. thus

$$L(\chi, s) = \sum_{j=1}^{m} \chi^{-1}(\alpha_j) n\alpha_j^{-s} \left(\sum_{\substack{\alpha \in \mathcal{P}_L(\mathcal{B}) \\ \alpha \alpha_j \subseteq \mathcal{O}_L}} n\alpha^{-s} \right).$$

Thus we are reduced to establishing that

$$L_j(s) := \left(\sum_{\substack{\alpha \in \mathcal{P}_L(\mathcal{B}) \\ a\alpha_j \subseteq \mathcal{O}_L}} n\alpha^{-s} \right) n\alpha_j^{-s}$$

is entire for all j. Fix $j = j_0$ for some j_0, $1 \leq j_0 \leq m$.

Let $t = \frac{r^{d_\infty}-1}{r-1}$ and let $\Phi = \{\theta_1, \ldots, \theta_t\}$ be the classes of $\mathbb{F}_\infty^*/\mathbb{F}_r^*$. The sign, $\mathrm{sgn}(I)$, of any principal ideal is invariantly given as an element of Φ. Thus

$$L_{j_0}(s) = \sum_{e=1}^{t} \left(\sum_{\substack{\alpha \in \mathcal{P}_L(\mathcal{B}) \\ a\alpha_{j_0} \subseteq \mathcal{O}_L \\ \mathrm{sgn}\, n\alpha = \theta_e}} n\alpha^{-s} \right) n\alpha_{j_0}^{-s}.$$

Fix $e = e_0$; so we are reduced, finally, to establishing that

$$L_{j_0,t_0}(s) = \left(\sum_{\substack{\alpha \in \mathcal{P}_L(\mathcal{B}) \\ a\alpha_{j_0} \subseteq \mathcal{O}_L \\ \mathrm{sgn}\, n\alpha = \theta_{e_0}}} n\alpha^{-s} \right) n\alpha_{j_0}^{-s}$$

is entire. For simplicity, set $L(s) := L_{j_0,e_0}(s)$.

The proof will now consist of using the Riemann-Roch Theorem to construct enough functions to establish the entirety of $L(s)$ as in the example following 8.8.2.

Set $\alpha_0 := \alpha_{j_0}$ and let $\hat{\alpha}_0$ be its inverse as a fractional ideal. Let a, b be two nonzero elements of $\hat{\alpha}_0$. We say $a \sim b$ if and only if there exists $u \in \mathcal{O}_L^*$ such that $a = ub$. Thus

$$a\alpha_0 = b\alpha_0 \iff a \sim b.$$

Let $[\hat{\alpha}_0]$ be the set of equivalence classes of $\hat{\alpha}_0 - \{0\}$ under $\sim$.

Let $[a] \in [\hat{\alpha}_0]$ be the class of a. Let $n[a]$ be the norm from L to $\mathbf{k}$ of the elements in $[a]$. As the norm of a unit $\in \mathbb{F}_r^*$, $n[a]$ uniquely describes a principal **A**-fractional ideal which is none other than $n(a\mathcal{O}_L)$. The sign of $n[a]$ is thus an element of Φ.

Let $-\sum_{i=1}^{\varepsilon} c_i^a(\infty_i)$ be the infinite part of the divisor of a taken over all the places of L. Thus the infinite part of the divisor of the norm $N_{\mathbf{k}}^L a$ is $-\left(\sum c_i^a f_i\right)(\infty)$ where f_i is the residue class degree of ∞_i over ∞. As the norm of a unit is in $\mathbb{F}_r^*$, we see that

$$\deg na = d_\infty \left(\sum c_i^a f_i \right)$$

depends only on $[a]$.

Set $D_\infty(a) = \sum_{i=1}^{\varepsilon} c_i^a(\infty_i)$. For each $i = 1, \ldots, \varepsilon$, let $\pi_i \in \hat{L}_i$ be a fixed uniformizer and for $\ell \geq 0$ let $U_i(\ell)$ be the group of units $\equiv 1$ (π_i^ℓ) in $\hat{L}_i$. Let ℓ_i be chosen so that $U_i(\ell_i) \subseteq U_i \cap K_i$ where $U_i := U_i(0)$ is the group of units. Put

$$D_a := \sum_{i=1}^{\varepsilon} (c_i^a - \ell_i)(\infty_i).$$

Let $\widehat{\alpha_0}(D_a)$ be those elements in $\widehat{\alpha_0}$ also in $\mathcal{B}_f$ and which have poles of order $\leq c_i^a - \ell_i$ at ∞_i. By standard algebraic geometry $\hat{\alpha}_0(D_a)$ is finite dimensional over $\mathbb{F}$ and also $\mathbb{F}_r$.

We now establish a lemma which is essential for the remainder of the proof.

Lemma 8.9.3. 1. Let $0 \neq a \in \hat{\alpha}_0$ with $a\mathcal{O}_L \in \mathcal{P}_L(\mathcal{B})$. Let $b \in \hat{\alpha}_0(D_a)$. Then $a + b \in \hat{\alpha}_0$ and $(a+b)\mathcal{O}_L \in \mathcal{P}_L(\mathcal{B})$.
2. $D_\infty(a) = D_\infty(a+b)$.
3. $\mathrm{sgn}(na) = \mathrm{sgn}(n(a+b))$.
4. Let $g \in \hat{\alpha}_0$ with $g\mathcal{O}_L \in \mathcal{P}_L(\mathcal{B})$ also. Then there exists $h \in \hat{\alpha}_0(D_g)$ with $[a+b] = [g+h] \iff$ there exists $b_1 \in \hat{\alpha}_0(D_a)$ with $[a+b_1] = [g]$.
5. Let $b_1, b_2 \in \hat{\alpha}_0(D_a)$ with $b_1 \neq b_2$. Then $(a+b_1)/(a+b_2) \notin \mathcal{O}_L^*$.

Proof. To see Part 1, notice that $a + b = a(1 + b/a)$ and $1 + b/a \in K_i$ at ∞_i by construction.

Parts 2 and 3 follow in a completely similar fashion.

To see Part 4, suppose $[a+b] = [g+h]$. Thus, by definition, there exists $u \in \mathcal{O}_L^*$ with

$$a + b = u(g + h).$$

From Part 2 we conclude that

$$D_\infty(a) = D_\infty(a+b) = D_\infty(u(g+h)) = D_\infty(ug).$$

Set $b_1 = b - uh$; it is easy to see that $b_1 \in \hat{\alpha}_0(D_a)$.

To see Part 5, notice that $D_\infty(a+b_1) = D_\infty(a+b_2)$. Thus if

$$(a+b_1)/(a+b_2) \in \mathcal{O}_L^*,$$

then $(a+b_1)/(a+b_2) \in \mathbb{F}^*$. Solving for a gives a contradiction and completes the proof of the lemma.

Fix $d_0 \in \mathbb{Z}$. Let $X = \{[a_1], \ldots, [a_i]\}$ be the finite, and possibly empty, collection of those $[a]$ such that
1. $a \in \hat{\alpha}_0 - \{0\}$;

2. $(a) \in \mathcal{P}_L(\mathcal{B})$;
3. $\mathrm{sgn}(na) \in \theta_{e_0}$;
4. $\deg(na) = d_0$.

By Lemma 8.9.3 we can decompose X into orbits under the action of translation by the various spaces $\widehat{\alpha}_0(D_{a_i})$. By 8.9.3.4 and 8.9.3.5 the isotropy vanishes.

Let $[a] \in X$ and let $X_a \subseteq X$ be its orbit under $\widehat{\alpha}_0(D_a)$. Let $\kappa \in \widehat{\alpha}_0(D_a)$ and let a_i, κ_i be the injections of these elements into $L_{\infty_i} = \hat{L}_i \subset \mathbf{C}_\infty$. Write

$$a_i = \pi^{n_i} \cdot u_i$$

where $n_i \in \mathbb{Q}$ (and we fix a root of π if necessary) and u_i is a unit in $\mathbf{C}_\infty$. Define

$$m_i \colon \widehat{\alpha}_0(D_n) \to \overline{\mathbf{K}} \subset \mathbf{C}_\infty$$

by

$$m_i(\kappa) := \kappa_i/\pi^{n_i}.$$

Obviously m_i is $\mathbb{F}_r$-linear and we let $\{M_h(\kappa)\}$ be the finite set of maps obtained by composing $m_i(\kappa)$ and $\mathrm{Aut}(\overline{\mathbf{K}}/\mathbf{K})$. As $\mathrm{Aut}(\overline{\mathbf{K}}/\mathbf{K})$ acts continuously, we also have $v_\infty(M_h(\kappa)) > 0$ for any h. Let $\beta \in X_a$. One can now express $\langle\!\langle n\beta \rangle\!\rangle$ in terms of $\langle\!\langle na \rangle\!\rangle$ and $M_h(\kappa)$ (keeping in mind the inseparability degree).

Recall that $t_{\mathbb{F}} = [\mathbb{F} \colon \mathbb{F}_r]$ where $\mathbb{F} \subset L$ is the field of constants. We have $d_\infty = \deg(\infty)$ and let us set $d_i := \deg_1(\infty_i)$. Thus

$$t_{\mathbb{F}} d_i = d_\infty f_i.$$

As $\deg(na) = d_\infty(\sum c_i^a f_i)$ we see $\sum c_i^a d_i = \deg(na)/t_{\mathbb{F}} = d_0/t_{\mathbb{F}}$. Thus the Riemann-Roch Theorem, for sufficiently large d_0, and Lemma 8.8.1, show that v_∞ of the coefficient of x^{-j} in $L(s)$ grows *quadratically* in j for any $s = (x,y) \in S_\infty$. The proof of the theorem is now complete. $\square$

Let L_2 be the compositum of L_1 and $\mathbf{H}^+$. So L_2/L is abelian and the norm of any ideal in the $\mathbf{A}$-integers of L_2 is principal and positive. As is classically done (see also our next subsection), $L(\chi, s)$ factors $\zeta_{\mathcal{O}_L}(s)$. Thus by 8.5.14 we deduce that $L(\chi, s)$ is also essentially algebraic. This can also be seen directly using the first part of 8.8.1.

The v-adic theory of $\zeta_{\mathcal{O}_L}(s)$ is handled in much the same way as the ∞-adic version. One also works with congruence classes mod v and 8.8.1 v-adically. The details are left to the reader.

Let ψ be a Drinfeld module of rank d over L with lattice M. We say that ψ has *sufficiently many complex multiplications* if and only if the set of $\alpha \in \mathbf{C}_\infty$ with $\alpha M \subseteq M$ is an order of rank d over $\mathbf{A}$. In this case the L-series $L(\psi, s)$ factors into a product of L-series associated to Hecke characters exactly as for elliptic curves. In turn, these L-series can be shown to be entire by the same technique as used here, for more see [G04]. See also Subsection 10.4.

8.10. The Question of Local Factors

Let L be a finite extension of $\mathbf{k}$ and let $G = \text{Gal}(L^{\text{sep}}/L)$ where $L^{\text{sep}} \subset \mathbf{C}_\infty$ is the separable closure. Let $\rho\colon G \to GL_n(\mathbf{C}_\infty)$ be a representation of Galois type; so ρ factors through $G_1 := \text{Gal}(L_1/L)$ with L_1/L finite Galois. There is an ambiguity in dealing with finitely many local finite factors (i.e., the ramified finite primes of L) that must now be addressed. The reader will immediately perceive the difficulty from the following example.

Example 8.10.1. Suppose that L_1/L is abelian of order p and totally-ramified above, say, one finite prime $\wp$ of L. Suppose also that $n = 1$; so $G_1 \simeq \mathbb{Z}/(p)$ and $\rho\colon G_1 \to \mathbf{C}_\infty^*$. But, of course, there are *no* non-trivial p-th roots of 1 in $\mathbf{C}_\infty$; thus $\rho \equiv 1$. So there are *no* bad primes for ρ and $L(\rho, s) = \zeta_{\mathcal{O}_L}(s)$, if we work in finite characteristic. However, we want to factor $\zeta_{\mathcal{O}_{L_1}}(s)$ as in classical arithmetic. But $\zeta_{\mathcal{O}_{L_1}}(s) \neq \zeta_{\mathcal{O}_L}(s)$, as the product of the factors above a good finite prime $\mathcal{B} \subset \mathcal{O}_L$ in $\zeta_{\mathcal{O}_{L_1}}(s)$ is $(1 - n\mathcal{B}^{-s})^p$. On the other hand, $\zeta_{\mathcal{O}_{L_1}}(s) \neq \zeta_{\mathcal{O}_L}(s)^p$ as the local factor above $\wp$ in $\zeta_{\mathcal{O}_{L_1}}(s)$ is $(1 - n\wp^{-s})^{-1}$.

As the reader will note, the problem lies in the use of characters into $\mathbf{C}_\infty^*$. However, as is standard in arithmetic, there is *no* difficulty with the local factors of *complex*-valued representations of G_1 [Ta1]. Moreover, as G_1 is finite we are free to work with *any* algebraically closed field of characteristic 0; in particular $\overline{\mathbb{Q}}_p$, a fixed algebraic closure of $\mathbb{Q}_p$.

Let $R_p \subset \overline{\mathbb{Q}}_p$ be the ring of integers and let M_p be its maximal ideal. So R_p/M_p is an algebraic closure of $\mathbb{F}_p$. Fix an embedding $\theta\colon R_p/M_p \hookrightarrow \mathbf{C}_\infty$; so θ has image the algebraic closure of $\mathbb{F}_r$ inside $\mathbf{C}_\infty$.

The solution to the problem of local factors is simply to work only with the local factors associated to $\overline{\mathbb{Q}}_p$-valued representations of G_1. As these are the "correct" factors for factorization of L-series, the local factor problem vanishes.

Thus let $\rho\colon G_1 \to GL(V)$, V a finite dimensional $\overline{\mathbb{Q}}_p$-vector space, be a representation of G_1. Let w be a finite prime of L with W a prime above w in L_1. Let D_W, I_W be the decomposition group and inertia groups, etc. Set

$$P_{\rho,w}(u) := \det(1 - u\rho(F_W) \mid V^{I_W})$$

where F_W is the geometric Frobenius at W. Since G_1 is finite, the eigenvalues of $\rho(F_W)$ are roots of unity and so

$$P_{\rho,w}(u) \in R_p[u]$$

and depends only on w, etc.

We now project $R_p[u]$ onto $R_p/M_p[u]$ and use θ to obtain

$$\overline{P}_{\rho,w}(u) \in \mathbf{C}_\infty[u].$$

Definition 8.10.2. We set for $s \in S_\infty$

$$L(\rho, s) := \prod_{w \text{ finite}} \overline{P}_{\rho,w}(nw^{-s})^{-1}.$$

As before, $L(\rho, s)$ converges on a half-plane of S_∞. Further, the v-adic version of $L(\rho, s)$ can now be easily defined and will be left to the reader.

Example 8.10.3. We now return to the example of 8.10.1. Let $\chi \colon G_1 \to \overline{\mathbb{Q}}_p^*$ be a non-trivial character; so the group of all characters is $\{1, \chi, \ldots, \chi^{p-1}\}$. Now

$$L(1, s) = \zeta_{\mathcal{O}_L}(s)$$

and

$$L(\chi^i, s) = (1 - n\wp^{-s})\zeta_{\mathcal{O}_L}(s),$$

as $\chi^i + (M_p)$ is the trivial character for $i = 1, \ldots, p-1$. Thus we obtain the correct factorization

$$\zeta_{\mathcal{O}_{L_1}}(s) = \zeta_{\mathcal{O}_L}(s)^p (1 - n\wp^{-s})^{p-1}.$$

In general, if G_1 is a p-group, we obtain a factorization completely similar to that of 8.10.3.

Therefore, the reader should ignore the temptation to find a Galois stable lattice in V and reduce modulo M_p *before* defining the local factors as this *precisely* gives the confusion of Example 8.10.1.

Finally, when we are dealing with an L-series, $L(\rho, s)$, where ρ is of Galois type, we shall always mean the functions of the type considered here unless otherwise specifically mentioned. Thus we will have no ambiguity at the bad primes in terms of the factors to be considered.

8.11. The Generalized Teichmüller Character

In this subsection we will construct a particular class of characteristic 0 valued characters of $\operatorname{Gal}(\mathbf{k}^{\operatorname{sep}}/\mathbf{k})$. This construction is based on our exponentiation of ideals.

Recall that in Subsection 8.1 we defined the elements $s_j \in S_\infty$ with the property that if $I = (i)$, i positive, then

$$I^{s_j} = i^j.$$

We then set $\mathbf{V} = \mathbf{k}(\{I^{s_1}\}) \subset \mathbf{C}_\infty$; so $[\mathbf{V} \colon \mathbf{k}] < \infty$ as we have seen.

Let $\mathcal{O}_\mathbf{V}$ be the ring of $\mathbf{A}$-integers in $\mathbf{V}$ and let $\sigma \colon \mathbf{V} \to \overline{\mathbf{k}}_\wp$ be an embedding over $\mathbf{k}$ where $\wp \in \operatorname{Spec}(\mathbf{A})$ is fixed and $\overline{\mathbf{k}}_\wp$ is an algebraic closure equipped with its canonical metric. This is completely similar to what was done in

8.11. The Generalized Teichmüller Character

Subsection 8.3. Let $\overline{\wp}$ be the finite prime of **V** over $\wp$ which is associated to σ.

As before let $\mathcal{I}$ be the group of **A**-fractional ideals of **k** with $\mathcal{I}(\wp)$ the subgroup of $\mathcal{I}$ generated by those primes $\neq \wp$. If $I \in \mathcal{I}(\wp)$ then $I^{s_1} = I^1 \in \mathbf{V}$ is prime to $\wp$ by Proposition 8.2.10. Let

$$\rho : \mathcal{I}(\wp) \to (\mathcal{O}_\mathbf{V}/\overline{\wp})^*$$

be the mapping

$$I \mapsto I^1 + \overline{\wp} = I^{s_1} + \overline{\wp}.$$

Lemma 8.11.1. *Let $\alpha \in \mathbf{k}^*$ with $\alpha \equiv 1 \pmod{\wp}$, α positive. Then*

$$\rho((\alpha)) = 1 \in \mathcal{O}_\mathbf{V}/\overline{\wp}.$$

Proof. As α is positive we see that $\rho((\alpha)) = \alpha + \overline{\wp}$. As $\alpha \equiv 1 \pmod{\overline{\wp}}$, the result follows. □

Recall that we set $\mathcal{P}_\wp^+$ to be group of principal ideals (α) such that α is positive *and* $\equiv 1 \pmod{\wp}$. From 8.11.1 we immediately deduce a homomorphism

$$\widehat{\rho} : \mathcal{I}(\wp)/\mathcal{P}_\wp^+ \to (\mathcal{O}_\mathbf{V}/\overline{\wp})^*.$$

Let $W = W_\sigma$ be the Witt ring of the finite field $\mathcal{O}_\mathbf{V}/\overline{\wp}$ which we consider as lying in $\overline{\mathbb{Q}}_p$ for our fixed algebraic closure $\overline{\mathbb{Q}}_p$ of $\mathbb{Q}_p$.

Let Teich: $(\mathcal{O}_\mathbf{V}/\overline{\wp})^* \to W^*$ be the usual map taking an element to its Teichmüller representative.

Definition 8.11.2. We set

$$\omega_\sigma := \omega_{\overline{\wp}} := \text{Teich} \circ \widehat{\rho} : \mathcal{I}(\wp)/\mathcal{P}_\wp^+ \to W^*.$$

The homomorphism ω_σ is the *generalized Teichmüller character*.

Note that if $\sigma_1, \sigma_2 : \mathbf{V} \to \overline{\mathbf{k}}_\wp$ differ by an automorphism of $\overline{\mathbf{k}}_\wp/\mathbf{k}_\wp$ then they give rise to the same ideal $\overline{\wp}$. Thus the notation above is indeed consistent. Moreover, if **A** is a principal ideal domain, then the generalized Teichmüller character *is* the Teichmüller character of $\mathbf{A}/\wp$.

Let ψ be a sgn-normalized Hayes-module for a fixed sign function sgn, and let, as before, $\mathbf{k}(\wp) \subset \mathbf{C}_\infty$ be the abelian extension obtained by adjoining to $\mathbf{k}$ the $\wp$-division points of ψ in $\mathbf{C}_\infty$. By 7.5.5 we know that there is an isomorphism

$$G := \text{Gal}(\mathbf{k}(\wp)/\mathbf{k}) \xrightarrow{\sim} \mathcal{I}(\wp)/\mathcal{P}_\wp^+,$$

via the Artin map. Thus we may consider ω_σ as a W-valued character of G.

Let $H_\sigma \subseteq (\mathcal{O}_\mathbf{V}/\overline{\wp})^*$ be the image of $\widehat{\rho}$. We now establish some general results on the size of H_σ. First of all, of course, H_σ is a cyclic group. Moreover, let $\widehat{\mathcal{P}} \subseteq \mathcal{I}(\wp)$ be the subgroup generated by *positive* α with $(\alpha, \wp) = 1$. Thus

$$\rho((\alpha)) = \alpha + \overline{\wp},$$

and

$$\rho(\widehat{\mathcal{P}}) = (\mathbf{A}/\wp)^* \subseteq (\mathcal{O}_\mathbf{V}/\overline{\wp})^*.$$

In particular, $r^{\deg \wp} - 1$ divides the order of H_σ.

Now let α be an element of $\mathbf{k}$ which is $\equiv 1 \pmod{\wp}$ but *not* necessarily positive. We have

$$(\alpha)^{s_1} = (\alpha)^1 = \alpha/\operatorname{sgn}(\alpha) \equiv \frac{1}{\operatorname{sgn}(\alpha)} \pmod{\overline{\wp}}.$$

Thus H_σ must also contain a subgroup, $\mathfrak{H}$, isomorphic to $\mathbb{F}_\infty^*$. We summarize all this in our next result.

Proposition 8.11.3. *The subgroup $H_\sigma \subseteq (\mathcal{O}_\mathbf{V}/\overline{\wp})^*$, which is the image of $\widehat{\rho}$, is cyclic of order t_σ where t_σ divides*

$$r^{f \deg \wp} - 1,$$

and where $f = f_{\overline{\wp}}$ is the residue class degree at $\overline{\wp}$. Moreover, t_σ is divisible by the l.c.m. of $r^{\deg \wp} - 1$ and $r^{d_\infty} - 1$. □

Question 8.11.4. What is the exact order of H_σ?

We note, of course, that the order of H_σ is the order of ω_σ. Note further that in general, there is *not* a canonical embedding of W into $\overline{\mathbb{Q}}_p$; however all such embeddings differ by some power of the Frobenius automorphism as $W/\mathbb{Z}_p$ is unramified.

8.12. Special-values at Negative Integers

Let $L \subset \overline{\mathbf{k}} \subset \mathbf{C}_\infty$ be a fixed finite extension of $\mathbf{k}$ as before. Let $\mathcal{O} := \mathcal{O}_L$ be the ring of $\mathbf{A}$-integers of L. We are interested here in the function $\zeta_{\mathcal{O}_L}(s)$ which we recall (see Subsection 8.6) is defined as

$$\zeta_{\mathcal{O}_L}(s) = \prod_{\substack{\wp \subset \mathcal{O}_L \\ \wp \text{ prime}}} (1 - n\wp^{-s})^{-1}$$

$$= \sum_{I \subseteq \mathcal{O}_L} nI^{-s}.$$

8.12. Special-values at Negative Integers

From the main result of Subsection 8.9 we know that $\zeta_{\mathcal{O}_L}(s)$ continues to an entire function on S_∞. Thus it is essentially algebraic, either by the argument at the end of Subsection 8.9 (which uses 8.5.14) or directly via 8.8.1.1. Therefore if π_* is our fixed root in $\mathbf{C}_\infty$ of

$$z^{d_\infty} - \pi,$$

and j is a non-negative integer, then

$$z(x, -j) := z_\zeta(x, -j) := h_\zeta(x, -j) = \zeta_{\mathcal{O}_L}(x\pi_*^j, -j)$$

is a *polynomial* in x^{-1} with coefficients in the ring $\mathcal{O}_\mathbf{V}$ of $\mathbf{A}$-integers of the value field $\mathbf{V}$. This subsection is devoted to the arithmetic interpretation of these polynomials.

Remarks. 8.12.1. 1. The vanishing result given in 8.8.1.1 is *not* always best possible. We illustrate this here with an example. Let $\mathcal{O} = \mathbf{A} = \mathbb{F}_r[T]$ for all of this remark and set

$$z(x, -j) = \sum_{h=0}^{\infty} s_h(j) x^{-h},$$

where

$$s_h(j) := \sum_{\substack{\deg n = h \\ n \text{ monic}}} n^j.$$

Thus by 8.8.1.1 we know that $s_h(j)$ vanishes whenever

$$j < (r-1)h. \tag{8.12.2}$$

On the other hand, put

$$e_h(x) := \prod_{\deg(\alpha) < h} (x + \alpha).$$

By Corollary 1.2.2 we know that $e_h(x)$ is an $\mathbb{F}_r$-linear polynomial. Note also that

$$e_h(T^h) = \prod_{\substack{n \text{ monic} \\ \deg(n) = h}} n = D_h$$

by 3.1.6.3. Thus $e_h(x - T^h) = e_h(x) - D_h = \prod_{\substack{n \text{ monic} \\ \deg n = h}} (x - n)$. By taking logarithmic derivatives and using Theorem 3.1.5, we find

$$(-1)^h \frac{D_h}{L_h} \cdot \frac{1}{e_h(x) - D_h} = \sum_{\substack{n \text{ monic} \\ \deg n = h}} (x - n)^{-1}.$$

Upon expanding out the sum in terms of $\frac{1}{x}$, one finds that $s_h(j)$ vanishes when $j < r^h - 1$; clearly a much better estimate than 8.12.2.

2. For more on $s_h(j)$ we refer the reader to [Th3, esp. Th. VIII] and the references given there. (The result of Carlitz, [Th3, Th. VIII.3] should be viewed with some suspicion for $r \neq p$; as of this writing, it is not clear if this result is valid when $r \neq p$.)

3. Because of the simplicity of $\mathbb{F}_r[T]$ as $\mathbb{F}_r$-algebra, it is possible to give a recursive definition of $z(x, -j)$. More precisely, we have $z(x, 0)$ is identically 1, and for $j > 0$ we have

$$z(x, -j) = 1 - \sum_{\substack{b=0 \\ (r-1)|(j-b)}}^{j-1} \binom{j}{b} T^b x^{-1} z(x, -b).$$

The proof follows from noting that every monic n of degree d can be written as $n = Th + \alpha$ where h is monic of degree $d - 1$ and $\alpha \in \mathbb{F}_r$; one then uses the binomial theorem. (See also [Go4, 5.1.2].)

We now want to relate $z(x, -j)$ to the generalized Teichmüller character ω_σ of Subsection 8.11. As mentioned in that subsection, we view ω_σ as a W-valued character of $\mathrm{Gal}(\mathbf{k}(\wp)/\mathbf{k})$ where W is the Witt ring of $\mathcal{O}_\mathbf{V}/\overline{\wp}$. Let

$$L(\wp) := L \cdot \mathbf{k}(\wp);$$

of course $L(\wp)/L$ is abelian with Galois group canonically isomorphic to a subgroup of $\mathrm{Gal}(\mathbf{k}(\wp)/\mathbf{k})$. Therefore, ω_σ also gives rise to a character $\omega_{\sigma,L}$ of $\mathrm{Gal}(L(\wp)/L)$. Let $\mathcal{B}$ be a finite prime of $\mathcal{O}_L$ not dividing $\wp$; thus $L(\wp)$ is unramified at $\mathcal{B}$. Class field theory implies that

$$\omega_{\sigma,L}((\mathcal{B}, L(\wp)/L)^{-1}) = \omega_\sigma((n\mathcal{B}, \mathbf{k}(\wp)/\mathbf{k})^{-1}).$$

Let u be a variable.

Definition 8.12.3. Let $L(\omega_{\sigma,L}^j, u)$ be the classical (characteristic zero valued) L-series of $\omega_{\sigma,L}^j$.

Thus let $\mathcal{B}$ be as above. The local factor at $\mathcal{B}$ in $L(\omega_{\sigma,L}^j, u)$ is

$$(1 - \omega_{\sigma,L}^j((\mathcal{B}, L(\wp)/L)^{-1}) u^{\deg_1 \mathcal{B}})$$

where $\deg_1 \mathcal{B}$ is the degree of $\mathcal{B}$ with respect to the full field $\mathbb{F}$ of constants of L (and where we have continued to use the geometric Frobenius). We note that $L(\omega_{\sigma,L}^j, u)$ is formed as a product over *all* places, finite or infinite, of L.

Let $L^{\mathrm{sep}} \subset \overline{\mathbf{k}}$ be the separable closure of L and let $\chi \colon \mathrm{Gal}(L^{\mathrm{sep}}/L) \to \overline{\mathbb{Q}}_p^*$ be a homomorphism that factors through a finite Galois extension L_1 of L. One says that χ is *principal* if and only if L_1 can be taken to be a constant

8.12. Special-values at Negative Integers 275

field extension (which may be trivial) of L. Let $L(\chi, u)$ be the L-series of χ. If χ is *not* principal, then standard theory (due to A. Weil) [Ta1] implies that $L(\chi, u) \in \overline{\mathbb{Q}}_p[u]$.

In particular, the above remark applies to all $\omega_{\sigma,L}^j$, where $\omega_{\sigma,L}^j$ is non-principal. For instance, if $\deg \infty = 1$ and L is unramified above $\wp$ and totally-split at ∞, then one sees from Subsection 7.5 that $\omega_{\sigma,L}^j$ is non-principal if it is non-trivial. Indeed, ramification considerations imply that $L(\wp)/L$ contains no non-trivial constant field extensions.

Let $\widehat{L}(\omega_{\sigma,L}^j, u)$ be the L-function of $\omega_{\sigma,L}^j$ but where we *only* use primes of $\operatorname{Spec}(\mathcal{O}_L)$ which do *not* lie over $\wp$. Thus $\widehat{L}(\omega_{\sigma,L}^j, u) = f(u) L(\omega_{\sigma,L}^j, u)$ where $f(u) \in \overline{\mathbb{Q}}_p[u]$ and is a product over the set of infinite primes and those above $\wp$. As such, $\widehat{L}(\omega_{\sigma,L}^j, u) \in \overline{\mathbb{Q}}_p[u]$ whenever $L(\omega_{\sigma,L}^j, u)$ does.

Remark. 8.12.4. Let L be unramified above $\wp$. One checks that $L(\omega_{\sigma,L}^j, u)$ will contain non-trivial Euler factors at $\wp$ if and only if $r^{\deg \wp} - 1$ divides j.

As before, let $\mathbb{F} \subset L$ be the full field of constants and set $t_{\mathbb{F}} = [\mathbb{F} : \mathbb{F}_r]$.

We can now present the basic relationship between $z(x, -j)$ and the classical function $\widehat{L}(\omega_{\sigma,L}^j, u)$. Of course, $z(x, -j) \in \mathcal{O}_\mathbf{V}[x^{-1}]$ and $\widehat{L}(\omega_{\sigma,L}^j, u) \in W(u)$ where $W =$ the Witt ring of $\mathcal{O}_\mathbf{V}/\overline{\wp}$. However, by reducing modulo $\overline{\wp}\mathcal{O}_\mathbf{V}[x^{-1}]$, the polynomial $z(x, -j)$ gives rise to an element $\pi_0(z(x, -j))$ in $\mathcal{O}_\mathbf{V}/\overline{\wp}[x^{-1}]$. Similarly, by reducing modulo $pW(u)$, $\widehat{L}(\omega_{\sigma,L}^j, u)$ is readily seen to give rise to a non-zero element $\pi_1(\widehat{L}(\omega_{\sigma,L}^j, u)) \in \mathcal{O}_\mathbf{V}/\overline{\wp}(u)$. We thus have the following *double congruence*.

Theorem 8.12.5. *Let $j > 0$. Consider $(\mathcal{O}_\mathbf{V}/\overline{\wp})(x^{-1})$ mapped isomorphically to $(\mathcal{O}_\mathbf{V}/\overline{\wp})(u)$ via $x^{-1} \leftrightarrow u$. Then, under this identification*

$$\pi_0(z(x, -j)) = \pi_1(\widehat{L}(\omega_{\sigma,L}^{-j}, u^{t_{\mathbb{F}}})).$$

Proof. This follows from the Euler products for both $z(x, -j)$ and $\widehat{L}(\omega_{\sigma,L}^{-j}, u)$, and the fact that $\deg = t_{\mathbb{F}} \deg_1$. □

Remarks. 8.12.6. 1. The above result differs from our earlier formulations of it (e.g., [Go4, 4.2.5]) because we are now using the geometric Frobenius.
2. The theorem implies that the reduction of $\widehat{L}(\omega_{\sigma,L}^{-j}, u)$ is always a polynomial for $j > 0$. This is also quite easy to see directly as the factors in the denominator (if there are any) will reduce to 1.
3. There is some ambiguity in the result arising from different injections of W into $\overline{\mathbb{Q}}_p$. However, applications of the theorem turn out to be independent of which embedding one uses.

8.13. Trivial Zeroes

We retain the notation of our previous two subsections. It is our purpose to show here how the double congruences of our last two sections give "trivial zeroes" to the functions $\zeta_{\mathcal{O}_L}(s)$. In fact we will see later on, in Subsection 8.17, how these double congruences give trivial zeroes in a very general context. Moreover, in Remark 8.17.6.1, we will present some *speculation* on possible trivial zeroes for the L-series of arbitrary Drinfeld modules. As of this writing, no general results are known in this area.

Recall that if $\wp \subseteq \mathbf{A}$ is prime (and σ is associated to a prime above $\wp$ as in Subsection 8.11), then $\mathcal{I}(\wp)/\mathcal{P}_\wp^+$ contains the subgroup $\mathfrak{H}$ generated by principal ideals (α) where $\alpha \equiv 1 \ (\wp)$ (α *not* necessarily positive). Thus, as we have seen, $\mathfrak{H}$ is isomorphic to $\mathbb{F}_\infty^*$ via our sign morphism sgn. Moreover, by Proposition 7.5.8, $\mathfrak{H}$ corresponds to both the decomposition *and* inertia groups at ∞.

A basic point about $\mathfrak{H}$ is that it is *independent* of $\wp$. Indeed the fields $\mathbf{k}(\wp)$, as $\wp$ varies, give rise to the *same* local extension of $\mathbf{K}$. This may be seen easily by local class field theory, or directly via the exponential function of a Hayes-module associated to sgn. Thus for L/k finite, the local extensions at the infinite primes of $L(\wp)/L$ are also independent of $\wp$.

Let $\{\infty_1, \ldots, \infty_t\}$ be the infinite primes of L. Let j be a positive integer and let $\{\infty_{i_1}, \ldots, \infty_{i_t}\}$ be the, possibly empty, subset of infinite primes where $\omega_{\sigma, L}^{-j}$ is unramified. From the discussion given above, and the definition of $\omega_{\sigma,L}^{-j}$, this subset is independent of the choice of $\wp$ and σ. Moreover, since L may be general, it can happen that $\omega_{\sigma,L}^{-j}$ may be unramified *and* non-trivial at these infinite primes. However, one sees from our description of $\mathfrak{H}$ that the local factor

$$(1 - \omega_{\sigma,L}^{-1}(\infty_{i_e}) t^{\deg_1(\infty_{i_e})})$$

is *independent* of σ and $\wp$ and depends only on j. Moreover, from 7.5.7, the value $\omega_{\sigma,L}^{-j}(\infty_{i_e})$ depends only on sgn and so we may, without ambiguity, use the same symbol to denote this value in $\overline{\mathbf{K}}$.

Lemma 8.13.1. *Let D be a Dedekind domain and let $g(u), h(u) \in D[u]$ with $g(0) = h(0) = 1$. Suppose that $g(u)/h(u)$ is a polynomial modulo an infinite number of primes of D. Then $g(u)/h(u) \in D[u]$.*

Proof. Assume the result is false. Without loss of generality, we may assume that the coefficients of $g(u)$ and $h(u)$ are units in D as these coefficients are divisible by only finitely many primes. We may also assume that $g(u)/h(u)$ is reduced. Let L be a common splitting field for both $g(u)$ and $h(u)$ and let $\{\alpha_i, \beta_j\} \subset L$ be such that $g(u) = \prod(1 - \alpha_i u)$, $h(u) = \prod(1 - \beta_j u)$. Finally, let $\mathfrak{B}$ be any prime not dividing $\{\beta_j - \alpha_i\}$. Thus modulo $\mathfrak{B}$, $g(u)/h(u)$ still has a pole. This is a contradiction. □

8.13. Trivial Zeroes 277

Definition 8.13.2. We set $\widetilde{z}(x,-j)$ equal to

$$z(x,-j)(1 - \omega_{\sigma,L}^{-j}(\infty_{i_1})x^{-\deg(\infty_{i_1})})^{-1} \cdots (1 - \omega_{\sigma,L}^{-j}(\infty_{i_t})x^{-\deg(\infty_{i_t})})^{-1},$$

where, as usual, deg means "degree with respect to $\mathbb{F}_r$."

We now want to present a small variation of our double congruence 8.12.5. Let us define $\widetilde{L}(\omega_{\sigma,L}^j, u)$ to be the L-function of $\omega_{\sigma,L}^j$ where we do *not* use primes lying above $\wp$ but where we *do* use the infinite primes. Thus $\widetilde{L}(\omega_{\sigma,L}^j, u)$ is again a polynomial in u whenever $L(\omega_{\sigma,L}^j, u)$ is. In the notation of 8.12.5, we then have the following result.

Theorem 8.13.3. *Let j be a positive integer. Consider $(\mathcal{O}_\mathbf{V}/\overline{\wp})(x^{-1})$ mapped isomorphically to $(\mathcal{O}_\mathbf{V}/\overline{\wp})(u)$ via $x^{-1} \leftrightarrow u$. Then*

$$\pi_0(\widetilde{z}(x,-j)) = \pi_1(\widetilde{L}(\omega_{\sigma,L}^{-j}, u^{t_\mathbb{F}})).$$

Proof. This follows as in 8.12.5. □

Corollary 8.13.4. *$\widetilde{z}(x,-j)$ is a polynomial in x^{-1}.*

Proof. The coefficients of $\widetilde{z}(x,-j)$ are also in the value field $\mathbf{V}$. Furthermore, it is easy to see that $\pi_1(L(\omega_{\sigma,L}^{-j}, u^{t_\mathbb{F}}))$ is a polynomial in u. Thus the result follows from Proposition 8.13.1. □

We have

$$z(x,-j) = \widetilde{z}(x,-j)(1 - \omega_{\sigma,L}^{-j}(\infty_{i_1})x^{-\deg \infty_{i_1}}) \cdots (1 - \omega_{\sigma,L}^{-j}(\infty_{i_t})x^{-\deg \infty_{i_t}}),$$

where the right hand side is a product of *polynomials* by the corollary.

Definition 8.13.5. The zeroes of $z(x,-j)/\widetilde{z}(x,-j)$ are the *trivial zeroes* of $z(x,-j)$.

Of course the trivial zeroes of $z(x,-j)$ immediately translate to trivial zeroes of $\zeta_{\mathcal{O}_L}(s)$, as, by definition, the value of $z(x,-j)$ at x equals $\zeta_{\mathcal{O}_L}(x\pi_*^j, -j)$. If $x=1$ is a trivial zero for $z(x,-j)$ then we say $-j \in S_\infty$ is a trivial zero for $\zeta_{\mathcal{O}_L}(s)$, etc. Moreover, the same techniques can be used to give trivial zeroes of abelian L-series, etc. See Subsection 8.17.

In order to illustrate the above result on trivial zeroes, we present two concrete examples. It is instructive to compare these examples to classical results on trivial zeroes.

Example 8.13.6. Let ∞ be a rational point over $\mathbb{F}_r$ and let $L = \mathbf{k}$; thus $\mathcal{O}_L = \mathbf{A}$. Let $\wp$ be a prime of $\mathbf{A}$. We know that the decomposition and inertia groups at ∞ inside $\mathrm{Gal}(\mathbf{k}(\wp)/\mathbf{k})$ correspond to $\mathbb{F}_r^* \subseteq \mathrm{Gal}(\mathbf{k}(\wp)/\mathbf{k})$. Let j be a positive integer; thus ω_σ^{-j} is ramified at ∞ if and only if j is *not* divisible by $r - 1$. We then have two cases:

1. $(r - 1) \nmid j$. In this case ∞ does not contribute to ω_σ^{-j}; in consequence $-j$ is not a trivial zero for $\zeta_\mathbf{A}(s)$. (**N.B.**: This does *not* imply that $\zeta_\mathbf{A}(-j) \neq 0$; only that such a zero is not "trivial.")
2. $(r - 1) \mid j$. Then ∞ contributes $1 - x^{-1}$, as above, and we find a trivial zero at $-j \in S_\infty$.

Example 8.13.7. Let ∞ continue to be rational over $\mathbb{F}_r$ and now let $L = \mathbf{H} =$ the Hilbert class field totally split at ∞. The field $\mathbf{H}$ is always contained in $\mathbf{k}(\wp)$ for any prime $\wp$; thus $L(\wp) = \mathbf{k}(\wp)$. Moreover, $\omega_{\sigma,L}$ is just the restriction of ω_σ to $\mathrm{Gal}(\mathbf{k}(\wp)/\mathbf{H}) \simeq (\mathbf{A}/\wp)^*$ and thus does not depend on σ. Since $\mathbf{k}(\wp)/\mathbf{H}$ totally ramifies at the primes above $\wp$, $\omega_{\sigma,L}^{-j}$ is ramified at the primes above $\wp$ whenever it is non-trivial. Moreover, the decomposition and inertia groups of the infinite primes of $\mathbf{H}$ still correspond to $\mathbb{F}_r^* \subseteq (\mathbf{A}/\wp)^*$.

Therefore, let $j > 0$. We then have two cases as above:

1. $(r - 1) \nmid j$. In this case the infinite primes of $\mathbf{H}$ do *not* contribute to the L-series of $\omega_{\sigma,L}^{-j}$; in consequence there are no trivial zeroes at $-j$ for $\zeta_{\mathcal{O}_L}(s)$.
2. $(r - 1) \mid j$. Then *each* infinite prime of $\mathbf{H}$ contributes $(1 - x^{-1})$ and we find a trivial zero at $-j \in S_\infty$ of order $\geq [\mathbf{H}:\mathbf{k}] = \#\mathrm{Pic}(\mathbf{A})$.

Remarks. 8.13.8. 1. The formula given for the trivial zeroes is *only* a lower bound for the order of zero. In the case $\mathbf{A} = \mathbb{F}_r[T]$, the recursive formula of 8.12.1.3 can be easily used to establish that $\zeta_\mathbf{A}(-j) \neq 0$ for $j > 0, j \not\equiv 0(r-1)$. Moreover, if $j > 0$ is $\equiv 0 \ (r-1)$, then $\zeta_\mathbf{A}(-j)$ has a *simple* trivial zero at $-j$.
2. The trivial zeroes of $\zeta_{\mathcal{O}_L}(s)$ at negative integers can be used to give trivial zeroes to

$$\zeta_{\mathcal{O}_L,\sigma,v}(s)$$

at *positive* integers via v-adic continuity. For more see [Go4, 4.2.7].
3. The lower bound on the order of the trivial zeroes is *not* always exact. For a counterexample see [Th4, §3] (this counterexample is also discussed in Subsection 8.24). It would be *very* interesting to have a formula and interpretation for the exact order of the trivial zeroes in general.

It would also be *very* interesting to have the techniques to handle trivial zeroes directly *without* using classical results in characteristic 0 (and a double congruence). The following is a primitive example of such techniques.

Example 8.13.9. We will show here how to deduce the results of Example 8.13.6 from the results of Subsection 8.8. First of all, by definition $\zeta_\mathbf{A}(s) = \sum_I I^{-s}$ where I runs through all the ideals of $\mathbf{A}$. Let $\{\alpha_1, \ldots, \alpha_h\}$ be representatives of the ideal classes of $\mathbf{A}$. Thus we have

$$\zeta_{\mathbf{A}}(s) = \sum_{t=1}^{h} \alpha_t^{-s} \left(\sum_{\substack{a \text{ monic} \\ a \in \alpha_t^{-1}}} a^{-s} \right).$$

Let j be a positive integer divisible by $r - 1$. We will therefore be finished if we show that for any t

$$\sum_{d=0}^{\infty} \left(\sum_{\substack{a \in \alpha_t^{-1} \\ a \text{ monic} \\ \deg a = d}} a^j \right) = 0,$$

where by 8.8.1.1 we know that the sum inside the parentheses vanishes for sufficiently large d. Now choose $e \gg 0$ sufficiently large so that, by 8.8.1.1,

$$\sum_{\substack{a \in \alpha_t^{-1} \\ \deg a \leq e}} a^j = 0.$$

Note that as $(r - 1) \mid j$, and ∞ is rational, the sum over *all* a is the opposite of the sum over *monics*. A little thought completes the computation.

Remark. 8.13.10. We finish this subsection with examining the trivial zeroes theory for $\zeta_{\mathbf{A}}(s) = \sum_I I^{-s}$ for *arbitrary* ∞. In this case, for any prime $\wp$ of $\mathbf{A}$ we know that the inertia and decomposition subgroups of $\mathrm{Gal}(\mathbf{k}(\wp)/\mathbf{k})$ are isomorphic to $\mathbb{F}_\infty^*$ where $\mathbb{F}_\infty$ is the field of constants of $\mathbf{K}$. Thus ω_σ^{-j} is unramified at ∞ if and only if j is divisible by $r^{d_\infty} - 1$ (and $d_\infty := [\mathbb{F}_\infty : \mathbb{F}_r]$) in which case ∞ contributes $1 - x^{-d_\infty}$. In particular, we deduce $\zeta_{\mathbf{A}}(-j) = 0$ for such j. On the other hand, we obtain *no* information for *arbitrary* $j \equiv 0 \bmod (r - 1)$, which is both intriguing and somewhat counter-intuitive.

8.14. Applications to Class Groups

As mentioned in the introduction to this section, the reader may profit by just assuming that $\mathbf{A}$ is a principal ideal domain (or even $\mathbb{F}_r[T]$) during a first reading. This may be especially true of this subsection where the use of general $\mathbf{A}$ will cause us to introduce a fair amount of notation. In any case, in this subsection we will use our double congruence (Theorem 8.13.3) to establish results on the divisibility of certain class numbers by p, where p is the characteristic of $\mathbf{C}_\infty$. While there may be some technicalities involved, the basic idea used here is very simple: a p-primary Galois component of the class group is non-trivial if and only if some classical L-value is not divisible by p. On the other hand, this L-value *modulo p* turns out to be the same (up to isomorphism) as some special zeta-value (in finite characteristic!) modulo the prime $\wp$ of $\mathbf{A}$. Thus the divisibility of this characteristic p value by $\wp$ determines the non-triviality of the p-primary Galois component we want to study.

We begin by examining more closely the generalized Teichmüller character ω_σ as a map of $G_\wp := \text{Gal}(\mathbf{k}(\wp)/\mathbf{k})$ to the Witt ring $W = W_\sigma$ of $\mathcal{O}_\mathbf{V}/\overline{\wp}$.

Definition 8.14.1. Let $G_\wp$ be as above and $G_\sigma \subset G_\wp$ the kernel of ω_σ. Let $\mathbf{k}(\wp)_\sigma$ be the subfield fixed by G_σ; we call $\mathbf{k}(\wp)_\sigma$ the *subfield cut out by ω_σ*.

Thus, if $\mathbf{A}$ is a principal ideal domain, then $\mathbf{k}(\wp)_\sigma = \mathbf{k}(\wp)$; in general, however, it will be a much smaller subfield.

Proposition 8.14.2. *The extension $\mathbf{k}(\wp)/\mathbf{k}(\wp)_\sigma$ is everywhere unramified.*

Proof. It is clear that we need only concentrate on the primes above $\wp$ and the infinite primes. We present the argument for the primes above $\wp$ with a similar argument working for the primes at infinity. Let $\mathfrak{B}$ be a prime of $\mathbf{k}(\wp)$ which lies above $\wp$. We need only show that the inertia group $I_\mathfrak{B}$ of $\mathfrak{B}$ inside G_σ is trivial. However $I_\mathfrak{B}$ is a subgroup of full inertia group I of $\mathfrak{B}$ in $G_\wp$ and this is mapped *injectively* by ω_σ. Thus $I_\mathfrak{B}$ has trivial intersection with G_σ. □

Now $L \subset \mathbf{C}_\infty$ be a finite extension of $\mathbf{k}$. Let $\wp \in \text{Spec}(\mathbf{A})$ be fixed. To simplify matters, we assume that $\wp$ is *unramified* in L; so, in particular, $L/\mathbf{k}$ is separable. Let $z(x, -j)$ be as in Subsection 8.12 and let $\tilde{z}(x, -j)$ be as in Definition 8.13.2. Set $L_1 := \mathbf{H}^+ \cdot L$. Thus $L_1 \subseteq L(\wp) = L \cdot \mathbf{k}(\wp)$, and is also unramified at the primes above $\wp$. Finally, $L(\wp)/L_1$ is totally *ramified* at the primes above $\wp$. Set $G_{\wp,L} := \text{Gal}(L(\wp)/L)$. Let $\omega_{\sigma,L}$ be as in Subsection 8.12 and let $o(\sigma, L)$ be its order. Thus $\omega_{\sigma,L}: G_{\wp,L} \to W_\sigma$; we denote the kernel by $G_{\sigma,L}$. One can now form, in the obvious fashion, $L(\wp)_\sigma$, etc. At the primes above $\wp$ there is also a version of the above proposition which we leave to the reader.

Lemma 8.14.3. *The p-primary component $G_{\wp,L}^{(p)}$ of $G_{\wp,L}$ is contained in the kernel $G_{\sigma,L}$.* □

In the rest of this subsection, and indeed, in the rest of this section we shall need to compose certain $W[G_{\wp,L}]$-modules under the powers of $\omega_{\sigma,L}$. Note that if $G_{\wp,L}^{(p)}$ is non-trivial, then $W[G_{\wp,L}]$ will *not* be semi-simple. In order to avoid this bad state of affairs, we make the following convention, which we will *always* use unless *explicitly* stated otherwise. Let $L(\wp)_p$ be the extension of $L(\wp)_\sigma$ which is fixed by $G_{\wp,L}^{(p)}$.

8.14. Applications to Class Groups

Convention on forming isotypic components: Let M be a $W[G_{\wp,L}]$-module such as a class group, Tate module, etc. The symbol "$M(\omega_{\sigma,L}^i)$" will always refer to the corresponding construction done over $L(\wp)_p$ viewed as a $G_{\wp,L}/G_{\wp,L}^{(p)}$-module.

Thus our convention will save us from always having to explicitly refer to the subfield $L(\wp)_p$ etc., and will at the same time always insure that we have semisimplicity.

Let $0 < i < o(\sigma, L)$. Let $\mathbb{F}$ be the constant field of L as before. Let $\mathbb{F}_1$ be the constant field of $L_1 =$ the constant field of $L(\wp)$. Let C be the smooth, projective, geometrically irreducible curve over $\mathbb{F}_1$ associated to $L(\wp)$. We set

$$\overline{C} := C \otimes_{\mathbb{F}} \overline{\mathbb{F}},$$

where $\overline{\mathbb{F}} \subset \mathbf{C}_\infty$ is the algebraic closure of $\mathbb{F}$. Note that $\overline{C}$ may possibly have many connected components.

Let $T_p := T_p(\overline{C})$ be the p-adic Tate module of $\overline{C}$ (and so is constructed in the usual fashion from the Jacobians of all the components). Note that $\text{Gal}(L(\wp)/L)$ acts as a group of automorphisms of $\overline{C}$ which commute with the action of the Frobenius morphism F associated to $\mathbb{F}$. Thus $T_p \otimes W$ can be decomposed into isotypic components under powers of $\omega_{\sigma,L}$ in line with the above convention. In particular, in line with the above convention, $T_p(\omega_{\sigma,L}^{-i})$ is just the submodule of $T_p \otimes W$ consisting of those t with $\lambda \cdot t = \omega_{\sigma,L}^{-i}(\lambda) \cdot t$ for Galois elements λ.

Let $L(\omega_{\sigma,L}^i, u)$ be the L-series of $\omega_{\sigma,L}^i$ over $\mathbb{F}$ as before. To simplify matters, we assume that $\omega_{\sigma,L}^{-i}$ is not principal; thus $L(\omega_{\sigma,L}^{-i}, u)$ is a polynomial in u.

Let $\overline{W}$ be the integral closure of W in some algebraic closure of its quotient field (or $\overline{\mathbb{Q}}_p$ if we view W as being embedded in $\overline{\mathbb{Q}}_p$). Over $\overline{W}$ we can write

$$L(\omega_{\sigma,L}^{-i}, u) = \prod_{\{\alpha_j\} \subset \overline{W}} (1 - \alpha_j u).$$

We set

$$L_{\text{un}}(\omega_{\sigma,L}^{-i}, u) := \prod_{\substack{\alpha_j \in \overline{W} \\ \alpha_j \text{ a unit}}} (1 - \alpha_j u) \in W[u].$$

This is the *unit-root* piece of $L(\omega_{\sigma,L}^{-i}, u)$.

Crystalline cohomology [Cr1] then tells us that

$$L_{\text{un}}(\omega_{\sigma,L}^{-i}, u) = \det(1 - Fu \mid T_p(\omega_{\sigma,L}^{-i})).$$

We now assume that $r^{\deg \wp} - 1$ does *not* divide i. Therefore, $\wp$ ramifies for $\omega_{\sigma,L}^{-i}$ and $\omega_{\sigma,L}^{-i}$ is consequently *not* principal. The L-series for $\omega_{\sigma,L}^{-i}$ will then have the Euler factor 1 at the primes above $\wp$.

Let $\text{Cl}_{L(\wp)}^{(p)}$ be the p-primary class group of the *field* $L(\wp)$, i.e., the p-primary component of the $\mathbb{F}_1$-rational points of the Jacobian of C (as curve over $\mathbb{F}_1$). Alternatively, this can be described as the p-primary subgroup of the divisors on C rational over $\mathbb{F}_1$ and of degree 0, modulo the subgroup of principal divisors of elements in $L(\wp)^*$. As before the Galois module $\mathrm{Cl}_{L(\wp)^{(p)}} \otimes W$ may be decomposed under powers of $\omega_{\sigma,L}$. One has

$$\mathrm{Cl}_{L(\wp)^{(p)}}(\omega_{\sigma,L}^{-i}) \simeq T_p(\omega_{\sigma,L}^{-i})/(1-F)T_p(\omega_{\sigma,L}^{-i}).$$

Thus we deduce that

$$\mathrm{ord}_p L(\omega_{\sigma,L}^{-i}, 1) = \mathrm{ord}_p L_{\mathrm{un}}(\omega_{\sigma,L}^{-i}, 1)$$

is the *length* of $\mathrm{Cl}_{L(\wp)^{(p)}}(\omega_{\sigma,L}^{-i})$ as W-module.

We then have the following result.

Theorem 8.14.4. *Under the above hypotheses on $\wp$ and i we have*

$$\mathrm{Cl}_{L(\wp)^{(p)}}(\omega_{\sigma,L}^{-i}) \neq \{0\} \iff \bar{\wp} \mid \tilde{z}(1, -i).$$

Proof. Note that the reduction modulo (p) of $L(\omega_{\sigma,L}^{-i}, u)$ and $L_{\mathrm{un}}(\omega_{\sigma,L}^{-i}, u)$ are the same. Note also that from the fact that $(r^{\deg \wp} - 1) \nmid i$, the L-series of $\omega_{\sigma,L}^{-i}$ contains no non-trivial factors above $\wp$. Thus the result follows from our double congruence 8.13.3. □

Remarks. 8.14.5. 1. We would also like such a result for those i, $0 < i < o(\sigma, L)$ where $r^{\deg \wp} - 1$ divides i. The difficulty here lies in handling the factors above $\wp$ as $\wp$ will then be *unramified* for $\omega_{\sigma,L}^{-i}$.

2. As a remarkable corollary, note that if $\tilde{z}(1, -i) = 0$, then $\mathrm{Cl}_{L(\wp)^{(p)}}(\omega_{\sigma,L}^{-i})$ is *always* non-trivial. One is then left with finding a natural interpretation of the first *non-zero* term in the expansion of $\tilde{z}(x, -i)$ at $x = 1$.

3. Let $L = \mathbf{k} = \mathbb{F}_r(T)$ and $\mathbf{A} = \mathbb{F}_r[T]$. Let $\wp \in \mathrm{Spec}(\mathbf{A})$. Thus in this example we may denote the Teichmüller character simply by ω and dispense with the subscripting of L, etc. In down to earth terms Theorem 8.14.4 says the following. Let $0 < i < r^{\deg \wp} - 1$. If $(r-1) \nmid i$ then

$$\mathrm{Cl}(\wp)^{(p)}(\omega^{-i}) \neq \{0\} \iff \wp \mid \sum_{t=0}^{\infty} \left(\sum_{\substack{n \in \mathbf{A} \\ n \text{ monic} \\ \deg n = t}} n^i \right),$$

where the sum inside parentheses vanishes for $t \gg 0$. If, $(r-1) \mid i$ then

$$\mathrm{Cl}(\wp)^{(p)}(\omega^{-i}) \neq \{0\} \iff \wp \mid \sum_{t=0}^{\infty} -t \left(\sum_{\substack{n \in \mathbf{A} \\ n \text{ monic} \\ \deg n = t}} n^i \right).$$

Furthermore, let $h(\wp)$ be the class number of the field $\mathbf{k}(\wp)$. Let $h(\wp)^+$ be the class of number of the maximal totally-real subfield $\mathbf{k}(\wp)^+$ of $\mathbf{k}(\wp)$. One easily sees that $h(\wp)^+ \mid h(\wp)$, and we set

$$h(\wp)^- := h(\wp)/h(\wp)^+ \,.$$

The prime $\wp$ is said to be *irregular of the first kind* if $p \mid h(\wp)^-$ and *irregular of the second kind* if $p \mid h(\wp)^+$. If $r = 2$, then $h(\wp)^- = 1$. If $r > 2$, then one can see that there are infinitely many irregular primes of the first kind. For general r one can also show that there are infinitely many irregular primes of the second kind. See [Go6] and [Fe1].

4. In classical cyclotomic theory one knows that

$$p \nmid h(p)^- \iff p \nmid h(p) \,.$$

No such result has been found in function fields. However, see the new work by Anderson summarized in Subsection 10.6.

8.15. "Geometric" Versus "Arithmetic" Notions

In this subsection we discuss a point that does *not* appear in classical algebraic number theory; in particular the theory of cyclotomic number fields. Indeed, let p be a prime and let $\mu_{p^n} \subset \mathbb{C}$ be the group of p^n-th roots of 1. Let $\zeta_{p^n} \in \mu_{p^n}$ be a generator and let $\mathbb{Q}(\zeta_{p^n}) \subset \mathbb{C}$ be the associated cyclotomic number field. It is very well known that $\mathbb{Q}(\zeta_{p^n})/\mathbb{Q}$ is *totally-ramified* above p. On the other hand, for function fields over finite fields one, obviously, never obtains such a ramified extension via roots of unity (= constant field extensions); all constant field extensions are everywhere unramified. In order to obtain ramified extensions we need to use other techniques such as those discussed in this book.

Let k be a function field in one variable over a constant field k_0. Let L be a finite extension of k with constants L_0. One says that L/k is *geometric* if and only if $L_0 = k_0$.

Let ψ be a Drinfeld module over a finite extension L of $\mathbf{k}$. Note that the extension of L obtained by adjoining all division points of ψ includes at most a finite constant field extension. Indeed, embed L into $\mathbf{C}_\infty$ over $\mathbf{k}$; now (as in Remark 7.1.9) the lattice of a Drinfeld module defined over an algebraic extension $\mathbf{K}_1$ of $\mathbf{K} \subset \mathbf{C}_\infty$ will always lie in a *finite* extension $\mathbf{K}_2 \supseteq \mathbf{K}_1$ of $\mathbf{K}$ inside $\mathbf{C}_\infty$. Now the exponential of ψ is defined over $\mathbf{K}_1 \subseteq \mathbf{K}_2$ and thus by continuity maps $\mathbf{K}_2$ to itself. As such, the torsion points of ψ will also lie in $\mathbf{K}_2$. But the constants of $\mathbf{K}_2$ will always be a *finite* field and the claim is established. Therefore, by only *slightly* abusing language, we call the (ramified) extensions obtained by adjoining torsion points of sign normalized Hayes modules *geometric cyclotomic extensions*. Constant field extensions are then the *arithmetic cyclotomic extensions*.

In a similar vein, concepts related to constant field extensions will be called "arithmetic" whereas those related to Drinfeld modules will be labeled "geometric." For instance, the theory of the classical L-series of function fields, due to Artin, Hasse, Weil, etc., is well-known to be intimately connected to constant field extensions (via Tate modules, etc.). Thus we refer these L-series as the *arithmetic L-series*. Those L-series associated to Drinfeld modules etc. are then *geometric L-series*.

The geometric/arithmetic dichotomy in the theory of function fields is found quite widely. For instance, in Theorem 8.14.4 we proved a criterion for the existence of certain class groups which depended on the $\overline{\wp}$-divisibility of $\widetilde{z}(1, -i)$. The element $\widetilde{z}(1, -i) \in \mathbf{V}$ is very much like a classical Bernoulli-number. Because it is connected with the arithmetic L-series, it is an *arithmetic Bernoulli-element* (or *arithmetic Bernoulli function*). Similarly, in future subsections, we will encounter *geometric Bernoulli-elements*.

One can, perhaps naively, view the *functional equation* of classical L-series of number fields as manifesting the identification of the "geometric" with the "arithmetic."

8.16. The Arithmetic Criterion for Cyclicity

In classical cyclotomic theory, one is very interested in whether certain class groups are cyclic or not. More precisely, let p be prime, ζ_p a primitive p-th root of unity. Let $C^{(p)}$ be the p-primary class group of $\mathbb{Q}(\zeta_p)$ and ω the standard Teichmüller character of $(\mathbb{Z}/p)^* \simeq \text{Gal}(\mathbb{Q}(\zeta_p)/\mathbb{Q})$. One forms $C^{(p)}(\omega^i)$ in the usual way and one is interested in whether it is a cyclic $\mathbb{Z}_p$-module or not.

The most important sufficient condition for cyclicity in classical theory is the *Kummer-Vandiver Conjecture*, $p \nmid h(p)^+$, where $h(p)^+$ is the class number of $\mathbb{Q}(\zeta_p + \zeta_p^{-1})$ (and, of course $h(p)^- = h(p)/h(p)^+$). One then deduces cyclicity via the *reflection theorem*, see [Wa1, §10].

In the function field setting we are also interested in cyclicity, and in this section we will present a sufficient criterion for it. In a following subsection (Subsection 8.22), we present a possible analog of the Kummer-Vandiver Conjecture. Another remarkable analog of the Kummer-Vandiver Conjecture is given in the new work of Anderson discussed in Subsection 10.6.

We use the set-up and decomposition conventions of Subsection 8.14. Thus $L \subset \mathbf{C}_\infty$ is a finite extension of $\mathbf{k}$, $\wp \in \text{Spec}(\mathbf{A})$ is unramified in L, and $\omega_{\sigma, L}$ is the Teichmüller character of $\text{Gal}(L(\wp)/L)$ associated to L and an embedding σ of the value field $\mathbf{V}$. Let $o(\sigma, L)$ be the order of $\omega_{\sigma, L}$ as before and let

$$0 < i < o(\sigma, L).$$

We assume that $r^{\deg \wp} - 1$ does *not* divide i (so we may use an appropriate double congruence).

As in Subsection 8.14, let $L_{\text{un}}(\omega_{\sigma,L}^{-i}, u)$ be the unit root L-series of $\omega_{\sigma,L}^{-i}$. Let W now be the Witt ring of $\mathcal{O}_{\mathbf{V}}/\overline{\wp}$ and let $\overline{W}$ be its integral closure in some algebraic closure of the fraction field of W. We write

$$L_{\text{un}}(\omega_{\sigma,L}^{-i}, u) = \prod_{\alpha_j \text{ a unit}} (1 - \alpha_j u)$$

in $\overline{W}[u]$.

Proposition 8.16.1. *Let i be positive integer not divisible by $r^{\deg \wp} - 1$. Let $\overline{M} \subset \overline{W}$ be the maximal ideal. Suppose that at most one of $\{\alpha_j\}$ is congruent to $1 \pmod{\overline{M}}$. Then $\operatorname{Cl}_L(\wp)^{(p)}(\omega_{\sigma,L}^{-i})$ is a cyclic W-module.*

Proof. This follows from the description given for $\operatorname{Cl}_L(\wp)^{(p)}(\omega_{\sigma,L}^{-i})$ in Subsection 8.14 in terms of the p-adic Tate module and elementary linear algebra. □

Let $\tilde{z}(x, -i)$ be as in Subsection 8.14. Let $\mathbf{V}_{\overline{\wp}}$ be the completion of $\mathbf{V}$ at $\overline{\wp}$ and let $\overline{\mathbf{V}}_{\overline{\wp}}$ be an algebraic closure of $\mathbf{V}_{\overline{\wp}}$ with integers $\overline{\mathcal{O}}_{\overline{\wp}}$ and maximal ideal $\overline{M}_{\overline{\wp}}$. Over $\overline{\mathcal{O}}_{\overline{\wp}}$ we write

$$\tilde{z}(x, -i) = \prod (1 - \theta_e x^{-t_{\mathbb{F}}}).$$

Our double congruence, Theorem 8.13.3, then immediately gives the following result.

Corollary 8.16.2. *Let i be as above. Then $\operatorname{Cl}_L(\wp)^{(p)}(\omega_{\sigma,L}^{-i})$ is a cyclic W-module if at most one of $\{\theta_e\}$ is $\equiv 1 \pmod{\overline{M}_{\overline{\wp}}}$.* □

Corollary 8.16.2 is the *arithmetic criterion for cyclicity*. It is "arithmetic" precisely since it relates to the theory of $\zeta_{\mathbf{A}}(s)$, $s = (x, y)$, when y is a negative integer and these values are related to arithmetic L-series.

Remark. 8.16.3. It is easy to reformulate 8.16.2 in terms of the value and derivative at $x = 1$ of $\tilde{z}(x, -i)$. For details see [Go4, §4.4].

8.17. The "Geometric Artin Conjecture"

In this subsection we will show how the work of A. Weil on the classical Artin conjecture for function fields (the *arithmetic Artin Conjecture*) can be used to establish a *geometric Artin conjecture* for function fields; i.e., for the characteristic p L-series associated to characteristic 0 representations as presented in Subsection 8.10. The result [Go7] will actually follow rapidly as *all* the needed ideas are now in place. At the end of this subsection we will present some results in classical theory suggested by the characteristic p theory.

To fix ideas let L be a finite extension of $\mathbf{k} \subset \mathbf{C}_\infty$ and let $G := \mathrm{Gal}(L^{\mathrm{sep}}/L)$ where L^{sep} is the separable closure of L. As in Subsection 8.10 we let $\rho\colon G \to GL(V)$ be a representation of Galois type. We form $L(\rho, s)$ as in Definition 8.10.2. We shall establish here that $L(\rho, s)$ is always an essentially algebraic entire function. The case where ρ is abelian (i.e., $\dim_{\overline{\mathbf{Q}}_p} V = 1$) is contained in Theorem 8.9.2.

We are thus reduced to considering the case where $\dim_{\overline{\mathbf{Q}}_p} V > 1$.

We now note that the function $L(\rho, s)$ satisfies *all* the usual formalism of L-series (see [Ta1]). Indeed this is *precisely* because of the way we have chosen the local factors. Therefore we can restrict our attention to the case where ρ is irreducible of degree ($= \dim_{\overline{\mathbf{Q}}_p} V) > 1$.

For the next step we invoke Brauer induction and abelian reciprocity to express $L(\rho, s)$ as
$$\frac{f_1(s)}{f_2(s)}$$
where both $f_1(s)$ and $f_2(s)$ are products of L-series of abelian representations. In particular both $f_1(s)$ and $f_2(s)$ are essentially algebraic entire functions. We shall simply show that $f_2(s)$ divides $f_1(s)$.

Let $t_{\mathbb{F}} = [\mathbb{F}\colon \mathbb{F}_r]$ where $\mathbb{F} \subset L$ is the field of constants. Let $L_0 \subseteq L$ be the separable closure of $\mathbf{k}$ and let $L \subseteq L_1$ where L_1 is finite Galois such that ρ factors through $\mathrm{Gal}(L_1/L)$. Note that almost all primes of $\mathbf{A}$ are unramified in L_0. Moreover, almost all finite primes of L are unramified in L_1. Thus almost all primes of $\mathbf{A}$ have the properties that they are unramified in L_0 and the primes above them in L are unramified in L_1. We let $\wp$ be one such prime of $\mathbf{A}$.

A straightforward argument involving ramification indices now shows that $L(\wp)/L \cdot \mathbf{H}^+$ is totally ramified at the primes above $\wp$. Now let $\omega_{\sigma, L}$ be the Teichmüller character.

Our proof that $f_2(s)$ divides $f_1(s)$ now has two parts. Let $s = (x, y)$ as before. We first prove that $f_2(x, -j)$ divides $f_1(x, -j)$ for $j > 0$. We then deduce from this that $f_2(x, y)$ divides $f_1(x, y)$ for all y which will complete the proof.

Let $h_{f_i}(x, -j)$, $i = 1, 2$, be as in Definition 8.5.12 (so, as usual, we have "removed π_*" from $f_i(x, -j)$). To show that $f_2(x, -j)$ divides $f_1(x, -j)$, it is clearly enough to show that $h_{f_2}(x, -j)$ divides $h_{f_1}(x, -j)$.

Note that for almost all primes $\mathfrak{P}$ of $\mathbf{A}$, $r^{\deg \mathfrak{P}} - 1$ does *not* divide j. Thus without loss of generality we may also assume that $r^{\deg \wp} - 1$ does not divide j. Under this hypothesis $\omega_{\sigma, L}^{-j}$ is ramified.

By Weil, the classical L-series, $L(\rho \otimes \omega_{\sigma, L}^{-j}, u)$ is a *polynomial* in u. Let $\widehat{L}(\rho \otimes \omega_{\sigma, L}^{-j}, u)$ be the classical L-series of $\rho \otimes \omega_{\sigma, L}^{-j}$ obtained by removing the

8.17. The "Geometric Artin Conjecture" 287

infinite factors; thus once again $\widehat{L}(\rho \otimes \omega_{\sigma,L}^{-j}, u)$ is a polynomial in u. Then, exactly as in 8.12.5, one deduces a *double congruence*

$$\pi_0(h_L(x,-j)) = \pi_0(h_{f_1}(x,-j)/h_{f_2}(x,-j)) = \pi_1(\widehat{L}(\rho \otimes \omega_{\sigma,L}^{-j}, u^{t_\mathbb{F}})),$$

where we work over the finite extension $\mathbf{V}_1$ of $\mathbf{V}$ containing all t^{th} roots of unity for t dividing the order of $\text{Gal}(L_1/L)$. Thus $h_L(x,-j)$ is a polynomial modulo an infinite number of primes of $\mathbf{V}_1$. By 8.13.1 we conclude that $h_{f_2}(x,-j)$ divides $h_{f_1}(x,-j)$ and the proof of the first part is complete.

To see that $f_2(s)$ divides $f_1(s)$ one now uses the continuity of the roots as in Subsection 8.5.

We recapitulate all of this in the following result.

Theorem 8.17.1. *Let L be a finite extension of $\mathbf{k}$. Let G be the Galois group over L of the separable closure of L, and let V be a finite dimensional $\overline{\mathbb{Q}}_p$-vector space. Let $\rho: G \to GL(V)$ be a representation of Galois type. Then the L-series $L(\rho,s)$ is an essentially algebraic entire function on S_∞.* □

Remark. 8.17.2. The obvious v-adic version of the theorem is also true. We leave the details to the reader.

Corollary 8.17.3. *Let $L/\mathbf{k}$ be as in 8.17.1. Let K be a finite separable extension of L. Let $\mathcal{O}_L$, $\mathcal{O}_K$ be the rings of $\mathbf{A}$-integers. Then $\zeta_{\mathcal{O}_K}(s)/\zeta_{\mathcal{O}_L}(s)$ is an essentially algebraic entire function.*

Proof. Let H be a Galois closure of K over L. Let $G_1 := \text{Gal}(H/L)$ and let $G_0 \subseteq G_1$ be the subgroup associated to K. Let ρ be the trivial representation of G_0. Induce ρ to G_1 and use the formalism of L-series. □

Let $\wp$ be as above with $r^{\deg(\wp)} - 1$ not dividing j, $j > 0$. Let $L(\rho \otimes \omega_{\sigma,L}^{-j}, u)$ be the arithmetic L-series of $\rho \otimes \omega_{\sigma,L}^{-j}$; note that under our assumptions on $\wp$, the primes above $\wp$ have trivial local Euler factor. Let $M_p \subset \overline{\mathbb{Q}}_p$ be the maximal ideal and assume that $L(\rho \otimes \omega_{\sigma,L}^{-j}, 1) \notin M_p$.

Corollary 8.17.4. *Under the above hypothesis, $L(\rho \otimes \omega_{\sigma,L}^{-j}, 1) \notin M_p$ for almost all primes $\wp$ with $\wp$ as above.*

Proof. We know that there is a double congruence between $L(\rho \otimes \omega_{\sigma,L}^{-j}, 1)$ and $h_L(1,-j)$. Thus as $L(\rho \otimes \omega_{\sigma,L}^{-j}, 1)$ does not reduce to zero we find $h_L(1,-j) \neq 0$. The result is now easy. □

We can also, finally, present the trivial zeroes theory for $L(\rho,s)$, ρ of Galois type as above, in its most general context. Let ρ factor through $\text{Gal}(L_1/L)$ where $[L_1:L] < \infty$. Let ∞_i be a fixed infinite prime of L and let $D_{\overline{\infty}_i}$ and

$I_{\overline{\infty}_i}$ be the decomposition and inertia groups at a prime $\overline{\infty}_i$ of L_1 above ∞_i. Set
$$P_{\rho,\infty_i}(u,-j) := \det(1 - Fu \mid \widetilde{V}^{I_{\overline{\infty}_i}})$$
where $\widetilde{V}$ is V equipped with the action $\rho \otimes \omega_{\sigma,L}^{-j}$. One checks that this polynomial is independent of both the choice of $\overline{\infty}_i$ and $\wp$.

We set
$$P_\rho(u,-j) := \prod_{\infty_i} P_{\rho,\infty_i}(u,-j).$$

One checks, as before, that the reduction $\overline{P}_\rho(x^{-t_\mathbb{F}},-j)$ divides $h_L(x,-j)$ and gives the trivial zeroes of $L(\rho,s)$, where $t_\mathbb{F}$ is as in Theorem 8.12.5.

Question 8.17.5. Let G be as in 8.17.1 and let W now be a finite dimensional $\mathbf{C}_\infty$-vector space; so we are taking representations in characteristic p. Let $\rho\colon G \to GL(W)$ be a representation of Galois type. We now define the L-series of ρ, $L(\rho,s)$, by using the inertia and decomposition groups at *all* finite primes in the usual fashion. Can $L(\rho,s)$ be continued to an essentially algebraic entire function? The answer is certainly "yes" if $\dim W = 1$ as we have seen. Moreover, Theorem 8.17.1 combined with [Se, §16] tells us that $L(\rho,s)$ *automatically* has a *meromorphic* continuation as a quotient $f_1(s)/f_2(s)$ with $f_i(s)$ essentially algebraic and entire for $i = 1, 2$. This is exactly similar to what was concluded via Brauer induction at the beginning of this section for representations in characteristic 0.

Remarks. 8.17.6. 1. Recall that in Subsection 8.6 we introduced the L-series, $L(\psi,s)$, of a Drinfeld module ψ over L. It is expected that $L(\psi,s)$ will continue to an essentially algebraic entire function on S_∞. Recall further that $L(\psi,s)$ is defined via the representation of G on
$$H_v^1(\psi,\mathbf{k}_v) = \mathrm{Hom}_{\mathbf{A}_v}(T_v,\mathbf{k}_v),$$
where T_v is the v-adic Tate module of ψ. Let ∞_i be an infinite prime of L and let $D_{\overline{\infty}_i}$ be the decomposition group at a prime, $\overline{\infty}_i$, of the separable closure of L which lies above ∞_i, etc. The group $D_{\overline{\infty}_i}$ is isomorphic to $\mathrm{Aut}(\overline{\mathbf{K}}/L_{\infty_i})$ via the embedding of L_{∞_i} into $\overline{\mathbf{K}}$ at ∞_i. Now note that T_v may now be expressed easily via the lattice $M \subset \overline{\mathbf{K}}$ associated to ψ at ∞_i; indeed $T_v \simeq M \otimes_\mathbf{A} \mathbf{A}_v$. The key point is to note that this is an isomorphism of $D_{\overline{\infty}_i}$-modules. As such the action of $D_{\overline{\infty}_i}$ factors on T_v factors through the action on M, and is *independent* of v and of Galois type (as M lives in a finite separable extension of $\mathbf{K}$)! Let j be as above. One now obtains a polynomial
$$P_{\psi,\infty_i}(u,-j) \in \mathbb{F}_r[u] \subset \mathbf{A}[u]$$
exactly as above (via $\omega_{\sigma,L}$, etc.) and independent of v. It is now reasonable to expect that
$$P_\psi(u,-j) := \prod_{\infty_i} P_{\psi,\infty_i}(u,-j)$$

8.17. The "Geometric Artin Conjecture"

can be used to provide *trivial zeroes* for $L(\psi, s)$. One can, for instance, verify this in the case $L = \mathbf{K} = \mathbb{F}_r(T)$ and $\psi = C$ is the Carlitz module as $L(C, s) = \zeta_{\mathbb{F}_r[T]}(s - 1)$. For general ψ and L, however, these trivial zeroes seem out of reach at the moment.

2. In the classical theory of motives over algebraic number fields, trivial zeroes are, of course, given via Γ-factors which, in turn, are associated to Galois representations of the Archimedean completions. These Galois groups are thus either trivial or of order 2. In contrast, for functions fields, the completions at infinity give rise to Galois groups of infinite order and thus one needs some sort of argument as in Part 1.

The basic flow of the proof of Theorem 8.17.1 is the following: We use Brauer induction to get a meromorphic continuation; then we use Weil's result for the collection $\{\rho \otimes \omega_{\sigma,L}^j\}$, where $j > 0$ and $\wp$ runs over a certain infinite list of primes, to conclude holomorphy at the positive integers; finally we use strong continuity to conclude the result.

Brauer induction is, of course, an essential tool of the classical theory of Artin L-series also. Moreover, the fundamental "yoga" of Iwasawa theory stipulates that the *Main Conjecture* is the cyclotomic (in the sense of algebraic numbers) version of Weil's results. Motivated by this, the author and W. Sinnott searched for a classical analog of our second step above. In [GS1] the following result was obtained. Let $\overline{\mathbb{Q}}$ be a fixed algebraic closure of $\mathbb{Q}$ and let $G := \text{Gal}(\overline{\mathbb{Q}}/\mathbb{Q})$. Let V be a finite dimensional $\mathbb{C}$-vector space and let $\rho: G \to GL(V)$ be a representation of Galois type. Let $\chi: G \to \mathbb{C}$ be the character of ρ. Let $L(\rho, s) = L(\chi, s)$ be the L-series of ρ. One knows (Brauer induction) that $L(\rho, s)$ has a meromorphic continuation and is *finite* at the negative integers; these special values are known to be in $L = \mathbb{Q}(\chi) := \{\text{smallest subfield of } \mathbb{C} \text{ containing the values of } \chi\}$.

Definition 8.17.7. 1. We set

$$g(\rho, x) := \sum_{k=0}^{\infty} L(\rho, -k) \cdot \frac{x^k}{k!}.$$

2. Let $\log(T) = \sum_{n=1}^{\infty} (-1)^{n-1} \frac{(T-1)^n}{n}$, as usual. We set

$$f(\rho, T) := g(\rho, \log(T)),$$

as a power series in $T - 1$.

Let D be the operator $T\frac{d}{dT}$. One then sees readily that

$$D^k f(\rho, T)|_{T=1} = L(\rho, -k), \quad k = 0, 1, 2, \ldots.$$

By checking Γ-factors, it is known that $g(\rho, x)$ is *not* constant if and only if ρ is totally-even ($\rho(\tau) = 1_V$ for all complex conjugations τ) or totally-odd

$(\rho(\tau) = -1_V$ for all τ). In this case $g(\rho, x)$ determines the isomorphism type of ρ.

Let $\mathcal{O} \subset L$ be the integers of L.

Theorem 8.17.8. *Assume the p-adic Artin Conjecture (which follows from the Main Conjecture of Iwasawa theory) then $f(\rho, T) - f(\rho, 1)$ has $\wp$-adically bounded denominators for all finite primes $\wp$ of L.* □

(Under the additional hypothesis that Iwasawa's μ-invariant vanishes, one actually obtains that $f(\rho, T) - f(\rho, 1) \in \mathcal{O}[[T-1]]$.)

In other words, under the p-adic Artin Conjecture, the function $g(\rho, T)$ is p-adically analytic in the domain $|T - 1|_p < 1$. Any connection, in general, between these functions, (or $\mathrm{Spec}(L[[T - 1]]/f(\rho, T))$, etc.) and the complex analytic theory of $L(\rho, s)$ would be very interesting. For instance, if $\dim \rho = 1$, then $f(\rho, T)$ can be used to complex-analytically continue $L(\rho, s)$.

8.18. Special Values at Positive Integers

We now turn our attention to the values at positive integers of $\zeta_{\mathcal{O}_L}(s)$ where L will be restricted to certain finite extensions of **k**. We will establish results for $\zeta_{\mathcal{O}_L}(s)$ that are analogous to the classical results in algebraic number theory of Euler and Siegel.

The basic observation (Lemma 8.18.1) was originally discovered by L. Carlitz in the 1930's [C1]. As the reader will no doubt observe, Carlitz's result uses the classical connection between Bernoulli numbers, zeta values and the exponential function as a model. The present author independently rediscovered these constructions in the context of q-expansions of Eisenstein series [Go8]. (In fact, these results spurred the development of the L-series of this section and, in particular, the exponentiation theory of our first few subsections.)

Let sgn be our fixed sign function. Let $I \subseteq \mathbf{A}$ be an ideal. From Hayes' theory (cf. Th. 7.4.8 and Cor. 4.9.5.2) we can always find a sgn-normalized Hayes module ψ with lattice $\Lambda = \xi \cdot I$, for some $\xi \in \mathbf{C}_\infty^*$. Thus

$$e_\Lambda(z) = e_\psi(z) = z \prod_{0 \neq \alpha \in \Lambda} (1 - z/\alpha),$$

and is an entire $\mathbb{F}_r$-linear mapping of $\mathbf{C}_\infty$ to itself. As ψ can be defined over the **A**-integers $\mathcal{O}^+$ of $\mathbf{H}^+$, $e_\psi(z)$ has coefficients in $\mathbf{H}^+$.

We now assume to begin with, for simplicity of exposition, that ∞ is rational ($d_\infty = 1$).

Lemma 8.18.1. *Let j be a positive integer with $(r-1) \mid j$. Then*

$$\xi^{-j} \sum_{\substack{i \in I \\ i \text{ monic}}} i^{-j} \in \mathbf{H}^+,$$

where $\mathbf{H}^+ = \mathbf{H}$ as $d_\infty = 1$.

Proof. Note that
$$\frac{d}{dz} e_\Lambda(z) = e'_\Lambda(z) = 1.$$
Thus by taking logarithmic derivatives, we find
$$\frac{e'_\Lambda(z)}{e_\Lambda(z)} = \frac{1}{e_\Lambda(z)} = \sum_{\alpha \in \Lambda} (z-\alpha)^{-1}.$$
Notice that for $\alpha \neq 0$ we have
$$\frac{1}{z-\alpha} = \frac{-1}{\alpha(1-z/\alpha)} = \frac{-1}{\alpha}(1 + z/\alpha + (z/\alpha)^2 + \cdots).$$
Thus
$$\frac{z}{e_\Lambda(z)} = 1 - \sum_{(r-1)|j} \left(\sum_{0 \neq \alpha \in \Lambda} \alpha^{-j} \right) z^j,$$
as $\sum_{\zeta \in \mathbb{F}_r^*} \zeta^{-j} = 0$ for $j \not\equiv 0 \ (r-1), j > 0$. Finally $\sum_{\zeta \in \mathbb{F}_r^*} \zeta^{-j} = -1$ for $j \equiv 0 \ (r-1)$, and so
$$\frac{z}{e_\Lambda(z)} = 1 + \sum_{(r-1)|j} \xi^{-j} \left(\sum_{\substack{i \in I \\ i \text{ monic}}} i^{-j} \right) z^j. \qquad \square$$

The situation when $d_\infty > 1$ is a little more complicated. Let J be a principal $\mathbf{A}$-ideal. Recall that, in the notation of Subsection 8.2, we have
$$J^{s_1} = \alpha/\text{sgn}(\alpha)$$
for *any* α with $J = (\alpha)$. So if $j \equiv 0 \ (r^{d_\infty} - 1)$, then
$$J^{s_j} = J^j = \alpha^j.$$
We can, therefore, rephrase 8.18.1 in the following way which works for *all* ∞.

Lemma 8.18.2. *Let e be a positive integer $\equiv 0 \ (r^{d_\infty} - 1)$. Then*
$$\xi^{-e} \sum_{\substack{J \subseteq I \\ J \text{ principal}}} J^{-e} \in \mathbf{H}^+.$$

Proof. This follows as in 8.18.1 upon noting that any two generators of the *same* **A**-ideal differ by an element of $\mathbb{F}_r^*$. □

In fact, 7.10.10 implies that the above sum actually lies in **H**.

One also knows that the sums in 8.18.1 and 8.18.2 are non-zero as observed by Thakur. Indeed by Theorem 7.4.8 and Corollary 4.9.5.2 we can always find a conjugate of our sums, with respect to $\text{Gal}(\mathbf{H}^+/\mathbf{k})$, where the ideal I equals **A**. In this case, up to $\mathbb{F}_r^*$, 1 is the unique non-zero element of minimal degree. The result now follows easily.

Next we use 8.18.1 and 8.18.2 to establish the analog of Euler's results for $\zeta_{\mathbf{A}}(s)$ at those positive integers $\equiv 0 \ (r^{d_\infty} - 1)$. Recall that in Subsection 8.2 we defined the value field **V**. We now set

$$\mathbf{T} := \mathbf{V} \cdot \mathbf{H}^+ ;$$

so **T** is the compositum of two "principal ideal fields" (i.e., fields where ideals of **A** become principal). Moreover, $[\mathbf{T}:\mathbf{k}] < \infty$.

Theorem 8.18.3. *Let j be a positive integer with $j \equiv 0 \ (r^{d_\infty} - 1)$. Let ξ be as above. Then*

$$0 \neq \zeta_{\mathbf{A}}(j)/\xi^j \in \mathbf{T}.$$

Proof. The fact that $\zeta_{\mathbf{A}}(j) \neq 0$ follows from the existence of the Euler product.

Let $\{\alpha_1, \ldots, \alpha_h\}$ be representatives of $\mathcal{I}/\mathcal{P}$. Thus we have

$$\zeta_{\mathbf{A}}(j) = \sum_{e=0}^{h} \alpha_e^{-j} \sum_{\substack{I \subseteq \alpha_e^{-1} \\ I \text{ principal}}} \mathcal{I}^{-j}.$$

The fractional ideal α_e is a $\mathbf{k}^*$ multiple of an **A**-integral ideal. Thus the result follows from 8.18.2. □

The reader should note that one only has trivial zeroes for $\zeta_{\mathbf{A}}(s)$ at $-j$, for $j \equiv 0 \ (r^{d_\infty} - 1)$ (Remark 8.13.10). Thus although the number of units in **A** is $r - 1$, the special values theory actually works modulo $r^{d_\infty} - 1$.

Example 8.18.4. Let $\mathbf{k} = \mathbb{F}_r(T)$, $\mathbf{A} = \mathbb{F}_r[T]$ and j a positive integer divisible by $r - 1$. Then 8.18.3 implies that

$$\zeta_{\mathbf{A}}(j)/\xi^j \in \mathbf{k}^*$$

which is Carlitz's original result.

We now turn toward establishing an analog of Siegel's results on values of Dedekind zeta functions associated to totally-real number fields. We say that an extension $L/\mathbf{k}$ of function fields is *totally-real* if and only if it is totally-split at ∞. Let L be finite totally-real extension of $\mathbf{k}$ that we also assume is *abelian*. This last assumption should, as with number fields, ultimately *not* be needed; see 8.18.7.1. We use it so as to be able to reduce to abelian L-series. Moreover, in order to avoid problems with $\mathbf{A}$ having non-trivial class group, we shall assume that the Hilbert class field $\mathbf{H}$ (totally split at ∞) is contained in L. Thus the norms of $\mathcal{O}_L$-ideals down to $\mathbf{A}$ will be principal (but *not* necessarily positively generated).

Let $I \subseteq \mathbf{A}$ be an ideal and let ψ be a sgn-normalized Hayes-module with lattice $\Lambda = \xi I$ as before. Let $B \subseteq \mathbf{A}$ be any ideal and $\alpha \in B^{-1}I - I$. Thus $e_\psi(\alpha\xi)$ is a non-zero B-division point of ψ.

Lemma 8.18.5. *Let j be a positive integer. Then we have*

$$\xi^{-j} \sum_{\beta \in I} (\alpha + \beta)^{-j} \in \mathbf{k}(B).$$

Proof. The element $e_\psi(\alpha\xi)$ is an B-division point of ψ; thus it belongs the field $\mathbf{k}(B)$ by definition. We know that

$$e_\psi(z) = z \prod_{0 \neq \beta \in I} \left(1 - \frac{z}{\beta\xi}\right).$$

As in Lemma 8.18.1,

$$\frac{1}{e_\psi(z)} = \sum_{\beta \in I} \frac{1}{z + \beta\xi},$$

and

$$\frac{1}{e_\psi(z + \alpha\xi)} = \frac{1}{e_\psi(z) + e_\psi(\alpha\xi)} = \sum_{\beta \in I} \frac{1}{z + \alpha\xi + \beta\xi} = \sum_{\beta \in I} \frac{1}{z + \xi(\alpha + \beta)}.$$

The coefficients of the exponential function of ψ are in $\mathbf{H}^+$. The result now follows directly by computing the Taylor expansion about the origin of $\frac{1}{e_\psi(z + \xi\alpha)}$ and using the above equalities. $\square$

Since we have assumed that L is both totally-real and abelian, it follows that there is some integral conductor $B \subseteq \mathbf{A}$ with

$$L \subseteq \mathbf{k}(B).$$

Set $G := \mathrm{Gal}(L/\mathbf{k})$ and let $\widehat{G}$ be the group of $\overline{\mathbb{Q}}_p^*$-valued characters of G. As was explained in Subsection 8.10, we have the factorization

$$\zeta_{\mathcal{O}_L}(s) = \prod_{\chi \in \widehat{G}} L(\chi, s).$$

Recall that we set $d_\infty := \deg(\infty)$. Put $t := [L:\mathbf{k}]$. Our main result here is then our next theorem.

Theorem 8.18.6. *Let j be a positive integer divisible by $r^{d_\infty} - 1$. Then*
$$\xi^{-tj}\zeta_{\mathcal{O}_L}(j)$$
is non-zero, contained in the maximal totally-real subfield of $\mathbf{k}(B)$, and
$$(\xi^{-tj}\zeta_{\mathcal{O}_L}(j))^2 \in \mathbf{H}.$$

Proof. We will use the well-known *Dedekind Determinant Formula* which states the following: let W be a finite abelian group and let $\widehat{W}$ be its $\overline{\mathbb{Q}}_p$-valued character group (where $\overline{\mathbb{Q}}_p$ is a fixed algebraic closure of $\mathbb{Q}$). Let f be any $\overline{\mathbb{Q}}_p$-valued function on W. Then, upon fixing an ordering O of W,

$$\prod_{\chi \in \widehat{W}} \sum_{w \in W} \chi(w) f(w^{-1}) = \det(f(w_0^{-1} w_1))_{(w_0,w_1) \in O \times O}.$$

Although the Dedekind Determinant Formula is established over $\mathbb{Q}_p$, it is clear that, in fact, it works over $\mathbb{Z}/p$ by simple reduction modulo p, and thus over $\mathbf{k}$ provided that we use $\overline{\mathbb{Q}}_p$-valued characters of W as explained in Subsection 8.10. In particular, we may use this formula to evaluate L-series in the set-up of this section.

We note first of all that we may reduce to considering the product $\prod L(\chi, j)_B$ where the subscript B means that we have removed all Euler factors at the primes dividing B. Indeed, we know that this product equals the value of $\zeta_{\mathcal{O}_L}(j)$ with the Euler factors above B removed. On the other hand, let $\wp$ be a prime of $\mathcal{O}_L$ dividing B. Then

$$(1 - n\wp^{-j}) \in \mathbf{k}$$

since $n\wp$ is principal and, as $j \equiv 0 \pmod{r^{d_\infty} - 1}$, $n\wp^j$ is positive.

Now let $\sigma \in G$ and set

$$f(\sigma) = \sum_{\substack{U \subseteq A \\ (U,B)=A \\ \sigma_U = \sigma}} U^{-j}$$

where σ_U is the Artin symbol at U. Then the determinant formula gives

$$L(\chi, j)_B = \det(f(\sigma^{-1}\sigma_1))$$

where (σ, σ_1) are in an ordering of G.

Since $\mathbf{H} \subseteq L$ there is a *surjective* map from G to the ideal classes of $\mathbf{A}$. Let $R = \{\mathcal{B}_1, \ldots, \mathcal{B}_h\}$ be a collection of representatives prime to B and where h is the class number of $\mathbf{A}$. For each $\sigma \in G$ let $g(\sigma) \in R$ be its representative.

The two determinants
$$\det(f(\sigma^{-1}\sigma_1)),$$
and
$$\det(f(\sigma^{-1}\sigma_1)g(\sigma^{-1})^j g(\sigma_1)^j),$$
differ by the factor
$$\prod g(\sigma)^{2j}.$$
But simple arguments in group theory tell us that, in the ideal class group, $\prod g(\sigma)^2$ is the identity; i.e.,
$$\prod g(\sigma)^2 = (a)$$
for some $a \in \mathbf{A}$. Again, since $j \equiv 0 \pmod{r^{d_\infty} - 1}$ we see that, by our definition of exponentiation of ideals,
$$\prod g(\sigma)^{2j} = a^j;$$
consequently, we are reduced to computing the second determinant.

Note that for α, β in $\mathbf{k}$, we have
$$(\zeta\alpha + \beta)^j = (\alpha + \zeta^{-1}\beta)^j$$
for $\zeta \in \mathbf{A}^*$. Using this elementary observation, the explicit class field theory of Hayes, and the fact that $j \equiv 0 \pmod{r^{d_\infty} - 1}$, the reader can now, in the spirit of Lemma 8.18.2, express $\xi^{-j} f(\sigma^{-1}\sigma_1)g(\sigma^{-1})g(\sigma_1)^j$ in terms of the sums of Lemma 8.18.5. In particular, we see that these elements are in $\mathbf{k}(B)$.

Let η be an element of $\text{Gal}(L/\mathbf{H})$. It is now easy to see that η permutes the rows of
$$(\xi^{-j} f(\sigma^{-1}\sigma_1)g(\sigma^{-1})^j g(\sigma_1)^j);$$
thus the determinant gets multiplied by ± 1. The result now follows. □

In classical theory, the base field is $\mathbb{Q}$ which is its own Hilbert class field. For function fields, of course, in general $\mathbf{k}$ is strictly contained in $\mathbf{H}$. One is thus naturally curious as to whether the result can be improved by replacing $\mathbf{H}$ with $\mathbf{k}$. In fact, one cannot; for a beautiful counterexample see [Th4, remarks at the end of §4.]

We may, therefore, learn something of a meta-mathematical nature about number fields from the function field theory: in the functional equation of classical Dedekind zeta functions, it seems that we should view "$\mathbb{Q}$ on the right hand side" (e.g., Euler and Siegel's results at positive integers) as being the Hilbert class field of $\mathbb{Q}$, while on the "left hand side" (i.e., the negative integers) "$\mathbb{Q}$" is indeed the rational field, $\mathbb{Q}$ itself.

Remarks. 8.18.7. 1. It should be possible to establish 8.18.6 for all totally-real L with $\mathbf{H} \subseteq L$ via *q-expansions* of rigid analytic Eisenstein series associated to L, [Go8]. Computing these expansions is a deep and interesting problem.

2. Let L be any finite Galois extension of $\mathbf{k}$ where $G := \mathrm{Gal}(L/\mathbf{k})$ is a p-group. We can, in the standard fashion of Artin L-series, express $\zeta_{\mathcal{O}_L}(s)$ as a product of L-series associated to the irreducible (char. 0) representations of G. Let $t := [L:\mathbf{k}] = \#G$. Note that if ρ is *any* representation of G, then the eigenvalues of $\rho(g)$ for $g \in G$ are p-power roots of unity and so reduce to 1. Therefore one sees that, up to finitely many Euler factors, the L-series will be a power of $\zeta_{\mathbf{A}}(s)$, and
$$\zeta_{\mathcal{O}_L}(s) = \zeta_{\mathbf{A}}(s)^t.$$
(The reader can also see this directly without L-series.) As such, we deduce rationality results for $\zeta_{\mathcal{O}_L}(s)$ trivially from those of $\zeta_{\mathbf{A}}(s)$.

3. If we put ourselves in the situation of "two T's" as in Subsection 7.11, then the above results would need to be phrased a little differently. We would first have to apply θ_{con} to the zeta-value and *then* multiply by a period, etc. Another way to accomplish the same end would be to use the fact that we have now chosen a Drinfeld module ψ. The "invariant differential" on ψ (i.e., the constant term of ψ_a for $a \in \mathbf{A}$) then allows one to pass from "T as operator" to "T as scalar."

Finally we end this section with some results on the transcendence of zeta values. First of all, Yu's general result on periods [Yu1], together with Theorems 8.18.3 and 8.18.6, establishes the transcendence over $\mathbf{k}$ of zeta values at *positive* integers divisible by $r^{d_\infty} - 1$. Now let $\mathbf{A} = \mathbb{F}_r[T]$ in which case much more can be said. Indeed, let ξ be the period of the Carlitz module. Then using the *special points* on the higher tensor powers of the Carlitz module found by Anderson and Thakur [AT1], Jing Yu [Yu3] established the following remarkable result which goes far beyond classical theory.

Theorem 8.18.8 (Yu). *Let* $\mathbf{A} = \mathbb{F}_r[T]$. *Let* $\zeta(s)$ *be the zeta function of* $\mathbf{A}$ *(as a function on S_∞).*
1. *We have $\zeta(i)$ is transcendental over $\mathbf{k}$ for all positive i.*
2. *Let i now not be divisible by $r-1$. Then $\zeta(i)/\xi^i$ is also transcendental over $\mathbf{k}$.*
3. *(v-adic version) Let i be as in Part 2 and let $\zeta_v(\mathbf{A}, e_v)$ be the v-adic interpolation as in Subsection 8.6 (there is no non-trivial σ in this case). Then*

$$\zeta_v(\mathbf{A}, 1, i, i) = \sum_{j=0}^{\infty} \left(\sum_{\substack{n \text{ monic} \\ v \nmid n \\ \deg n = j}} n^{-i} \right)$$

is transcendental over $\mathbf{k}$. □

Theorem 8.18.8.1 was also obtained independently in [DamH1]. For another approach to 8.18.2, we refer the reader to [Ber1], [Ber1].

8.19. The Functional Equation of the Special-values

In classical number theory, Euler discovered the functional equation for the Riemann zeta function $\zeta(s) := \zeta_{\mathbb{Q}}(s)$ by computing the values of $\zeta(s)$ at both positive even and negative odd integers and then noting that both expressions can be described in terms of Bernoulli numbers. See [Ay1] for a very readable account of this.

As we will need the functional equation for $\zeta(s)$ for comparison, we now recall it. The reader should note, of course, that a very similar phenomenon takes place with arbitrary Dedekind zeta functions (as well as L-functions of more general motives). Thus let $\Gamma(s)$ be Euler's gamma function and set

$$\xi(s) := \Gamma(s/2)\pi^{-s/2}\zeta(s).$$

Then, as is universally known, $\xi(s)$ has a meromorphic continuation to $\mathbb{C}$ with simple poles at $s = 0$ and $s = 1$, *and*

$$\xi(s) = \xi(1-s).$$

This is the *symmetric form* of the functional equation. A *non-symmetric form* states that

$$\zeta(s) = \pi^{s-1/2}\zeta(1-s)\Gamma\left(\frac{1-s}{2}\right) \Big/ \Gamma\left(\frac{s}{2}\right);$$

various other forms are possible using standard Γ-identities. Finally if one wants to work with *entire* functions, as opposed to meromorphic ones, simply set

$$\hat{\xi}(s) := s(1-s)\xi(s);$$

so $\hat{\xi}(s)$ obviously satisfies the same functional equation as $\xi(s)$.

Remark. 8.19.1. Let **k** be our base function field, as usual, and let $\zeta_{\mathbf{k}}(s)$ be its classical (complex) arithmetic zeta function in the sense of Artin and Weil. So

$$\zeta_{\mathbf{k}}(s) = \frac{P_{\mathbf{k}}(r^{-s})}{(1-r^{-s})(1-r^{1-s})},$$

where $P_{\mathbf{k}}(u) \in \mathbb{Z}[u]$ has degree $2g$, g = genus of **k**. As is very well-known, $P_{\mathbf{k}}(u)$ satisfies the functional equation

$$P_{\mathbf{k}}(u) = r^g u^{2g} P_{\mathbf{k}}\left(\frac{1}{ru}\right),$$

which, as the reader may readily see, endows $\zeta_{\mathbf{k}}(s)$ with a functional equation under $s \mapsto 1-s$. Set

$$\xi_{\mathbf{k}}(s) := r^{s(g-1)}\zeta_{\mathbf{k}}(s);$$

Then one verifies directly that

8. L-series

$$\xi_{\mathbf{k}}(s) = \xi_{\mathbf{k}}(1-s).$$

If one wants to deal with "entire" functions, one sets

$$\hat{\xi}_{\mathbf{k}}(s) := \frac{r^{s(g-1)}\zeta_{\mathbf{k}}(s)}{r^{-s}\zeta_{\mathbb{P}^1}(s)}$$

$$= r^{sg} P_{\mathbf{k}}(r^{-s}).$$

We now return to studying the zeta function

$$\zeta_{\mathbf{A}}(s) = \sum_{I \subseteq \mathbf{A}} I^{-s}$$

as before. The special values of $\zeta_{\mathbf{A}}(s)$ at the integers are similar to those of $\zeta_{\mathbb{Q}}(s)$. For instance, as $\zeta_{\mathbf{A}}(s)$ is essentially algebraic, we see immediately that

$$\zeta_{\mathbf{A}}(-j) \in \mathcal{O}_{\mathbf{V}}$$

where $\mathbf{V}$ is the value field, $\mathcal{O}_{\mathbf{V}}$ its $\mathbf{A}$-integers, and $j \geq 0$. Moreover, as in our last subsection, the values of $\zeta_{\mathbf{A}}(s)$ at the positive integers satisfy a form of Euler's Theorem. *However, all attempts to link these two types of special-values in the classical fashion of $\zeta_{\mathbb{Q}}(s)$ have so far been unsuccessful.* We, therefore, see that the arithmetic/geometric dichotomy of subsection 8.15 also carries over to special-values. We have called the special-values at negative integers *arithmetic Bernoulli-elements*. Since the values at the positive integers arise from exponentials of Drinfeld modules, we call them *geometric Bernoulli-elements*. (Strictly speaking, we should also introduce some Γ-functions here, but we will ignore this point for the moment. However, see Subsection 9.2.)

In this subsection we will establish a "functional equation" for the special-values of $\zeta_{\mathbf{A}}(s)$ at the positive integers which are divisible by $r^{d_\infty} - 1$. Recall that we set

$$\mathbf{T} := \mathbf{H}^+ \cdot \mathbf{V}.$$

Let ψ be a sgn-normalized rank one Drinfeld module defined over the $\mathbf{A}$-integers $\mathcal{O}^+$ of $\mathbf{H}^+$. Without loss of generality we can and shall assume that the lattice Λ of ψ is $\xi \cdot \mathbf{A}$.

For each ideal $I \subseteq \mathbf{A}$ we have the $\mathbb{F}_r$-linear polynomial $\psi_I(\tau)$ which is monic with roots $\psi[I] \subset \mathbf{C}_\infty$. Let J be another ideal. Then by 7.6.1 we know that

$$D\psi_{IJ} = D(J * \psi)_I D\psi_J$$

$$= D\psi_I^{\sigma_J} \cdot D\psi_J,$$

where σ_J is the Artin symbol at J and the action is written multiplicatively.

Lemma 8.19.2. *The element $I^{s_1}/D\psi_I$ is a unit in the ring of $\mathbf{A}$-integers of $\mathbf{T}$ for all integers I of $\mathbf{A}$.*

8.19. The Functional Equation of the Special-values

Proof. We have shown in 8.2.10 that

$$I\mathcal{O}_\mathbf{V} = (I^{s_1}),$$

and, in 7.6.2 that

$$I\mathcal{O}^+ = (D\psi_I).$$

Thus the result follows immediately. □

Note that $\mathbf{T}/\mathbf{V}$ is Galois with group canonically isomorphic to a subgroup of $\mathrm{Gal}(\mathbf{H}^+/\mathbf{k}) \simeq \mathcal{I}/\mathcal{P}^+$.

Let $\{\alpha_1, \ldots, \alpha_h\}$ be our representatives of $\mathcal{I}/\mathcal{P}$ and let $(r^{d_\infty} - 1) \mid j$. Then we have

$$\zeta_\mathbf{A}(j) = \sum_{e=0}^{h} \alpha_e^{-j} \sum_{\substack{I \subseteq \alpha_e^{-1} \\ I \text{ principal}}} I^{-j}$$

$$= -\sum_{e=0}^{h} \alpha_e^{-j} \sum_{0 \neq \beta \in \alpha_e^{-1}} \beta^{-j},$$

(since $\mathbb{F}_r^* \cdot \beta$ corresponds to the same principal ideal (β)). Thus

$$\zeta_\mathbf{A}(j) = -\sum_{e=0}^{h} \left(\frac{D\psi_{\alpha_e}}{\alpha_e}\right)^j \left(\sum_{0 \neq \beta \in \alpha_e^{-1}} (D\psi_{\alpha_e} \cdot \beta)^{-j}\right). \quad (8.19.3)$$

Theorem 8.19.4. *Let j be as above. Then the element $\zeta_\mathbf{A}(j)/\xi^j$ is in $\mathbf{T}$ and generates a $\mathrm{Gal}(\mathbf{T}/\mathbf{V})$ invariant fractional ideal.*

Proof. We have assumed that the lattice associated to ψ is $\xi \mathbf{A}$. Let J be an ideal of $\mathbf{A}$ with Artin symbol σ_J. By 4.9.5.2 we know that the lattice associated to ψ^{σ_J} is $D\psi_J \cdot J^{-1} \cdot \xi$. Notice that if we divide the sums inside parenthesis in 8.19.3 by ξ^j, we see that they are all Galois *conjugate* as elements of $\mathbf{H}^+$. The first part of the result follows.

Suppose now that J is chosen so that $\sigma_J \in \mathrm{Gal}(\mathbf{T}/\mathbf{V})$. We then deduce, with a little thought (and the basic formula 7.6.1) that

$$\sigma_J(\zeta_\mathbf{A}(j)/\xi^j) = \frac{\zeta_\mathbf{A}(j)}{\xi^j} \cdot \frac{J^j}{D\psi_J^j}.$$

The result now follows from 8.9.2. □

The idea of searching for Galois invariant ideals comes from L. Shu, [Shu1].

Notice that we can now descend $(\zeta_\mathbf{A}(j)/\xi^j)$ to a fractional ideal of the integers of $\mathbf{V}$ as $\mathbf{T}/\mathbf{V}$ is unramified at finite primes. We denote this fractional ideal by "$[\zeta_\mathbf{A}(j)/\xi^j]$."

We can now deduce the "functional equation" for the values $\zeta_{\mathbf{A}}(j)/\xi^j$. Recall that 7.6.1 implies the *cocycle relation*

$$D\psi_{IJ} = D\psi_I^{\sigma_J} \cdot D\psi_J,$$

and that, by definition, $J^1 = J^{s_1}$.

Lemma 8.19.5. *Let J be chosen so that the Artin symbol $\sigma_J \in \mathrm{Gal}(\mathbf{T}/\mathbf{V})$. Then the mapping*

$$\sigma_J \mapsto \frac{J^1}{D\psi_J} = \frac{J^{s_1}}{D\psi_J}$$

is a 1-cocycle on $\mathrm{Gal}(\mathbf{T}/\mathbf{V})$ with values in $\mathbf{T}^*$.

Proof. This now follows immediately. □

Corollary 8.19.6. *Let J be as above. Then there exists $\gamma \in \mathbf{T}^*$ such that*

$$\frac{J^1}{D\psi_J} = \frac{\gamma}{\gamma^{\sigma_J}} = \frac{\gamma}{\sigma_J(\gamma)}.$$

Proof. $H^1(\mathrm{Gal}(\mathbf{T}/\mathbf{V}), \mathbf{T}^*)$ vanishes. □

Corollary 8.19.7. *Let j, J, γ, etc., be as above. Then we have*

$$\sigma_J(\zeta_{\mathbf{A}}(j)/\xi^j) = (\zeta_{\mathbf{A}}(j)/\xi^j) \cdot \gamma^j/\sigma_J(\gamma)^j.$$ □

Corollary 8.19.7 is the *unsymmetric functional equation for the special-values*; in it, $\mathrm{Gal}(\mathbf{T}/\mathbf{V})$ plays the role of the group $\{1, \beta\}$, $\beta(s) = 1 - s$, and where all "Γ-factors" are grouped together.

Corollary 8.19.8. *Let j, J, γ, etc., be as above. Then*

$$\sigma_J(\gamma^j \zeta_{\mathbf{A}}(j)/\xi^j) = \gamma^j \xi_{\mathbf{A}}(j)/\xi^j.$$

Thus

$$\gamma^j \zeta_{\mathbf{A}}(j)/\xi^j \in \mathbf{V}^*.$$

Proof. Use a little algebra. □

Corollary 8.19.8 is the *symmetric form* of the functional equation and is, in fact, simply a statement of Galois theory. The reader should note, of course, that these results are vacuous when $\mathbf{A}$ is a principal ideal domain.

Remark. 8.19.9. Note that $\gamma/\sigma_J(\gamma) = J^{s_1}/D\psi_J$ is a unit in $\mathbf{T}$. Thus the principal $\mathcal{O}_{\mathbf{T}}$-ideal (γ), where $\mathcal{O}_{\mathbf{T}}$ is the ring of $\mathbf{A}$-integers of $\mathbf{T}$, is also $\mathrm{Gal}(\mathbf{T}/\mathbf{V})$ invariant.

Questions 8.19.10. 1. The element γ is only defined up to elements in $\mathbf{V}^*$ by Galois theory. Is there a way to define it canonically?
2. Does the statement, Corollary 8.19.8, actually lift to a statement about *the function* $\zeta_{\mathbf{A}}(s)$?

One can prove similar results in other situations. For instance, in the setting and notation of Theorem 8.18.6 one can show via the Dedekind determinant that
$$(\gamma^{tj}\zeta_{\mathcal{O}_L}(j)/\xi^{tj})^2 \in \mathbf{V}^*\,;$$
etc. Presumably such a formula would also hold for arbitrary totally-real fields, etc.

Classically, with the obvious definitions, we have $\mathbf{H} = \mathbf{V} = \mathbf{T} = \mathbb{Q}$ as is obvious. However, it seems unreasonable to expect that $\mathbf{V}$ can be replaced with $\mathbf{k}$ in the above results just as $\mathbf{H}$ could not be replaced in Theorem 8.18.6. Perhaps a counter-example may also be found in the setting of [Th4].

8.20. Applications to Class Groups

In this subsection we explain how the special-values of the previous two subsections *should* be connected to class groups. We then describe what is known at present.

We recall that the convention of Subsection 8.14 is in effect. We also drop the subscript "L" on class groups as everything will refer only to $\mathbf{k}(\wp)$ (i.e., $L = \mathbf{k}$) itself.

In Theorem 8.19.4 we established that for $j \equiv 0 \ (r^{d_\infty} - 1)$, the element $\zeta_{\mathbf{A}}(j)/\xi^j$ generates a $\mathrm{Gal}(\mathbf{T}/\mathbf{V})$-invariant fractional ideal of the $\mathbf{A}$-integers $\mathcal{O}_T$ of $\mathbf{T}$. It thus arises from a unique fractional ideal
$$[\zeta_{\mathbf{A}}(j)/\xi^j]$$
of $\mathcal{O}_{\mathbf{V}}$, the ring of $\mathbf{A}$-integers of $\mathbf{V}$.

Recall further how the construction of the generalized Teichmüller character was given in Subsection 8.11. We fixed a prime $\wp$ of $\mathbf{A}$ and we let $\overline{\mathbf{k}}_\wp$ be an algebraic closure of $\mathbf{k}_\wp$. We let $\sigma\colon \mathbf{V} \to \overline{\mathbf{k}}_\wp$ be an embedding over $\mathbf{k}$ corresponding to some prime $\overline{\wp}$ of $\mathcal{O}_{\mathbf{V}}$ which lies above $\wp$. We then used our exponentiation of ideals and Teichmüller representatives to obtain
$$\omega_\sigma = \omega_{\overline{\wp}}\colon \mathcal{I}(\wp)/\mathcal{P}_\wp^+ \to W_\sigma^*$$
where W_σ is the Witt ring of $\mathcal{O}_{\mathbf{V}}/\overline{\wp}$.

Let $\mathbf{k}(\wp)$ be the abelian extension of $\mathbf{k}$ obtained by adjoining the $\wp$-division points of any sgn normalized Hayes module to $\mathbf{H}^+$. So, of course, $\mathbf{k}(\wp)$ is our general analog of the field $\mathbb{Q}(\zeta_p)$. Let $\mathbf{k}(\wp)^+ \subseteq \mathbf{k}(\wp)$ be the maximal totally-real subfield. Thus if $G := \mathrm{Gal}(\mathbf{k}(\wp)/\mathbf{k})$ and $D_\infty \subseteq G$ is the decomposition group at ∞, then $D_\infty \simeq \mathbb{F}_\infty^*$ and $\mathbf{k}(\wp)^+$ is the fixed field of D_∞. Let $\mathcal{O}(\wp)$, $\mathcal{O}(\wp)^+$ be the rings of $\mathbf{A}$-integers.

Definition 8.20.1. 1. Let $R(\wp)$ (resp. $R(\wp)^+$) be the group of divisors of degree 0 supported on the infinite primes of $\mathbf{k}(\wp)$ (resp. $\mathbf{k}(\wp)^+$).
2. Let $U(\wp) \subseteq R(\wp)$ (resp. $U(\wp)^+$) be the divisor of the units in $\mathcal{O}(\wp)$ (resp. $\mathcal{O}(\wp)^+$).
3. Let $r(\wp)$ (resp. $r(\wp)^+$) be $R(\wp)/U(\wp)$ (resp. $R(\wp)^+/U(\wp)^+$).

The groups $r(\wp)$ and $r(\wp)^+$ are the *regulator groups* of the fields $\mathbf{k}(\wp)$ and $\mathbf{k}(\wp)^+$. Let $\mathrm{Cl}(\mathcal{O}(\wp))$ be the class group of $\mathcal{O}(\wp)$ and $\mathrm{Cl}(\mathcal{O}(\wp))^{(p)}$ its p-primary part, etc. Let P (resp. P^+) be the projection which takes a divisor of degree 0 to its part supported on the finite primes. We have two exact sequences

$$0 \to r(\wp) \to \mathrm{Cl}(\wp) \xrightarrow{P} \mathrm{Cl}(\mathcal{O}(\wp)) \to 0, \qquad (8.20.2)$$

and

$$0 \to r(\wp)^+ \to \mathrm{Cl}(\wp)^+ \xrightarrow{P^+} \mathrm{Cl}(\mathcal{O}(\wp)^+) \to 0. \qquad (8.20.3)$$

(Indeed, the primes at infinity are rational over $\mathbb{F}$, where $\mathbb{F} \simeq \mathbb{F}_{r^{d_\infty}}$ is the full field of constants of $\mathbf{k}(\wp)$.) Note that the elements in these sequences are all natural G-modules. Upon tensoring with W_σ, we can form the p-primary components of these modules under ω_σ.

Proposition 8.20.4. 1. *Let $u \in U(\wp)$. Then u^e, $e = r^{d_\infty} - 1$, is in $U(\wp)^+$.*
2. *Let i be a positive integer not divisible by $r^{d_\infty} - 1$. Then $r(\wp)^{(p)}(\omega_\sigma^{\pm i}) = \{0\}$.*
3. *There is an isomorphism of p-primary parts*

$$(r(\wp)^+)^{(p)} \simeq r(\wp)^{(p)}.$$

Proof. 1. Let $\sigma \in D_\infty$ and $u \in U(\wp)$. As the divisor of u is supported on $R(\wp)$, we see that $\sigma(u)/u$ must be a global unit; i.e.; $\sigma(u)/u \in \mathbb{F}^*$. Thus Part 1 is now obvious.
2. This follows since the primes at ∞ are $\mathbb{F}$-rational and $\mathbb{F} \subset \mathbf{H} \subseteq \mathbf{k}(\wp)^+$.
3. The primes at infinity in $\mathbf{k}(\wp)$ are totally ramified over $\mathbf{k}(\wp)^+$ and $r^{d_\infty} - 1$ is prime to p. $\square$

If $d_\infty = 1$, then one can refine Part 1 of 8.20.4 to see that $u \in U(\wp)^+$. Indeed $\mathbf{k}(\wp)^+(u)/\mathbf{k}(\wp)^+$ is étale above $\wp$. However $\mathbf{k}(\wp)/\mathbf{k}(\wp)^+$ is now totally-ramified above $\wp$.

Corollary 8.20.5. *Let i be as in 8.20.4.2. Then*

$$\mathrm{Cl}(\wp)^{(p)}(\omega_\sigma^{\pm i}) \simeq \mathrm{Cl}(\mathcal{O}(\wp))^{(p)}(\omega_\sigma^{\pm i}). \qquad \square$$

Corollary 8.20.5 is important in that it tells us that our criterion in Subsection 8.14 (e.g., 8.14.4) and 8.16 for the non-triviality of $\mathrm{Cl}(\wp)^{(p)}(\omega_\sigma^{\pm i})$ *automatically* apply to $\mathrm{Cl}(\mathcal{O}(\wp))^{(p)}(\omega_\sigma^{\pm i})$. Thus we may turn our attention to the "plus parts." More precisely, we turn our attention to $\mathrm{Cl}(\mathcal{O}(\wp)^+)^{(p)}$ and $(r(\wp)^+)^{(p)}$. Set $G^+ := \mathrm{Gal}(\mathbf{k}(\wp)^+/\mathbf{k})$. Let $I^+ \subset \mathbb{Z}[G^+]$ be the augmentation ideal, i.e., those $\sum n_\sigma \sigma$ with $\sum n_\sigma = 0$.

Proposition 8.20.6. $R(\wp)^+ \simeq I^+$ *as G^+-modules.* □

We will review below what is known in the prototypical $\mathbb{F}_r[T]$-case. Based on this, and also the work of L. Shu [Shu1], we now pose the following problem. First of all, recall that we set $o(\sigma)$ to be the order of ω_σ and have shown that $o(\sigma) \equiv 0 \ (r^{d_\infty} - 1)$.

Problems 8.20.7. 1. Let $0 < i < o(\sigma)$ with $(r^{d_\infty} - 1) \mid i$. Show that $v_{\overline{\wp}}[\zeta_\mathbf{A}(i)/\xi^i] \geq 0$ where $v_{\overline{\wp}}[\zeta_\mathbf{A}(i)/\xi^i]$ is the order of $\overline{\wp}$ in this $\mathcal{O}_\mathbf{V}$ fractional ideal.
2. Let i be as above. Show that
$$\mathrm{Cl}(\mathcal{O}(\wp)^+)^{(p)}(\omega_\sigma^i) \neq 0 \Rightarrow v_{\overline{\wp}}[\zeta_\mathbf{A}(i)/\xi^i] > 0.$$

Remark. 8.20.8. Let $\mathbf{A} = \mathbb{F}_r[T]$. In this case $\mathbf{V} = \mathbf{k} = \mathbf{H}^+$. Thus the fractional ideal $[\zeta_\mathbf{A}(i)/\xi^i]$ of $\mathbf{k}$ just corresponds to the fractional ideal generated by
$$\zeta_\mathbf{A}(i)/\xi^i \in \mathbf{k}^*.$$
If we multiply $\zeta_\mathbf{A}(i)/\xi^i$ by the appropriate "Γ-factor," (see Subsection 9.2) we obtain the *i-th Bernoulli-Carlitz number B_i*. However, it is easy to see that the Γ-factors in the range $0 < i < o(\sigma) = r^{\deg \wp} - 1$ are *prime to* $\wp$. As such, we are justified in ignoring them in posing the problems.

We now turn to a remarkable problem which is inverse to 8.20.7.2. This problem is just the straightforward generalization of the special case $\mathbf{A} = \mathbb{F}_r[T]$ where it is due to E.-U. Gekeler. Gekeler begins with the following observation: the Frobenius automorphism of W_σ permutes the isotypic components of $\mathrm{Cl}(\wp)^{(p)} \otimes W_\sigma$ and the same for the *p*-class groups of $\mathcal{O}(\wp)$. Thus the components of an orbit will all live or die together. More formally, let us define the following equivalence relation: let i and j be two positive numbers. We define $i \overset{\overline{\wp}}{\sim} j$ by
$$i \overset{\overline{\wp}}{\sim} j \iff i \equiv p^e j \mod o(\sigma),$$
for some e (and where, we recall, $o(\sigma)$ is the order of ω_σ). Then
$$i \overset{\overline{\wp}}{\sim} j \implies \mathrm{Cl}(\wp)^{(p)}(\omega_\sigma^i) \simeq \mathrm{Cl}(\wp)^{(p)}(\omega_\sigma^j),$$
via the Frobenius, etc.

Let I be an ideal of $\mathbf{A}$ which is prime to $\wp$. Let i and j be two positive integers which are congruent modulo $o(\sigma)$. It is easy to see that

$$I^i \equiv I^j \pmod{\overline{\wp}},$$

where $I^i = I^{s_i}$, etc. Since the sums describing values of $\zeta_{\mathbf{A}}(s)$ at negative integers are always *finite* sums, we deduce that

$$\zeta_{\mathbf{A}}(-i) \equiv \zeta_{\mathbf{A}}(-j) \pmod{\overline{\wp}},$$

as elements of $\mathcal{O}_{\mathbf{V}}$. Moreover, we have trivially

$$\zeta_{\mathbf{A}}(-pi) = \zeta_{\mathbf{A}}(-i)^p.$$

We leave it to the reader to verify that, in this case (i.e., $L = \mathbf{k}$) these relationships also carry over to the special values $\tilde{z}(1, -i)$ of Theorem 8.14.4. Therefore, we see

$$i \overset{\overline{\wp}}{\sim} j \Longrightarrow \overline{\wp} \mid \tilde{z}(1, -i) \Longleftrightarrow \overline{\wp} \mid \tilde{z}(1, -j).$$

In other words, divisibility of $\tilde{z}(1, -j)$ by $\overline{\wp}$ respects the equivalence relation $\overset{\overline{\wp}}{\sim}$. However, for $[\zeta_{\mathbf{A}}(i)/\xi^i]$ it is another matter altogether. Indeed, clearly $\zeta_{\mathbf{A}}(pi) = \zeta_{\mathbf{A}}(i)^p$, thus $[\zeta_{\mathbf{A}}(pi)/\xi^{pi}] = [\zeta_{\mathbf{A}}(i)/\xi^i]^p$. However, for congruences modulo $o(\sigma)$ nothing is known.

In any case, we have the following problem of E.-U. Gekeler for which there is some evidence in the $\mathbf{A} = \mathbb{F}_r[T]$-case [Ge4].

Problem 8.20.9. Let i be an integer $< o(\sigma)$. Suppose that for all $j < o(\sigma)$, with $j \overset{\overline{\wp}}{\sim} i$, we have $\overline{\wp} \mid [\zeta_{\mathbf{A}}(j)/\xi^j]$. Show that

$$\mathrm{Cl}\,(\mathcal{O}(\wp)^+)^{(p)}(\omega_\sigma^i) \neq \{0\}.$$

Should 8.20.9 be established, we would then have a weak *functional equation* relating the values of $\zeta_{\mathbf{A}}(s)$ at positive and negative integers. Indeed, for $i < o(\sigma)$ with $(r^{d_\infty} - 1) \mid i$, we deduce from 8.14.4 that

$$\overline{\wp} \mid [\zeta_{\mathbf{A}}(j)/\xi^j] \text{ for all } j \overset{\overline{\wp}}{\sim} i \Longrightarrow \overline{\wp} \mid \tilde{z}(1, o(\sigma) - j) \text{ for all } j \overset{\overline{\wp}}{\sim} i;$$

at least for those i not divisible by $r^{\deg \wp} - 1$.

We now briefly describe how the solutions of Problems 8.20.7.1 and 8.20.7.2 go in the $\mathbf{A} = \mathbb{F}_r[T]$-case. The reader will note that they are modeled on the classical work of Kummer in cyclotomic fields.

Let $\mathbf{A} = \mathbb{F}_r[T]$, $\mathbf{k} = \mathbb{F}_r(T)$, etc. Let $\wp = (f)$ be a prime of $\mathbf{A}$ and let C be the Carlitz module for $\mathbf{A}$. Let λ be a non-zero root of $C_f(x) = 0$; so $\mathbf{k}(\wp) = \mathbf{k}(\lambda)$ with Galois group $G := \mathrm{Gal}(\mathbf{k}(\wp)/\mathbf{k}) = (\mathbf{A}/\wp)^*$. Note that $\mathbf{V} = \mathbf{H}^+ = \mathbf{T} = \mathbf{k}$ and so we may dispense with the subscript "σ" in ω_σ.

Let $\tau \in G$; so we have seen (Subsection 7.5) that $\tau(\lambda)/\lambda$ is a unit in $\mathcal{O}(\wp)^+$. These units generate the *group of cyclotomic units* in $\mathbf{k}(\wp)$. We will denote the group of cyclotomic units by $\mathrm{Cy}\,(\wp)$. It is trivial to see that $\mathbb{F}_r^* \subseteq \mathrm{Cy}\,(\wp)$.

Definition 8.20.10. 1. We define $Q(\wp)^+$ to be $R(\wp)^+$ (cf. 8.20.1) modulo divisors of cyclotomic units.
2. We set $H(\wp)^+ := U(\wp)^+/\mathrm{Cy}\,(\wp)$.

Proposition 8.20.11. *The following sequence of abelian groups is exact:*

$$0 \to H(\wp)^+ \to Q(\wp)^+ \to r(\wp)^+ \to 0\,. \qquad \square$$

By computing explicitly the divisors of the cyclotomic units and comparing them to class number formulae (as in classical theory) Galovich and Rosen have shown the following basic result, [GR1], [GR2].

Theorem 8.20.12. $Q(\wp)^+$ *and* $\mathrm{Cl}\,(\wp)^+$ *have the same order.* $\qquad \square$

Corollary 8.20.13. $H(\wp)^+$ *and* $\mathrm{Cl}\,(\mathcal{O}(\wp)^+)$ *have the same order.*

Proof. Use 8.20.12, 8.20.11, and 8.20.3. $\qquad \square$

The above result can be refined. Indeed, if we pass to p-primary components, we obtain the next result. Recall that $W = $ the Witt ring of $\mathbf{A}/\wp$.

Proposition 8.20.14. *Let $0 < i < o(\sigma) = r^{\deg f} - 1$ with $(r-1) \mid i$. Then the W-modules $(H(\wp)^+)^{(p)}(\omega^i)$ and $\mathrm{Cl}\,(\mathcal{O}(\wp)^+)^{(p)}(\omega^i)$ have the same length.* $\qquad \square$

We now discuss the work of S. Okada [Ok1] which introduces an analog of the *Kummer map* into the theory. We have seen that (λ) is the unique prime above $\wp$ in $\mathbf{k}(\wp)$; we denote the completion of $\mathbf{k}(\wp)$ at this prime by $\mathbf{k}_\wp^1$ and view it as an extension of $\mathbf{k}_\wp$. Let $\alpha \in \mathbf{k}_\wp^1$ be a unit. It is clear that we can write

$$\alpha = f(u)$$

with $f(u) \in \mathbf{A}[[u]]$. Let $e(z)$ be the exponential of the Carlitz module. In the next definition "′" denotes the derivative.

Definition 8.20.15. Let $f'(e(z))/f(e(z)) = \sum_{i=0}^{\infty} a_i z^i$ (recall $f(0) \neq 0$). We set

$$\Phi_i(\alpha) = a_{i-1} \pmod{\wp},$$

for $i = 1, \ldots, r^{\deg f} - 2$.

We know that $e(z) = \sum z^{r^i}/D_i$ where D_i is the product of the monics of degree i. Thus, as we will see in our next section, D_i is a (piece of a) "factorial" with *many* properties similar to those of $n!$. As such, one can show that a_i is $\wp$-integral in the above range. Using the fact that $C_f(z)/z$ is irreducible, one can show that Φ_i is well-defined. And using the fact that $(fg)'/fg = f'/f + g'/g$, as usual, one sees that Φ_i is a homomorphism from the units $E \subset \mathbf{k}_\wp^1$ to $\mathbf{A}/\wp$.

Let $\mathrm{Cy}\,(\wp)(\omega^i)$ be the isotypic component associated to ω^i which is obtained upon tensoring $\mathrm{Cy}\,(\wp)$ with W. If ω^i is non-trivial, then $\mathrm{Cy}\,(\wp)(\omega^i)$ is generated by

$$\lambda^{\sum \omega^i(\beta)\beta^{-1}}$$

where β runs over $\mathrm{Gal}(\mathbf{k}(\wp)^+/\mathbf{k})$. Now if

$$(H(\wp)^+)^{(p)}(\omega^i) \neq 0,$$

then $\lambda^{\sum \omega^i(\beta)\beta^{-1}}$ is a p-th power and so

$$\Phi_i(\lambda^{\sum \omega^i(\beta)\beta^{-1}}) = 0 \pmod{\wp}.$$

Next, let M be a monic element and $\sigma_M \in \mathrm{Gal}(\mathbf{k}(\wp)^+/\mathbf{k})$ its Artin symbol. Thus

$$\sigma_M(\lambda) = C_M(\lambda),$$

and we may take $f(u) = C_M(u)/u$. A little computation gives

$$\Phi_i(\lambda) = (M^i - 1)\zeta_{\mathbf{A}}(i)/\xi^i$$

(as in the proof of 8.18.1). Thus

$$\Phi_i(\lambda^{\sum \omega^i(\beta)\beta^{-1}}) = \frac{\zeta_{\mathbf{A}}(i)}{\xi^i} \pmod{\wp},$$
$$= 0 \pmod{\wp},$$

from above. Combining this with 8.20.14 completes the proof of 8.20.7.1 and 8.20.7.2 in the $\mathbf{A} = \mathbb{F}_r[T]$-case.

Overcoming a number of technical difficulties, L. Shu in [Shu1] establishes Problems 8.20.7.1 and 8.20.7.2 when $\mathbf{A}$ is a principal ideal domain. She also has results related to 8.20.7.1 and 8.20.7.2 when $d_\infty = 1$. The reader should also see the papers [Yi1], [Yi2] of L. Yin as well as [A9].

8.21. The Geometric Criterion for Cyclicity

We now explore the implication of the results and problems in Subsection 8.20 for cyclicity of class groups. We continue with the same notation (and, as usual, with the *convention* of Subsection 8.14 in effect). We also make the assumption that Problems 8.20.7.1 and 8.20.7.2 are established *affirmatively* (as is known when **A** is a principal ideal domain).

Proposition 8.21.1. *Let* $0 < i < o(\sigma)$ *with* $(r^{d_\infty} - 1) \mid i$. *Suppose that* $v_{\overline{\wp}}[\zeta_{\mathbf{A}}(i)/\xi^i] = 0$. *Then*
$$\mathrm{Cl}\,(\wp)^{(p)}(\omega_\sigma^i)$$
is cyclic as a W-module.

Proof. By 8.20.7.2 we know that $\mathrm{Cl}\,(\mathcal{O}(\wp)^+)^{(p)}(\omega_\sigma^i) = 0$. Thus, by 8.20.3, we have $\mathrm{Cl}\,(\wp)^{(p)}(\omega_\sigma^i) \simeq (\mathrm{Cl}\,(\wp)^+)^{(p)}(\omega_\sigma^i)$ is supported on the primes at infinity. Thus the result follows from Proposition 8.20.6. □

Proposition 8.21.1 is the *geometric criterion for cyclicity*. It is "geometric" as it relates to $\zeta_{\mathbf{A}}(s)$ where s is a positive integer.

Let $\widetilde{z}(x, -j) = \prod(1 - \theta_e x^{-t_{\mathbb{F}}})$ be as in Subsection 8.16, where $\{\theta_e\}$ are in some algebraic closure of $\mathbf{V}_{\overline{\wp}}$ and $t_{\mathbb{F}}$ is as in Theorem 8.12.5. Let $\overline{M}_{\overline{\wp}}$ be the maximal ideal in $\mathbf{V}_{\overline{\wp}}$. Let $0 < i < o(\sigma)$ with $i \equiv 0 \bmod (r^{d_\infty} - 1)$ but $i \not\equiv 0 \bmod (r^{\deg \wp} - 1)$. Set $j = o(\sigma) - i$ (so j satisfies the same congruences as i).

Theorem 8.21.2. *The geometric criterion for cyclicity implies the arithmetic criterion for cyclicity. That is, at most one θ_e is $\equiv 1 \pmod{\overline{M}_{\overline{\wp}}}$.*

Proof. We present the proof under the simplifying assumption that $d_\infty = 1$. We leave the general case to the reader.

Let $T_p(\wp)^+$ be the p-adic Tate module of the field $\mathbf{k}(\wp)^+$. Note that as $d_\infty = 1$, the constant field of $\mathbf{k}(\wp)$ is $\mathbb{F}_r$. Let F be the Frobenius endomorphism of the Jacobian of $\mathbf{k}(\wp)$ with respect to $\mathbb{F}_r$.

Without loss of generality, we may assume that $(\mathrm{Cl}\,(\wp)^+)^{(p)}(\omega_\sigma^i)$ is non-trivial (else the result is vacuous). We shall show that on $(T_p(\wp)^+)^{(p)}(\omega_\sigma^i)$ the mapping $F - 1$ has a *unique* non-unit eigenvalue. The result will then follow by the double congruence 8.13.3.

In this proof, for simplicity, we shall drop the symbol "$\wp$" in our notation for various groups when no confusion will result. Thus "$(\mathrm{Cl}\,(\mathcal{O})^+)^{(p)}$" will be the p-class group of $\mathcal{O}(\wp)^+$ etc.

Let $\overline{\mathbb{F}}_r$ be a fixed algebraic closure of $\mathbb{F}_r$. Let $(\overline{\mathrm{Cl}}^+)^{(p)}$ be the p-primary class group of the field $\overline{\mathbb{F}}_r \cdot \mathbf{k}(\wp)^+$ and let $(\overline{\mathrm{Cl}}\,(\mathcal{O})^+)^{(p)}$ be the p-primary class group of $\overline{\mathbb{F}}_r \cdot \mathcal{O}(\wp)^+$. Recall that the primes at ∞ on $\mathbf{k}(\wp)_+$ are $\mathbb{F}_r$-rational.

Thus, as in 8.20.3, we have the exact sequence:

$$0 \to (r^+)^{(p)}(\omega_\sigma^i) \to (\overline{\text{Cl}}^+)^{(p)}(\omega_\sigma^i) \to (\overline{\text{Cl}}(\mathcal{O})^+)^{(p)}(\omega_\sigma^i) \to 0.$$

By taking Tate modules ($= \text{Hom}(\mathbb{Q}_p/\mathbb{Z}_p, ?)$) of the above sequence, we obtain

$$0 \to \text{Hom}(\mathbb{Q}_p/\mathbb{Z}_p, (r^+)^{(p)}(\omega_\sigma^i)) \to T_p^+(\omega_\sigma^i) \to \text{Hom}(\mathbb{Q}_p/\mathbb{Z}_p, (\overline{\text{Cl}}(\mathcal{O})^+)^{(p)}(\omega_\sigma^i))$$
$$\to \text{Ext}^1(\mathbb{Q}_p/\mathbb{Z}_p, (r^+)^{(p)}(\omega_\sigma^i)) \to \text{Ext}^1(\mathbb{Q}_p/\mathbb{Z}_p, (\overline{\text{Cl}}^+)^{(p)}(\omega_\sigma^i)).$$

As $(r^+)^{(p)}(\omega_\sigma^i)$ is finite, it is easy to see that the first term vanishes. As $(\overline{\text{Cl}}^+)^{(p)}(\omega_\sigma^i)$ is divisible, the last term also vanishes. Let us set $\widetilde{T}_p^+(\omega_\sigma^i) = \text{Hom}(\mathbb{Q}_p/\mathbb{Z}_p, (\overline{\text{Cl}}(\mathcal{O})^+)^{(p)}(\omega_\sigma^i))$. We, therefore, have

$$0 \to T_p^+(\omega_\sigma^i) \to \widetilde{T}_p^+(\omega_\sigma^i) \to \text{Ext}^1(\mathbb{Q}_p/\mathbb{Z}_p, (r^+)^{(p)}(\omega_\sigma^i)) \to 0.$$

From the exact sequence

$$0 \to \mathbb{Z}_p \to \mathbb{Q}_p \to \mathbb{Q}_p/\mathbb{Z}_p \to 0,$$

we deduce that

$$\text{Ext}^1(\mathbb{Q}_p/\mathbb{Z}_p, (r^+)^{(p)}(\omega_\sigma^i)) \simeq (r^+)^{(p)}(\omega_\sigma^i).$$

We then have

$$0 \to T_p^+(\omega_\sigma^i) \to \widetilde{T}_p^+(\omega_\sigma^i) \to (r^+)^{(p)}(\omega_\sigma^i) \to 0. \tag{8.21.3}$$

Note that $\widetilde{T}_p^+(\omega_\sigma^i)$ is *torsion free*; thus

$$T_p^+(\omega_\sigma^i) \otimes \mathbb{Q}_p \simeq \widetilde{T}_p^+(\omega_\sigma^i) \otimes \mathbb{Q}_p.$$

Consequently, $T_p^+(\omega_\sigma^i)$ is a *sublattice* of $\widetilde{T}_p^+(\omega_\sigma^i)$ of maximal rank.

The endomorphism $F - 1$ clearly acts on the exact sequence 8.21.3 and kills $(r^+)^{(p)}(\omega_\sigma^i)$. Thus

$$(F - 1)\widetilde{T}_p^+(\omega_\sigma^i) \subseteq T_p^+(\omega_\sigma^i).$$

We shall show that this is, in fact, an equality.

By the assumption that $v_{\overline{p}}[\zeta_\mathbf{A}(i)/\xi^i] = 0$, we find

$$T_p^+(\omega_\sigma^i)/(F - 1)T_p^+(\omega_\sigma^i) = (r^+)^{(p)}(\omega_\sigma^i).$$

Thus, $\text{ord}_p(\det(F - 1))$ is the W-length of $(r^+)^{(p)}(\omega_\sigma^i)$. But then

$$\widetilde{T}_p^+(\omega_\sigma^i)/((F - 1)\widetilde{T}_p^+(\omega_\sigma^i))$$

must have the same W-length. As this is also the W-length of

$$\widetilde{T}_p^+(\omega_\sigma^i)/T_p^+(\omega_\sigma^i),$$

we conclude
$$(F-1)\widetilde{T}_p^+(\omega_\sigma^i) = T_p^+(\omega_\sigma^i).$$

Next, choose a basis for $\widetilde{T}_p^+(\omega_\sigma^i)$, say $\{e_1,\ldots,e_h\}$, so that $\{qe_1,\ldots,e_h\}$ is a basis for $T_p^+(\omega_\sigma^i)$, for some $q \in W$. (Recall that $(r^+)^{(p)}(\omega_\sigma^i)$ is a cyclic W-module.) Let $A = (a_{i,j})$ be the matrix of $F-1$ with respect to this basis. Let B be the diagonal matrix with $\frac{1}{q}$ in the first row and all other diagonal elements $=1$. One finds that BA is an automorphism of $T_p^+(\omega_\sigma^i)$ and AB is an automorphism of $\widetilde{T}_p^+(\omega_\sigma^i)$. Thus one sees that both the first row and column of A must have coefficients divisible by q.

Let A_1 be the minor obtained from A by deleting the first row and column of A. If we now expand the determinant along the first row we find
$$\det(A) \equiv a_{1,1} \det(A_1) \pmod{q^2}.$$

But $\mathrm{ord}_p(q) = \mathrm{ord}_p(\det(A)) > 0$, thus $\det(A_1)$ is a W-unit.

If we reduce $\det(tI - A)$ modulo (p), we obtain $t\widetilde{P}(t)$, where $\widetilde{P}(t)$ is the characteristic polynomial of the reduction of A_1. The result now follows. □

8.22. Magic Numbers

Magic numbers are certain integers with particularly simple p-adic digits. Judging from the $\mathbf{A} = \mathbb{F}_r[T]$-case, they *appear* to play an important, if little understood, role. We will state some problems about them in full generality as a guide for further research. Perhaps, as time goes by, the true importance of these integers for the theory will become clear.

Let $\mathbf{A}$ be as before and set $d = d_\infty$ for simplicity.

Definition 8.22.1. 1. Let $0 < c \leq r^d - 1$. Set
$$i_\mathbf{A}(c,j) := cr^{dj} + (r^{dj} - 1).$$

2. Set
$$i_\mathbf{A}(0,0) := 0.$$

The set $\{i_\mathbf{A}(c,j)\}$ is the set of $\mathbf{A}$-*magic numbers*. A positive integer i is an $\mathbf{A}$-magic number if and only if all its r^d-adic digits, except *perhaps* the highest, equal $r^d - 1$.

Remarks. 8.22.2. 1. Every non-negative integer has a canonical minimal expansion as an alternating sum of $\mathbf{A}$-magic numbers such that there is *no* carry over of r^d-adic digits. Indeed, let α be a positive integer with

$$\alpha = \sum_{i=0}^{\nu} \beta_i r^{di}, \quad 0 \leq \beta_i < r^d.$$

We may suppose $\beta_\nu \neq 0$. Set $i_1 := i_\mathbf{A}(\beta_\nu, \nu)$; so

$$i_1 - \alpha = \sum_{i=0}^{\nu-1} (r^d - 1 - \beta_i) r^{di}.$$

Now iterate the procedure. This process clearly eventually stops and gives rise to the required sum.

2. Let $0 < i_\mathbf{A}(c,j)$, then $i_\mathbf{A}(c,j) \equiv 0 \ (r^d - 1)$ if and only if $c = r^d - 1$.

3. Let i, j be two non-negative integers, with $i \equiv j \ (ed)$. Then

$$i_\mathbf{A}(c,i) \equiv i_\mathbf{A}(c,j) \pmod{r^{ed} - 1}.$$

The importance of 8.22.2.3 is the following. Recall that in Subsection 8.11 we defined the generalized Teichmüller character ω_σ associated to a prime $\bar{\wp}$ of $\mathbf{V}$ lying over $\wp$. We let $o(\sigma)$ be the order of ω_σ and showed in 8.11.3 that $o(\sigma)$ is $\equiv 0 \ (r^d - 1)$ and $\equiv 0 \ (r^{\deg \wp} - 1)$. Thus *if* one knew that $o(\sigma)$ is of the form $r^t - 1$, then, at least, $t \equiv 0 \ (d)$ and thus by 8.22.2.3 the residue mod $(o(\sigma))$ of an $\mathbf{A}$-magic number is an $\mathbf{A}$-magic number. This may be unlikely in general, but can be seen to hold when the class number of $\mathbf{A}$ is a power of p (e.g., $\mathbf{A}$ is a principal ideal domain). For instance, if $d = 1$, then $\mathbf{V}/k$ is purely inseparable and so $o(\sigma) = r^{\deg(\wp)} - 1$, etc.

We next present an important lemma of Carlitz from his original paper [C1]. Let R be a finite extension of the complete field $\mathbf{K}$. Let $N \subset R$ be a discrete $\mathbb{F}_r$-vector space (e.g., an $\mathbf{A}$-lattice) and form the function $f(z) = z \prod_{0 \neq \alpha \in N} (1 - \frac{z}{\alpha})$ (so if N is a lattice then $f(z)$ is its exponential function); as N is the union of finite additive groups, we know that $f(z)$ is an entire $\mathbb{F}_r$-linear function. Let $g(z)$ be the inverse function of f, which is also additive, and

$$h(z) = \frac{z}{f(z)}.$$

We write

$$f(z) = \sum_{i=0}^{\infty} \alpha_i z^{r^i},$$

$$g(z) = \sum_{i=0}^{\infty} \beta_i z^{r^i},$$

and

$$h(z) = \sum_{m=0}^{\infty} \gamma_m z^m.$$

By taking logarithmic derivatives, as in Subsection 8.18, we see that $\gamma_m = 0$ unless $(r-1) \mid m$ and that $\gamma_{pm} = \gamma_m^p$. Moreover, we have the evident recursive formula

$$\sum_{r^i \leq m} \gamma_{m-r^i} \alpha_i = \begin{cases} 0 & \text{for } m > 1, \\ 1 & \text{for } m = 1. \end{cases} \tag{8.22.3}$$

Lemma 8.22.4 (Carlitz). *Let $m = p^t(r^k - 1)$ for $t \geq 0$ and $k > 0$. Then*

$$\gamma_m = \beta_k^{p^t}.$$

Proof. Without loss of generality, we may assume that $t = 0$. We next use induction and note that the case $k = 0$ reduces to $\gamma_0 = \beta_0 = 1$ which follows trivially by 8.22.3. Thus we assume that the result is true up to and including $k-1$. From 8.22.3, we know that

$$0 = \sum_{i \leq k} \gamma_{r^k - r^i} \alpha_i,$$

or

$$\gamma_{r^k - 1} = -\sum_{i=1}^{k} \alpha_i \gamma_{r^k - r^i}$$

$$= -\sum_{i=1}^{k} \alpha_i \gamma_{r^{k-i} - 1}^{r^i}$$

$$= -\sum_{i=1}^{k} \alpha_i \beta_{k-i}^{r^i},$$

by the induction hypothesis.

On the other hand, $\{\beta_t\}$ and $\{\alpha_t\}$ satisfy the obvious relations that come from

$$z = g(f(z)) = f(g(z)).$$

These state that if $k > 0$ then

$$0 = \sum_{i=0}^{k} \alpha_i \beta_{k-i}^{r^i},$$

or

$$\beta_k = -\sum_{i=1}^{k} \alpha_i \beta_{k-i}^{r^i}.$$

The result now follows. □

With Lemma 8.22.4 in hand, we can now present another problem to go along with 8.20.7.1, 8.20.7.2 and 8.20.9. We will assume that 8.20.7.1 and 8.20.7.2 are given *affirmative* solutions.

Let $i := i_\mathbf{A}(r^d - 1, j) = r^{d(j+1)} - 1$ be a magic number divisible by $r^d - 1$. Assume that $0 < i < o(\sigma)$ where $o(\sigma)$ is the order of the generalized Teichmüller character ω_σ.

Problem 8.22.5. Let i be as above. Show that the valuation

$$v_{\wp}[\zeta_\mathbf{A}(i)/\xi^i] = 0.$$

As before, the evidence arises from the $\mathbf{A} = \mathbb{F}_r[T]$-case. We apply 8.22.4 to the case where $f(z) = e(z)$ is the exponential of the Carlitz module. Then as in the proof of 8.18.1, we have

$$\frac{z}{f(z)} = \frac{z}{e(z)} = 1 + \sum_{(r-1)|j} \xi^{-j}\zeta_\mathbf{A}(j)z^j$$

where ξ is the period of the Carlitz-module. On the other hand, by Subsection 3.4, the logarithm of the Carlitz module is

$$\sum_{i=0}^{\infty}(-1)^i \frac{z^{r^i}}{L_i}.$$

Lemma 8.22.4 implies that for $i = r^{h+1} - 1 = i_\mathbf{A}(r-1, h)$ we have *exactly*

$$\zeta_\mathbf{A}(i)/\xi^i = (-1)^{h+1}/L_{h+1}.$$

If now $\wp$ is a prime of degree δ and $i < o(\sigma) = r^\delta - 1$, then $h + 1 < \delta$. As it is *obvious* that $v_\wp(L_{h+1}) = 0$ in this range, we have solved 8.22.5 affirmatively! Moreover, by Theorem 8.21.2, we now conclude that the arithmetic criterion for cyclicity also holds for $(\mathrm{Cl}\,(\wp)^+)^{(p)}(\omega^i)$ where $\omega = \omega_\sigma$ is the Teichmüller character.

We should also point out the remarkable results of Gekeler [Ge5] on special values of $\zeta_\mathbf{A}(i)$ where $\mathbf{A} = \mathbb{F}_r[T]$ and i is positive and divisible by $r - 1$. These results are somewhat in the spirit of Lemma 8.22.4; they are established by using the *cohomology* of Drinfeld modules; see [Go5]. It would be very interesting to generalize these results to arbitrary $\mathbf{A}$. It would also be very interesting to know if Gekeler's results are an indication of some deeper underlying phenomenon, etc.

It may turn out that 8.22.5 will follow from the techniques of Anderson's Proposition 7.11.6.

An affirmative solution to 8.22.5 has the following curious corollary when the class number of $\mathbf{A}$ is a pure p-th power (so that $o(\sigma)$ is of the form $r^e - 1$, $e = e_0 d$). Let

$$0 < i = i_\mathbf{A}(r^d - 1, j) < o(\sigma);$$

thus

$$o(\sigma) - i = r^{e_0 d} - 1 - (r^{d(j+1)} - 1)$$
$$= r^{d(j+1)}(r^{e_1} - 1), \quad \text{where } e_1 = d(e_0 - j - 1).$$

Thus $r^{e_1} - 1$ is also an **A**-magic number $\equiv 0 \ (r^d - 1)$; therefore, we find

$$v_{\overline{\wp}}[\zeta_{\mathbf{A}}(o(\sigma) - i)/\xi^{o(\sigma)-i}] = 0$$

also, via the r^d-th power mapping. Whether this holds for *general* **A** is unknown and we state it as a problem.

Problem 8.22.6. Let **A** be general and let $0 < i = i_{\mathbf{A}}(r^d - 1, j) < o(\sigma)$. Put $t = o(\sigma) - i$, so $t \equiv 0 \ (r^d - 1)$. Show that

$$v_{\overline{\wp}}[\zeta_{\mathbf{A}}(t)/\xi^t] = 0.$$

Note that $(\mathrm{Cl}\,(\wp)^+)^{(p)}(\omega_\sigma^t)$ corresponds to $\widetilde{z}(x, -i)$ as $i = o(\sigma) - t$. Thus if 8.22.6 is shown affirmatively, then we deduce information on the reductions of the roots of $\widetilde{z}(x, -i)$ by 8.21.2. Moreover, we also deduce that

$$(\mathrm{Cl}\,(\wp)^+)^{(p)}(\omega_\sigma^{-i})$$

is a cyclic W_σ-module (as, again, this component is then supported on the primes at infinity).

It, therefore, appears natural to ask what happens for the other **A**-magic numbers. We state this as a problem.

Problem 8.22.7. Let **A** be general and let

$$0 < i = i_{\mathbf{A}}(c, j) < o(\sigma)$$

where $c < r^d - 1$. Show that the arithmetic criterion for cyclicity holds for $\mathrm{Cl}\,(\wp)^{(p)}(\omega_\sigma^{-i})$.

Problem 8.22.7 has been checked affirmatively in some cases where $\mathbf{A} = \mathbb{F}_r[T]$ by computer. It is one analog for function fields of the Kummer-Vandiver Conjecture of classical cyclotomic theory. In Subsection 10.6 we will present another analog of a very different kind.

An affirmative solution to 8.22.7 would have the following nice corollary when the class number of **A** is a pure p-th power. Recall that, at the end of Subsection 8.16 we remarked that the arithmetic criterion for cyclicity may be described in terms of certain derivatives. We now make that precise when $L = \mathbf{k}$. Let i be as in 8.22.7 and let $\widetilde{z}(x, -i)$ be as before. Thus $\widetilde{z}(x, -i)$ has coefficients in **A**. Let $\overline{\wp}$ be the prime of **V** corresponding to σ; then the arithmetic criterion asserts that 1 is at most a simple root of $\widetilde{z}(x, -i)$ modulo $\overline{\wp}$. Set

$$\beta_{\mathbf{A}}(i) := \widetilde{z}(1, -i);$$

thus if $\bar{\wp} \nmid \beta_\mathbf{A}(i)$ there is nothing to establish as 1 is *not* a root. If $\bar{\wp} \mid \beta_\mathbf{A}(i)$, then $\bar{\wp}$ should *not* divide $\beta'_\mathbf{A}(i) := \frac{d}{dx}\tilde{z}(x, -i)\mid_{x=1}$. Set $\mathcal{O}_\mathbf{V} = \mathbf{A}$-integers of $\mathbf{V}$. Now if 8.22.7 is true for all $\mathbf{A}$-magic $i = i_\mathbf{A}(c, u)$, $c < r^d - 1$, and all $\bar{\wp}$, then the congruences of 8.22.2.3 imply that

$$\mathcal{O}_\mathbf{V} = (\beta_\mathbf{A}(i), \beta'_\mathbf{A}(i))$$

for *all* $\mathbf{A}$-magic $i = i_\mathbf{A}(c, j)$, $c < r^d - 1$. As above, for small r, $\wp$, i, this has been checked on a computer for $\mathbf{A} = \mathbb{F}_r[T]$. See also [Go4, §5.7].

For $\mathbf{A} = \mathbb{F}_r[T]$ there is yet another place where $\mathbf{A}$-magic numbers arise. All attempts to generalize this *directly* to other $\mathbf{A}$ (even when $\mathbf{A}$ is a principal ideal domain) have failed up to now. As such, we will be somewhat brief and refer the reader to [Go9], [Go10], [Go11], and [Go5] for details. However, it does appear that some sort of "calculus" is underlying the phenomenon that we will now discuss and that these techniques should ultimately be of great importance to the theory. As the reader will come to see, one obvious difficulty is the lack of a good "Haar measure" in finite characteristic (with of course values in finite characteristic!).

Thus let $\mathbf{A} = \mathbb{F}_r[T]$ and let v be a prime of $\mathbf{A}$ with completion $\mathbf{A}_v$. Recall that in Subsection 3.5 we discussed the polynomials

$$E_j(t) = \frac{e_j(t)}{D_j}$$

of Carlitz, and in 3.5.5, we showed that

$$E_j(a) \in \mathbf{A}$$

for all $a \in \mathbf{A}$. Thus, by continuity E_j maps $\mathbf{A}_v$ to itself and is, therefore, analogous to the classical functions $\binom{t}{k}$. *However*, by definition

$$e_C(t\log(x)) = \sum_{j=0}^{\infty} E_j(t) x^{r^j}.$$

whereas, classically,

$$e^{(t\log(1+x))} = (1+x)^t$$
$$= \sum_{k=0}^{\infty} \binom{t}{k} x^k.$$

In other words, there are big "gaps" in the definition of $E_j(t)$. The function $E_j(t)$ corresponds to x^{r^j} not to x^j. In order to find the "correct" coefficient of x^j, L. Carlitz made the following wonderful *digit expansion* construction which is basic for *all* the gamma functions of our next section, and which is still quite mysterious. Let i be *any* non-negative integer with

$$i = \sum_{e=0}^{m} c_e r^e, \qquad 0 \le c_e < r.$$

Definition 8.22.8. We set
$$G_i(t) := \prod_{e=0}^{m} E_e(t)^{c_e}.$$

Clearly $G_i(t)$ is also a continuous function from $\mathbf{A}_v$ to itself. It is easy to see that
$$\deg G_i(t) = i.$$
Thus any polynomial $p(t)$ of degree d can be written uniquely as
$$\sum_{i=0}^{d} c_i G_i(t).$$

In [C4], Carlitz develops a certain *integral formalism* (actually finite sums) for the coefficients $\{c_i\}$ in terms of $p(t)$. One thus sees easily the following result.

Lemma 8.22.9. *A polynomial $p(t) \in \mathrm{k}[t]$ takes $\mathbf{A}$ to itself if and only if $\{c_i\} \subseteq \mathbf{A}$.* □

Now let $\{c_i\}_{i=0}^{\infty} \subset \mathbf{A}_v$ with $c_i \to 0$ as $i \to \infty$. Clearly
$$f(t) = \sum_{i=0}^{\infty} c_i G_i(t) \qquad (8.22.10)$$
converges to a continuous function from $\mathbf{A}_v$ to itself. Using ideas as in Theorem 8.4.6, for example, one can prove conversely [Go9] that *all* continuous functions from $\mathbf{A}_v$ to itself can be expanded uniquely as in 8.22.10; a result originally established by C. Wagner [Wag1]. This is analogous, of course, to Theorem 8.4.6.

Definition 8.22.11. An $\mathbf{A}_v$-*valued measure* (or v-*adic measure*) on $\mathbf{A}_v$ is a finitely additive $\mathbf{A}_v$-valued function on the compact open subsets of $\mathbf{A}_v$.

Let μ be one such measure and let $f(t): \mathbf{A}_v \to \mathbf{A}_v$ be continuous. By using step-function approximations, one can form the integral $\int_{\mathbf{A}_v} f(t) \, d\mu(t)$. If $f(t) = \sum_{i=0}^{\infty} c_i G_i(t)$, then
$$\int_{\mathbf{A}_v} f(t) \, d\mu(t) = \sum_{i=0}^{\infty} c_i \left(\int_{\mathbf{A}_v} G_i(t) \, d\mu(t) \right).$$

Thus μ is uniquely determined by the collection $\{\int_{\mathbf{A}_v} G_i(t) \, d\mu(t)\} \subset \mathbf{A}_v$. Conversely, *any* such collection $\mu = \{b_i\}$ can easily be seen to define a *unique* v-adic measure.

The reader can now easily see the impossibility of finding a non-trivial Haar measure on $\mathbf{A}_v$ with values in finite characteristic.

Recall now the *algebra of divided power series*, $\mathbf{A}_v\{\{z\}\}$, consisting of all formal sums $\sum_{i=0}^{\infty} c_i \frac{z^i}{i!}$, $\{c_i\} \subset \mathbf{A}_v$, with, as usual,

$$\frac{z^i}{i!} \cdot \frac{z^j}{j!} = \binom{i+j}{i} \frac{z^{i+j}}{(i+j)!}.$$

As $\binom{i+j}{i} \in \mathbb{Z}$, this definition works in *any* characteristic.

To the measure $\mu = \{b_i\}$ one assigns the divided power series $f_\mu(z)$ with

$$f_\mu(z) := \sum_{i=0}^{\infty} b_i \frac{z^i}{i!}.$$

Remarkably, the mapping $\mu \mapsto f_\mu(z)$ is readily checked to be an *isomorphism* between the algebra of measures with the convolution product (in the usual sense), and the algebra of divided power series.

Recall that $z(x, -j)$ is defined by

$$z(x, -j) = \sum_{h=0}^{\infty} \left(\sum_{\substack{n \text{ monic} \\ \deg n = h}} n^j \right) x^{-h},$$

where the sum in parentheses vanishes for $h \gg 0$; thus

$$z(1, -j) = \zeta_{\mathbf{A}}(-j).$$

It is very simple to give a v-adic measure μ_x on $\mathbf{A}_v$ such that

$$\int_{\mathbf{A}_v} t^j \, d\mu_x(t) = z(x, -j), \qquad j \geq 0.$$

As $G_i(t)$ is a polynomial in t, and so one can compute $\int G_i(t)$ in terms of $\{\int t^j\}$, one sees that μ_x is unique. Finally, we have the wonderful computation by Thakur [Th3] of $f_{\mu_x}(z)$ which we present in our next result.

Theorem 8.22.12 (Thakur). *Let $f_{\mu_x}(z) = \sum_{i=0}^{\infty} m_x(i) \frac{z^i}{i!}$ be the divided power series to μ_x.*
1. *We have $m_x(i) \neq 0$ if and only if i is an $\mathbf{A}$-magic number.*
2. *We have $1 = m_x(0)$.*
3. *Let $i = i_{\mathbf{A}}(c, j)$ be a positive $\mathbf{A}$-magic number. Then we have*

$$m_x(i) = \begin{cases} (-1)^j x^{-j} & \text{if } c < r - 1 \\ (-1)^j (1 - x^{-1}) x^{-j} & \text{if } c = r - 1. \end{cases}$$

□

Remark. 8.22.13. Note that $f_{\mu_x}(z)$ is *independent* of v. Thus, presumably, it contains global information; see [Go10]. What sort of information is a *major* and intriguing problem. Recall that we do not have any sort of Haar measure (with finite characteristic values) at ∞ for the same reasons there is not one v-adically, so one cannot easily imitate classical theory!

Now let **A** be arbitrary and let $\wp \in \text{Spec}(\mathbf{A})$. By standard structure theorems, the completed ring $\mathbf{A}_\wp$ is isomorphic to $\mathbb{F}_r[T]_{\mathfrak{P}}$ for some $\mathfrak{P} \in \text{Spec}(\mathbb{F}_r[T])$. Thus we can always carry over the theory of measures *locally* from $\mathbb{F}_r[T]$ to arbitrary **A**. The problem is to find a *canonical* global set of functions (like $\{G_i(t)\}$) which everywhere gives the local theory on **A**. At this point all efforts towards this have failed.

One natural attempt towards finding analogs of the $G_i(t)$ is to just use arbitrary sign normalized rank one modules instead of the Carlitz module. Thus let sgn be a fixed sign function and let ψ be a sgn-normalized Hayes module, defined over the ring of **A**-integers $\mathcal{O}^+$ of $\mathbf{H}^+$, which we still assume has lattice $\xi \cdot \mathbf{A}$. Let $e(x)$ and $\log(x)$ be the exponential and logarithm function of ψ. Finally we define $E_j(t) := E_{\psi,j}(t)$ by $e(t\log(x)) = \sum_{j=0}^{\infty} E_j(t) x^{r^j}$, as in [AT1], [Th5].

Let $e(x) = \sum_{i=0}^{\infty} a_i x^{r^i}$ and $\log(x) = \sum_{i=0}^{\infty} b_i x^{r^i}$. Then one sees easily that

$$E_j(t) = \sum_{i=0}^{j} t^{r^i} a_i b_{j-i}^{r^i}.$$

Thus $\deg E_j(t) \leq j$. If e.g., $d_\infty = 1$, then there is equality (see Prop. 7.11.7.2). As ψ is defined over the integers $\mathcal{O}^+$ of $\mathbf{H}^+$, we see immediately that

$$E_j(t) \in \mathbf{H}^+[t].$$

Moreover, by definition

$$\psi_a(x) = \sum_{j=0}^{\infty} E_j(a) x^{r^j}.$$

We therefore see immediately our next result.

Proposition 8.22.14. *Let $a \in \mathbf{A}$. Then $E_j(a) \in \mathcal{O}^+$.* □

As the degree of $\psi_a(x)$ is $r^{\deg(a)}$, we deduce $E_j(a) = 0$ for $j > \deg(a)$. Thus $E_j(t)$ behaves much like in the case $\mathbf{A} = \mathbb{F}_r[T]$. However, even when **A** is a general principal ideal domain, we still do *not* get an integral basis for continuous functions as in the $\mathbb{F}_r[T]$-case.

Let us return finally to $\mathbf{A} = \mathbb{F}_r[T]$ for a short moment. Let $\alpha \in \mathbf{A}_v$; one then has the *Dirac measure* δ_α supported at α with

$$\int_{\mathbf{A}_v} f(t)\, d\delta_\alpha(t) = f(\alpha)$$

for all continuous $f(t)$. Let $f_\alpha(z) := \sum_{i=0}^{\infty} G_i(\alpha) \frac{z^i}{i!}$; so f_α is the divided power series associated to δ_α. It is quite remarkable that $f_\alpha(z)$ is also related to digit expansions in that it can be expanded in the following infinite product. Let $j \geq 0$ and set

$$f_{j,\alpha}(z) := \sum_{b=0}^{r-1} E_j^b(\alpha) \frac{z^{br^j}}{br^j!}.$$

Then the reader can check that Lucas' result on the p-adic residues of binomial coefficients implies that

$$f_\alpha(z) = \prod_j f_{j,\alpha}(z).$$

In any case, general Dirac formalism then implies that the mapping $\alpha \mapsto f_\alpha(z)$ satisfies an *exponential formalism*:

$$f_{\alpha+\beta}(z) = f_\alpha(z) \cdot f_\beta(z).$$

Therefore, it may be the case that for general $\mathbf{A}$ the connection with measures is not essential. Indeed, we may *always* form $G_{\psi,i}(t) = G_i(t)$ *exactly* as for $\mathbb{F}_r[T]$; i.e., if $i = \sum_{e=0}^{m} c_e r^e$, $0 \leq c_e < r$, then

$$G_{\psi,i}(t) := \prod_{e=0}^{m} E_{\psi,e}(t)^{c_e}.$$

Moreover, in [Go11] it is shown that the mapping

$$t \mapsto f_t(z) := \sum_{i=0}^{\infty} G_i(t) \frac{z^i}{i!}$$

also always satisfies the exponential formalism for arbitrary $\mathbf{A}$. Thus, keeping Subsection 8.19 in mind, we are lead to the following construction. Let $\{\alpha_1, \ldots, \alpha_{h^+}\}$ be representatives for $\mathcal{I}/\mathcal{P}^+$ with Artin symbols $\{\sigma_{\alpha_e}\}$. Set

$$f_t(z) := \sum_{e=1}^{h^+} \left(\frac{D\psi_{\alpha_e}}{\alpha_e} \right) \left(\sum_{i=0}^{\infty} G_i^{\sigma_{\alpha_e}}(t) \frac{z^i}{i!} \right),$$

where σ_{α_e} acts on the coefficients of $G_i(t)$, and finally

$$f_{\mu_x}(z) := \sum_{\substack{a \in A \\ a \text{ positive}}} x^{-\deg(a)} f_a(z).$$

Note that for all $a \in \mathbf{A}$, $f_a(z)$ is a divided polynomial. Moreover, if $\mathbf{A} = \mathbb{F}_r[T]$, it is easy to see that $f_{\mu_x}(z)$ is *precisely* the same divided power series as in 8.22.12. It thus may happen that $f_{\mu_x}(z)$ plays some important role in the theory for general $\mathbf{A}$ *and* that the actual computation of $f_{\mu_x}(z)$ à la 8.22.12 is of interest.

8.23. Finiteness in Local and Global Fields

It is well known that local fields come in three "sizes." Indeed, the "small" size consists of $\{\mathbb{R}, \mathbb{C}\}$ and is characterized as those local K with $[\overline{K}:K] < \infty$, where $\overline{K}$ will always denote an algebraic closure of K. The "medium" size local fields are those K such that K is finite over $\mathbb{Q}_p$. Thus medium size fields are characterized as those local K such that $[\overline{K}:K] = \infty$, *but K has only finitely many extensions of bounded degree* (it is well known that finite extensions of $\mathbb{Q}_p$ have this property; below, in 8.23.1, we show that local fields in finite characteristic do *not*). Finally the "large" size fields are those local K with finite characteristic; thus $K \simeq F((t))$ for some finite F. A local field is thus large if and only if $[\overline{K}:K] = \infty$ *and* there may be infinitely many extensions of bounded degree. There are many ways to see that $F((t))$ has this property, and we present one in 8.23.1. We choose this method because of its importance in Subsection 8.24.

Example 8.23.1. Let $F \simeq \mathbb{F}_r$, $r = p^{m_0}$, $K = \mathbf{K} = \mathbb{F}_r((t))$, $t = \frac{1}{T}$. Let $\{S_\infty, \mathbf{C}_\infty\}$ be as in Subsection 8.1. We define

$$f(s): S_\infty \to \mathbf{C}_\infty$$

by

$$f(s) = f(x, y) := x^{-p} + \left(\left(1 + \frac{1}{T}\right)^y - 1\right) x^{-1} - \frac{1}{T}.$$

It is trivial to see that $f(s)$ is entire in S_∞. Let $y \in \mathbb{Z}_p$ be fixed; then $f(x, y)$ is an *Eisenstein polynomial* in x^{-1} for the t-valuation and so irreducible. Moreover, for $y \neq 0$, $f(x, y)$ is separable. Let $\mathbf{K}_y \subseteq \mathbf{C}_\infty$ be the field obtained by adjoining the roots in x of $f(x,y)$ to $\mathbf{K}$. Let $\mathcal{F}$ be the compositum of all $\mathbf{K}_y$ for $y \neq 0$; thus $\mathcal{F}/\mathbf{K}$ is separable. *If there are only finitely many extensions of bounded degree*, then L would be finite over $\mathbf{K}$, thus closed in the canonical topology. Continuity then implies that

$$\mathbf{K}_0 \subseteq \mathcal{F}$$

also. But $\mathbf{K}_0 = \mathbf{K}(T^{1/p})$ and is totally inseparable. Thus we have a contradiction.

Let $B \subset \mathbb{Z}_p - \{0\}$ be compact. One may use *Krasner's Lemma* to show that the compositum of all $\mathbf{K}(y)$, $y \in B$, is of finite degree. One can not do so for all $y \in \mathbb{Z}_p$, precisely because $f(x,0)$ is *not* separable.

Finiteness for $F((t))$ may be restored as in our next result.

Proposition 8.23.2. *Let F be a finite field and let $K \simeq F((t))$.*
1. *K has only finitely many totally inseparable extensions of bounded degree.*
2. *K has only finitely many unramified extensions of bounded degree.*
3. *K has only finitely many separable extensions of bounded degree and bounded discriminant.*

Proof. Part 1 follows from noting that
$$K^{1/p^j} \simeq F((t^{1/p^j})).$$
Part 2 is standard. For Part 3, note that Part 2 reduces us immediately to where the extension K_1/K is totally ramified. The result then follows from [Se6]. □

The following statement works for *all* non-Archimedean local fields.

Proposition 8.23.3. *Let K be a local non-Archimedean field with integers $\mathcal{O}_K$. Let $\mathfrak{d}$ be a non-trivial ideal of $\mathcal{O}_K$. Then K has only finitely many extensions with discriminant $\mathfrak{d}$.*

Proof. Let L be an extension with integers $\mathcal{O}_L$ and different $\mathfrak{D}$; so $\mathfrak{d}$ is the norm of $\mathfrak{D}$. Suppose now that M_K is the maximal ideal in $\mathcal{O}_K$ with
$$\mathfrak{d} = M_K^t, \qquad t > 0,$$
and let e, f be the standard ramification and residual degree numbers for L. If $\mathfrak{D} \simeq M_L^a$, then we find
$$\mathfrak{d} = M_K^{af}.$$
Thus f must be bounded. As there are only finitely many unramified extensions of bounded degree, we are reduced to having L/K totally ramified. In this case we find
$$\mathcal{O}_L = \mathcal{O}_K[\pi_L]$$
where π_L is a uniformizer in L and satisfies an Eisenstein polynomial $f(x)$ of degree e. Moreover, $\mathfrak{D} = M_L^t$ if $\mathfrak{d} = M_K^t$. We find that
$$\mathfrak{D} = (f'(\pi_L))$$
and that $f'(\pi_L) \equiv 0 \ (\pi_L^{e-1})$. Thus t bounds $e-1$ from above and we are reduced to having both bounded discriminant *and* degree in which case we are finished. □

We now turn our attention to global fields. Of course there are two kinds, number fields and finitely generated function fields of transcendence degree 1 over a finite field. These are obviously distinguished by their characteristic.

Let K be a number field. The following statement contains two well-known and basic finiteness results.

Theorem 8.23.4. 1. *Only a finite collection of number fields can have the same discriminant.*
2. *Let S be a finite set of finite places of K. Then there are only finitely many extensions of K which are unramified outside of S and which have bounded degree.* □

Let K now be a function field as above. The constant field extensions immediately give a counterexample to 8.23.4.1 for K. To see that 8.23.4.2 is also false for K we can, for instance, use the extension $\mathbf{k}(\wp^n)$, $n \geq 0$, for $\mathbf{k} = \mathbb{F}_r(T)$ obtained by adjoining the $\wp^n$-th division points of the Carlitz module. We have seen that $\mathrm{Gal}(\mathbf{k}(\wp^n)/\mathbf{k}) \simeq (\mathbf{A}/\wp^n)^*$; so if $\mathbf{k}(\wp^\infty) = \cup \mathbf{k}(\wp^n)$, then $\mathrm{Gal}(\mathbf{k}(\wp^\infty)/\mathbf{k}) \simeq \mathbf{A}_\wp^*$, where $\mathbf{A}_\wp$ is the completion. Note that $\mathbf{k}(\wp^\infty)/\mathbf{k}$ is ramified only at $\wp$ and ∞. Finally

$$\mathbf{A}_\wp^* \simeq (\mathbf{A}/\wp)^* \times U_1$$

where U_1 is the group of 1-units. It is well-known that U_1 is topologically an infinite product of $\mathbb{Z}_p$'s. As such we can construct infinitely many extensions of $\mathbf{k}$ which have degree p and are unramified outside $\{\wp, \infty\}$. Finiteness for global function fields of characteristic p may be restored as in our next results.

Theorem 8.23.5. *Let K be a finitely generated function field of transcendency one over a finite field F.*
1. *K has only finitely many totally inseparable extensions of bounded degree.*
2. *K has only finitely many constant field extensions of bounded degree.*
3. *K has only finitely many separable extensions of bounded degree and discriminant.*

Proof. Part 1 follows immediately from 8.2.12. Part 2 is obvious. To see Part 3 we use the following nice argument (told to us by Y. Taguchi). First of all, by Part 2, we are reduced to looking at geometric extensions L/K of bounded degree and discriminant (so L and K have the same fields of constants by definition). Next observe that there are only finitely many isomorphism classes of curves over a finite field of bounded genus. Indeed [Mu3, §7.4] the moduli space of proper smooth curves of fixed genus over $\mathbb{Z}/(p)$ (perhaps equipped with a level n structure on their Jacobians) is of finite type over $\mathbb{Z}/(p)$. For $n \gg 0$ this space is a fine moduli scheme. Now given a curve C/F, one finds a level n structure over a finite extension F_1 of F of bounded degree. Thus, as our varieties are of finite type over F_1, there are only finitely many isomorphism classes of curves with level n structure. Now if the genus is > 1 then such a curve has only finitely many (absolute) automorphisms. Thus as the forms of a curve C will be measured by $H^1(\mathrm{Gal}(F_1/F), \mathrm{Aut}(C))$ the result follows. If the genus is 1 or 0, one argues directly with Weierstrass models or quadrics. (Alternatively, one can use the classical Riemann hypothesis results of Weil to show that there is exists a rational point in an extension of bounded degree and then use Galois descent again.)

Let L/K be a geometric extension. Hurwitz's formula then gives the genus of L in terms of $[L:K]$ and the degree of the different. However, if we bound $[L:K]$ and the discriminant we also bound the different and thus the genus of L.

Part 3 will now follow if we show that for fixed curves X and Y over F, there are only finitely many morphisms

$$f: X \to Y$$

of fixed degree. One reduces to having $Y \simeq \mathbb{P}^1/F$. Now if we fix the degree then there are only finitely many possible polar and zero divisors of f (as, again, this result is clearly true for $\mathbb{P}^1/F$ which is finitely covered by every curve). The result now follows.

Finally we have the following.

Proposition 8.23.6. *Let K be as in 8.23.5. Let δ be a fixed non-trivial effective divisor on the curve associated to K. Then K has only finitely many Galois extensions L of discriminant divisor equal to δ.*

Proof. Let $\wp$ be a place occurring non-trivially in δ as $c(\wp)$, $c > 0$. Let Δ be the different divisor of L. As L is Galois, *all* the places above $\wp$ occur in Δ with the same multiplicity, say c_0. Let e be the ramification at $\wp$, f the residue degree and r the splitting number so that $[L:K] = efr$. Moreover, $c = c_0 fr$, so f and r are bounded.

The result will follows from 8.23.5.3 if we can also bound e. However, this follows immediately from 8.23.3. □

8.24. Towards a Theory of the Zeroes

In this subsection we will describe some results that *appear* to be a sort of Riemann hypothesis for $\zeta_{\mathbf{A}}(s)$, $\mathbf{A} = \mathbb{F}_r[T]$. However, in order to place these ideas in the proper perspective, we must first discuss some classical theory.

Recall that in Subsection 8.19 we defined $\xi(s)$ as

$$\xi(s) := \Gamma(s/2)\pi^{-s/2}\zeta(s),$$

and $\hat{\xi}(s)$ as

$$\hat{\xi}(s) := s(1-s)\xi(s).$$

Thus $\hat{\xi}(s)$ is *entire* and satisfies the functional equation

$$\hat{\xi}(s) = \hat{\xi}(1-s).$$

Of course, the classical Riemann hypothesis (*Riemann's* Riemann hypothesis!) is that the zeroes of $\hat{\xi}(s)$ are of the form $\frac{1}{2} + i\beta$, $\beta \in \mathbb{R}$. This may be reformulated in the following well-known fashion. First of all, set $t := s - 1/2$ and

$$w(t) := \hat{\xi}(s) = \hat{\xi}(t + 1/2).$$

As $1 - s = 1/2 - t$, we see

$$w(t) = \hat{\xi}(s) = \hat{\xi}(1-s) = \hat{\xi}(-t + 1/2) = w(-t).$$

Upon expanding $w(t)$ about $t = 0$, we find

$$w(t) = \sum_{j=0}^{\infty} e_j t^j$$

with $\{e_j\} \subset \mathbb{R}$ and $e_j = 0$ for j odd. Thus the Riemann hypothesis now becomes the statement that the zeroes of $w(t)$ are of the form $i\beta$, $\beta \in \mathbb{R}$. Finally we put $u := it$, and $\theta(u) := w(t) = w(u/i) = w(-iu)$. Thus

$$\theta(u) = \sum_{j=0}^{\infty} e_{2j}(-1)^j u^{2j}.$$

Clearly $\{(-1)^j e_{2j}\} \subset \mathbb{R} = \mathbb{Q}_\infty$ and the Riemann hypothesis is now the statement that the zeroes of $\theta(u)$ are *real*. Similar statements can be made, e.g., for Dedekind zeta functions of arbitrary number fields. We shall see a little later that this formulation of the Riemann hypothesis will be echoed in the theory of $\zeta_A(s)$.

Next we need to discuss some of the theory of Artin L-functions for number fields. Let $\overline{\mathbb{Q}}$ be a fixed algebraic closure of the rational field $\mathbb{Q}$. Let $k/\mathbb{Q}$ be a finite extension lying in $\overline{\mathbb{Q}}$ and let $G := \text{Gal}(\overline{\mathbb{Q}}/k)$. Let H be a subgroup of finite index in G and let $\rho \colon H \to \text{Aut}(V)$ be a representation of Galois type of H, where V is now a finite dimensional complex vector space (N.B.: for representations of Galois type, one may assume, without loss of generality, that V is finite dimensional over *any* algebraically closed field of characteristic 0). Let $\text{ind}(\rho)$ be the induction of ρ from H to G. Recall that the formalism of L-series implies that $L(\rho, s) = L(\text{ind}(\rho), s)$. Recall also that representations into fields of characteristic 0 are determined by their characters; as such there is no ambiguity in denoting $L(\rho, s)$ by $L(\chi, s)$ where χ is the character of ρ, etc. The classical *Artin Conjecture* then asserts that $L(\chi, s)$ is *entire* when χ is irreducible of degree > 1.

We are interested, in particular, when one can reasonably expect that the zeroes of $L(\chi,s)$ are *simple*. The first observation along these lines is that one should restrict to having $k = \mathbb{Q}$. Indeed, suppose, for instance, that k was itself Galois over $\mathbb{Q}$ and set $G_1 := \text{Gal}(k/\mathbb{Q})$. Let $G \subset \text{Gal}(\overline{\mathbb{Q}}/\mathbb{Q})$ be the normal subgroup fixing k and suppose also that G_1 is *not* abelian. Then there is an irreducible character χ of G_1 with degree $d := d_\chi > 1$. Let 1_G be the trivial character on G and let $\text{ind}(1_G)$ be its induction to $\text{Gal}(\overline{\mathbb{Q}}/\mathbb{Q})$; of course $L(\text{ind}(1_G), s) = \zeta_k(s)$ the Dedekind zeta function of k. The basic theory of group representations implies that χ occurs in $\text{ind}(1_G)$ with multiplicity d. Thus the Artin Conjecture implies that *any* zero of $L(\chi, s)$ will have multiplicity at least d in $\zeta_k(s)$.

Note that it is always possible to apply the theory of L-series over $\mathbb{Q}$ to the theory of L-series over arbitrary fields by simply inducing the character of a given representation to $\text{Gal}(\overline{\mathbb{Q}}/\mathbb{Q})$. One can then reduce to considering only L-series of irreducible characters because every character has a unique expansion over the irreducible ones.

We now present a previously unpublished conjecture of J.-P. Serre related to the simplicity of zeroes; we thank him for his kind permission to quote it. In it, we shall say that a zero s of an L-series is "non-trivial" if and only if $\Re(s) > 0$.

Conjecture 8.24.1 Let $G := \text{Gal}(\overline{\mathbb{Q}}/\mathbb{Q})$ and let χ be an irreducible character of G.
1. All the non-trivial zeroes of $L(\chi, s)$ should be simple.
2. $s = 1/2$ should be a zero of $L(\chi, s)$ only if χ is real and the functional equation of $L(\chi, s)$ has a minus sign.
3. If χ and χ' are distinct, then the zeroes of $L(\chi, s)$ not equal to $s = 1/2$ should be distinct from those of $L(\chi', s)$.

It is known that if χ is the real character corresponding to an orthogonal representation, then the constant in the functional equation is $+1$; if the representation is symplectic, then the constant can be either 1 or -1. Moreover, there is not much evidence for the part of the conjecture related to $s = 1/2$; thus perhaps this part of 8.24.1 may be viewed more as a guide than a conjecture. In any case, it would be interesting to have computer experiments carried out in order to have more evidence one way or the other. (The reader should also see [RS1, p. 195] for more along these lines.)

We return now to the function field setting. We set $\mathbf{A} = \mathbb{F}_r[T]$, $\mathbf{k} = \mathbb{F}_r(T)$, etc.

Recall that the holomorphy of $\zeta_\mathbf{A}(s)$, $s \in S_\infty$, was established through the basic estimates of Lemma 8.8.1. Our next example shows that these estimates are sometimes *far* from the best possible.

Example 8.24.2. Let $s = (x, y) \in S_\infty$ and set $y = 1$. Thus

$$\zeta_{\mathbf{A}}(x,1) = \sum_{j=0}^{\infty} x^{-j} \left(\sum_{\substack{n \text{ monic} \\ \deg n = j}} \langle n \rangle^{-1} \right);$$

set

$$\widetilde{s}_j := \sum_{\substack{n \text{ monic} \\ \deg n = j}} \langle n \rangle^{-1} = \pi^{-j} s_j,$$

where

$$s_j := \sum_{\substack{n \text{ monic} \\ \deg n = j}} n^{-1},$$

and π is our fixed positive uniformizer (e.g., $\pi = \frac{1}{T}$). Thus

$$v_\infty(\widetilde{s}_j) = v_\infty(s_j) - j,$$

and we may compute $v_\infty(\widetilde{s}_j)$ via $v_\infty(s_j)$. To find $v_\infty(s_j)$ we now establish a nice closed form expression for s_j as in [C1]. Recall that we set

$$e_j(x) := \prod_{\substack{\alpha \in \mathbf{A} \\ \deg(\alpha) < j}} (x - \alpha),$$

and found, in the notation of Subsection 3.1, that

$$e_j(x) = \sum_{i=0}^{j} (-1)^{j-i} x^{r^i} \frac{D_j}{D_i L_{j-i}^{r^i}}.$$

As $e_j(x)$ is $\mathbb{F}_r$-linear, we see

$$\frac{e_j'(x)}{e_j(x)} = \frac{(-1)^j D_j}{L_j} \frac{1}{e_j(x)} = \sum_{\deg(\alpha) < j} \frac{1}{x - \alpha}.$$

As $e_j(T^j) = D_j$, we see that

$$\frac{(-1)^j D_j}{L_j} \frac{1}{e_j(x + T^j)} = \frac{(-1)^j D_j}{L_j} \frac{1}{e_j(x) + D_j}$$
$$= \sum_{\substack{n \text{ monic} \\ \deg n = j}} (x + n)^{-1}. \qquad (8.24.3)$$

Thus, upon setting $x = 0$, we deduce

$$\frac{(-1)^j}{L_j} = \sum_{\substack{n \text{ monic} \\ \deg n = j}} \frac{1}{n} = s_j. \qquad (8.24.4)$$

As $\deg(L_j) = r^j + r^{j-1} + \cdots + r = \frac{r(r^j - 1)}{(r-1)}$, we see that

$$v_\infty(s_j) = \frac{r(r^j - 1)}{r - 1}.$$

Thus $v_\infty(s_j)$ and $v_\infty(\tilde{s}_j) \to \infty$ *exponentially* as opposed to quadratically as given by 8.8.1!

As Daqing Wan pointed out, this calculation has other more interesting consequences. Recall from Section 2 that the Newton polygon of $\zeta_\mathbf{A}(x,1)$ (as a power series x^{-1}) is the lower convex hull of the points

$$\{(j, v_\infty(\tilde{s}_j))\} = \left\{\left(j, \frac{r(r^j - 1)}{r - 1} - j\right)\right\}.$$

As
$$\frac{r(r^j - 1)}{r - 1} - j - \left(\frac{r(r^{j-1} - 1)}{r - 1} - (j - 1)\right) = r^j - 1,$$

we deduce that each side of the Newton polygon has projection onto the x-axis of length 1. Thus for each slope there is *one* root of $\zeta_\mathbf{A}(x,1)$ associated to it. Moreover, this root *must* be in $\mathbf{K}$ and be simple! As $\{\tilde{s}_j\} \subset \mathbf{K} = \mathbf{k}_\infty$, the analogy with the classical theory, and conjecture, as presented above is quite clear!

Inspired by this, Wan attempted to calculate the Newton polygon of $\zeta_\mathbf{A}(x,y)$ for all y. We state his conclusions [Wan1] next. The reader will see that they are in accordance with the theory presented at the beginning of this section as well as Conjecture 8.24.1.1. At the end of this section, we will present a possible generalization of 8.24.5 that seems to have a chance of being true for all $\zeta_{\mathcal{O}_L}(s)$ where L is a finite extension of $\mathbf{k}$ and $\mathcal{O}_L$ is its ring of $\mathbf{A}$-integers.

Theorem 8.24.5 (Wan). *Let $y \in \mathbb{Z}_p$ have the property that each r-adic digit of $-y$ is less than p or equal to $r - 1$. Then the zeroes of $\zeta_\mathbf{A}(x,y)$ are in $\mathbf{K}$ and are simple.*

Corollary 8.24.6. *Let $r = p$. Then all zeroes of $\zeta_\mathbf{A}(s)$ are in $\mathbf{K}$ and are simple.* □

Note that if all r-adic digits of a p-adic integer are $< p$, then the r-adic expansion *is* the p-adic expansion.

We now present some results leading towards a proof of 8.24.5 *under the simpler assumption that* all r-adic digits of $-y$ are $< p$. We follow the elegant exposition of Diaz-Vargas [D-V1] and refer the reader to [Wan1] for the general result. First of all, it is clear in this very simple case that without loss of generality we may take $\pi = 1/T$; the general dependence upon uniformizers will be discussed later on in this subsection. Now let $j \geq 0$ and $i \in \mathbb{Z}$. Set

$$s_j(i) := S_j(i) := \sum_{\deg n = j} n^i;$$

so
$$s_j(-1) = S_j(-1) = s_j.$$
(We use both $s_j(i)$ and $S_j(i)$ to be consistent with the notation of [D-V1].)

Definition 8.24.7. Let i be a positive integer. We write
$$i = m_0 \oplus \cdots \oplus m_j$$
if and only if
1. for $t > 0$, we have $m_t > 0$ and $m_t \equiv 0 \ (r-1)$;
2. $i = m_0 + \cdots + m_j$;
3. in the above sum there is *no* carry over of p-adic digits.

Proposition 8.24.8. *Let $i > 0$. Then*
$$S_j(i) \neq 0.$$
only if there exists $m_1, \ldots, m_j$ such that
$$i = m_1 \oplus \cdots \oplus m_j.$$

Proof. Let $n = T^j + a_1 T^{j-1} + \cdots + a_j$, $\{a_t\} \subseteq \mathbb{F}_r$; thus
$$n^i = \sum_{i=m_0+\cdots+m_j} \frac{i!}{m_0!\cdots m_j!} (T^j)^{m_0} (a_1 T^{j-1})^{m_1} \cdots a_j^{m_j}.$$
By the well-known result of Lucas,
$$\frac{i!}{m_0! m_1! \cdots m_j!} \neq 0 \ (p)$$
if and only if $i = m_0 + \cdots + m_j$ and there is no carry over of p-adic digits. If such an expression is impossible, we deduce that
$$S_j(i) = 0.$$

On the other hand,
$$S_j(i) = \sum_{a_1,\cdots,a_j \in \mathbb{F}_r} \sum_{i=m_0+\cdots+m_j} \frac{i!}{m_0!\cdots m_j!} a_1^{m_1} \cdots a_j^{m_j} T^{jm_0+\cdots+m_{j-1}}.$$
Note that for $h \geq 0$
$$\sum_{a \in \mathbb{F}_r} a^h = 0 \iff (r-1) \nmid h \text{ or } h = 0.$$
Therefore, $S_j(i) \neq 0$ implies that i can be written
$$i = m_0 \oplus \cdots \oplus m_j$$
and the proof is finished. □

Proposition 8.24.9. *Let i be a positive integer with the property that each r-adic digit of i is less than p. Then $S_j(i) \neq 0$ if we can find $m_0, \ldots, m_j$ such that*

$$i = m_0 \oplus \cdots \oplus m_j.$$

Proof. Suppose $i = m_0 \oplus \cdots \oplus m_j$. We may clearly choose $\{m_t\}$ in such a way as to maximize the degree $jm_0 + \cdots + m_{j-1}$. We will thus be finished if we show that this maximum is unique.

Note that

$$\max\{jm_0 + \cdots + m_{j-1}\} = ij - \min\{m_1 + \cdots + jm_j\}.$$

Thus to find the maximum on the left hand side, it is clearly enough to minimize the *weight* $w := m_1 + \cdots + jm_j$ subject to $i = m_0 \oplus m_1 \oplus \cdots \oplus m_j$. Of *all* such decompositions, we first choose those with minimum m_j, then the ones with minimum m_{j-1} and so on. We end up with a solution called the *greedy solution*. The result clearly follows if we show that any *non*-greedy solution $\{m'_t\}$ with $i = m'_0 \oplus \cdots \oplus m'_j$ can be modified in such a fashion as to obtain another solution of smaller weight *and* such that we ultimately end up at the greedy solution.

Without loss of generality, we can assume that

$$m'_j \leq m'_{j-1} \leq \cdots \leq m'_1.$$

As $(r-1)$ divides m'_t for $t > 0$ by definition, we have that each m'_t, $t > 0$ is a sum of $\alpha(r-1)$ powers of r (the sum of the r-adic digits of m'_t is $\alpha(r-1)$). If $\alpha > 1$, then, as each r-adic digit of i is $< p$, we may move $(\alpha-1)(r-1)$ of these r-th powers into m'_0 and decrease the weight. Thus we may assume that $\alpha = 1$. (The reader should note that this step is *not* valid without our assumptions on the digits of i.)

Let t_0 be the largest index such that $m'_{t_0} \neq m_{t_0}$. By construction

$$m'_{t_0} > m_{t_0}.$$

If

$$m_{t_0} = \sum_{e=1}^{r-1} r^{i_e}, \qquad i_e \leq i_{e+1},$$

and

$$m'_{t_0} = \sum_{e=1}^{r-1} r^{i'_e}, \qquad i'_e \leq i'_{e+1},$$

then choose the least e such that

$$a := i_e < b := i'_e.$$

Thus, for some $t_1 < t_0$, r^a must occur in m'_{t_1}.

Now define $\{\widetilde{m}_k\}$, $k = 0, \ldots, j$ by

$$\tilde{m}_k := m'_k \qquad \text{for } k \neq t_0, t_1,$$
$$\tilde{m}_{t_0} := m'_{t_0} - r^b + r^a$$
$$\tilde{m}_{t_1} := m'_{t_1} - r^a + r^b.$$

Note that $\tilde{m}_t$ is still $\equiv 0$ $(r-1)$ for all t. But it has weight $w' - (t_0 - t_1)(r^b - r^a)$ where w' is the weight of $\{m'_i\}$. Moreover, this process clearly converges to the greedy solution. □

Remarks. 8.24.10. 1. Results 8.24.8 and 8.24.9 are due to Carlitz [C3], where they (e.g., 8.24.9) are claimed for all i; however, as of this writing, nobody has been able to substantiate this. Carlitz works directly with the sum $dm_0 + \cdots + m_{d-1}$ and claims that the maximum occurs uniquely when one first maximizes m_0 (subject to $m_0 \oplus \cdots \oplus m_d = i$), then m_1, etc. The method we use of finding the greedy solution does not work in general to give the maximum of $dm_0 + \cdots + m_{d-1}$; moreover, there are no known counterexamples to Carlitz's method. Clearly this is an area deserving of attention.

2. If there is no carry over modulo p, there is certainly none modulo r. However, the difficulty arises in carrying out Diaz-Vargas' method for arbitrary r in that preserving congruences modulo $(r-1)$ *may* lead to carry over of p-adic digits when $r \neq p$. Moreover, it is necessary for this method to have the r-adic digits $< p$. Indeed, even though a number i may have *each* r-adic digit $< p$ or equal to $r - 1$, there may be decompositions (8.24.7) where this is *not* true. (I am indebted to Diaz-Vargas for showing me such an example.)

3. Let i be as in 8.24.9. As Thakur points out, if $\{m_t\}$ is the greedy solution, then
$$\deg S_j(i) = jm_0 + \cdots + m_{j-1}.$$

Let i be as in 8.24.9. Then, of course,
$$\zeta_A(x, -i) = \sum_{j=0}^{\infty} x^{-j} \left(\sum_{\substack{n \text{ monic} \\ \deg n = j}} \langle n \rangle^i \right)$$
$$= \sum_{j=0}^{\infty} x^{-j} (\pi^{ji} S_j(i)).$$

Define $v_j(i) := v_\infty \left(\sum_{\substack{n \text{ monic} \\ \deg n = j}} \langle n \rangle^i \right)$. Thus
$$v_j(i) = ji - \deg S_j(i)$$
$$= m_1 + \cdots + jm_j$$

where $\{m_t\}$ is the greedy solution (cf. 8.24.10.3). Thus $v_j(i)$ is precisely the *weight* of the greedy solution (as in the proof of 8.24.9).

In order to understand the Newton polygon of $\zeta_{\mathbf{A}}(x,-i)$ we need to understand the greedy solutions where the degree j varies over all non-zero integers. Thus we let $\{m_t(j);\, t=0,\ldots,j\}$ be the greedy solution associated to j and with our fixed i.

Lemma 8.24.11. *Let i be as in 8.24.9. Then all the roots of $\zeta_{\mathbf{A}}(x,-i)$ are in $\mathbf{K}$ and are simple.*

Proof. Let $\sum_{e=0}^{m} c_e r^e$, be the r-adic expansion of i. Then the greedy solution may be constructed in the following fashion:
$$m_j(j) = c_0 + \cdots + c_\alpha^* r^\alpha,$$
where $0 \le c_\alpha^* \le c_\alpha$ and $c_0 + \cdots + c_\alpha^* = r - 1$. Next, $m_{j-1}(j)$ is obtained in the same way from the r-adic expansion of $i - m_j(j)$, etc., until
$$m_0(j) = i - (m_j(j) + \cdots + m_1(j)).$$
The reader should note that this construction precisely uses our assumptions on the r-adic digits of i. If such a construction is not possible, then $S_j(i) = 0$ and $v_j(i) = \infty$ by 8.24.8. The important point is to notice that by our construction
$$m_{j-t}(j) = m_{j-t-1}(j-1)$$
for $t = 0, \ldots, j-2$.

Now the Newton polygon is the lower convex hull of $\{(j, v_j(i))\}$. As $v_j(i) = m_1(j) + \cdots + jm_j(j)$ and $m_{j-t}(j) = m_{j-t+1}(j-1)$, for $t = 0, \ldots, j-2$, we see that the slope $\lambda_j(i)$ from $(j-1, v_{j-1}(i))$ to $(j, v_j(i))$ equals
$$m_1(j) + m_1(j-1) + \cdots + m_{j-1}(j-1) = m_1(j) + m_2(j) + \cdots + m_j(j).$$
Consequently
$$\lambda_j(i) - \lambda_{j-1}(i) = m_1(j) > 0$$
for $j > 0$. Thus the projections onto the X-axis of the Newton polygon have length 1 and the proof is finished. $\square$

Note that the set of positive integers with the property that each r-adic digit is $< p$ is *dense* in all such p-adic integers. Thus continuity of the roots implies the first part of 8.24.5 under our simplifying assumption. We now finish the proof of all of 8.24.5 under this assumption.

Proof of 8.24.5. By assumption $-y$ has the property that each of its r-adic digits is $< p$. If $-y$ is a non-negative integer, then the result is precisely 8.24.11. Thus we may assume that $-y$ is not a non-negative integer, in which case its r-adic expansion is infinite; say

$$-y = \sum_{e=0}^{\infty} c_e r^e, \qquad 0 \le c_e < p.$$

Let $j \ge 0$ and choose $k > 0$ so that

$$i := i_k = \sum_{e=0}^{k} c_e r^e$$

has enough digits to determine a greedy solution $\{m_t(j)\}$ with $i = m_0(j) \oplus \cdots \oplus m_j(j)$. Note that if k works for j, it works for $j-1$ also. Note also that if k works for j, then so does $k+1$.

Now

$$\lim_{k \to \infty} v_j(i_k) = v_\infty \left(\sum_{\substack{\deg n = j \\ n \text{ monic}}} \langle n \rangle^{-y} \right)$$

by continuity. As k increases, only $m_0(j)$ increases by construction; but $m_0(j)$ does *not* appear in $v_j(i_k)$. Thus if k works for j, we see that

$$v_j(i_k) = v_j(i_{k+1}) = \cdots = v_\infty \left(\sum_{\substack{\deg n = j \\ n \text{ monic}}} \langle n \rangle^{-y} \right).$$

As such, all the results that we concluded when $-y$ is a positive integer *automatically* work when $-y$ is *general* (subject to the constraint on its r-adic digits) and the result is established. □

Note also that if all the r-adic digits of $-y$ equal $r-1 \implies -y = -1$ and $y = 1$. Thus the result in this case also follows by the discussion before the statement of the theorem.

Remarks. 8.24.12. 1. Suppose $r = p$. It should be relatively easy to compute the *distribution of zeros* of $\zeta_A(s)$ from the theorem and its proof. It would be interesting to compare these results to the classical theory of Riemann's zeta function.

2. The proof of 8.24.5 shows the following. Let $-y = \sum_{e=0}^{\infty} c_e r^e$, $0 \le c_e < p$ and $c_e \ne 0$ infinitely often. Let

$$\zeta_A(x, y) = \sum_{j=0}^{\infty} x^{-j} \tilde{s}_j(-y).$$

Then $\tilde{s}_j(-y) \ne 0$ for $j \ge 0$.

Let $y \in \mathbb{Z}_p$ and let r be arbitrary. As above, write

$$\zeta_{\mathbf{A}}(x,y) = \sum_{j=0}^{\infty} x^{-j}\tilde{s}_j(-y).$$

If it happens that $\tilde{s}_{j-1}(-y) = 0$ but $\tilde{s}_j(-y) \neq 0$, then the Newton polygon of $\zeta_{\mathbf{A}}(x,y)$ will have sides with projection down to the x-axis of length > 1. Thus we can no longer conclude simplicity or the fact that the zeroes are in $\mathbf{K}$ from the Newton polygon alone. In this regard, the following elegant lemma of H. Lee [Lee1] is somewhat reassuring.

Lemma 8.24.13. *Let $i > 0$. Then*

$$\tilde{s}_j(-i) = \sum_{\substack{\deg n = j \\ n \text{ monic}}} \langle n \rangle^{-i}$$

is non-zero for all j.

Proof. We are immediately reduced to showing

$$s_j(-i) = \sum_{\substack{\deg n = j \\ n \text{ monic}}} n^{-i}$$

is non-zero. Let L_j be as in 3.1.4.3; by 3.1.6.4 L_j is the l.c.m. of all monics of degree j.

Thus

$$\alpha_j^{(i)} := L_j^i s_j(-i)$$

is a sum of over elements in $\mathbf{A}$. Let f be a monic prime of degree j. Then one sees easily that

$$\alpha_j^{(i)} \equiv L_j^i/f^i \pmod{f}.$$

But $L_j^i/f^i \not\equiv 0 \pmod{f}$ and the result follows. □

Let $f(u) = 1 + \sum_{j=1}^{\infty} c_j u^j$ be an entire power series with the property that every side of the Newton Polygon projects to the X-axis onto a segment of unit length. We will then say that the Newton polygon of $f(u)$ is *simple*.

Let $y \in \mathbb{Z}_p$. If the Newton polygon of $\zeta_{\mathbf{A}}(x,y)$ (as a function of x^{-1}) is simple, then we say that y is *simple*. It turns out that there are other instances of simple y besides those given in Wan's result. We discuss these briefly now.

Lemma 8.24.14. *Let y be simple and $t \geq 0$. Then $p^t y$ is also simple.*

Proof. Just use the p^t-th power mapping. □

Corollary 8.24.15. *Let y be as in* Theorem 8.24.5. *Then $p^t y$ is simple.* □

8.24. Towards a Theory of the Zeroes 333

Proposition 8.24.16. *Let $0 < m \leq r$. Then $p^t m$ is simple.*

Proof. We are now in the situation of Example 8.24.2. Recall that we showed in 8.24.3 that

$$\frac{(-1)^j D_j}{L_j} \frac{1}{e_j(x) + D_j} = \sum_{\substack{n \text{ monic} \\ \deg n = j}} (x+n)^{-1}.$$

Now expand both sides about the origin and use the expression for $e_j(x)$ given above in 8.24.2. We conclude immediately [C1] that for $0 < m \leq r$ we have

$$\sum_{\substack{n \text{ monic} \\ \deg n = j}} \frac{1}{n^m} = (-1)^{jm} \frac{1}{L_j^m}.$$

Thus the result for $0 < m \leq r$ follows as in 8.24.2. An application of 8.24.14 finishes the proof. $\square$

Next we turn to some examples at the negative integers. Before presenting the results we need a definition. Recall that we have set

$$[i] = T^{r^i} - T.$$

We now set for $0 < j \leq i$

$$[i]_j := [i][i-1]^r \cdots [i-j+1]^{r^{j-1}},$$

and $[i]_0 := 1$. Thus

$$[i]_j = \frac{D_i}{D_{i-j}^{r^j}}$$

and $\deg[i]_j = jr^i$. If $j > i$, we set $[i]_j = 0$. In [Lee1], Lee shows the following.

Proposition 8.24.17 (Lee). *Let $i = r^{k_1} + \cdots + r^{k_s}$, $s < r$. Then*

$$S_j(i-1) = (-1)^j \frac{[k_1]_j \cdots [k_s]_j}{L_j}. \qquad \square$$

Corollary 8.24.18. *Let i be as above. Then $p^t(1-i)$ is simple.*

Proof. Since we can compute the degrees and so the valuation at infinity, the result follows as before. $\square$

Similarly, in [Lee1], the following is shown.

Proposition 8.24.19 (Lee). *In the set-up of 8.24.17 let $s = r$. Then*

$$S_j(i-1) = (-1)^j \frac{[k_1]_j \cdots [k_s]_j}{L_j} + (-1)^j \frac{[k_1]_{j-1} \cdots [k_s]_{j-1}}{L_{j-1}}. \qquad \square$$

It turns out that the degree of the sum in 8.24.19 is easily computed as the two elements in the sum have different degrees.

Corollary 8.24.20. *Let i be as in 8.24.19. Then $p^t(1-i)$ is simple.* □

Remarks. 8.24.21. 1. There are more such results in Lee and they are all quite mysterious. As s increases, they get more complicated, but perhaps the degrees may be readily found as in 8.24.19. These results are proved by using the additive harmonic analysis of the functions $\{G_k(t)\}$ of Subsection 8.22 (but Lee uses $E_j(t) = e_j(t)$ instead of $e_j(t)/D_j$). It would be very interesting to put all of these results in the proper context (whatever that might be), and perhaps generalize to arbitrary **A**.
2. Similarly, sums of the type occurring in the proof of Proposition 8.24.16 are also poorly understood. Any progress towards understanding them for all m would be very useful. Perhaps Anderson's theory [A2], [A3] of *solitons* will be crucial here; see also Subsection 10.2.
3. A final note on Lee's paper. His proof of Theorem 3.2 in [Lee1] is suspect. The reader should refer to the papers of Wan [Wan1] or Gekeler [Ge1] for a better exposition.

It is natural to wonder about applications of the above results on zeroes, and we now present a few. However, as the reader will see, these results do *not* appear to be at the heart of the matter. Indeed one would hope for some implications of the above results for the arithmetic of Drinfeld modules, etc. We first deal with the negative integers.

Let $i > 0$ be chosen so that $-i$ is simple; thus $\zeta_\mathbf{A}(x, -i)$ has simple roots which all belong to **K**. As before, let $z(x, -i) = \sum_{j=0}^{\infty} x^{-j} s_j(i)$, where $s_j(i) = S_j(i) = \sum_{\substack{\deg n = j \\ n \text{ monic}}} n^i$, and set

$$\widetilde{z}(x, -i) = \begin{cases} z(x, -i) & \text{if } i \not\equiv 0 \ (r-1) \\ z(x, -i)/(1 - x^{-1}) & \text{if } i \equiv 0 \ (r-1) \end{cases}$$

(Thus in $\widetilde{z}(x, -i)$ we have removed the trivial zero *when* there is one.) Let $\wp$ be a prime with injection $\sigma: \mathbf{k} \to \mathbf{k}_\wp$ and Teichmüller character $\omega := \omega_\sigma$. Theorem 8.13.3 gives a double congruence between $\widetilde{z}(x, -i)$ and $\widetilde{L}(\omega_\sigma^{-i}, u)$ where $\widetilde{L}(\omega_\sigma^{-i}, u)$ is the usual L-series but *without* the factor above $\wp$ (which can only happen if $(r^{\deg \wp} - 1) \mid i$). For $\deg \wp \gg 0$, we see that $\widetilde{L}(\omega_\sigma^{-i}, u) = L(\omega_\sigma^{-i}, u)$.

Theorem 8.24.22. *Let $L_{\mathrm{un}}(\omega_\sigma^{-i}, u)$ be the unit-root L-series associated to $\wp$ and i (cf. Subsection 8.14) where i is simple. As $\deg \wp \to \infty$, $L_{\mathrm{un}}(\omega_\sigma^{-i}, u)$ eventually becomes, and stays, a separable polynomial.*

Proof. We know that i is simple; thus $z(x,-i)$ and $\tilde{z}(x,-i)$ are separable polynomials. As $\deg \wp \to \infty$, their reductions modulo $\wp$ eventually become separable as is easy to see. Thus the reduction of the unit root polynomials eventually becomes separable by the double congruence. The result now follows easily. □

Question 8.24.23. Is $L_{\mathrm{un}}(\omega_\sigma^{-1}, u)$ always separable?

Let $\hat{\rho}$ be a strictly compatible system of representations with L-series $L(\hat{\rho}, s)$, $s \in S_\infty$, as in Subsection 8.6. Set $f(s) := L(\hat{\rho}, s)$ and assume that $f(s)$ can be shown to be entire in the sense of Subsection 8.5. Then the zeroes of $f(x,y)$ flow continuously in the $\mathbb{Z}_p$-variable y. Let $\Lambda_y \subset \mathbf{C}_\infty$ be set of zeroes of $f(x,y)$ counted with their multiplicities as usual. Note that $0 \notin \Lambda_y$. Let $N_y \in \{0, 1, \ldots, \infty\}$ be the number of element in Λ_y. Thus we may write (Theorem 2.14)

$$f(x,y) = \prod_{i=1}^{N_y} (1 - \beta_i(y)/x)$$

where $\Lambda_y = \{\beta_i(y)\}$, and where we set $f(x,y) = 1$ if $N_y = 0$. If we set $\beta_i(y) = 0$ for $i > N_y$, then we can write

$$f(x,y) = \prod_{i=1}^{\infty} (1 - \beta_i(y)/x).$$

Upon perhaps reordering the sets $\{\beta_i(y)\}$, the reader may use the inverse function theorem [Se7, LG 2.10] to show that the elements $\beta_i(y)$ can be taken to form continuous sections of the canonical projection $\mathbf{C}_\infty \times \mathbb{Z}_p \to \mathbb{Z}_p$. These continuous sections are the best way to view the flow of the zeroes as y varies.

Let $\mathbf{A} = \mathbb{F}_r[T]$ and $f(s) = \zeta_\mathbf{A}(s)$. Thus $\{\beta_i(y)\}$ gives the factorization of $\zeta_\mathbf{A}(x, y)$. Let N_n be the norm map from $\mathbb{F}_{r^n}\left(\left(\frac{1}{T}\right)\right)$ to $\mathbb{F}_r\left(\left(\frac{1}{T}\right)\right) = \mathbf{K}$. Let $\pi = \frac{1}{T}$. Our next result and its proof are due to W. Sinnott.

Proposition 8.24.24 (Sinnott). *Let $n \geq 1$. Then*

$$\sum_{i=0}^{\infty} \beta_i(y)^n = \sum_{\alpha \in \mathbb{F}_{r^n}} N_n(1 - \alpha/T)^{-y}.$$

Proof. Fix y and apply logarithm differentiation (with respect to x) to

$$\zeta_\mathbf{A}(x,y) = \prod_{i=1}^{\infty} (1 - \beta_i(y)/x).$$

We find, with a little algebra, that

$$x \cdot \frac{\zeta'_A(x,y)}{\zeta_A(x,y)} = \sum_{n=1}^{\infty} x^{-n} \left(\sum_{i=1}^{\infty} \beta_i(y)^n \right).$$

On the other hand, from the Euler product for $\zeta_A(x,y)$ we have

$$\zeta_A(x,y) = \prod_{f \text{ monic prime}} (1-f^{-s})^{-1}$$

$$= \prod_f (1 - x^{-\deg f} \langle f \rangle^{-y})^{-1}$$

$$= \prod_f \left(1 - x^{-\deg f} \left(\frac{f}{T^{\deg f}} \right)^{-y} \right)^{-1}.$$

We rewrite this as

$$\prod_{d=1}^{\infty} \prod_{\substack{f \text{ monic prime} \\ \deg f = d}} \left(1 - x^{-d} \left(\frac{f}{T^d} \right)^{-y} \right)^{-1}.$$

Taking the logarithmic derivative of $\zeta_A(x,y)$ with respect to x gives

$$x \cdot \frac{\zeta'_A(x,y)}{\zeta_A(x,y)} = -\sum_{d=1}^{\infty} \sum_{\substack{f \text{ monic prime} \\ \deg f = d}} \frac{x}{1-x^{-d}(T^{-d}f(T))^{-y}} \cdot \frac{d(T^{-d}f(T))^{-y}}{x^{d+1}}$$

$$= -\sum_{d=1}^{\infty} \sum_{\substack{f \text{ monic prime} \\ \deg f = d}} \sum_{m=1}^{\infty} \frac{d}{x^{md}} (T^{-d}f(T))^{-my}.$$

We can rewrite this formula for $x \cdot \frac{\zeta'_A(x,y)}{\zeta_A(x,y)}$ as

$$-\sum_{n=1}^{\infty} x^{-n} \sum_{\substack{d|n \\ d \geq 1}} \sum_{\substack{f \text{ monic prime} \\ \deg f = d}} d(T^{-d}f(T))^{-\frac{n}{d}y}. \qquad (8.24.25)$$

Note that

$$\sum_{\alpha \in \mathbb{F}_{r^n}} N_n(1-\alpha/T)^{-y} = \sum_{d|n} \sum_{\substack{\alpha \in \mathbb{F}_{r^n} \\ \deg \alpha = d}} (T^{-n}N_n(T-\alpha))^{-y},$$

where "$\deg \alpha = d$" means α is of degree d over $\mathbb{F}_r$. If we let $f_\alpha(T)$ be the monic minimal polynomial of α over $\mathbb{F}_r$, we find

$$\sum_{\alpha \in \mathbb{F}_{r^n}} N_n(1-\alpha/T)^{-y} = \sum_{d|n} \sum_{\substack{\alpha \in \mathbb{F}_{r^n} \\ \deg \alpha = d}} (T^{-n} f_\alpha(T)^{n/d})^{-y}.$$

If $\deg \alpha = d$, then $f_\alpha(T)$ will occur d times. Thus we find

$$\sum_{\alpha \in \mathbb{F}_{r^n}} N_n(1-\alpha/T)^{-y} = \sum_{d|n} d \sum_{\substack{f \text{ monic prime} \\ \deg f = d}} (T^{-n} f(T)^{n/d})^{-y}$$

$$= \sum_{d|n} d \sum_{\substack{f \text{ monic prime} \\ \deg f = d}} (T^{-d} f(T))^{-\frac{n}{d} y} .$$

The result follows by plugging this formula into 8.24.25 and then comparing the coefficients of both expressions for

$$x \cdot \frac{\zeta'_\mathbf{A}(x,y)}{\zeta_\mathbf{A}(x,y)} . \qquad \square$$

Thus, if y is simple, we may obtain an exact formula for

$$\left| \sum_{\alpha \in \mathbb{F}_{r^n}} N_n(1-\alpha/T)^{-y} \right|$$

for all $n \geq 1$.

Remarks. 8.24.26. 1. Let $f(s) = L(\widehat{\rho}, s)$ as before where $f(s)$ is assumed entire; so

$$f(x,y) = \prod_{i=1}^\infty \left(1 - \frac{\beta_i(y)}{x}\right) .$$

As in the proof of 8.24.24, or directly, one sees that the logarithmic derivative of $f(x,y)$ with respect to x depends only on the residue mod p of the multiplicity that each non-zero $\beta_i(y)$ occurs. Thus, of course, we lose information by taking logarithmic derivatives in finite characteristic.

2. Let $v \in \text{Spec}(\mathbf{A})$ where $\mathbf{A} = \mathbb{F}_r[T]$. Thus we know that the v-adic interpolation of $\zeta_\mathbf{A}(s)$ (cf. Definition 8.6.8) is also entire. It would be very interesting to have calculations of the Newton polygons of these functions. As of now, one has no idea what to expect.

Let $\mathbf{A} = \mathbb{F}_r[T]$ and let $i > 0$ be chosen so that $-i$ is simple (e.g., all such i if $r = p$). Let

$$z(x, -i) = \sum_{j=0}^\infty x^{-j} s_j(i)$$

as before; thus $z(x, -i)$ is *separable*. Let $\mathbf{k}_{-i} \subset \mathbf{C}_\infty$ be the Galois extension generated by its roots. As these roots are contained in $\mathbf{K}$ we see that $\mathbf{k}_{-i} \subset \mathbf{K}$ and $\mathbf{k}_{-i}/\mathbf{k}$ is *geometric*.

Questions 8.24.27. 1. What is $\text{Gal}(\mathbf{k}_{-i}/\mathbf{k})$?

2. What are the primes that split completely in $\mathbf{k}_{-i}/\mathbf{k}$?
3. (Weaker than Question 2.) What is the discriminant of $\mathbf{k}_{-i}/\mathbf{k}$?

Perhaps the v-adic theory will be helpful with 8.24.27.3.

Note that, in one sense, Theorem 8.24.5 is more remarkable than the classical Riemann Hypothesis. Obviously, of course, the classical Riemann hypothesis is a problem of spectacular depth and difficulty, but $\mathbb{R}$ is still a "small" local field; thus the zeroes of $\theta(u)$ ($\theta(u)$ as in the beginning of this subsection) must either be in $\mathbb{R}$ or $\mathbb{C}$ = the extension of $\mathbb{R}$ of degree 2. On the other hand, for $\zeta_\mathbf{A}(s)$, all we know a priori is that for fixed y, $\zeta_\mathbf{A}(x,y)$ is entire in x^{-1} with coefficients in $\mathbf{K}$. Thus a-priori the roots could be in an extension of $\mathbf{K}$ of *arbitrary degree* d, $1 \leq d \leq \infty$. Thus it is all the more remarkable that, at least for $\mathbb{F}_p[T]$, they are in $\mathbf{K}$ itself. The big question is what, if anything, is actually behind this phenomenon...

Next we make some general remarks on what *may* be expected in general. The first one is that the obvious analog of Theorem 8.24.5 does not hold in general, even when $\mathbf{A}$ is a principal ideal domain. Indeed, let

$$\mathbf{A} = \mathbb{F}_2[T, u]/(u^2 + u = T^5 + T^3 + 1).$$

One knows that $\mathbf{A}$ is the affine ring of a genus 2, separable, hyperelliptic extension of $\mathbf{k} = \mathbb{F}_2(T)$ minus a rational place at ∞. Moreover, $\mathbf{A}$ has class number 1 (cf. Example 7.11.9.4). Let $y_0 = -2^j - 1$ for $j > 0$; in [Th4, Th.8]. D. Thakur provides the tools necessary to compute $\zeta_\mathbf{A}(x,y_0)$ and we find

$$\zeta_\mathbf{A}(x,y_0) = \zeta_\mathbf{A}(x,-2^j-1) = (1 - \pi^{2^{j+1}+2}x^{-2})(1 - \pi^{2^{j+1}+2}(T^{2^j}+T)x^{-2}).$$

Thus the zeroes are

$$\{\pi^{2^j+1}, \pi^{2^j+1}(T^{2^{j-1}} + \sqrt{T})\},$$

each with multiplicity 2. The first zero is the trivial zero $(\pi^{2^j+1}, -2^j-1) \in S_\infty$ at y_0 (so we see that a trivial zero may have a *larger* order zero than predicted by Corollary 8.13.4; it would be very interesting to have a formula for the *exact* order of zero). The second zero generates the field $\mathbf{K}(\sqrt{T}) \neq \mathbf{K}$; thus not only does it not belong to $\mathbf{K}$, it is even *inseparable*.

For each $y \in \mathbb{Z}_p$, let $\mathbf{K}_y$ be the normal (possibly infinite) extension generated by the roots of $\zeta_\mathbf{A}(x,y)$. It is not yet known if the zero $\pi^{2^j+1}(T^{2^{j-1}}+\sqrt{T})$ of $\zeta_\mathbf{A}(x,y_0)$ is isolated or not. If it *is* isolated, and is the limit of separable roots $\beta_i(y_n)$ for $y_n \to y_0$, then, as in Example 8.23.1, *infinitely* many $\mathbf{K}_{y_n}$ must be different from $\mathbf{K}$, and their union must be infinite over $\mathbf{K}$. In this way, inseparable zeroes may play havoc on the fields $\mathbf{K}_y$!

Still, all is not yet lost. At this time it appears reasonable to expect the following: Let $\mathbf{A}$ be a principal ideal domain and let $\{\beta_i(y)\}$ be the roots of $\zeta_\mathbf{A}(x,y)$ as above. Then there exists some integer n, independent of y, such that for $i > n$, we have $\beta_i(y) \in \mathbf{K}$.

It turns out that a similar phenomenon is even found when $\mathbf{A} = \mathbb{F}_r[T]$. Indeed, as in Subsection 8.10, let $L/\mathbf{k}$ be a finite abelian extension which is ramified at one finite prime f (f monic) and where $[L\colon\mathbf{k}] = p$. Let χ be non-trivial character of the Galois group into $\overline{\mathbb{Q}}_p^*$. From our definitions, we find as before that
$$L(\chi, s) = (1 - f^{-s})\zeta_{\mathbf{A}}(s).$$
Thus the zeroes of $1 - f^{-s}$ are also zeroes of the L-series. This is unavoidable, and we have no *a priori* control over such zeroes.

We will shortly pose some questions on the zeroes which seem, as of this writing to have a reasonable chance of being true. However, before doing so we need to make a few definitions as well as recall a basic fact of non-Archimedean analysis. Let $\mathbf{A}$ be totally general and let $L/\mathbf{k}$ be a finite extension with $\mathbf{A}$-integers $\mathcal{O}_L$. Let $\zeta_{\mathcal{O}_L}(s)$ be the zeta function of its ring of integers (which is analogous to a classical Dedekind zeta function). Let $\{\beta_i(y)\}$ now denote its root sections as before 8.24.24. Let G be the Galois group of the separable closure of $\mathbf{k}$ over $\mathbf{k}$. From Subsection 8.17 we know that $\zeta_{\mathcal{O}_L}(s)$ can be factored into a product of L-series associated to $\overline{\mathbb{Q}}_p$ representations of G of Galois type. Thus, as in the reasoning before 8.24.1, questions about the zeroes of $\zeta_{\mathcal{O}_L}(s)$ are reduced to questions about L-series where we have Theorem 8.17.1. Moreover we need only discuss irreducible such representations ρ.

Recall that the value field was defined as a subfield of $\mathbf{C}_\infty$. Let $\mathbf{K}_\mathbf{V}$ be the smallest extension of $\mathbf{K}$ containing $\{\langle I \rangle\}$ for $I \in \mathcal{I}$ ($= \mathbf{K} \cdot \mathbf{V}$ if $d_\infty = 1$). Let $\mathbf{K}_\mathbf{V}(\rho)$ be the smallest extension of $\mathbf{K}_\mathbf{V}$ containing the coefficients of the characteristic polynomials of the Frobenius elements as discussed before 8.10.2. Thus, for each $y \in \mathbb{Z}_p$, the function $x \mapsto L(\rho, (x, y))$ is given by a power series in x^{-1} with coefficients precisely in $\mathbf{K}_\mathbf{V}(\rho)$. Note that by construction, as ρ is of Galois type, the field $\mathbf{K}_\mathbf{V}(\rho)$ is an unramified finite extension of $\mathbf{K}_\mathbf{V}$. Similarly we let $\mathbf{V}(\rho)$ be the smallest extension of $\mathbf{V}$ containing the above coefficients. Again, $\mathbf{V}(\rho)$ is a finite constant field extension of $\mathbf{V}$.

Finally, let $f(u) = \sum_{j=0}^{\infty} a_j u^j$, $a_0 = 1$, be an entire power series over some non-Archimedean field. As presented in Section 2, $f(u)$ *always* has a factorization over its roots like an infinite polynomial. As such, any closed field containing the roots *must* also contain the coefficients $\{a_j\}$. This is not true in classical analysis (look at $\frac{e^{2\pi i z}-1}{2\pi i z}$) and is essential for what we state next.

Let $L(\rho, s)$ be as above and let $\{\beta_i(y)\}$ be its root sections. With all of the above in mind we now pose the following questions.

Questions 8.24.28. 1. Does there exist an integer $n \geq 1$ depending only on ρ, so that for $i \geq n$, and all $y \in \mathbb{Z}_p$ the root $\beta_i(y)$ belongs to $\mathbf{K}_\mathbf{V}(\rho)$?

2. Let y be a non-positive integer and let $h_L(x, y)$ be as in 8.5.12. Thus $h_L(x, y)$ will be a polynomial in x^{-1} with coefficients in the field $\mathbf{V}(\rho)$. Let $\mathbf{V}(\rho)_y$ be the finite normal extension obtained by adjoining the roots

of $h_L(x,y)$. What is the inseparability degree of $\mathbf{V}(\rho)_y$ over $\mathbf{V}(\rho)$? What are the primes that split totally in the maximal separable subfield of $\mathbf{V}(\rho)_y$, etc?

Recall from our last subsection, if we know the answers to both parts of 8.24.28.2, then we know $\mathbf{V}(\rho)_y$ *exactly*!

Question 8.24.28.1 is modeled on Wan's result for $\zeta_{\mathbb{F}_p[T]}(s)$. It may turn out that it is the correct form of the Riemann hypothesis for these characteristic p functions. It can be stated succinctly as follows: *outside of a finite number of spurious cases, the zeroes of $L(\rho,s)$ should be as small as possible (i.e., are on the "line" $\mathbf{K_V}(\rho)$)*. We leave it to the reader to formulate a version of this question for, say, Drinfeld modules defined over $\mathbf{k}$, etc. (but see Subsection 10.4). More computations are needed before we know whether these questions need modification or not. We also point out that, occasionally, one might be able to specify that the roots lie in a proper subfield of $\mathbf{K_V}(\rho)$. For instance, let $\mathbf{A} = \mathbb{F}_r[T]$ where r is strictly bigger than p etc., and let $\pi = 1/T$. With these hypotheses, and a little thought, the reader will see that the zeta function of $\mathbf{A}$ is a family of power series with coefficients in $\mathbb{F}_p((1/T)) \subset \mathbf{K}$. Thus we can also expect the roots to lie in this subfield for our particular choice of π.

Next we discuss what can be expected in terms of the simplicity of the non-spurious zeroes. The main problem here is to avoid those irreducible representations ρ which factor through p-groups as in 8.18.7.2. Indeed for such ρ the eigenvalues will always reduce to 1. Thus if the degree of ρ is > 1, then $L(\rho,s)$ will, up to finitely many Euler factors, be divisible by the square of $\zeta_\mathbf{A}(s)$. As such, the zeroes will never be simple. Thus, to be safe let ρ be a representation which either is of degree 1 or factors through a group of order *prime* to p.

Question 8.24.29. Does there exist an integer $n_1 \geq 1$, depending only on ρ, such that for $i \geq n_1$ and all $y \in \mathbb{Z}_p$ the root $\beta_i(y)$ is simple?

Questions 8.24.29 and 8.24.28.1 may very well ultimately be attacked together via Newton polygons as in Wan's Theorem. It is unclear to us, as of this writing, how to phrase a version of 8.24.9 for those representations of higher degree which factor through groups which are *not* p-groups but have non-trivial p-Sylow subgroups.

Finally we discuss the effect on the above fields of changing the parameter at ∞. Recall that d_∞ is the degree of ∞ with respect to $\mathbb{F}_r$.

Proposition 8.24.30. *Let $L(\rho,s)$ be as above and let $y \in \mathbb{Z}_p$. Let $\mathbf{K_V}(\rho)_y \subseteq \mathbf{C}_\infty$ be the extension obtained by adjoining the roots of $L(\rho,(x,y))$ to $\mathbf{K}$.*
1. *If $p \nmid d_\infty$, then $\mathbf{K_V}(\rho)_y$ is independent of the choice of parameter.*
2. *For any d_∞, the maximal separable subfield of $\mathbf{K_V}(\rho)_y$ is always independent of the choice of parameter.*

3. *Whether* $[\mathbf{K_V}(\rho)_y\colon \mathbf{K}] < \infty$ *or not is independent of the choice of parameter.*

Proof. Let u be as in 8.2.15, and let β be a root of $L(\rho, (x,y))$ defined with respect to π_1. Changing π_1 to π_2 multiplies β by u^{y/d_∞}. If $p \nmid d_\infty$, then $u^{1/d_\infty} \in \mathbf{K}$ and Part 1 follows. If $p \mid d_\infty$, set

$$d_\infty = p^t q, \qquad (p,q) = 1.$$

Note that $u^{1/q} \in \mathbf{K}$. Thus we may assume $q = 1$. Let $\{\beta_1(y), \ldots, \beta_n(y)\}$ be the first n roots of $L(\rho, (x,y))$ defined with respect to π_1; thus

$$\{\beta_1(y)u^{y/p^t}, \ldots, \beta_n(y)u^{y/p^t}\}$$

are the first n roots with respect to π_2. Let $L_n \subseteq \mathbf{C}_\infty$ be the field obtained by adjoining $\{\beta_1(y), \ldots, \beta_n(y)\}$ and let L_n^1 be the field obtained by adjoining $\{\beta_1(y)u^{y/p^t}, \ldots, \beta_n(y)u^{y/p^t}\}$. Thus $L_n^1 \subseteq L_n((u^y)^{1/p^t})$. Therefore, the maximal separable subfield of L_n^1 is contained in the maximal separable subfield of L_n. Reversing the argument gives equality and Part 2. Part 3 follows easily. □

Of course, one need not work with all $\beta_i(y)$ in the above result, but only with those for i sufficiently large, etc.

Proposition 8.24.31. *Let y and $h_L(x,y)$ be as in 8.24.28.2. Let*

$$\{\widetilde{\beta}_i(y), \ldots, \widetilde{\beta}_n(y)\}$$

be its roots. Then changing from one positive uniformizer to another, multiplies $\widetilde{\beta}_i(y)$ by a d_∞-th root of unity.

Proof. Use Proposition 8.2.16. □

Thus if d_∞ is a p-th power, the extension of $\mathbf{V}(\rho)$ obtained by adjoining $\{\widetilde{\beta}_i(y)\}$ is independent of the choice of parameter. In general, the fact that this extension is separable or not, is always independent of the parameter.

To repeat, more calculations in all these areas would be very useful.

8.25. Kapranov's Higher Dimensional Theory

At the beginning of Subsection 8.23 we discussed the "classical" local fields in terms of their "sizes." Of course these local fields occur as the completions of global fields (= number fields or finitely generated function fields of transcendency one over finite fields). Moreover, they possess a (local) class field theory consistent with such completions and global class field theory.

All of this has been well known for decades. More recently, it has become apparent that class field theory, in both its local and global forms, can be established for much general fields and schemes. Those fields with a local class field theory are the *n-dimensional (non-Archimedean) local fields* of A. Parshin.

Definition 8.25.1. A 0-*dimensional field* is a finite field. For $n \geq 1$, an n-*dimensional local field* is a complete discrete valuation field whose residue field is an $(n-1)$ dimensional local field.

Thus the non-Archimedean local fields contained in Subsection 8.23 are "1-dimensional" in Parshin's definition.

Example 8.25.2. Let K be a finite extension of $\mathbb{Q}_p$ or $\mathbb{F}_r((t))$. Then, the field $K((x))$ is a 2-dimensional local field.

Local class field theory may then be established for such n-dimensional local fields through the use of Milnor K-groups. As we will not need this here, we refer the reader to [Ka1], [Ra1] for this local class field theory, as well as the *global* class field theory of integral schemes of finite type over $\mathbb{Z}$.

It would be interesting to have a characterization of n-dimensional local fields via their "size" as in Subsection 8.23.

In [Ka1], M. Kapranov uses such n-dimensional local fields in a very elegant fashion to define zeta-functions generalizing those discussed in this section. We will content ourselves with presenting the basic definitions and properties of these functions as well as posing some basic problems about them. The interested reader can consult [Ka1] for more information. As the theory of these functions is in its infancy, one may expect that much more of interest about them will eventually be discovered.

The basic set-up of Kapranov is the following: let X be a smooth projective geometrically irreducible variety over $\mathbb{F}_r$ of dimension n. Let

$$\infty = \{X = X_n \supset X_{n-1} \supset \cdots \supset X_0\}$$

be a maximal flag of smooth irreducible subvarieties of X such that $\dim(X_i) = i$ and X_{i-1} is *ample* in X_i for all i. Thus $X_i - X_{i-1}$ is an *affine* variety. Of course, the basic example of this section is $X =$ a curve and $X_0 = \infty$ is a closed point of X.

Remark. 8.25.3. The ampleness requirement is not as restrictive as it might a priori seem; see Bertini's Theorem [Har1, Th. 8.18].

We set $\mathbf{A} := \Gamma(X - X_{n-1}, \mathcal{O}_X)$; so $X - X_{n-1} = \mathrm{Spec}(\mathbf{A})$. We will follow Kapranov and use the flag ∞ to define an n-dimensional completion, $\mathbf{K} = \mathbf{k}_\infty$, of $\mathbf{k} = \mathbb{F}_r(X) =$ the field of rational functions on X, as well as a good notion of "monic" (or "positive") element in **A**. With these notions in hand, we then define a zeta function

$$\zeta_\mathbf{A}(s) := \zeta_{\mathbf{A},\infty}(s)$$

of the form

$$\zeta_\mathbf{A}(s) = \sum_{a \text{ monic}} a^{-s},$$

for appropriate s. In particular, we will concentrate on the prototypical case $\mathbf{A} = \mathbb{F}_r[T_1, \ldots, T_n]$.

Remarks. 8.25.4. 1. As will be evident below, the choice of the flag is essential in determining which elements are monic. Thus $\zeta_\mathbf{A}(s)$ does indeed depend on the flag ∞ and not just on **A** itself.

2. Kapranov's function actually corresponds to a partial zeta function in the curve case. Indeed, in the curve case Kapranov's $\zeta_\mathbf{A}(s)$ can be written as

$$\sum_{\substack{I \subseteq \mathbf{A} \\ I=(i) \\ i \text{ positive}}} I^{-s};$$

so we are missing the ideals which are not principal and positively generated. In contrast, in earlier subsections, we worked with the function $\zeta_\mathbf{A}(s) = \sum_{I \subseteq \mathbf{A}} I^{-s}$; a sum over *all* ideals.

We now need to define our completion **K**. The rigorous construction is a bit involved, so we shall rather give an informal presentation. The details can be found in [Hub1] which follows Beilinson. First of all, as X is smooth, the hypersurface $X_{n-1} \subset X$ corresponds to a discrete valuation on X. Completing **k** with respect to this valuation gives a complete field $\mathbf{k}_{X_{n-1}}$. Note that the field $\mathbb{F}_r(X_{n-1})$ of rational functions on X_{n-1} appears as the residue field of $\mathbf{k}_{X_{n-1}}$. Moreover, X_{n-1} is smooth and X_{n-2} gives rise to a discrete valuation on it. We can then complete with respect to this valuation, etc. In this fashion, the field **K** emerges when we complete, finally, at X_0.

One can actually make the construction of **K** more precise (but coordinate dependent) as follows. A set $\{\pi_1, \ldots, \pi_n\}$ of elements of **k** is a *system of parameters adjusted to the flag* ∞ if all X_i near X_0 are given by the equations

$$\pi_{i+1} = \pi_{i+2} = \cdots = \pi_n = 0.$$

Thus $\mathbf{k}_{X_{n-1}} \simeq \mathbb{F}_r(X_{n-1})((\pi_n))$. We now use the parameter π_{n-1} to embed $\mathbb{F}_r(X_{n-1})$ into $\mathbb{F}_r(X_{n-2})((\pi_{n-1}))$, and so $\mathbb{F}_r(X_{n-1})((\pi_n))$ will embed into $\mathbb{F}_r(X_{n-2})((\pi_{n-1}))((\pi_n))$, and so on. Clearly $\mathbb{F}_r(X_0)$ is a finite extension of $\mathbb{F}_r$ and $\mathbf{K} \simeq \mathbb{F}_r(X_0)((\pi_1)) \cdots ((\pi_n))$.

Example 8.25.5. Let $X = \mathbb{P}^n_{\mathbb{F}_r}$ where we use homogeneous coordinates

$$[y_0, \ldots, y_n].$$

We let X_{n-1} be the "infinite" hyperplane $y_0 = 0$. Thus $X - X_{n-1}$ is isomorphic to affine n-space with coordinates

$$T_i := y_i/y_0, \qquad i = 1, \ldots, n;$$

so $\mathbf{A} = \mathbb{F}_r[T_1, \ldots, T_n]$. We define the flag ∞ by

$$\infty := \{X = X_n \supset X_{n-1} \supset X_{n-2} \supset \cdots \supset X_0\}$$

where X_i is given by $y_0 = \cdots = y_{n-1-i} = 0$. Note that X_0 is the rational point $[0, \ldots, 0, 1]$ in homogeneous coordinates. We set

$$\pi_1 := \frac{y_{n-1}}{y_n} = \frac{T_{n-1}}{T_n}, \quad \pi_2 := \frac{y_{n-2}}{y_n} = \frac{T_{n-2}}{T_n}, \ldots, \pi_n := \frac{y_0}{y_n} = T_n^{-1}.$$

Clearly $\{\pi_i\}$ is a system of parameters adjusted for the flag ∞ and $\mathbf{K} \simeq \mathbb{F}_r((\pi_1)) \cdots ((\pi_n))$. Finally

$$T_i = y_i/y_0 = \frac{\pi_{n-i}}{\pi_n}, \qquad i = 1, \ldots, n-1,$$
$$T_n = \pi_n^{-1}.$$

Let $\overline{\mathbf{K}}$ be a fixed algebraic closure of $\mathbf{K}$ equipped with the canonical topology arising from the parameter π_n. Let $\mathbf{C}_\infty$ be its completion, etc.

Let $a \in \mathbf{K}^*$. We now define a canonical sequence of elements $a_{(i)} \in \mathbf{K}$ associated to $\{\pi_i\}$. We begin by setting $a_{(0)} := a$. We then set $a_{(1)} \in \mathbb{F}_r(x_0)((\pi_1)) \cdots ((\pi_{n-1}))$ to be the coefficient of the minimal degree (in π_n) of the expansion of $a_{(0)}$. We set $a_{(2)} \in \mathbb{F}_r(X_0)((\pi_1)) \cdots ((\pi_{n-2}))$ to be the coefficient of minimal degree (in π_{n-1}) of $a_{(1)}$ and so on.

Definition 8.25.6. 1. We set $\mathrm{sgn}(a) := a_{(n)} \in \mathbb{F}_r(X_0)$.
2. The element a is *positive* or *monic* if and only if $\mathrm{sgn}(a) = 1$.

With these definitions, the proof of our next result is easy.

Lemma 8.25.7. 1. *Let $\{a, b\} \subset \mathbf{K}^*$. Then*

$$\mathrm{sgn}(ab) = \mathrm{sgn}(a)\mathrm{sgn}(b).$$

2. *Suppose a is positive and $b \in \mathbf{K}$ is an element with $v_{\pi_n}(b) > v_{\pi_n}(a)$. Then $a + b$ is also positive.* □

Example 8.25.8. We return to the case of $\mathbb{P}^2_{\mathbb{F}_r}$ as in Example 8.25.5 to work out a specific example as in [Ka1]. Let

$$f(T_1, T_2) := T_1^3 + T_1^2 T_2 + T_2^3 + T_1^2 + T_2.$$

We have $T_1 = \pi_1/\pi_2$ and $T_2 = \pi_2^{-1}$. Thus

$$f = \left(\frac{\pi_1}{\pi_2}\right)^3 + \left(\frac{\pi_1}{\pi_2}\right)^2 \frac{1}{\pi_2} + \frac{1}{\pi_2^3} + \left(\frac{\pi_1}{\pi_2}\right)^2 + \pi_2^{-1}.$$

Consequently,
$$f_{(1)} = \pi_1^3 + \pi_1^2 + 1$$

and
$$f_{(2)} = 1.$$

Thus $\operatorname{sgn}(t) = 1$ and $f(T_1, T_2)$ is monic.

Definition 8.25.9. Let $a \in \mathbf{K}^*$.
1. We say that a is *absolutely integral* if and only if a is positive and integral for the π_n-valuation, its residue class is integral for the π_{n-1}-valuation, etc. (Note that the last residue is 1 as a is positive.)
2. Set
$$\langle a \rangle := a \cdot \pi_n^{-v_{\pi_n}(a)} \cdot \pi_{n-1}^{-v_{\pi_{n-1}}(a_{(1)})} \cdots \pi_1^{-v_{\pi_1}(a_{(n-1)})}.$$

Lemma 8.25.10. 1. *Let $a, b \in \mathbf{K}^*$. Then*
$$\langle ab \rangle = \langle a \rangle \langle b \rangle.$$

2. *$a \in \mathbf{K}^*$ is absolutely integral if and only if $\operatorname{sgn}(a) = 1$ and $\langle a \rangle = a$.* □

Let $g(T_1, \ldots, T_n)$ be a polynomial in n variables. Recall that $\deg(g)$ is the maximum of the degrees of the monomials of g and the degree of a monomial is the sum of its exponents.

Example 8.25.11. Let $f(T_1, T_2)$ be as in 8.25.8. So

$$f(T_2, T_2) = T_1^3 + T_1^2 T_2 + T_2^3 + T_1^2 + T_2 = \frac{\pi_1^3 + \pi_1^2 + 1}{\pi_2^3} + \frac{\pi_1^2}{\pi_2^2} + \frac{1}{\pi_2}.$$

Thus $v_{\pi_2}(f) = -3 = -\deg(f)$. (The reader will note that, in general, $v_{\pi_n}(f) = -\deg(f)$.) Moreover, as in 8.25.8, $f_{(1)} = \pi_1^3 + \pi_1^2 + 1$; so $v_{\pi_1}(f) = 0$. Therefore,
$$\langle f \rangle = \pi_2^3 \cdot f = \pi_1^3 + \pi_1^2 + 1 + \pi_2 \pi_1^2 + \pi_2^2.$$

Let F be any field of characteristic p and $h(t) = 1 + tE[[t]]$. Then, as before, the binomial theorem tells us how to define $h(t)^y$ for $y \in \mathbb{Z}_p$.

Definition 8.25.12. let $b \in \mathbf{K}^*$ be absolutely integral, and let $y \in \mathbb{Z}_p$. We then define b^y by

$$b^y := (b/b_{(1)})^y (b_{(1)}/b_{(2)})^y \cdots (b_{(n-1)}/b_{(n)})^y,$$

where $b_{(n)} = 1$ as b is positive. (This definition makes sense as $b_{(i)}/b_{(i+1)}$ will be congruent to 1 modulo π_{n-1}.)

Definition 8.25.13. We set $S_\infty := S_\infty^{(n)} = (\mathbf{C}_\infty^*)^n \times \mathbb{Z}_p$.

Thus, as before, S_∞ is the topological group whose operation will be written additively. Let $a \in \mathbf{K}^*$ be positive and $s = (x_1, \ldots, x_u, y) \in S_\infty$.

Definition 8.25.14. We set

$$a^s := \langle a \rangle^y \left(\frac{1}{x_n}\right)^{v_{\pi_n}(a)} \left(\frac{x_1}{x_n}\right)^{v_{\pi_{n-1}}(a_{(1)})} \cdots \left(\frac{x_{n-1}}{x_n}\right)^{v_{\pi_1}(a_{(n-1)})}.$$

We leave it to the reader to check that a^s satisfies the usual rules of exponentiation. Moreover, and quite curiously, we see that s has n "real" parts corresponding to $\{x_1, \ldots, x_n\}$ and one "imaginary" part corresponding to $y \in \mathbb{Z}_p$.

The usual integral powers, $\mathbb{Z}$, are embedded into S_∞ by the rule

$$i \mapsto ((\pi_{n-1}/\pi_n)^i, (\pi_{n-2}/\pi_n)^i, \ldots, (\frac{1}{\pi_n})^i, i),$$

as the reader may easily check. If X_0 is *not* a rational point, then this definition will not agree with our original one when X is a curve. Indeed, Kapranov is using the valuation at X_0 instead of the degree which is what was used earlier; however, this difference is minor (and Kapranov may as well have used degree of divisors, etc.).

Example 8.25.15. In the prototypical case of Example 8.25.5, we see that the above embedding of $\mathbb{Z}$ is just

$$i \mapsto (\pi_1^i, \ldots, \pi_n^i, i).$$

With all of the above techniques, we can now discuss zeta functions.

Definition 8.25.16. Let $s \in S_\infty$. Then we set

$$\zeta_\mathbf{A}(s) := \sum_{\substack{a \text{ monic} \\ a \in \mathbf{A}}} a^{-s}.$$

The reader should be aware that the definition of $\zeta_\mathbf{A}(s)$ depends on our uniformizers $\{\pi_i\}$ as well as $\mathbf{A}$. Moreover, the series for $\zeta_\mathbf{A}(s)$ will converge in the topology of $\mathbf{C}_\infty$ in a "half-space" of S_∞.

Remarks. 8.25.17. 1. The topology that $\zeta_{\mathbf{A}}(s)$ converges in depends only on π_n. However, we needed the *full* n-dimensional local field $\mathbf{K}$ in order to even define "a^s."

2. In the situation of Example 8.25.5, $\zeta_{\mathbf{A}}(s)$ has an Euler product

$$\zeta_{\mathbf{A}}(s) = \prod_{f \text{ monic prime}} (1 - f^{-s})^{-1}.$$

By an *entire function* on S_∞ we shall mean a continuous function $g(x_1, \ldots, x_n, y)$ such that for each $y \in \mathbb{Z}_p$ we obtain an everywhere convergent (on $(\mathbf{C}_\infty^*)^n$) Laurent series in $x_1, \ldots, x_n$. (In fact, one should also require uniform convergence on bounded subsets as in Subsection 8.5, etc., but we shall not worry about that here.)

Kapranov then proves the following result.

Theorem 8.25.18. *The function $\zeta_{\mathbf{A}}(s)$ analytically continues to an entire function on S_∞.* □

For the proof we refer the interested reader to [Ka1]. We point out, however, that the proof uses intersection theory (which is why the ampleness requirements are crucial) and Hilbert polynomials to establish estimates on certain vector space dimensions. The result then follows from Lemma 8.8.1 as in the curve case discussed earlier in this section.

Let $\mathbb{Z}$ be embedded in S_∞ as above. Then one has the following result on special values, [Ka1].

Theorem 8.25.19. 1. *For $i \geq 0$ the value $\zeta_{\mathbf{A}}(-i) \in \mathbf{A}$.*
2. *If X_0 is a rational point, then $\zeta_{\mathbf{A}}(-i) = 0$ for $i \equiv 0 \ (r-1)$, $i > 0$.* □

The proofs of these results are actually quite standard. The non-constant elements of $\mathbf{A}$ all have poles at X_{n-1}. So one can "sum by degree" as we did in the curve case to obtain 8.25.19.1; in particular the sum for $\zeta_{\mathbf{A}}(-i)$ is finite. The second part follows as in Example 8.13.9.

Let v be a *closed point* of $\mathrm{Spec}(\mathbf{A})$ corresponding to a maximal ideal M_v of $\mathbf{A}$. Let $a \in \mathbf{A}$ be positive with $a \notin M_v$, and let $\mathbf{A}_v$ be the completion at v. Let $\mathbf{A}/M_v$ have r^m elements and set

$$S_v = \mathbb{Z}/(r^m - 1) \times \mathbb{Z}_p.$$

The function $i \mapsto a^i$ clearly interpolates to a continuous function $S_v \to \mathbf{A}_v$. Since the sums in $\zeta_{\mathbf{A}}(-i)$ are all **finite**, they satisfy the same congruences as a^i, $i \neq 0$. As such, we obtain the following result.

Theorem 8.25.20. *$\zeta_{\mathbf{A}}(i)$ interpolates to a continuous function from S_v to $\mathbf{A}_v$.* □

Remark. 8.25.21. In [Ka1] it is pointed out that one can obtain a somewhat stronger result than 8.25.20. However, these results are still weaker than the v-adic interpolations in the curve case which was presented earlier in this section.

We finish with a number of interesting and potentially extremely fruitful questions related to $\zeta_\mathbf{A}(s)$. The first and most basic is the following.

Question 8.25.22. Does there exist a "good" theory of objects similar to Drinfeld modules to go along with $\zeta_\mathbf{A}(s)$?

Note that the sum $\zeta_\mathbf{A}(i) = \sum\limits_{a \text{ monic}} a^{-i}$, $i > 0$, converges to an element of $\mathbf{K}$.

Question 8.25.23. Does there exist an "Euler Theorem" for the values of $\zeta_\mathbf{A}(s)$ at positive integers (satisfying an appropriate congruence)?

We saw above that $\zeta_\mathbf{A}(-i) \in \mathbf{A}$.

Question 8.25.24. Is there an arithmetic interpretation for $\zeta_\mathbf{A}(-i)$ similar to the one presented in the curve case?

The generalized class field theory *suggests* a connection with the Chow group, $CH_0(X)$, of zero cycles on X modulo rational equivalence.

Question 8.25.25. Does there exist a good theory of "v-adic measures" on $\mathbf{A}_v$, etc.?

Question 8.25.26. Can one describe the zeroes of $\zeta_\mathbf{A}(s)$, where $X = \mathbb{P}^n_{\mathbb{F}_r}$, in a fashion similar to that of Subsection 8.24?

The above list of questions is certainly not exhaustive, but represents a good start!

9. Γ-functions

In this section we will introduce Γ-functions into the arithmetic of function fields. We do this by building on a basic, and still quite mysterious, construction of L. Carlitz in the $\mathbf{A} = \mathbb{F}_r[T]$-case. Recall that in Section 3.3 we introduced the Carlitz exponential

$$e_C(z) = \sum_{j=0}^{\infty} z^{r^j}/D_j$$

where $D_0 = 1$, $D_j = [j][j-1]^r \cdots [1]^{r^{j-1}}$, for $j > 1$, and $[j] = T^{r^j} - T$. In Proposition 3.1.6 we showed that

$$D_j = \prod_{\substack{g \in \mathbf{A} \\ g \text{ monic} \\ \deg g = j}} g\,.$$

Thus D_j looks very much like a *piece* of a factorial. Carlitz's wonderful idea is that one can define a *full* factorial by using D_j "linearly over r-adic expansions." More precisely, let i be a non-negative integer which we expand r-adically as

$$i = \sum_{j=0}^{m} c_j r^j, \qquad 0 \le c_j < r\,.$$

Carlitz then sets (in our notation)

$$\Pi(i) := \prod_{j=0}^{m} D_j^{c_j}\,;$$

thus $\Pi(r^j) = D_j$. This definition then has properties very analogous to $i!$. In turn, these formulas led the present author to introduce continuous Γ-functions into the theory; first an *arithmetic Γ-function* based on Carlitz's factorial, and later a *geometric Γ-function* based on the product formula for the Carlitz exponential.

The major breakthrough in the study of these functions came with the seminal Harvard thesis of Dinesh Thakur, [Th6]. Here it was established that these continuous Γ-functions possess the "correct" properties in analogy with

Euler's Γ-function and Morita's p-adic Γ-function. In fact, the theory of the Γ-functions described in this section has been so successful that virtually the only thing missing is the connection with L-series. As the reader will observe, here we only have some interesting hints. One such hint (see Remark 9.9.13) is that, as mentioned before, from the point of view of the "two T's" (e.g., Remark 5.4.3.1) the Γ-functions arise from one T ("T as constant" = the theory of exponential functions) whereas the L-functions arise from the *other* T ("T" as operator = the theory of characteristic polynomials of Frobenius morphisms associated to Drinfeld modules). Thus, from the point of view of completions, (as in Subsection 7.11) the Γ-functions will be seen to be functions from $\mathbf{C}_\infty \to \mathbf{C}_\infty$ whereas the L-functions (of Section 8) take $\mathbf{C}_\infty \to \mathbf{C}_\infty$. Now there is a canonical isomorphism between these two fields, but it may ultimately prove to be more natural to view these fields as *distinct* and combine them as in the *non-commutative* ring $L[T, \tau]$ of Subsection 5.4. In any case, as of this writing, it is certainly not clear what the final role of the Γ-functions is in the theory of the L-series (but see the very recent paper [Zh2]). More development is clearly called for.

It is our purpose here to describe the basic set-up of the theory. For more information we refer the reader to [Go12], [Th6], [Th7].

9.1. Basic Properties of the Carlitz Factorial

The first result that we present is ironically *not* due to Carlitz, but is due to W. Sinnott. Let i be a positive integer. Recall that classically one has the factorization
$$i! = \prod_{p \text{ prime}} p^{\alpha_p},$$
where
$$\alpha_p = \sum_{e \geq 1} [i/p^e],$$
and [?] is the greatest integer. Sinnott proved that a perfectly similar result holds for $\Pi(i)$, as defined in the introduction to this section, for $\mathbf{A} = \mathbb{F}_r[T]$.

Theorem 9.1.1 (Sinnott). *Let i be a non-negative integer. Then*
$$\Pi(i) = \prod_{f \text{ monic prime}} f^{\alpha_f},$$
where $\alpha_f = \sum_{e \geq 1} [i/r^{e \deg(f)}]$.

Proof. Let $i = \sum_{j=0}^{m} c_j r^j$, $0 \leq c_j < r$, as before; so

9.1. Basic Properties of the Carlitz Factorial   351

$$\Pi(i) = \prod_{j=0}^{m} D_j^{c_j}.$$

From the definition of D_j we find

$$\begin{aligned}\Pi(i) &= \prod_{j=0}^{m}\left(\prod_{e=1}^{j}(T^{r^j}-T^{r^{j-e}})\right)^{c_j}\\ &= \prod_{j=0}^{m}\prod_{e=1}^{j}(T^{r^e}-T)^{c_j r^{j-e}}\\ &= \prod_{e=1}^{m}\prod_{j=e}^{m}(T^{r^e}-T)^{c_j r^{j-e}}\\ &= \prod_{e=1}^{m}(T^{r^e}-T)^{c_e+c_{e+1}r+\cdots}.\end{aligned} \qquad (9.1.2)$$

Now notice $c_e + c_{e+1}r + \cdots = [i/r^e]$. Moreover,

$$T^{r^e} - T = \prod_{\substack{f \text{ monic prime}\\ \deg(f)|e}} f.$$

Thus we see that 9.1.2 equals

$$\prod_{e=1}^{m}\left(\prod_{\deg f | e} f\right)^{[i/r^e]}.$$

Collecting terms for each f gives the result. □

Remark. 9.1.3. Both the classical formula and Theorem 9.1.1 may be put into the same form. Indeed, let $\wp$ be a prime of a global field and set, as before, $N\wp := \#\mathcal{O}_\wp/\wp$ where $\mathcal{O}_\wp$ is the local ring at $\wp$. Let Fact (i) be either $i!$ or $\Pi(i)$. Then we have

$$(\text{Fact}\,(i)) = \prod_{\wp} \wp^{\alpha_\wp},$$

where

$$\alpha_\wp = \sum_{e \geq 1}[i/N\wp^e].$$

The Carlitz factorial satisfies many divisibility results analogous to those of $i!$. We refer the interested reader to [Go2], [C5] and [C6] for a complete list. For the purpose of illustrating these results, we will be content with the following examples.

9. Γ-functions

Definition 9.1.4. Let $\{a_m\} \subseteq \mathbf{A}$ be any collection of elements. Any series of the form $\sum_{m=0}^{\infty} \frac{a_m}{\Pi(m)} z^m$ is an **A**-*Hurwitz series* (or just *Hurwitz series*).

Most of the divisibility properties of $\Pi(i)$ may be summed up in the statement that **A**-Hurwitz series have properties similar to classical divided power series (= "**Z**-Hurwitz series"). More precisely we have the next result.

Proposition 9.1.5. 1. *The set of **A**-Hurwitz series forms an **A**-algebra under the usual addition and multiplication of power series.*
2. *Let $s(z)$ be an **A**-Hurwitz series with constant term 1. Then $1/s$ is also an **A**-Hurwitz series.*
3. *Let $s(z)$ be an **A**-Hurwitz series without constant term and let $s_1(z)$ be an arbitrary **A**-Hurwitz series. Then $s_1(s(z))$ is also an **A**-Hurwitz series.*

Proof. For the full proof, see, e.g. [Go2]. We will be content to show Parts 1 and 2. For this it is obvious that the **A**-Hurwitz series form an **A**-module. Thus we need only concentrate on multiplication, and for this we need only establish that

$$\frac{z^j}{\Pi(j)} \cdot \frac{z^j}{\Pi(i)} = \frac{\alpha z^{i+j}}{\Pi(i+j)}$$

where $\alpha = \frac{\Pi(i+j)}{\Pi(i)\Pi(j)} \in \mathbf{A}$, (i.e., $\Pi(i)\Pi(j)$ divides $\Pi(i+j)$). To see that $\alpha \in \mathbf{A}$, let

$$i = \beta_0 + \beta_1 r + \cdots + \beta_s r^s, \qquad 0 \le \beta_t < r \text{ all } t,$$

and

$$j = \gamma_0 + \gamma_1 r + \cdots + \gamma_s r^s, \qquad 0 \le \gamma_t < r \text{ all } t.$$

Thus

$$i + j = (\beta_0 + \gamma_0) + (\beta_1 + \gamma_1)r + \cdots + (\beta_s + \gamma_s)r^s$$

with $0 < \beta_t + \gamma_t < 2r$. Let $\beta_0 + \gamma_0 = \alpha_0 + \delta_0 r$, where $\delta_0 = 0$ or 1 and $0 \le \alpha_0 < r$. Now define inductively

$$\delta_0 + \beta_1 + \gamma_1 = \alpha_1 + \delta_1 r, \qquad 0 \le \alpha_1 < r, \quad \delta_1 = 0 \text{ or } 1,$$
$$\delta_1 + \beta_2 + \gamma_2 = \alpha_2 + \delta_2 r, \qquad 0 \le \alpha_2 < r, \quad \delta_2 = 0 \text{ or } 1,$$

etc., until we obtain

$$\delta_{s-1} + \beta_s + \gamma_s = \alpha_s + \delta_s r, \qquad 0 \le \alpha_{s-1} < r, \quad \delta_s = 0 \text{ or } 1.$$

Thus

$$i + j = \alpha_0 + \alpha_1 r + \cdots + \alpha_s r^s + \delta_s r^{s+1}$$

and

$$\frac{\Pi(i+j)}{\Pi(i)\Pi(j)} = D_{s+1}^{\delta_s} \prod_{t=1}^{s} D_t^{\alpha_t - \beta_t - \gamma_t}.$$

As $D_{s+1} = [s+1]D_s^r$, we see

$$\frac{\Pi(i+j)}{\Pi(i)\Pi(j)} = [s+1]^{\delta_s} D_s^{\delta_s r + \alpha_s - \beta_s - \gamma_s} \prod_{t=1}^{s-1} D_t^{\alpha_t - \beta_t - \gamma_t}$$

$$= [s+1]^{\delta_s} D_s^{\delta_s - 1} \prod_{t=1}^{s-1} D_t^{\alpha_t - \beta_t - \gamma_t}.$$

Similarly

$$D_s^{\delta_s - 1} D_{s-1}^{\alpha_{s-1} - \beta_{s-1} - \gamma_{s-1}} = [s]^{\delta_{s-1}} D_{s-1}^{\delta_{s-2}},$$

and so continuing this process we find

$$\frac{\Pi(i+j)}{\Pi(i)\Pi(j)} = [s+1]^{\delta_s}[s]^{\delta_{s-1}} \cdots [1]^{\delta_0},$$

which gives Part 1.

For Part 2, write $s(z) = 1 + s_0(z)$; so $\frac{1}{s(z)} = \frac{1}{1+s_0(z)} = 1 - s_0(z) + s_0^2(z) \cdots$. Now use Part 1. □

Remark. 9.1.6. As we have seen, 9.1.5.1 is equivalent to the statement $\Pi(i)\Pi(j)$ divides $\Pi(i+j)$. In particular we deduce that

$$\Pi(i)^j \mid \Pi(ij).$$

Let $\ell_r(i)$ be the sum of the r-adic digits of i. Then [Go2] Carlitz shows that, in fact,

$$\Pi(i)^j \Pi(j)^{\ell_r(i)} \mid \Pi(ij).$$

9.2. Bernoulli-Carlitz Numbers

In this subsection we will use the Carlitz factorial to define Carlitz's analog of Bernoulli numbers. We will state, but not prove, Carlitz's von-Staudt type theorem for these elements and, finally, we will present a table of them for $\mathbb{F}_3[T]$. In Subsection 8.20 the Bernoulli-Carlitz numbers were mentioned but not defined. The reason was that from the point of view of arithmetic applications the Bernoulli-Carlitz numbers give no new information (i.e., in the relevant range the Bernoulli-Carlitz numbers and the zeta-values have the same divisibility properties). On the other hand, the Bernoulli-Carlitz numbers have the advantage that their denominators may be computed.

As in the previous subsection, we set $\mathbf{A} := \mathbb{F}_r[T]$.

Definition 9.2.1. Let ξ be the period of the Carlitz module. If $i > 0$, $i \equiv 0$ $(r-1)$, then we set $BC_i := \Pi(i)\zeta_{\mathbf{A}}(i)/\xi^i$. If $i > 0$, $i \not\equiv 0$ $(r-1)$, then we set $BC_i := 0$.

The element BC_i is the *Bernoulli-Carlitz number of order i*. Thus if $e_C(z)$ is the exponential of the Carlitz module, then

$$\frac{z}{e_C(z)} = \sum_{i=0}^{\infty} \frac{BC_i}{\Pi(i)} z^i.$$

Using properties of the **A**-Hurwitz series, Carlitz then establishes the following von-Staudt result for his numbers. We will state it in the case $r \neq 2$. Recall that $r = p^{m_0}$.

Theorem 9.2.2. *Let i be written in its p-adic expansion as $\sum_{t=0}^{\nu} \beta_t p^t$. There are two conditions on i:*

1. $(p-1)m_0$ *divides* $\sum_{t=0}^{\nu} \beta_t$; *if this is so, set* $h := h_i := \left(\sum_{t=0}^{\nu} \beta_t \right) / ((p-1)m_0)$.
2. $(r^h - 1)$ *divides i.*

If i satisfies both conditions then the denominator of BC_i (in reduced form) is $\prod_{\substack{f \text{ monic prime} \\ \deg f = h}} f$. *If i does not satisfy both these conditions, then BC_i is an element of* **A**. □

Remark. 9.2.3. For $r = 2$, the result as originally stated in [C5], [C6], [Go2] is not correct. A corrected version is given in [Ge5], [C7]. We refer the interested reader there.

Finally, we present a table of the Bernoulli-Carlitz numbers for $\mathbb{F}_3[T]$ for $i = 2$ to $i = 80$; of course we need only mention those associated to *even* i. These computations were put into TEX *by hand* from computer printouts. The Bernoulli-Carlitz numbers are elements of $\mathbb{F}_3(T)$ and are presented in *reduced* form. Every effort has been made to make sure that all is correct; the author apologizes if unknown errors have been introduced in passing from the printout to TEX.

$i = 2$
$$\frac{2}{T^3 + 2T}$$

$i = 4$
$$\frac{1}{T^3 + 2T}$$

$i = 6$
$$\frac{2}{T^3 + 2T}$$

$i = 8$
$$\frac{1}{T^6 + T^4 + T^2 + 1}$$

$i = 10$
$$\frac{2T^6 + 2T^4 + 2T^2 + 1}{T^3 + 2T}$$

9.2. Bernoulli-Carlitz Numbers

$i = 12$
$$\frac{T^6 + T^4 + T^2 + 1}{T^3 + 2T}$$

$i = 14$
$$2$$

$i = 16$
$$\frac{T^6 + T^4 + T^2}{T^6 + T^4 + T^2 + 1}$$

$i = 18$
$$\frac{2T^{12} + T^{10} + 2T^6 + T^2 + 2}{T^3 + 2T}$$

$i = 20$
$$T^6 + T^4 + T^2 + 1$$

$i = 22$
$$2T^6 + 2T^4 + 2T^2$$

$i = 24$
$$\frac{T^{12} + 2T^{10} + 2T^6 + T^4}{T^6 + T^4 + T^2 + 1}$$

$i = 26$
$$\frac{2T^{27} + 2T^{21} + T^{19} + 2T^{15} + T^{13} + T^7}{T^{24} + T^{22} + T^{20} + T^{18} + T^{16} + T^{14} + T^{12} + T^{10} + T^8 + T^6 + T^4 + T^2 + 1}$$

$i = 28$

$T^{42} + T^{40} + T^{38} + 2T^{34} + 2T^{32} + 2T^{30} + 2T^{24} + 2T^{22} +$
$2T^{20} + 2T^{18} + T^{16} + T^{14} + T^{12} + 2T^{10} + 2T^2 + 1$

over

$T^3 + 2T$

$i = 30$

$2T^{42} + 2T^{40} + 2T^{38} + T^{36} + T^{34} + T^{32} + T^{24} + T^{22} + T^{20} +$
$T^{18} + 2T^{16} + 2T^{14} + 2T^{12} + T^4 + T^2 + 1$

over

$T^3 + 2T$

$i = 32$

$$T^{42} + T^{40} + T^{38} + 2T^{36} + 2T^{34} + 2T^{32} +$$
$$2T^{24} + 2T^{22} + 2T^{20} + T^{18} + T^{16} + T^{14} + 2T^6 + 2T^4 + 2T^2 + 2$$

over

$$T^6 + T^4 + T^2 + 1$$

$i = 34$

$$2T^{36} + T^{28} + 2T^{18} + T^2$$

$i = 36$

$$T^{48} + 2T^{46} + 2T^{42} + T^{38} + T^{34} + 2T^{32} + T^{30} + 2T^{28} + T^{24} + 2T^{20} +$$
$$T^{18} + 2T^{16} + T^{14} + T^{10} + 2T^6 + 2T^2 + 1$$

over

$$T^3 + 2T$$

$i = 38$

$$2T^{42} + 2T^{40} + 2T^{38} + T^{36} + T^{34} + T^{32} + 2T^{24} + 2T^{22} + 2T^{20} +$$
$$T^{12} + T^{10} + T^8 + 2T^6 + 2T^4 + 2T^2 + 2$$

$i = 40$

$$T^{48} + 2T^{46} + T^{42} + 2T^{40} + T^{36} + 2T^{34} + T^{30} + 2T^{28} +$$
$$T^{24} + 2T^{20} + T^{18} + 2T^{14} + T^{12} + 2T^8 + 2T^4 + T^2$$

over

$$T^6 + T^4 + T^2 + 1$$

$i = 42$

$$2T^{42} + 2T^{40} + 2T^{38} + T^{34} + T^{32} + T^{30} + T^{16} + T^{14} + T^{12} + 2T^8 + 2T^6 + 2T^4$$

$i = 44$

$$T^{39} + 2T^{37} + 2T^{33} + T^{31} + 2T^{15} + T^{13} + T^9 + 2T^7$$

$i = 46$

$$2T^{48} + T^{46} + 2T^{42} + T^{40} + 2T^{36} + T^{34} + 2T^{24} + T^{20} + 2T^{18} + T^{14} + 2T^{12} + T^8$$

$i = 48$

$$\frac{T^{54} + 2T^{36} + 2T^{28} + T^{10}}{T^6 + T^4 + T^2 + 1}$$

$i = 50$

$$2T^{45} + T^{39} + T^{37} + 2T^{31} + T^{27} + 2T^{21} + 2T^{19} + T^{13}$$

9.2. Bernoulli-Carlitz Numbers   357

$i = 52$
$$\frac{\begin{array}{c}T^{69} + T^{67} + T^{65} + 2T^{63} + 2T^{61} + 2T^{59} + 2T^{51} + \\ 2T^{49} + 2T^{47} + T^{39} + T^{37} + T^{35} + T^{27} + T^{25} + T^{23} + 2T^{21} + 2T^{19} + 2T^{17}\end{array}}{T^{24} + T^{22} + T^{20} + T^{18} + T^{16} + T^{14} + T^{12} + T^{10} + T^8 + T^6 + T^4 + T^2 + 1}$$

$i = 54$
$$\frac{\begin{array}{c}2T^{84} + T^{82} + 2T^{66} + T^{64} + 2T^{60} + T^{56} + 2T^{48} + T^{46} + \\ 2T^{42} + T^{38} + 2T^{36} + T^{28} + 2T^{24} + T^{20} + 2T^{18} + T^2 + 2\end{array}}{T^3 + 2T}$$

$i = 56$
$$\frac{\begin{array}{c}T^{84} + 2T^{82} + T^{66} + 2T^{64} + 2T^{58} + T^{56} + T^{48} + 2T^{46} + 2T^{40} + \\ T^{38} + T^{30} + 2T^{28} + T^{24} + 2T^{20} + T^{18} + T^{16} + T^{14} + T^{12} + T^{10} + T^8 + T^6 + 2T^2 + 1\end{array}}{T^6 + T^4 + T^2 + 1}$$

$i = 58$
$$\begin{array}{c}2T^{78} + 2T^{76} + 2T^{74} + T^{72} + 2T^{70} + 2T^{68} + T^{66} + 2T^{64} + 2T^{62} + T^{60} + 2T^{58} + 2T^{56} + \\ T^{54} + 2T^{52} + 2T^{50} + T^{48} + 2T^{46} + 2T^{44} + T^{42} + 2T^{40} + \\ 2T^{38} + T^{36} + 2T^{34} + 2T^{32} + T^{30} + 2T^{28} + 2T^{26} + \\ T^{24} + 2T^{22} + 2T^{20} + 2T^{18} + 2T^{16} + 2T^{14} + 2T^{12} + 2T^{10} + 2T^8 + 2T^6 + 2T^4 + 2T^2\end{array}$$

$i = 60$
$$\begin{array}{c}T^{78} + T^{76} + T^{74} + T^{72} + T^{70} + T^{68} + T^{66} + T^{64} + T^{62} + T^{60} + T^{58} + T^{56} + T^{52} + \\ T^{50} + T^{48} + T^{46} + T^{44} + T^{42} + T^{40} + T^{38} + T^{36} + T^{34} + \\ T^{32} + T^{30} + T^{26} + T^{24} + T^{22} + T^{20} + T^{18} + T^{16} + T^{14} + T^{12} + T^{10} + T^8 + T^6 + T^4\end{array}$$

$i = 62$
$$\begin{array}{c}2T^{75} + T^{73} + 2T^{67} + T^{65} + 2T^{59} + T^{57} + T^{51} + 2T^{47} + \\ T^{43} + 2T^{39} + T^{35} + 2T^{31} + 2T^{25} + T^{23} + 2T^{17} + T^{15} + 2T^9 + T^7\end{array}$$

$i = 64$
$$\frac{T^{90} + 2T^{84} + 2T^{24} + 2T^{22} + 2T^{20} + T^{18} + 2T^{16} + 2T^{14} + 2T^{10} + 2T^8}{T^6 + T^4 + T^2 + 1}$$

$i = 66$
$$\begin{array}{c}2T^{84} + T^{82} + 2T^{78} + 2T^{76} + 2T^{74} + 2T^{72} + 2T^{70} + 2T^{68} + \\ T^{64} + 2T^{62} + 2T^{58} + T^{56} + 2T^{52} + 2T^{50} + 2T^{48} + 2T^{46} + 2T^{44} + 2T^{42} + \\ T^{38} + 2T^{36} + 2T^{32} + T^{30} + 2T^{26} + 2T^{24} + 2T^{22} + 2T^{20} + 2T^{18} + 2T^{16} + T^{12} + 2T^{10}\end{array}$$

$i = 68$
$$T^{81} + 2T^{63} + 2T^{57} + 2T^{55} + T^{39} + T^{37} + T^{31} + 2T^{13}$$

$i = 70$
$$\begin{array}{c}2T^{81} + T^{75} + 2T^{73} + T^{67} + 2T^{65} + T^{63} + T^{59} + T^{55} + \\ 2T^{51} + T^{47} + 2T^{43} + 2T^{39} + 2T^{35} + T^{33} + 2T^{31} + T^{25} + 2T^{23} + T^{17}\end{array}$$

$i = 72$

$T^{96}+T^{94}+T^{92}+2T^{90}+2T^{88}+2T^{86}+2T^{78}+2T^{76}+2T^{74}+T^{72}+2T^{70}+2T^{68}+$
$T^{66}+2T^{64}+2T^{62}+2T^{60}+2T^{52}+2T^{50}+2T^{48}+T^{46}+2T^{44}+2T^{42}+$
$T^{40}+2T^{38}+2T^{36}+2T^{34}+2T^{26}+2T^{24}+2T^{22}+T^{20}+T^{18}+T^{16}$

over

$T^6+T^4+T^2+1$

$i = 74$

$2T^{87}+2T^{85}+2T^{83}+2T^{81}+2T^{69}+2T^{67}+2T^{65}+2T^{63}+T^{61}+T^{59}+$
$T^{57}+T^{55}+2T^{51}+2T^{49}+2T^{47}+2T^{45}+T^{43}+T^{41}+T^{39}+T^{37}+T^{25}+T^{23}+$
$T^{21}+T^{19}$

$i = 76$

$T^{87}+T^{85}+T^{83}+T^{69}+T^{67}+T^{65}+2T^{63}+2T^{61}+$
$2T^{59}+T^{51}+T^{49}+T^{47}+2T^{45}+2T^{43}+2T^{41}+2T^{27}+2T^{25}+2T^{23}$

$i = 78$

$2T^{111}+T^{109}+T^{105}+T^{103}+T^{101}+T^{97}+T^{95}+T^{91}+$
$T^{89}+T^{87}+T^{83}+2T^{81}+T^{57}+2T^{55}+2T^{51}+2T^{49}+2T^{47}+2T^{43}+2T^{41}+2T^{37}+$
$2T^{35}+2T^{33}+2T^{29}+T^{27}$

over

$T^{24}+T^{22}+T^{20}+T^{18}+T^{16}+T^{14}+T^{12}+T^{10}+T^8+T^6+T^4+T^2+1$

$i = 80$

$T^{156}+2T^{154}+T^{150}+2T^{148}+T^{144}+2T^{142}+T^{138}+2T^{136}+2T^{132}+$
$T^{128}+2T^{126}+T^{122}+2T^{120}+T^{116}+2T^{114}+T^{110}+T^{106}+2T^{104}+2T^{102}+$
$2T^{100}+2T^{98}+2T^{96}+2T^{94}+2T^{92}+2T^{90}+2T^{88}+2T^{86}+2T^{84}+$
$T^{82}+T^{78}+2T^{74}+T^{72}+2T^{68}+T^{66}+2T^{62}+T^{60}+2T^{56}+2T^{52}+$
$T^{50}+2T^{46}+T^{44}+2T^{40}+T^{38}+2T^{34}+T^{32}$

over

$T^{72}+T^{64}+T^{56}+T^{48}+T^{40}+T^{32}+T^{24}+T^{16}+T^8+1$

9.3. The Γ-ideal

In this subsection we will present an ideal-theoretic version of the Carlitz factorial. This was introduced in [Go13] which contained some errors (cf. [Go13, 4.3.2, 4.3.3]). The presentation given here, which uses local properties, is due to D. Thakur as given in Shu's paper [Shu1].

Let $\mathbf{A}$, $\mathbf{k}$, etc., be general. Recall that we set $[j] = T^{r^j} - T \in \mathbb{F}_r[T]$ for $j > 0$.

Definition 9.3.1. Let $x \in \mathbf{k}$.
1. We set $[j](x) := x^{r^j} - x$ for $j > 0$.
2. For such j, set $\widetilde{[j]}$ to be the ideal of $\mathbf{A}$ generated by $[j](a)$ for all $a \in \mathbf{A}$.

Lemma 9.3.2. We have
$$\widetilde{[j]} = \prod_{\substack{\wp \text{ prime} \\ \deg(\wp) | j}} \wp.$$

Proof. Let $\wp$ be a prime with $\deg(\wp) \mid j$. Let $a \in \mathbf{A}$. Clearly
$$[j](a) = a^{r^j} - a \equiv 0 \; (\wp).$$
Thus $\prod_{\deg \wp | j} \wp$ divides $\widetilde{[j]}$. Conversely, let a be an element of order 1 at $\wp$. Then $a^{r^j} - a = a(a^{r^j - 1} - 1)$ also has order 1 at $\wp$. The result follows readily. $\square$

Note that for $\mathbf{A} = \mathbb{F}_r[T]$, $\widetilde{[i]} = ([i](T)) = (T^{r^i} - T)$.

Definition 9.3.3. Let $j > 0$.
1. We set
$$D_j(x) := [j](x)[j-1]^r(x) \cdots [1]^{r^{j-1}}(x)$$
$$= (x^{r^j} - x)(x^{r^j} - x^r) \cdots (x^{r^j} - x^{r^{j-1}}).$$

2. We set $\widetilde{D_j}$ to be the ideal of $\mathbf{A}$ generated by $D_j(a)$ for all $a \in \mathbf{A}$.

Note, of course, that $D_j(x) = [j](x) D_{j-1}^r(x)$.

Lemma 9.3.4. *Let $\wp$ be a prime of $\mathbf{A}$ and let $g(x)$ be a polynomial with $\mathbf{A}$-coefficients. Then*
$$\min\{v_\wp(g(a)) \mid a \in \mathbf{A}\} = \min\{v_\wp(g(a)) \mid a \in \mathbf{A}_\wp\}.$$

Proof. $\mathbf{A}$ is dense in $\mathbf{A}_\wp$. $\square$

Proposition 9.3.5. *We have*
$$\widetilde{D}_j = \widetilde{[j]}\widetilde{[j-1]}^r \cdots \widetilde{[1]}^{r^{j-1}}.$$

Proof. Let $\wp$ be a prime of $\mathbf{A}$. It is clear that
$$v_\wp(\widetilde{[j]}\widetilde{[j]}^r \cdots \widetilde{[1]}^{r^{j-1}}) \leq v_\wp(\widetilde{D}_j).$$
Thus we need only show the reverse inequality.

Suppose $\deg(\wp) = d$. Without loss of generality we may assume $d \leq j$. Let
$$j = ed + f, \qquad 0 \leq f < d.$$
Then one sees that
$$v_\wp(\widetilde{[j]}\widetilde{[j-1]}^r \cdots \widetilde{[1]}^{r^{j-1}}) = r^{j-d} + r^{j-2d} + \cdots + r^{j-ed}.$$
Note that this formula only depends on r, j and d. For $\mathbf{A} = \mathbb{F}_r[T]$, it agrees with Sinnott's formula for $v_\wp(D_j) = v_\wp(\Pi(r^j))$. Thus for $\mathbb{F}_r[T]$ we deduce
$$\widetilde{[j]}\widetilde{[j-1]}^r \cdots \widetilde{[1]}^{r^{j-1}} = (D_j) = \widetilde{D}_j.$$

From Lemma 9.3.4 we see that we may compute $v_\wp(\widetilde{D}_j)$ locally. However $\mathbf{A}_\wp$ is isomorphic to $\mathbb{F}_r[T]_f$ where f is some prime polynomial of degree d. Thus the result follows from the special case $\mathbf{A} = \mathbb{F}_r[T]$. $\square$

As mentioned in the proof, for $\mathbf{A} = \mathbb{F}_r[T]$ we have $\widetilde{D}_j = (D_j)$. Moreover, the techniques of 9.3.5 give a different proof of 9.3.2.

Definition 9.3.6. Let $i = \sum_{j=0}^{m} c_j r^j$, $0 \leq c_j < r$.

1. We set $\Pi(i)(x) := \prod_{j=0}^{m} D_j(x)^{c_j}$.

2. We set $\widetilde{\Pi}(i)$ to be the ideal of $\mathbf{A}$ generated by $\Pi(i)(a)$ for all $a \in \mathbf{A}$.

The ideals defined in 9.3.6.2 are, with perhaps some historically bad notation, the Γ-*ideals*. As in Proposition 9.3.5 we have the following result.

Theorem 9.3.7. 1. $\widetilde{\Pi}(i) = \prod_{j=0}^{m} \widetilde{D}_j^{c_j}$.

2. *We have*
$$\widetilde{\Pi}(i) = \prod_{\wp \text{ prime}} \wp^{\alpha_\wp},$$
where $\alpha_\wp = \sum_{e \geq 1} [i/r^{e \deg \wp}]$. $\square$

Corollary 9.3.8. *Let* $\mathbf{A} = \mathbb{F}_r[T]$. *Then* $\widetilde{\Pi}(i) = (\Pi(i)(T))$. $\square$

Let **A** be general. Let d_∞ be the degree of infinity. Let **T**, **V**, ξ be as in Subsection 8.18. Finally let $j > 0$ be $\equiv 0 \ (r^{d_\infty} - 1)$. Let $\widetilde{\Pi}(j) \subset \mathbf{A}$ be the Γ-ideal. From 8.18.3 we know that

$$\zeta_{\mathbf{A}}(j)/\xi^j \in \mathbf{T}^*.$$

Question 9.3.9. Let $(\widetilde{\Pi}(j)\zeta_{\mathbf{A}}(j)/\xi^j)$ be the $\mathcal{O}_T$-fractional ideal generated by $\widetilde{\Pi}(j)\zeta_{\mathbf{A}}(j)/\xi^j$. Can one prove a von-Staudt result for this ideal generalizing Theorem 9.2.2?

9.4. The Arithmetic Γ-function

In this subsection we show how the Carlitz factorial, and some generalizations, can be fit into continuous Γ-functions as in [Go3], [Go2] and [Th6]. This construction is heavily influenced by the classical Morita p-adic Γ-function.

Let $\mathbf{A} = \mathbb{F}_r[T]$ and, for now, let $\pi = 1/T \in \mathbf{K} = \mathbb{F}_r\left(\left(\frac{1}{T}\right)\right)$; later we will allow π to be an arbitrary uniformizer of the form $\pi = 1/T + \{\text{higher terms}\}$. As in Subsection 8.1, for monic $a \in \mathbf{A}$ we set

$$\langle a \rangle := \pi^{\deg(a)} a = T^{-\deg(a)} \cdot a.$$

Thus $\langle a \rangle$ is a 1-unit $\in \mathbf{K}$.

Lemma 9.4.1. $\langle D_i \rangle \to 1 \in \mathbf{K}$ as $i \to \infty$.

Proof. By definition we have

$$D_i = (T^{r^i} - T^{r^{i-1}}) \cdots (T^{r^i} - T).$$

Thus

$$\langle D_i \rangle = (1 - T^{r^{i-1}-r^i}) \cdots (1 - T^{1-r^i}).$$

Therefore, in particular,

$$\langle D_i \rangle \equiv 1 \ (T^{(r^{i-1}-r^i)}).$$

However as $i \to \infty$, $r^{i-1} - r^i = r^{i-1}(1-r) \to -\infty$ and the result follows. $\square$

Definition 9.4.2. Let $y = \prod\limits_{j=0}^{\infty} c_j r^j$, $0 \leq c_j < r$ for all j. Then we set

$$\Pi_\infty(y) := \prod_{j=0}^{\infty} \langle D_j \rangle^{c_j}.$$

It is easy to see that $\Pi_\infty(y) \colon \mathbb{Z}_p \to \mathbf{K}^*$ is continuous.

Next we present Thakur's beautiful calculation of $\Pi_\infty(-1)$ in terms of the period ξ of the Carlitz module. In Subsection 3.2 we showed that

$$\xi = \sqrt[r-1]{T - T^r}\,\xi_*$$

where

$$\begin{aligned}
\xi_* &= \prod_{j=1}^{\infty} \left(1 - \frac{T^{r^j} - T}{T^{r^{j+1}} - T}\right) \\
&= \prod_{j=1}^{\infty} \left(\frac{T^{r^{j+1}} - T - T^{r^j} + T}{T^{r^{j+1}} - T}\right) \\
&= \prod_{j=1}^{\infty} [1]^{r^j}/[j+1] \\
&= \prod_{j=0}^{\infty} [1]^{r^j}/[j+1].
\end{aligned}$$

Thus

$$\begin{aligned}
\xi_* &= \lim_{k \to \infty} \prod_{j=0}^{k-1} [1]^{r^j}/[j+1] \\
&= \lim_{k \to \infty} \left(\prod_{j=0}^{k-1} [1]^{r^j}\right) / ([1] \cdots [k]) \\
&= \lim_{k \to \infty} [1]^{\frac{r^k-1}{r-1}}/([1] \cdots [k]).
\end{aligned}$$

Recall that in Subsection 3.2 we let θ be the positive (= 1-unit) $(r-1)$-st root of $1 - T^{1-r}$ where $1 - T^{1-r} = \langle[1]\rangle$. We then set

$$\xi_u := \theta \xi_*;$$

so ξ_u is the 1-*unit part* of ξ. From above we see that

$$\xi_u = \lim_{k \to \infty} T^{-r}[1]^{r^k/(r-1)}/([1] \cdots [k]),$$

where, again, we take the positive $(r-1)$-st root. Thus, as $\langle \xi_u \rangle = \xi_u$, we see

$$\begin{aligned}
\xi_u &= \lim_{k \to \infty} \langle[1]\rangle^{r^k/(r-1)}/\langle[1] \cdots [k]\rangle \\
&= \lim_{k \to \infty} \theta^{r^k}/\langle[1] \cdots [k]\rangle.
\end{aligned}$$

However, $\lim_{k \to \infty} \theta^{r^k} = 1$. So, finally, we have

$$\xi_u = \lim_{k \to \infty} \langle[1] \cdots [k]\rangle^{-1}.$$

Theorem 9.4.3 (Thakur). *We have*
$$\Pi_\infty(-1) = \xi_u.$$

Proof. By definition,
$$\Pi_\infty(-1) = \lim_{k\to\infty} \langle (D_0 \cdots D_k)^{r-1} \rangle.$$

Note that
$$D_{i+1} = [i+1]D_i^r$$
as is easily seen. Thus
$$D_i^{r-1} = D_{i+1}/([i+1]D_i),$$
and so
$$(D_0 \cdots D_{k-1})^{r-1} = D_k/([1]\cdots[k]).$$
Therefore
$$\langle (D_0 \cdots D_{k-1})^{r-1} \rangle = \langle D_k \rangle / \langle [1]\ldots[k] \rangle.$$
But as $k \to \infty$, we know that $\langle D_k \rangle \to 1$. If we combine this with the remark just before the statement of the theorem, the proof is complete. □

Since $\langle D_i^{r-1} \rangle = \prod_{\deg g = i} \langle g \rangle$, this result is in agreement with Theorem 7.10.10.

Corollary 9.4.4. *Let $0 < a < r$. then*
$$\Pi_\infty(a/(1-r)) = \xi_u^{\frac{a}{r-1}},$$
where we take the positive root.

Proof. This follows exactly as before. □

Next we discuss the effect of changing the parameter at ∞. We first need a lemma. Note that $\deg(D_j) = jr^j$ as a polynomial in T.

Lemma 9.4.5. *The function $j \mapsto \deg(\Pi(j))$ interpolates p-adically to a continuous function $\deg_\Pi(y): \mathbb{Z}_p \to \mathbb{Z}_p$.*

Proof. If $y = \sum_{j=0}^{m} c_j r^j$, $0 \leq c_j r$, then $\deg(\Pi(y)) = \sum_{j=0}^{m} c_j j r^j$. This clearly interpolates to the continuous function $\mathbb{Z}_p \to \mathbb{Z}_p$ which takes $y = \sum_{j=0}^{\infty} c_j r^j$ to $\sum_{j=0}^{\infty} c_j j r^j$. □

It might be of interest to find the Mahler coefficients of the above function. Now let π_1 be another monic of the form $\pi_1 = 1/T + \{\text{higher terms}\}$. Set

$$v := \pi_1/\pi \in \mathbf{K};$$

so v is a 1-unit. We now define $\langle D_j \rangle_1$ by

$$\langle D_j \rangle_1 = \pi_1^{\deg D_j} D_j$$
$$= \pi_1^{jr^j} D_j$$
$$= v^{jr^j} \langle D_j \rangle.$$

Note that $\langle D_j \rangle_1 \to 1$ as $j \to \infty$ and form $\Pi_{\infty,1}(y) \colon \mathbb{Z}_p \to \mathbf{K}^*$ just as for $\Pi_\infty(y)$; clearly $\Pi_{\infty,1}(y)$ is also continuous, etc.

Proposition 9.4.6. $\Pi_{\infty,1}(y)/\Pi_\infty(y) = v^{\deg_\Pi(y)}$.

Proof. Immediate from the definitions. □

In particular,

$$\Pi_{\infty,1}(-1) = v^{\deg_\Pi(-1)} \Pi_\infty(-1)$$
$$= v^{\deg_\Pi(-1)} \xi_u$$

by Theorem 9.4.3. Note that

$$\deg_\Pi(-1) = \sum_{j=0}^\infty (r-1)jr^j$$
$$= (r-1)\frac{r}{(1-r)}$$
$$= \frac{r}{r-1}.$$

On the other hand this is precisely the degree of the period ξ and $Z'_{0,\mathbf{A}}(1)$ as in Example 7.10.12. This may be viewed as analogous to the classical Formula of Lerch.

Since

$$\xi = {}^{r-1}\!\!\sqrt{-T^r}\xi_u.$$

We see that

$$\xi^{r-1} = -T^r \xi_u^{r-1} \qquad (\in \mathbf{K})$$
$$= -\pi_1^{-r} \cdot \left(\frac{\pi_1}{\pi}\right)^r \xi_u^{r-1}$$
$$= -\pi_1^{-r}(v^r \xi_u^{r-1}),$$

or

$$\xi = {}^{r-1}\!\!\sqrt{-\pi_1^{-r} v^{\frac{r}{r-1}} \xi_u}$$
$$= {}^{r-1}\!\!\sqrt{-\pi_1^{-r} \cdot \Pi_{\infty,1}(-1)};$$

9.4. The Arithmetic Γ-function

thus giving a formula valid for any positive uniformizer. (We shall see later on that this also will follow from Gekeler's calculations of Subsection 7.9.)

The above calculations imply that one may do somewhat better in terms of expressing ξ in terms of a gamma function. Let $U_1 \subset \mathbf{K}^*$ be the subgroup of 1-units and let us use the sign function, sgn, given by $\pi = 1/T$ to decompose $\mathbf{K}^*$ as

$$\mu \times T^{\mathbb{Z}} \times U_1$$

where $\mu \subset \mathbf{K}^*$ is the subgroup of roots of unity. Define $\widehat{\mathbf{K}}^*$ to be the p-adic completion of $\mathbf{K}^*$ which, via the above decomposition, is isomorphic to

$$\mathbb{Z}_p \times U_1$$

(as U_1 is already p-adically complete).

Definition 9.4.7. Let $y \in \mathbb{Z}_p$. We define $\Pi(y) \in \widehat{\mathbf{K}}^*$ by

$$\Pi(y) := T^{\deg_\Pi(y)} \Pi_\infty(y).$$

Clearly $\Pi(y)$ is continuous. Note that $\xi \notin \mathbf{K}^*$ but that $\xi^{(r-1)}$ obviously does, as we have seen. Let $\widehat{\xi}^{(r-1)}$ be the image of $\xi^{(r-1)}$ in $\widehat{\mathbf{K}}^*$. One then sees directly that

$$\Pi(-1) = (\widehat{\xi}^{(r-1)})^{\frac{1}{r-1}},$$

where the root is unique as $(r-1) \in \mathbb{Z}_p^*$. Note also that $\Pi(y)$ is *independent* of the choice of parameter and may be expanded in terms of the images of the various D_j in $\widehat{\mathbf{K}}^*$.

Definition 9.4.8. We set

$$\Gamma_\infty(y) := \Pi_\infty(y - 1),$$

and

$$\Gamma(y) := \Pi(y - 1).$$

The function $\Gamma_\infty(y)$ is the *arithmetic gamma function*. The reason for the use of "arithmetic" in this name will be apparent from our next result. The reason for the shift from y to $y - 1$ lies in the functional equation for Γ presented in our next subsection. However, the shift also puts certain special values in line with the classical gamma function. For instance, suppose that $2 \nmid r$, then

$$\Gamma_\infty(1/2) = \Pi_\infty\left(-\frac{1}{2}\right) = \Pi_\infty\left(\frac{r-1}{2} \Big/ (1-r)\right) = \xi_u^{1/2}$$

by Corollary 9.4.4.

It is clear that for $y \in \mathbb{N}$, $\Pi_\infty(y) \in \mathbf{k}$. On the other hand, Wade has shown that ξ, and thus, ξ_u is transcendental. As such we deduce that $\Pi_\infty\left(-\frac{a}{r-1}\right)$ is transcendental for $0 < a < r$. Building on some work of D. Thakur, J. Yu, A. Thiery, and L. Denis, J.-P. Allouche [Al1] has recently established that $\Pi_\infty(y)$ is transcendental for all $y \in (\mathbb{Q} \cap \mathbb{Z}_p)\setminus\mathbb{N}$. For more history on this result, we refer to [Th8] and [Al1].

Let t be a positive integer and let $\mathbf{A}_t = \mathbb{F}_{r^t}[T]$. For each such $\mathbf{A}_t$ let $_tC$ be the Carlitz module given by $_tC_T(\tau_t) = T\tau_t^0 + \tau_t$, where $\tau_t(x) = x^{r^t}$. Let $_t\xi$ be the period of $_tC$ with 1-unit part, $_t\xi_u$, as before.

Theorem 9.4.9 (Thakur). *We have*

$$_t\xi_u = \Pi_\infty(r^{t-1}/(1-r^t))^r/\Pi_\infty(1/(1-r^t)).$$

Proof. By definition, we have

$$\Pi_\infty\left(\frac{1}{1-r^t}\right)\Pi_\infty\left(\frac{r^{t-1}}{1-r^t}\right)^{-r} = \lim_{n\to\infty}\left\langle\frac{D_{tn}\cdots D_0}{D_{tn-1}^r\cdots D_{t-1}^r}\right\rangle$$

$$= \lim_{n\to\infty}\langle[tn]\cdots[t]\rangle$$

$$= {_t\xi_u^{-1}}.$$

The result follows. □

In [Th6], it is explained that Theorem 9.4.9 is the *Chowla-Selberg formula for constant field extensions*. This provides justification for calling $\Gamma_\infty(y)$ the "arithmetic Γ-function."

We now explain how to generalize the above Γ-construction to arbitrary $\mathbf{A}$; for more details, see [Th6]. Recall that d_∞ is the degree of the place ∞ of $\mathbf{k}$. Let $I \subseteq \mathbf{A}$ be an ideal. As in Subsection 7.8, for $j \geq 0$ we set

$$I_j := \{\alpha \in I \mid \deg(\alpha) \leq j\}.$$

Of course $\deg(\alpha) \equiv 0 \ (d_\infty)$. Moreover the Riemann-Roch theorem tells us that for $j \gg 0$,

$$\#I_{jd_\infty} = r^{jd_\infty + c},$$

for some fixed constant c.

Choose now a sign function sgn of $\mathbf{K}$ and let ψ be a sgn normalized rank one Drinfeld module with lattice $\Lambda = I\xi$.

Definition 9.4.10. Let $j \geq 0$. We set

$$D_j := D_j^I := \prod_J J^{s_1}$$

where $s_1 \in S_\infty$ is defined as in Subsection 8.1 and J runs over all *principal* ideals such that $J \subseteq I$ and $\deg(J) = jd_\infty$. If there are no such elements then we set $D_j := 1$.

Remark. 9.4.11. In practice, D_j is computed as follows. Let I_{jd_∞} be as above. Let $\widetilde{I}_{jd_\infty} \subseteq I_{jd_\infty}$ be those elements of degree equal to jd_∞. Let $\{\alpha_1, \ldots, \alpha_e\}$ be representatives of the orbits of the obvious action of $\mathbb{F}_r^*$ on $\widetilde{I}_{jd_\infty}$. Then

$$D_j = \prod_i \alpha_i / \mathrm{sgn}(\alpha_i).$$

Let $\langle D_j \rangle$ also be defined as in Subsection 8.1. From the above equality we conclude that, while our definition of D_j differs from that of [Th6] when $d_\infty > 1$, our definition of $\langle D_j \rangle$ agrees with [Th6] always. It also agrees with our earlier definition for $\mathbf{A} = \mathbb{F}_r[T]$.

Lemma 9.4.12. *As $j \to \infty$, $\langle D_j \rangle \to 1$ in* $\mathbf{K}$.

Proof. Clearly $\langle D_j \rangle^{r-1} = U_j$ where U_j is given in Definition 7.9.6. Thus the result follows from Proposition 7.9.7. □

The reader may now easily define $\Pi_\infty(y)$ and $\Gamma_\infty(y)$ for general $\mathbf{A}$. Theorem 7.10.10 immediately implies that $\Pi_\infty(-1) = \Gamma_\infty(0) = u_{0,I}$ where $u_{0,I}$ is given in 7.9.8.1. Clearly 7.10.10 implies that $u_{0,I}$ is the *1-unit part ξ_u of ξ*; so our formula agrees with the earlier one for $\mathbf{A} = \mathbb{F}_r[T]$.

We leave it to the reader to complete $\mathbf{K}^*$ (as above and in [Th6]) to obtain a formula which also includes the degree part of ξ (as we did earlier for $\mathbb{F}_r[T]$).

In Subsection 9.3 we defined the Γ-ideal $\widetilde{\Pi}(i)$. Let $\mathbf{V}$ be the value field (of Section 8) with $\mathbf{A}$-integers $\mathcal{O}_\mathbf{V}$. In Theorem 9.3.7 we showed

$$\widetilde{\Pi}(i) = \prod_{\wp \text{ prime}} \wp^{\alpha_\wp}$$

where

$$\alpha_\wp = \sum_{e \geq 1} [i/r^{e \deg(\wp)}].$$

In $\mathcal{O}_\mathbf{V}$ we have

$$\widetilde{\Pi}(i)\mathcal{O}_\mathbf{V} = \left(\prod_{\wp \text{ prime}} (\wp^{s_1})^{\alpha_\wp} \right);$$

set

$$\widehat{\Pi}(i) := \prod_{\wp \text{ prime}} (\wp^{s_1})^{\alpha_\wp}$$

thus defining an *element* of $\mathcal{O}_\mathbf{V}$.

Question 9.4.13. Does the function $\mathbb{N} \to \mathbf{C}_\infty^*, i \mapsto \widehat{\Pi}(i)$, interpolate to continuous functions generalizing $\Pi_\infty(y)$, $\Pi(y)$, etc.? If so, what is the relationship of these new functions to $\Pi_\infty(y)$, $\Pi(y)$, etc.?

9.5. Functional Equations

We will now present Thakur's functional equations [Th6] for $\Pi_\infty(y)$ and $\Gamma_\infty(y)$, etc. As the reader will see, these results are of a purely formal nature involving only manipulations of p-adic digits.

Thus let f be a function defined on $\mathbb{Z}_p$ of the form

$$f(y) = \prod_{j=0}^{\infty} A_j^{a_j}$$

where $y = \sum_{j=0}^{\infty} a_j r^j$, $0 \leq a_j < r$, $\{A_j\}$ is a collection of non-zero elements in a complete non-Archimedean field, and where we assume convergence of the (possibility) infinite product.

Lemma 9.5.1. *Let $0 \neq y \in \mathbb{Z}_p$. Then $f(y)/f(y-1)$ depends only on $\mathrm{ord}_r(y)$.*

Proof. This is obvious from the definitions. □

Proposition 9.5.2. *We have $f(y-1)f(-y) = f(-1)$.*

Proof. Let $-y = \sum_{j=0}^{\infty} c_j r^j$, $0 \leq c_j < r_j$ thus

$$y - 1 = -1 - (-y)$$
$$= \sum_{j=0}^{\infty}((r-1) - c_j)r^j.$$

Therefore, $f(y-1) = \prod A_j^{r-1-c_j}$ and

$$f(y-1)f(-y) = (\prod A_j^{r-1-c_j})(\prod A_j^{c_j})$$
$$= \prod A_j^{r-1}$$
$$= f(-1).$$
□

The "f-function" is obviously a generalization of $\Pi_\infty(y)$, etc. To obtain a "Γ-version" we simply set

$$g(y) := f(y-1).$$

Corollary 9.5.3. *We have $g(y)g(1-y) = g(0)$.*

Proof. By the proposition

$$f(y-1)f(-y) = f(-1).$$

But $f(-1) = g(0)$, $f(y-1) = g(y)$ and $f(-y) = g(1-y)$. Thus the result follows immediately. □

Corollary 9.5.3 is *precisely* the reason we have defined $\Gamma_\infty(y)$, etc., in the manner that we have.

Next we shall turn to Thakur's analog of the classical *multiplication formula* for the Γ-function. Let n be a positive number *prime* to p. If $p = 2$, then n is odd; thus $\frac{n-1}{2} \in \mathbb{N}$. If $p > 2$, then $(2,p) = 1$ and $g(0)^{1/2} = f(-1)^{1/2} = \prod A_j^{\frac{r-1}{2}}$. In any case, the element

$$g(0)^{\frac{n-1}{2}}$$

is unambiguously defined.

Proposition 9.5.4. *Let $y \in \mathbb{Z}_p$ and n be prime to p as above. Then*

$$g(y)g\left(y + \frac{1}{n}\right)\cdots g\left(y + \frac{n-1}{n}\right) = g(0)^{\frac{n-1}{2}} g(ny).$$

Proof. As $(n,p) = 1$, $\text{ord}_r(ny) = \text{ord}_r(y)$. Moreover by Lemma 9.5.1 we know that $f(y)/f(y-1) = g(y+1)/g(y)$ depends only on $\text{ord}_r(y)$.

Next, let us check the formula for $y = \frac{1}{n}$. In this case, we must show

$$g\left(\frac{1}{n}\right)\cdots g\left(\frac{n-1}{n}\right) = g(0)^{\frac{n-1}{2}}.$$

The result then follows by pairing $g(a/n)$ with $g(1 - a/n) = g\left(\frac{n-a}{n}\right)$ and using 9.5.3.

We now assume that we know that the result for some y and show that it is also true for $y + \frac{1}{n}$. But note that $n(y + \frac{1}{n}) = ny + 1$ and

$$\frac{g(y)g(y+\frac{1}{n})\cdots g(y+\frac{n-1}{n})}{g(ny)} \cdot \frac{g(ny+1)}{g(y+\frac{1}{n})g(y+\frac{2}{n})\cdots g(y+1)}$$

$$= \frac{g(y)}{g(y+1)} \cdot \frac{g(ny+1)}{g(ny)}.$$

As $\text{ord}_r y = \text{ord}_r ny$, the above remark gives

$$\frac{g(y)}{g(y+1)} \cdot \frac{g(ny+1)}{g(ny)} = 1,$$

and we see the result for $y + \frac{1}{n}$. Thus, by induction it is true for all elements in $\mathbb{Z}_p$ of the form a/n, where $a \geq 1$. These are *dense* in $\mathbb{Z}_p$ and so the result follows by continuity. □

The above results obviously apply immediately to $\Pi_\infty(y)$, $\Pi(y)$, $\Gamma_\infty(y)$, etc. For instance, if $\mathbf{A} = \mathbb{F}_r[T]$ and ξ is the period of the Carlitz module, then
$$\Gamma_\infty(y)\Gamma_\infty(1-y) = \Gamma_\infty(0) = \xi_u,$$
etc.

In [Th6] the interested reader will find a more general functional equation than the one presented here.

9.6. Finite Interpolations

In this subsection we describe how the Carlitz factorial and its generalizations can be interpolated at finite primes of $\mathbf{A}$ to continuous analogs of Morita's p-adic Γ-function.

As we did at ∞, we will begin with $\mathbf{A} = \mathbb{F}_r[T]$ and the Carlitz factorial. Let $v \in \operatorname{Spec}(\mathbf{A})$. Recall that $D_i = \prod_{\substack{n \text{ monic} \\ \deg n = i}} n$.

Definition 9.6.1. We set $D_{0,v} := 1$, and for $i > 0$,
$$D_{i,v} := \prod_{\substack{n \text{ monic} \\ \deg n = i \\ v \nmid n}} n.$$

Lemma 9.6.2. *We have* $-D_{i,v} \to 1$ *in* $\mathbf{k}_v$ *as* $i \to \infty$.

Proof. Let $\deg v = d$ and write for $i \gg 0$, $i = ed + j$ where $0 \leq j < d$. We consider $D_{i,v}$ modulo v^e. Let n be monic of degree i and write $n = f \cdot v^e + h$, where we abuse language and let v^e also denote a monic generator of the ideal v^e, and where $\deg f = j$ and $\deg h < ed$. Thus $v \nmid n \Leftrightarrow v \nmid h$. Therefore,
$$\prod_{\substack{\deg n = i \\ n \text{ monic} \\ v \nmid n}} n \equiv (\prod h)^{r^j} \pmod{v^e},$$
where h runs through *all* polynomials of degree $< ed$ prime to v.

By pairing an element with its inverse, it is very well-known that the product of *all* elements in a finite abelian group equals the product of the elements of order 2. If $p \neq 2$, then the elements in $(\mathbf{A}/v^e)^*$ of order 2 form a vector space over $\mathbb{Z}/(2)$ which will have dimension > 1 for $e \gg 0$. It is then easy to see that the product of the elements of order 2 vanishes. The result now follows directly since $e \to \infty$ as $i \to \infty$. □

Definition 9.6.3. Let $y \in \mathbb{Z}_p$ be written r-adically as $\sum_{j=0}^{\infty} c_j r^j$, $0 \le c_j < r$. Then we set
$$\Pi_v(y) := \prod_j (-D_{j,v})^{c_j},$$
and
$$\Gamma_v(y) := \Pi_v(y-1).$$
These are continuous functions from $\mathbb{Z}_p$ to $\mathbf{A}_v^*$.

Thus $\Pi_v(y)$ and $\Gamma_v(y)$ are defined exactly as $\Pi_\infty(y)$ and $\Gamma_\infty(y)$. They therefore have functional equations, etc., by our last subsection. Clearly all that is left to do is compute $\Gamma_v(0) = \Pi_v(-1)$.

Theorem 9.6.4 (Thakur). *We have*
$$\Pi_v(-1) = \Gamma_v(0) = (-1)^{\deg(v)-1}.$$

Proof. This proof is modeled on the proof of 9.6.2, the difference being that we do *not* just focus on one $D_{j,v}$. More precisely let $\deg(v) = d$ and let $e \ge 1$. Clearly, $\Pi_v(-1) = \lim_{e \to \infty} \prod_{j=0}^{ed-1} (-D_{j,v})^{(r-1)}$. Thus we focus on computing $\prod_{j=0}^{ed-1} (-D_{j,v})^{(r-1)} = \prod_{j=0}^{ed-1} D_{j,v}^{(r-1)}$ modulo v^e. But if n is monic, then clearly $n^{(r-1)} = -\prod_{\zeta \in \mathbb{F}_r^*} (\zeta n)$. Moreover, as in the proof of 9.6.2
$$\prod_{\substack{\deg(h) < ed \\ (h,v)=1}} h \equiv (-1) \pmod{v^e}$$
(i.e., we have removed the monicity assumption). Thus we find
$$\prod_{j=0}^{ed-1} (D_{j,v})^{(r-1)} \equiv (-1)^m (-1) \pmod{v^e}$$
where m is the number of numbers of monics of degree $< de$ prime to v. This is clearly the number of all such elements divided by $(r-1)$. A simple calculation then implies that $m \equiv d \pmod 2$ giving the result. □

Now let $\mathbf{A}$ be arbitrary and let $I \subseteq \mathbf{A}$ be an ideal. Recall that in 9.4.10 we defined

$$D_j = D_j^I = \prod_J J^{s_1}$$

with $s_1 \in S_\infty$ as in Subsection 8.1 and where J runs over all *principal* ideals such that $J \subseteq I$ and $\deg(J) = jd_\infty$. Thus, if $\{\alpha_1, \ldots, \alpha_e\}$ are representatives of the action of $\mathbb{F}_r^*$ on the elements in I of degree jd_∞ we have

$$D_j = \prod_i \alpha_i / \mathrm{sgn}(\alpha_i).$$

Thus D_j depended only on the choice of parameter π at infinity and I.

Now let $v \in \mathrm{Spec}(\mathbf{A})$ be prime to I. We want to interpolate the elements D_j at v. To do this we need to specify more carefully our choice of $\{\alpha_i\}$ via a device due to Thakur [Th6]. Let

$$S = \{1 = \lambda_1, \ldots, \lambda_t\}, \qquad t = \frac{r^{d_\infty} - 1}{r - 1},$$

be a fixed set of representatives of $\mathbb{F}_\infty^*/\mathbb{F}_r^*$. We now choose $\{\alpha_i\}$ by simply mandating that their signs lie in S. We denote the set $\{\alpha_i\}$ by S_j^I.

Definition 9.6.5. We set

$$D_{j,v} := D_{j,v}^I = \prod_{v \nmid \alpha_i} \alpha_i$$

where $\alpha_i \in S_j^I$.

Notice that $D_{j,v}$ depends on I, j and π. If $d_\infty > 1$, it also depends on S.

Lemma 9.6.6. *We have* $(-1)^{d_\infty} D_{j,v}^I \to 1$ *in* $\mathbf{k}_v$ *as* $j \to \infty$.

Proof. Let $(I/v^m I)^* := I/v^m I - vI/v^m I$. As $(v, I) = 1$, representatives of $(I/v^m I)^*$ are *also* representatives of $(\mathbf{A}/v^m)^*$. As such, and as in the proof of 9.6.2, we see their product is $\equiv -1 \ (v^m)$, for $m \gg 0$.

Now let m be fixed and let $j \gg 0$ be chosen so that the elements in S_j^I of *fixed* sign cover all of $(I/v^m I)^*$ (as can be done by Riemann-Roch for instance). Any two elements of the same sign in S_j^I differ by an element in $v^m I$ of degree $< jd_\infty$, and, conversely, adding such elements does not effect the sign *or* the coset in $(I/v^m I)^* \simeq (\mathbf{A}/v^m)^*$ (as $\mathbf{A}$-modules). Let r^t be the number of elements of $v^m I$ of degree $< jd_\infty$; then the number of elements of fixed sign is

$$r^t \cdot \#(I/v^m I)^* = r^t \cdot r^{(m-1)\deg v}(r^{\deg v} - 1),$$

and modulo v^m these elements product to

$$(-1)^{r^t} = -1.$$

There are $\frac{r^{d_\infty}-1}{r-1}$ possible signs; so the product of *all* elements in S_j^I mod v^m is
$$(-1)^{(r^{d_\infty}-1)/(r-1)} = (-1)^{d_\infty}.$$
The result follows immediately. □

Using 9.6.6 we deduce the existence of continuous functions $\Pi_v(y) = \Pi_v^I(y): \mathbb{Z}_p \to \mathbf{A}_v^*$ and $\Gamma_v(y) = \Gamma_v^I(y) = \Pi_v^I(y-1)$, etc., as for $\mathbf{A} = \mathbb{F}_r[T]$. As before, we need only compute $\Pi_v(-1) = \Gamma_v(0)$.

Theorem 9.6.7. *We have $\Pi_v(-1) = \Gamma_v(0) = (-1)^{\deg v - 1}$.*

Proof. Let m be fixed and let $j \gg 0$ be chosen so that the elements in $I - vI$ of degree $\leq j d_\infty$ cover $(I/v^m I)^*$. If α, β project to the same class then they differ by an element in $v^m I$ of degree $< j d_\infty$. Conversely, adding such an element does not change the residue class in $v^m I$. Thus each residue class will be hit a power of r times; say r^t times. Now the product of all elements $< j d_\infty$ in $I - vI$ is found as in the proof of 9.6.4 to be
$$(-1)^a \prod_{i=0}^{j} (D_{i,v})^{(r-1)} = (-1)^a \prod_{i=0}^{j} ((-1)^{d_\infty} D_{i,v})^{(r-1)},$$
where a is the number of elements with signs in S.

As in the proof of 9.6.4 we see that $(-1)^a = (-1)^{\deg v}$. Thus modulo $v^m I$, we see
$$\prod_{i=0}^{j} ((-1)^{d_\infty} D_{i,v})^{(r-1)} = (-1)^{r^t} \cdot (-1)^a \pmod{v^m I}$$
$$= (-1)^{\deg v - 1},$$
which implies the result upon letting m go to infinity. □

Remark. 9.6.8. Let $\mathbf{A} = \mathbb{F}_r[T]$. Let $v \in \mathrm{Spec}(\mathbf{A})$ have degree d. Let $\pi = \frac{1}{T}$ and suppose that $\pi_1 = \zeta/T$ for some $1 \neq \zeta \in \mathbb{F}_r^*$. Thus π and π_1 are positive for different choices of sign. Let $D_i = \prod_{\substack{n \text{ monic} \\ \deg n = i}} n$ be chosen as before using monics for π, and let $\widetilde{D}_i$ be chosen using monics for π_1; similarly we define $D_{i,v}$ and $\widetilde{D}_{i,v}$. Thus $D_{i,v}$ and $\widetilde{D}_{i,v}$ differ by a power of ζ. If $i \geq d$, then a simple counting implies that this power is divisible by $(r-1)$ and so $D_{i,v} = \widetilde{D}_{i,v}$. Thus there is an open subset U of $\mathbb{Z}_p$ such that $\Pi_v(y)$ on U is independent of the sign. A similar phenomenon happens for arbitrary $\mathbf{A}$.

In our next section we will construct a different v-adic Γ-function.

9.7. Another v-adic Γ-function

In this subsection we present the remarkable construction of another type of v-adic Γ-function by Thakur [Th1]. Let **A** be arbitrary and let sgn be our fixed sign function. Let ψ be a sgn-normalized rank one Hayes module defined over the **A**-integers of the extension $\mathbf{H}^+$ of **k**. Let

$$e(z) := e_\psi(z) := \sum_{j=0}^{\infty} e_j z^{r^j}$$

be the exponential function of ψ. In Proposition 7.11.7 we mentioned a result due to Thakur that assures that $e_j \neq 0$ for all j if **k** has a place of degree 1. Our first result is to prove a special case of this here.

Lemma 9.7.1. *Suppose $d_\infty = 1$. Then $e_j \neq 0$ for all $j \geq 0$.*

Proof. Let $\Lambda = \xi I$, where $I \subseteq \mathbf{A}$ is an ideal. Thus

$$e(z) = z \prod_{0 \neq \alpha \in I} (1 - z/\xi\alpha).$$

One can now expand this product out to obtain a sum formula for e_j. As $d_\infty = 1$, one is able to use the Riemann-Roch Theorem to show that, *for precisely the sums occurring in the formulas for e_j*, one has a unique non-zero element of highest degree in each such sum. Thus the sum can never vanish. □

In [Th1] the reader will find other criterion for the non-vanishing of e_j, $j \geq 0$. It *may* be the case that it is always true that $e_j \neq 0$ for $j \geq 0$; not enough is known one way or the other as of this writing. In any case, *for the rest of this section we will assume that $e_j \neq 0$ for all $j \geq 0$.*

With the above assumption we can write

$$e(z) = \sum_{j=0}^{\infty} z^{r^j}/d_j$$

where $d_j := 1/e_j$. By definition $\{d_j\} \subseteq \mathbf{H}^+$.

In order to proceed further, we now place ourselves in the situation of "two T's" as given in Subsection 7.11. Thus we view ψ as being defined over H^+, etc., where C_∞ is a second copy of $\mathbf{C}_\infty$ and $H^+ \subset C_\infty$ is the copy of $\mathbf{H}^+$, etc. Recall that we set $\overline{X} := C_\infty \otimes_{\mathbb{F}_r} X$ where X is the smooth, projective curve associated to **k**.

Let $v = \wp$ be a prime of **A** of degree d and let η be an $\overline{\mathbb{F}}_r$-valued point of $\overline{X}$ lying over $\wp$. Let $w \in \mathbf{k}$ be a local parameter at $\wp$ which gives rise to a parameter at η. Via θ_{con}, the tautological isomorphism between $\mathbf{C}_\infty$ and C_∞, we can view w as lying in H^+.

9.7. Another v-adic Γ-function

Definition 9.7.2. We set
$$d_{j,w} := d_j/(d_{j-d}w^{l_j})$$
where l_j is chosen so that $d_{j,w}$ is a unit at η and $d_{j-d} := 1$ if $j < d$.

Example 9.7.3. Let $\mathbf{A} = \mathbb{F}_r[T]$ and $\psi = C$ (the Carlitz module) as in Example 7.11.8. Thus we have $\theta = \bar{T} = \theta_{\mathrm{con}}(T) \in C_\infty$. Let $v = \wp$ correspond to a monic irreducible element also denoted $\wp$. Let $j \geq 0$. By definition,
$$d_j = D_j = (\theta^{r^j} - \theta) \cdots (\theta^{r^j} - \theta^{r^{j-1}}) \in A,$$
as the Carlitz exponential is $\sum z^{r^j}/D_j$. Recall that $D_j = \prod_{\substack{n \text{ monic} \\ \deg n = j}} n$ and $D_{j,v} = \prod_{\substack{n \text{ monic} \\ \deg n = j \\ v \nmid n}} n$. Now a monic n of degree j is divisible by $\wp$ if and only if $n = \wp n_0$ where n_0 is monic of degree $j - d$ ($d = \deg \wp$). Thus to obtain $D_{j,v}$, we first divide D_j by D_{j-d} and then remove the factors of $\wp$ that are in the quotient. This is *precisely* Definition 9.7.2.

It is also possible to give a proof of the fact that $(-1)D_{j,v} \to 1$ v-adically from the above definition as follows. We have
$$D_j/D_{j-1}^r = \theta^{r^j} - \theta.$$
Set $y_j := D_{j,v}/D_{j-1,v}^r = (\theta^{r^j} - \theta)/(\theta^{r^{j-d}} - \theta)w^{h_j}$ where h_j is chosen so that y_j is a unit at η. Let $j = j_0 + td$ where $0 \leq j_0 < d$ and $t \geq 0$.

Without loss of generality, we may assume that θ is a unit at $\wp$ (or replace θ by $\theta + 1$). Let $\theta = \zeta \cdot u$, where $\zeta \in A_v^*$ is a root of unity and u is a 1-unit; so $u^{r^j} \to 1$ as $t \to \infty$. Thus $\theta^{r^j} \to \zeta^{r^{j_0}}$ as $t \to \infty$, and so $y_j \to 1$ as $t \to \infty$. This immediately implies that $y_j \to 1$ for *all* j as $j \to \infty$.

Let j_0 be chosen so that for $j \geq j_0$, the unit y_j is a 1-unit. Now
$$y_{j+1} = D_{j+1,v}/D_{j,v}^r,$$
and so
$$y_{j+1}D_{j,v}^r = D_{j+1,v}. \tag{9.7.4}$$

Suppose $D_{j,v} = \alpha u_1$ where $\alpha \in A_v^*$ is a root of unity and u_1 is a 1-unit. Then we see $D_{j+1,v} = \alpha^r \cdot u_2$ where u_2 is also a 1-unit from 9.7.4. Thus if we multiply w by c, for some root of unity c in A_v^* so that $D_{j_0,v}$ is a 1-unit, we conclude that $D_{j,v}$ is a 1-unit for *all* $j \geq j_0$. With a little thought one sees that 9.7.4 now implies that $D_{j,v} \to 1$ v-adically. (In our previous approach we multiplied by (-1). Here we obtain the same goal by changing the parameter.)

The key to the above example is the calculation of the limit of $\theta^{r^j} - \theta$ as $j \to \infty$ where j is of the form $j_0 + td$, *and* seeing that the limit is non-zero. Thakur's elegant observation is that the *same* proof works in general *once* one calculates a similar limit (which will turn out to be a *Jacobi sum*). This will be discussed in our next subsection. One, therefore, has a new v-adic Γ-function as when $\mathbf{A} \neq \mathbb{F}_r[T]$, $d_{j,v}$ is *distinct* from $D_{j,v}$. We denote these functions by $_2\pi_w(y)$ and $_2\Gamma_w(y)$, etc. In private communication Thakur informed us that the value at 0 of these new v-adic gamma functions may be computed in terms of the Jacobi sums of our next subsection.

In a similar fashion, one can also interpolate $\{d_j\}$ at ∞ [Th1]. It is an interesting open question to compute the values of these new ∞-adic gamma functions at 0.

It is thus remarkable that for general $\mathbf{A}$, the arithmetic Γ-function itself breaks up into *smaller* pieces. We will have more to say about this in our next subsection.

9.8. Gauss Sums

Let $\mathbf{A}$ be general and let sgn be a fixed sign function. We continue to place ourselves in the situation of "two T's" and we let ψ be a sgn-normalized Hayes module defined over the abelian extension H^+ of k. We present here Thakur's "Gauss sums" associated to $(\psi, \wp)$ for $\wp \in \mathrm{Spec}(\mathbf{A})$ [Th9], [Th10] and [Th11] following a question raised in [Go3]. We shall see that, in analogy with classical theory, Thakur's Gauss sums involve a "multiplicative character" and an "additive character."

Let $d = \deg(\wp)$ and let $\psi[\wp] \subset C_\infty$ be the $\mathbf{A}$-module of $\wp$-division points of ψ. Let $e(z) := e_\psi(z)$ be the exponential of ψ. As we have seen, there is an isomorphism of $\mathbf{A}$-modules $\mathbf{A}/\wp \xrightarrow{\sim} \psi[\wp]$; we fix one such and denote it by $\widetilde{\psi}$. Thus $\widetilde{\psi}$ is our "additive" character.

Next let χ be an $\mathbb{F}_r$-algebra injection of the field $\mathbf{A}/\wp$ into C_∞. Set $\chi_0 := \chi$ and for $j = 1, \ldots, d-1$, set

$$\chi_j := \chi_{j-1}^r.$$

Thus $\{\chi_j\}$ is the set of all $\mathbb{F}_r$-algebra injections of $\mathbf{A}/\wp$ into C_∞. These are our "multiplicative" characters.

Definition 9.8.1. We set

$$g(j) := g(\chi_j) := -\sum_{\alpha \in (\mathbf{A}/\wp)^*} \chi_j(\alpha^{-1})\widetilde{\psi}(\alpha).$$

These are the *basic Gauss sums*.

9.8. Gauss Sums

The elements of Definition 9.8.1 are also referred to as the *basic Gauss-Thakur sums*.

Let χ be an *arbitrary* homomorphism from $(\mathbf{A}/\wp)^*$ to C_∞^*. It is easy to see that there is a unique i with $0 < i < r^d$ such that $\chi = \chi_0^i$. Let $i = \alpha_0 + \alpha_1 r + \cdots + \alpha_{d-1} r^{d-1}$ where $0 \leq \alpha_j < r$ for all j. Then clearly $\chi = \prod_j \chi_j^{\alpha_j}$. Thus we *define* the general Gauss sum by

$$g(\chi) := \prod g(j)^{\alpha_j}.$$

This is *exactly* similar to the definition of the Carlitz factorial using $\{D_j\}$ and is another example of the use of r-adic digits in fundamental definitions in the theory. The reason that a definition using the digit expansion is necessary is given in Remark 9.8.3.

Since $\widetilde{\psi} \colon \mathbf{A}/\wp \to \psi[\wp] \subset C_\infty$ is $\mathbb{F}_r$-linear, it can be expanded as

$$\widetilde{\psi}(\alpha) = \sum_{j=0}^{d-1} c_j \chi_j(\alpha),$$

for constants $\{c_j\}$ in C_∞. By restricting $\widetilde{\psi}$ to $(\mathbf{A}/\wp)^*$, and using standard Fourier theory, we see that

$$c_j = g(j).$$

We summarize this as our next result.

Proposition 9.8.2. *We have* $\widetilde{\psi}(\alpha) = \sum\limits_{j=0}^{d-1} g(j)\chi_j(\alpha)$. $\square$

Remark. 9.8.3. Let $\chi \colon (\mathbf{A}/\wp)^* \to C_\infty^*$ be an *arbitrary* homomorphism. For the moment let us define $\widetilde{g}(\chi)$ via Definition 9.8.1, so $\widetilde{g}(\chi_j) = g(j)$, etc. Then Proposition 9.8.2 immediately implies that $\widetilde{g}(\chi) = 0$ unless $\chi = \chi_j$ for some j. This is the reason for the above definition of $g(\chi)$ for *general* χ.

Theorem 9.8.4 (Thakur). *We have*

$$g(j) \neq 0$$

for $j = 0, \ldots, d-1$.

Proof. The pairing $\mathbf{A}/\wp \times \mathbf{A}/\wp \to \psi[\wp] \subset C_\infty$ given by

$$(\alpha, \beta) \mapsto \widetilde{\psi}(\alpha\beta)$$

is $\mathbb{F}_r$-bilinear and non-degenerate. Therefore, any $\mathbb{F}_r$-linear function on $\mathbf{A}/\wp$ can be expressed as

$$\widetilde{\psi}_\mu(\alpha) := \widetilde{\psi}(\mu\alpha)$$

for some $\mu \in \mathbf{A}/\wp$. Now

$$\widetilde{\psi}_\mu(\alpha) = \sum_{j=0}^{d-1}(\chi_j(\mu)g(j))\chi_j(\alpha).$$

Thus if $g(t) = 0$ for some t, then $\{\widetilde{\psi}_\mu(\alpha)\}$ spans at most a $(d-1)$-dimensional space. But $\dim_{\mathbb{F}_r} \mathbf{A}/\wp = d$ giving a contradiction. □

Definition 9.8.5. We set $J_0 := g(d-1)^r/g_0$, and
$$J_j := g(j-1)^r/g(j)$$
for $j = 1, \ldots, d-1$. The elements $\{J_j\}$ are *Jacobi sums*.

The above definition is due to Thakur. Note that, unlike classical Jacobi sums, the J_j are *not* just made up of sums of multiplicative characters (else we would have $J_j \in \overline{\mathbb{F}}_r$ for all j). However, we have the following result.

Proposition 9.8.6. *The Jacobi sums are independent of the choice of additive character $\widetilde{\psi}$.*

Proof. Let $\widehat{\psi}$ be another additive character. Then $\widehat{\psi}(\alpha) = \widetilde{\psi}(\beta\alpha)$ for some $\beta \in (\mathbf{A}/\wp)^*$. The result follows from rewriting the sums for $g(j)$, $g(j-1)^r$, and the fact that $\chi_{j-1}^r = \chi_j$. □

Example 9.8.7. Let $\mathbf{A} = \mathbb{F}_r[T]$ and $\psi = C$, the Carlitz module. By definition (using the "two T's" notation)
$$\widetilde{\psi}(T\alpha) = \theta\widetilde{\psi}(\alpha) + \widetilde{\psi}(\alpha)^r, \qquad \theta = \imath(T).$$

Thus
$$g(j)\chi_j(T) = g(j)\theta + g(j-1)^r.$$
As $g(j) \neq 0$, we have $J_j = g(j-1)^r/g(j) = -(\theta - \chi_j(T))$. (If $\wp = (T)$, use $T+1$ instead.) Multiplying all the J_j together gives
$$\prod_{j=0}^{d-1} g(j)^{(r-1)} = (-1)^d \wp$$
where "$\wp$" also denotes the monic generator of the ideal $\wp$. Thus from our definitions, we see
$$g(\mathrm{id}) = (-1)^d \wp$$
where "id" denotes the trivial character. It is known classically that Jacobi sums occur as eigenvalues of the Frobenius operator on the cohomology of Fermat curves. For $\mathbf{A} = \mathbb{F}_r[T]$, one can see that $\{J_j\}$ occur as eigenvalues for the Frobenius operator acting on the cohomology of the Drinfeld module ρ with $\rho_T(\tau) = \theta\tau^0 + \tau^d$, ($\tau = r^{\text{th}}$-power mapping) over the field $\mathbf{A}/\wp$. Indeed, ρ

9.8. Gauss Sums   379

has complex multiplication by $\mathbf{A}/\wp[T]$ where it becomes the "Carlitz module" for $\mathbf{A}/\wp[T]$. The result now follows with a little calculation.

We now want to generalize the above example to arbitrary $\mathbf{A}$. From Subsection 7.11 we see that the shtuka function for the Carlitz module is $T - \theta$. Thakur's remarkable observation is that the shtuka function should be related to Jacobi sums in complete generality.

Thus let f be the shtuka function associated to ψ and let $\widehat{f} = \zeta f$ as in Subsection 7.11 and Proposition 7.11.4. Moreover our $\mathbb{F}_r$-algebra homomorphism of $\mathbf{A}/\wp$ into C_∞ corresponds to a C_∞ point $\mathfrak{p}$ of $\mathrm{Spec}(\mathbf{A})$ lying above $\wp$. Thus, if $a \in \mathbf{A}$, then $\chi(a) = a(\mathfrak{p})$.

As in Subsection 7.11 we let V be the Drinfeld divisor associated to ψ.

Proposition 9.8.8. *A C_∞-valued point of $\mathrm{Spec}(\mathbf{A})$ cannot lie in the support of V.*

Proof. This follows as in 7.11.3 but where we work with the reduction mod $\wp$ of ψ. □

Our next result uses the notation of Subsection 7.11.

Theorem 9.8.9 (Thakur). *We have*
$$\widehat{f}\big|_{\mathfrak{p}^{(j)}} = J_j.$$

Proof. We begin by using the shtuka function to construct an $\mathbf{A}$-module homomorphism $\phi\colon \mathbf{A}/\wp \to \psi[\wp]$. This mapping will be of the form

$$\phi(z) = \sum_{j=0}^{d-1} h_j \chi_j(z) \tag{9.8.10}$$

for certain, specially chosen, h_j. From the fact that all $\chi_j(z)$ are morphisms on $\mathbf{A}/\wp$ we deduce immediately that $\phi(z)$ will be an additive mapping from $\mathbf{A}/\wp$ to C_∞. If moreover, we know that for $a \in \mathbf{A}$, $\phi(az) = \psi_a(\phi(z))$ (as the $\{h_j\}$ will be precisely chosen to insure) we will then have our homomorphism. Now let h_0 be a fixed $(r^d - 1)$-st root of $(\widehat{f}\widehat{f}^{(1)} \cdots \widehat{f}^{(d+1)})|_{\mathfrak{p}}$. Define h_i inductively by
$$h_i := (h_{i-1}^r / \widehat{f})\big|_{\mathfrak{p}^{(i)}},$$
as Proposition 9.8.8 allows. One checks that h_i depends only on i modulo d. It now follows from 7.11.5 that $\phi(az) = \psi_a(\phi(z))$ and the construction is finished.

As the $\{h_i\}$ are non-trivial the linear independence of characters immediately implies that ϕ is non-trivial and thus an isomorphism. Thus

$\phi(z) = \tilde{\psi}(uz)$ for some $u \in (\mathbf{A}/\wp)^*$. From Proposition 9.8.2 we deduce that $h_j = \chi_j(u)g(j)$. Therefore,

$$\hat{f}|_{\mathbf{p}(j)} = \frac{h_{j-1}^r}{h_j} = \frac{g(j-1)^r}{g(j)} = J_j,$$

and the theorem is established. □

Theorem 9.8.9 finishes the construction of the v-adic ($v = \wp$) Γ-function started in Subsection 9.7.

Our next and last result of this subsection is a version of the Gross-Koblitz Theorem for the gamma function whose construction was just completed (and is denoted $_2\Pi_w(y)$). Let Ξ be as in Subsection 7.11 and set

$$\hat{f}_j := \hat{f}|_{\Xi(j)}$$

in the notation of that subsection. By 7.11.5, we see that

$$d_j = \hat{f}_j d_{j-1}^r$$

(e.g., for $\mathbf{A} = \mathbb{F}_r[T]$, $D_j = [j]D_{j-1}^r$, etc.).

Let $\mathbf{k}_\wp$ be the completion of $\mathbf{k}$ at $\wp$. Let $\overline{\mathbf{k}}_\wp$ be a fixed algebraic closure of $\mathbf{k}_\wp$; similarly we construct the field $\overline{k}_\wp$ via θ_{con}. The mapping

$$\mathbf{A} \to \mathbf{k} \to \overline{\mathbf{k}}_\wp \xrightarrow{\sim} \overline{k}_\wp$$

gives a $\overline{k}_\wp$-valued point of Spec($\mathbf{A}$) and X which we denote $\Xi_\wp$.

Let $i: \overline{k} \to \overline{k}_\wp$ be a fixed embedding. Thus we can view our Gauss sums as being defined over $\overline{k}_\wp$, etc. The Teichmüller mapping $\mathbf{A}/\wp \to \mathbf{k}_\wp \to \overline{\mathbf{k}}_\wp$ gives us a geometric point lying over $\wp$ which we still denote by $\mathbf{p}$. Moreover, of course, via i, ψ and $\psi[\wp]$ are defined over $\overline{k}_\wp$, etc. Let w be as in Definition 9.7.2.

Theorem 9.8.11 (Thakur). *Let $0 \leq j < d \; (= \deg(\wp))$. Let μ be the valuation of $g(j)$ in $\overline{k}_\wp$. We then have*

$$g(j) = \zeta_1 w^\mu /_2\Pi_w\left(\frac{r^j}{1-r^d}\right)$$

for some $(r^d - 1)$-st root of unity ζ_1.

Proof. Set

$$M_j := {}_2\Pi_w\left(\frac{r^j}{1-r^d}\right)$$

$$= \lim_{m \to \infty} T_m$$

where $T_m := d_{j,w} d_{j+d,w} \cdots d_{j+md,w}$.

From Definition 9.7.2, we see that

$$T_m = \frac{d_j \cdots d_{j+md}}{d_j \cdots d_{j+md-d}} w^{e_1}$$

for some e_1 (recall that $j < d$). Thus $T_m = d_{j+md} w^{e_1}$. Since $d_j = \widehat{f}_j d_{j-1}^r$, we see

$$T_m = \widehat{f}_{j+md} \widehat{f}_{j+md-1}^r \cdots \widehat{f}_{j+1+(m-1)d}^{r^{d-1}} T_{m-1}^{r^d} w^{e_2},$$

for some e_2. As $m \to \infty$, reasoning as in Example 9.7.3, we see that $\widehat{f}_{j+md} \mapsto \widehat{f}|_{\mathfrak{p}(j)} = J_j$. Thus

$$M_j^{1-r^d} = J_j J_{j-1}^r \cdots J_{j-d+1}^{r^{d-1}} w^{e_3} = g(j)^{r^d-1} w^{e_3},$$

for some e_3, which gives the result. □

Remarks. 9.8.12. 1. The injection i is, of course, not unique. Any other such injection will change the additive character up to multiplication by an element in $(\mathbf{A}/\wp)^*$ and thus change the Gauss sum by an $(r^d - 1)$-st root of unity. This change by a root of unity is built into the statement of the theorem.
2. For a more precise statement when $\mathbf{A}$ is a principal ideal domain, see [Th9], [Th1] and [Th10].
3. Recall that for an arbitrary homomorphism from $(\mathbf{A}/\wp)^*$ to C_∞^* we extend the definition of Gauss sums multiplicatively through r-adic digits. One sees immediately that Theorem 9.8.11 carries over directly in the same fashion and provides further evidence for the naturality of Thakur's general definition via digit expansions.

It is very interesting that 9.8.11 shows that the Gauss sum is related to the coefficients of the exponential and *not* to "factorials." This is another example of how the function field theory comments on classical results.

For more on Gauss sums, see the references at the beginning of this subsection.

9.9. The Geometric Γ-function

The early and exciting results of D. Thakur on the arithmetic Γ-function inspired the present author to define some "geometric" Γ-functions. We will concentrate on the basic $\mathbf{A} = \mathbb{F}_r[T]$ case and explain how things *may* be generalized. As with the arithmetic Γ-functions, there is more than one way to proceed in general. Future research should shed more light on these constructions.

Let sgn be the sign function associated with the usual notion of monicity. We will continue to place ourselves in the case of "two T's." Let C be the Carlitz module defined over $A = \mathbb{F}_r[\theta]$, etc. Let $e(x) := e_C(x)$ be the exponential function of C and let $e_A(x) := x \prod_{0 \neq \alpha \in A} (1 - x/\alpha)$.

The idea behind the geometric Γ-function is to find a function connected to $e_A(x)$ in a fashion similar to the connection between Euler's Γ-function and $\sin(z)$.

Definition 9.9.1. Let $y \in \mathbb{Z}_p$ be written $\sum_{j=0}^{\infty} \alpha_j r^j$, $0 \le \alpha_j < r$ and let $x \in C_\infty$. We set

$$g(x,y) := \prod_{j=0}^{\infty} \left(\prod_{\substack{n \text{ monic} \in A \\ \deg(n)=j}} \left(1 + \frac{x}{n}\right) \right)^{\alpha_j}.$$

It is easy to see that $g(x,y)$ is entire in x for each fixed $y \in \mathbb{Z}_p$ and that the resulting function on $C_\infty \times \mathbb{Z}_p$ is continuous, etc.

Definition 9.9.2. 1. We put $\Pi_0(x,y) := g(x,y)^{-1}$.
2. We put
$$\Pi(x,y) := \Pi_0(x,y)\Pi_\infty(y)$$
(where $\Pi_\infty(y)$ is defined in Subsection 9.4).
3. We put $\Gamma_0(x,y) := \Pi_0(x, y-1)$.
4. We put
$$\Gamma(x,y) := \frac{\Pi(x,y-1)}{x}.$$

The function $g(x,y)$ is called the *inverse (geometric) gamma function* and $\Gamma_0(x,y)$ is called the *two-variable geometric gamma function*. The function $\Gamma(x,y)$ is called the *total gamma function*.
The 1-variable function
$$\Gamma(x) := \frac{1}{x}\Pi_0\left(x, \frac{1}{1-r}\right) = \frac{1}{x}\Gamma_0\left(x, 1 - \frac{1}{r-1}\right) = \frac{1}{x} \prod_{\substack{n \in A \\ n \text{ monic}}} \left(1 + \frac{x}{n}\right)^{-1}$$
has simple poles at $\{0\}$ and $\{-n \mid n \text{ is monic}\}$. As such, it is particularly similar to Euler's gamma function. It is called the *1-variable geometric gamma function* (or just the *geometric gamma function*). It has been studied very extensively by Thakur, [Th6], and Sinha, [Si1] and [Si2].
The connection between these functions and $e_A(x)$ is given by our next result. Let G be the group of automorphisms of $C_\infty \times \mathbb{Z}_p$ generated by $(x,y) \mapsto (\zeta x, y)$, $\zeta \in \mathbb{F}_r^*$, and $(x,y) \mapsto 1 - y$. Thus $G \simeq \mathbb{F}_r^* \times \mathbb{Z}/(2)$. We then have the following *functional equation/reflection formula*.

Proposition 9.9.3. *Let $s = (x,y) \in C_\infty$, let $\sigma \in G$ and let s^σ be the action of σ on s. Then*

9.9. The Geometric Γ-function

$$\prod_{\sigma \in G} \Gamma_0(s^\sigma) = \frac{x^{r-1}}{e_A(x)^{r-1}} = \frac{(\xi x)^{r-1}}{e_C(\xi x)^{r-1}}$$

where ξ is the period of the Carlitz module.

Proof. This follows immediately from the definitions. $\square$

Since $\Pi_\infty(y-1) = \Gamma_\infty(y)$ and $\Gamma_\infty(y)\Gamma_\infty(1-y) = \xi_u =$ the 1-unit period of the Carlitz module, 9.9.3 also leads immediately into a reflection formula for the function $\Gamma(x,y)$ etc. For the 1-variable gamma function, if we set $\Pi(x) := x\Gamma(x)$, then

$$\prod_{\zeta \in \mathbb{F}_r^*} \Pi(\zeta x) = \frac{x}{e_A(x)}.$$

This should be compared to the classical formula

$$\prod_{\zeta \in \mathbb{Z}^*} \Pi(\zeta x) = \frac{\pi x}{\sin(\pi x)}$$

where $\Pi(x) := \Gamma(x+1)$.

The geometric gamma functions also satisfy multiplication formula. Let $h \in A$ be monic of degree d.

Lemma 9.9.4. *We have for all $j \geq 0$*

$$\prod_{\deg(\alpha)<d} \prod_{\substack{n \text{ monic } \in A \\ \deg(n)=j}} \left(1 + \frac{x+\alpha/h}{n}\right) = \prod_{\substack{n \text{ monic } \in A \\ \deg(n)=j+d}} \left(1 + \frac{hx}{n}\right) \cdot \frac{D_{j+d}}{h^{r^{j+d}} D_j^{r^d}}.$$

Proof. We have

$$\prod_{\deg(\alpha)<d} \prod_{\substack{n \text{ monic in} A \\ \deg(n)=j}} \left(1 + \frac{x+\alpha/h}{n}\right) = \prod_{\deg(\alpha)<d} \prod_{\substack{n \text{ monic} \\ \deg(n)=j}} \left(\frac{hn + \alpha + hx}{hn}\right).$$

The result now follows via the division algorithm. $\square$

The quotient

$$\frac{D_{j+d}}{h^{r^{j+d}} D_j^{r^d}}$$

is a 1-unit. Thus, in the notation of Subsection 9.4 (obviously carried over to $\mathbb{F}_r[\theta]$) we have

$$\left\langle \frac{D_{j+d}}{h^{r^{j+d}} D_j^{r^d}} \right\rangle = \frac{\langle D_{j+d} \rangle}{\langle h \rangle^{r^{j+d}} \langle D_j \rangle^{r^d}}.$$

This leads to the multiplication formula of our next result.

Proposition 9.9.5. *Let* $y = \sum \alpha_i r^i$, $0 \leq \alpha_i < r$ *be in* $\mathbb{Z}_p$. *With the above notation, we have*

$$\prod_{\deg(\alpha)<d} g(x+\alpha/h, y) = g(hx, r^d y)\Pi_\infty(r^d y)\Pi_\infty(y)^{-r^d}\langle h\rangle^{-r^d y}.$$

Proof. This follows immediately from Lemma 9.9.4. □

Corollary 9.9.6. *We have*

$$\Pi_\infty(r^d y) \prod_{\deg(\alpha)<d} \Pi_0(x+\alpha/h, y) = \Pi_0(hx, r^d y)\Pi_\infty(y)^{r^d}\langle h\rangle^{r^d y}.$$

Proof. Use $\Pi_0(x, y) = g(x, y)^{-1}$ in 9.9.5. □

Next we establish another, more simple, expression for $g(x, y)$ and thus for $\Pi_0(x, y)$, etc. As before, let $e_j(x) := \prod_{\substack{\deg(\alpha)<j \\ \alpha \in A}} (x+\alpha)$ and $E_j(x) := \frac{e_j(x)}{D_j}$.

Both $e_j(x)$ and $E_j(x)$ are $\mathbb{F}_r$-linear.

Lemma 9.9.7. *Let* $j \geq 0$. *Then we have*

$$\prod_{\substack{n\in A \\ n \text{ monic} \\ \deg(n)=j}} \left(1+\frac{x}{n}\right) = 1+E_j(x).$$

Proof. Let h be monic of degree j. Then $e_j(h) = D_j$. Thus, $E_j(-h) = -1$ and both sides have the same roots and constant term ($=1$). □

Corollary 9.9.8. *Let* $y = \sum_{i=0}^{\infty} \alpha_i r^i$, $0 \leq \alpha_i < r$. *Then*

$$g(x, y) = \prod_{i=0}^{\infty}(1+E_i(x))^{\alpha_i}.$$

□

Lemma 9.9.9. *Let* L *be a complete topological field of characteristic* p. *Let* $\{e_i\} \subseteq L^*$ *be a sequence tending to* 0. *Let* $y = \sum_{i=0}^{\infty} \alpha_i r^i \in \mathbb{Z}_p$ *be written* r-*adically. Set*

$$h(y) := \prod_{i=0}^{\infty}(1+e_i)^{\alpha_i}.$$

Then $h(y)$ *is continuous and* $h(y) = \sum_{i=0}^{\infty} \binom{y}{i} f_i$, *where if then*

$$i = \sum_{t=0}^{v} \beta_t r^t, \quad 0 \leq \beta_t < r,$$

$$f_i = \prod_{t=0}^{v} e_t^{\beta_t}.$$

Proof. This follows from Lucas' formula. □

Corollary 9.9.10. *Let y be as in 9.9.8. Then we have*

$$g(x,y) = \sum_{j=0}^{\infty} \binom{y}{j} G_j(x)$$

when $\{G_j(x)\}$ is given in Definition 8.22.8. □

Remark. 9.9.11. As mentioned before, the construction of gamma and zeta functions has a very interesting feature when viewed under the optic of the "two T's;" i.e., the rings $\mathbf{A}$ and A which are canonically isomorphic via θ_{con}. The zeta functions deal with the action of $\mathbf{A}$ on additive groups via Drinfeld modules or T-modules. That is, they are given by Euler products created out of polynomials with $\mathbf{A}$-coefficients (= the characteristic polynomials of Frobenius morphisms). We thus obtain functions with values in $\mathbf{C}_\infty$. On the other hand, gamma functions are concerned with the exponential functions of these $\mathbf{A}$-actions. They, therefore, are made up out of elements in k and we obtain functions with values in C_∞. Only via the canonical isomorphism θ_{con} can we *directly* relate the two constructions. We will return to this theme a little later on.

Corollary 9.9.10 provides the only link we have between the theory of the geometric gamma functions and the theory of the L-functions. Recall that in Subsection 8.22 we introduced the algebra of divided power series $\{\sum a_j \frac{z^j}{j!}\}$. We pointed out how the Dirac measure at β corresponds to the divided power series

$$f_\beta(z) := \sum_{j=0}^{\infty} G_j(\beta) \frac{z^j}{j!}.$$

Now the algebra of divided power series can also be viewed as the algebra of hyperderivatives $\sum_{j=0}^{\infty} a_j \frac{D^j}{j!}$ where one sets

$$\frac{D^j}{j!} X^i := \binom{i}{j} X^{i-j}.$$

Let $f_\beta(D) := \sum_{j=0}^{\infty} G_j(\beta) \frac{D^j}{j!}$. We then have the formal transform for $y \in \mathbb{Z}_p$

$$f_\beta(D)(1+X)^j\big|_{X=0} = g(\beta,y). \tag{9.9.12}$$

(*N.B.*: The above statement is not quite correct as it is not given in the language of the "two T's;" we leave it to the reader to work out the appropriate restatement in this language.) Thus we see that the theory of v-adic measures leads, at least formally, to a relationship between the theory of the geometric gamma functions and the theory of the zeta functions.

In [Go9], [Go10] the above transform is studied, and in [Go11] a *harmonic analysis* is developed (or at least defined) which tries to "unravel" the differential operator $f_\beta(D)$; some success is obtained at lower order terms. More work is still needed to develop tools of sufficient power to understand the higher order terms, etc. Moreover, of course, as a goal one wants ultimately to apply such a calculus towards understanding the arithmetic contained in L-series, etc., with the classical theory as a model.

Recall that in Theorem 9.4.9 we presented a computation of the period of the Carlitz module over $\mathbb{F}_{r^t}[T]$ in terms of the arithmetic gamma functions. Now let $L = \mathbf{k}(C[T]) = \mathbf{k}$ adjoined with the T-division points of C. Let $\mathcal{O}$ be the ring of $\mathbf{A}$-integers in L; so $\mathcal{O} = \mathbb{F}_r[y]$, $y^{r-1} = -T$. We then have a Carlitz module for $\mathcal{O}$ and, in the optic of the "two T's," Thakur [Th6] shows that its period is an algebraic multiple of $\Gamma(1/\theta)$. This is a *geometric Chowla-Selberg formula* and gives further credence to our use of "geometric" for these functions.

In [Si2] the transcendence of the values of $\Gamma(x)$ ($\mathbf{A} = \mathbb{F}_r[T]$) at $x = a/f$, where a, f are monic and $\deg a < \deg f$ is established.

We can now discuss possible ways of generalizing the functions $g(x,y)$, etc., to arbitrary $\mathbf{A}$. As with the arithmetic Γ-functions, there are really two possible ways. The first is to use signs as in Subsection 9.6, choose an ideal I, and then simply generalize Definition 9.9.1. The second, and more subtle, approach is based on Corollary 9.9.8 and Subsection 8.22; it was also suggested in [Th7]. Recall that in Subsection 8.22 we suggested replacing $f_\beta(z)$, in general, by the divided power series

$$f_\beta(z) := \sum_{e=1}^{h^+} \left(\frac{D\psi_{\alpha_e}}{\alpha_e}\right) \left(\sum_{i=0}^\infty G_i^{\sigma_{\alpha_e}}(\beta) z^i/i!\right).$$

Thus 9.9.12 *suggests* defining the general inverse geometric gamma function

$$g_\mathbf{A}(\beta,y) := f_\beta(D)(1+X)^y\big|_{X=0}$$

$$= \sum_{e=1}^{h^+} \left(\frac{D\psi_{\alpha_e}}{\alpha_e}\right) \left(\sum_{i=0}^\infty G_i^{\sigma_{\alpha_e}}(\beta)\binom{y}{i}\right),$$

where again the formulas should be considered in the "two T's" language. Such an inverse geometric gamma function would fit with the theory of [Go11].

9.9. The Geometric Γ-function

Remark. 9.9.13. We can now return to the theme of 9.9.11. If one wants to try to use "two T's" classically one is led to consider, say, $\mathbb{G}_m$ as an object with "multiplication by $\mathbb{Z}$" and also "having coefficients in $\mathbb{Z}$." Thus we would be looking at an object with multiplication by the algebra "$\mathbb{Z} \otimes_{\mathbb{Z}} \mathbb{Z}$." *But* of course $\mathbb{Z} \otimes_{\mathbb{Z}} \mathbb{Z} \simeq \mathbb{Z}$, as indeed $R \otimes_{\mathbb{Z}} \mathbb{Z} \simeq R$ for any commutative ring R, and so one never sees "two $\mathbb{Z}$'s" classically; the ring of operators *coincides* with the ring of scalars. However, for function fields this is manifestly not true. This therefore suggests looking for possible functional relationships based *not* on "gamma · zeta," but rather of the form "gamma $\otimes$ zeta." Still, all of this "two T" theory awaits further clarification and development.

Finally, we return to $\mathbf{A} = \mathbb{F}_r[T]$ and briefly discuss the v-adic interpolations of the geometric gamma functions as in [Th6] and [Th7]. Recall that we set

$$g(x,y) = \prod_{j=0}^{\infty} \left(\prod_{\substack{n \text{ monic} \\ \deg(n)=j}} \left(\frac{x+n}{n} \right) \right)^{a_j}$$

where $y = \sum_{j=0}^{\infty} a_j r^j$, $0 \leq a_j < r$. Now let $v \in \mathrm{Spec}(A)$ and let $\alpha \in A_v$. Set

$$\overline{\alpha} := \begin{cases} \alpha & \text{if } \alpha \in A_v^* \\ 1 & \text{if not} \end{cases}.$$

Definition 9.9.14. We set

$$g_v(\alpha, y) := \prod_{j=0}^{\infty} \left(\prod_{\substack{n \text{ monic} \\ \deg(n)=j}} \frac{\overline{\alpha+n}}{\overline{n}} \right)^{a_j}.$$

It is easy to use congruences modulo powers of v, as in Subsection 9.6, to show that $g_v(\alpha, y)$ is a continuous function from $A_v \times \mathbb{Z}_p \to A_v^*$. Finally we set

$$\Pi_{0,v}(\alpha, y) := g_v(\alpha, y)^{-1},$$

etc. Thus we have one v-adic interpolation of the geometric Γ-functions. In the Remarks (4) of §V of [Th7], yet another v-adic interpolation of $g(x,y)$ is given in the spirit of Subsection 9.7 and certain arithmetic applications are hinted at. How these two interpolations at v are related is not yet clear.

In conclusion the results of this section show just how rich is the theory of gamma functions for function fields. This theory is in a state of rapid and exciting development with the promise of further connections with the theory of L-series.

10. Additional Topics

In this last section, we will briefly describe *some* current areas of research into the arithmetic described in the previous nine sections. We emphasize the word "some" as this section is not meant to be considered exhaustive.

10.1. The Geometric Fermat Equation

Let n be a positive integer. Then, of course, the n-torsion points of the standard $\mathbb{Z}$-action on $\mathbb{G}_m$ is just the group of n-th roots of unity. Inside the complex numbers $\mathbb{C}$, the points of this group are the roots of the equation $u^n - 1 = 0$. If we substitute x/y for u and multiply by y^n we obtain $x^n - y^n$; upon setting this equal to z^n we obtain the equation $x^n - y^n = z^n$ which is a form of the classical Fermat equation.

Let $\mathbf{A} = \mathbb{F}_r[T]$, $\mathbf{k} = \mathbb{F}_r(T)$, $r = p^{m_0}$, etc. One can clearly consider the classical Fermat equation over $\mathbf{k}$; this is well-known and due to Greenleaf [Ri1, p. 264]. Following our general arithmetic/geometric philosophy, we call the classical Fermat equation over $\mathbf{k}$, the *arithmetic Fermat equation*. Thus we now seek to find the *geometric Fermat equation* which amounts to redoing the above construction with the Carlitz module C. Thus let $f \in \mathbf{A}$ be an element of degree d; then the f-division points are precisely the roots to the equation $C_f(u) = 0$ in $\mathbf{C}_\infty$. We are therefore lead to the *geometric Fermat equation* [Go18]
$$y^{r^d} C_f(x/y) - z^{r^d} = 0.$$
This equation is written $\mathcal{F}_0(f)(x, y, z) = 0$; an inhomogeneous version is
$$\mathcal{F}_1(f)(x, y, z) := y^{r^d} C_f(x/y) - z^p = 0.$$

In [Den6] these equations are studied by L. Denis and, in particular, a form of Fermat's Last Theorem is established for them.

Theorem 10.1.1 (Denis). *Let $\mathbf{A}$ and f, etc., be as above.*
1. *Let $r \neq 2$ and $d > 1$, or $d > 2$ and $r = 2$. Then both $\mathcal{F}_0(x, y, z) = 0$ and $\mathcal{F}_1(x, y, z) = 0$ have only a finite number of rational (over $\mathbf{k}$) solutions with $\gcd(x, y) = \gcd(y, z) = 1$.*

2. Let $r \geq 3$ and $d \geq 2$. Then $\mathcal{F}_0(x, y, z) = 0$ has no solutions in **k** with $xyz \neq 0$.
3. Let $r \geq 3$, $p > 2$, and $d \geq 2$. Then $\mathcal{F}_1(x, y, z) = 0$ has no solutions in **k** with $xyz \neq 0$. □

The proof of the above result is *far* more elementary than the proof of the classical Fermat's Last Theorem by A. Wiles (as completed by Wiles and R. Taylor). Indeed, in the function field case our geometric equation is vulnerable to a *geometric* attack; in fact, by simply differentiating (with respect to T) the above equations simplify drastically and the proof follows without too much difficulty. Of course such an approach is *not* possible classically. Still it is remarkable how close the statement of Denis' result above is the statement of the classical result. Therefore it might prove very instructive to study the above equations over extensions of **k** for some ideas on how the classical equations *should* behave. It might also be very interesting to see what happens when one works with arbitrary **A** instead of just the polynomial ring.

In [Go25] an adjoint Fermat equation is studied and in [Den6] higher dimensional *Fermat families* are studied.

10.2. Geometric Deligne Reciprocity and Solitons

Let **A** continue to be $\mathbb{F}_r[T]$, etc. Let

$$\Gamma(x) := \frac{1}{x} \prod_{\substack{n \in \mathbf{A} \\ n \text{ monic}}} \left(1 + \frac{x}{n}\right)^{-1},$$

be the 1-variable geometric Γ-function of Subsection 9.9. (In this subsection we will use the notation of Subsection 9.9.) Let $f \in \mathbf{A}$ have degree d and let $\mathrm{Sy}(f)$ be the set of symbols

$$\left\{ \left[\frac{a}{f}\right] : a \in \mathbf{A} \text{ with } \deg a < d \right\}.$$

Let $\mathbf{F}(f)$ be the free abelian group generated by $\mathrm{Sy}(f)$ and let

$$\mathbf{a} := \sum_{\deg a < d} m_a \left[\frac{a}{f}\right]$$

be an element of $\mathbf{F}(f)$. Following [Si1] we set $\left\langle \frac{a}{f} \right\rangle$ to be 1 if a is monic, and 0 otherwise. Moreover we set

$$m(\mathbf{a}) := \sum_{\deg a < d} m_a \left\langle \frac{a}{f} \right\rangle.$$

10.2. Geometric Deligne Reciprocity and Solitons

Let $k(f)$ be the extension of k obtained by adjoining the f-division points of C inside C_∞. By the results of Section 7, we know that $G := \text{Gal}(k(f)/k) \simeq (\mathbf{A}/(f))^*$ via the Artin symbol. Thus G acts on $\mathbf{F}(f)$ simply via the obvious action of $(\mathbf{A}/(f))^*$ on $\text{Sy}(f)$. This action is denoted by $(\sigma, \mathbf{a}) \mapsto \mathbf{a}^\sigma$. Let $H(f) \subset \mathbf{F}(f)$ be the subgroup of elements fixed by this action. Let $\xi \in C_\infty$ be the period of the Carlitz module (which is defined up to an element in $\mathbb{F}_r^*$) and set

$$\Gamma(\mathbf{a}) := \xi^{-m(\mathbf{a})} \prod_{\deg a < d} \Gamma(a/f)^{m_a}.$$

It is clear that $\Gamma(\mathbf{a})$ is nonzero. Using his knowledge of the various functional equations satisfied by $\Gamma(x)$, Thakur [Th6] establishes the following result.

Lemma 10.2.1. *For* $\mathbf{a} \in H(f)$, *the element* $\Gamma(\mathbf{a}) \in C_\infty$ *is algebraic over* k. $\square$

In [Si1] Sinha shows that, in fact, $k(f)(\Gamma(\mathbf{a}))$ is a *Kummer extension* of $k(f)$ which is unramified at the finite primes not lying above f. Let $\mathfrak{p}$ be such a prime with Artin symbol $\sigma_\mathfrak{p}$. The main result of [Si1] provides an explicit construction of the character $\chi_\mathbf{a}(\mathfrak{p}) := \sigma_\mathfrak{p}(\Gamma(\mathbf{a}))/\Gamma(\mathbf{a})$ in terms of *solitons* which we will now discuss.

Solitons are yet another major player in the arithmetic of function fields related to Drinfeld modules. A complete discussion of them would of necessity be quite involved, so we shall be content with the goal of spiking the reader's curiosity. For more on solitons, we refer the reader to [A2], [A3], [A4] and [Th14].

Recall that at the end of Section 6 we mentioned a *differential* version of the shtuka correspondence (as reported on in [Mu2]). This differential version is due in modern times to I. Krichever. Let $D := \frac{d}{dx}$; then as with the shtuka correspondence, there is a dictionary between commutative subrings of $\mathbb{C}[[x]][D]$ and complete $\mathbb{C}$-curves Z equipped with certain sheaf data. This dictionary allows one to find soliton (=*periodic* or *quasi-periodic*) solutions to the famous non-linear 2-variable *Korteweg-deVries equation* (KdV):

$$\frac{\partial \phi}{\partial t} = \frac{\partial^3 \phi}{\partial x^3} + 6\phi \frac{\partial \phi}{\partial x}.$$

This construction proceeds roughly as follows. There exists a certain morphism from affine 2-space to the Jacobian of Z; our soliton solutions are then the pull back, via this morphism, of a particular derivative of the classical *Riemann theta function*.

Anderson made the elegant observation that via the analogy between the r-th power mapping τ and D, which has been stressed in this book (cf. Subsections 1.9 and 4.14), the above formalism of *differential equations* should translate into powerful techniques in function field *arithmetic*. So differential polynomials correspond to $\mathbb{F}_r$-linear polynomials, pseudo-differential operators correspond to formal operators $\sum_{i \gg -\infty} c_i \tau^{-i}$, the "certain morphism" above is replaced by the Abel mapping (or $F - 1$ where F is the Frobenius morphism) and so on. This translation was carried out in ad hoc manner in [A2] and more systematically in [A3]. Let X_f be the smooth, geometrically connected, projective curve associated to $k(f)$ equipped with the fixed C_∞-valued point Ξ giving the injection of $k(f)$ into C_∞. (This geometric point lies above the geometric point with the same symbol given in Section 6.) Moreover X_f clearly comes equipped with a functorial action of $(\mathbf{A}/(f))^* \simeq \mathrm{Gal}(k(f)/k)$. The action of $a + (f)$ for $\deg a < d$ and $(a, f) = 1$ is denoted $[a]$.

Anderson [A2] then uses this Frobenius calculus to prove the following remarkable result.

Theorem 10.2.2 (Anderson). *There exists a meromorphic function ϕ (the soliton) on $\mathcal{X}_f := X_f \times X_f$ with the following properties.*
1. *ϕ is regular on the open subset of $\mathcal{X}_f$ which is the fiber product of the open set of places not lying above f and the open set of places not lying above ∞.*
2. *Let a be as above and let N be any positive integer. Let F be the Frobenius morphism of $X_f/\mathbb{F}_r$. Then* $1 - \phi(F^n \Xi, [a]\Xi) = \displaystyle\prod_{\substack{n \in A \\ \deg n = N-1 \\ n \text{ monic}}} \left(1 + \frac{a}{fn}\right).$ □

Thus the soliton ϕ provides a sort of universal interpolation of the types of products used in the definition of $\Gamma(x)$. The construction of ϕ is functorial and thus allows us to construct solitons $\phi_{[\frac{a}{f}]}$ (where $\phi_{[\frac{a}{f}]}$ is again a function on $\mathcal{X}_f$) for any a with $\deg a < d$. The main result of [Si1] can now be stated.

Theorem 10.2.3 (Sinha). *Let $\mathbf{a}$, $\mathfrak{p}$, $\chi_{\mathbf{a}}(\mathfrak{p})$, etc., be as above. Then*

$$\chi_{\mathbf{a}}(\mathfrak{p}) = \frac{\prod \chi_{[\frac{a}{f}]}(\mathfrak{p})^{m_a}}{n\mathfrak{p}^{m_a}},$$

where

$$\chi_{[a/f]}(\mathfrak{p}) := \prod \left(1 - \phi_{[\frac{a}{f}]}(\alpha, \Xi)\right),$$

α runs over the C_∞ points of X_f lying over $\mathfrak{p}$, and $n\mathfrak{p}$ is the ideal norm of $\mathfrak{p}$ to $\mathbf{A}$. □

Theorem 10.2.3 is Sinha's version for our $\Gamma(x)$ of Deligne Reciprocity (Theorem I.7.15 of [DMOS1]) for the classical gamma function. The analog for the classical gamma function of Lemma 10.2.1 was originally shown by Koblitz and Ogus in the appendix to [De3]. Note that any point lying above $\mathfrak{p}$ must have coefficients in the algebraic closure of $\mathbb{F}_r$ in C_∞.

10.3. The Tate Conjecture for Drinfeld Modules

Finally, we refer the reader to [A9] for computations related to the gamma functions studied here as well as to classical gamma functions.

10.3. The Tate Conjecture for Drinfeld Modules

Let $\mathbf{A}$, $\mathbf{k}$, etc., now be general as in Section 4 (so for this subsection we will not need to work in the set-up of "two T's" as we did e.g., in Section 9). Let $\mathcal{F}$ be an $\mathbf{A}$-field and let ψ be a Drinfeld module over $\mathcal{F}$. Given a prime v of $\mathbf{A}$ which is different from the characteristic of $\mathcal{F}$, we have the Tate module $T_v(\psi)$ as defined in Definition 4.10.9. Let $\mathcal{F}^{\text{sep}}$ be a separable closure and put $G := \text{Gal}(\mathcal{F}^{\text{sep}}/\mathcal{F})$. Clearly G operates $\mathbf{A}_v$-linearly and continuously on $T_v(\phi)$. Moreover, if ϕ is another Drinfeld module defined over $\mathcal{F}$ and $f\colon \phi \to \psi$ is a $\mathcal{F}$-morphism, then we deduce a G-equivariant mapping $T_v(\phi) \to T_v(\psi)$. In other words there is a mapping

$$\text{Hom}_{\mathcal{F}}(\phi,\psi) \otimes_{\mathbf{A}} \mathbf{A}_v \to \text{Hom}_{\mathbf{A}_v[G]}(T_v(\phi), T_v(\psi)).$$

Of course all of this is totally similar to what happens with abelian varieties and was discussed in Subsection 4.11 when $\mathcal{F}$ is finite. As mentioned in Proposition 4.12.11, the above mapping is injective (essentially because an additive polynomial can have at most finitely many zeroes). The *Tate Conjecture for Drinfeld modules* over $\mathcal{F}$ is the assertion that this mapping is actually an isomorphism. This has been proved in the case of a finite field $\mathcal{F}$ in Theorem 4.12.12. The proof of the next result is due independently to Y. Taguchi [Tag7] and A. Tamagawa [Tam2], [Tam4].

Theorem 10.3.1 (Taguchi, Tamagawa). *The Tate Conjecture is true when $L = \mathcal{F}$ is a finite extension of* $\mathbf{k}$. □

We will briefly describe Taguchi's approach to the proof of the above fundamental result though both approaches are rather similar. In [Tag8] the reader will find an *excellent* discussion of the matters briefly touched on here.

The first important step is to realize that standard arguments in complex multiplication theory allow us to reduce the theory to having $\mathbf{A} = \mathbb{F}_r[T]$. The next basic element in the proof of the Tate Conjecture was given by Anderson [A5]. In this paper, using arguments *native* to the characteristic p theory (in particular, the ability to tensor two T-modules, cf. Section 5), a version of the Tate Conjecture was shown for *formal T-modules*. This was done via a tensor product reduction to the case where one of the objects was a higher tensor power of the Carlitz module. In this fashion the *global* Tate Conjecture is eventually reduced to a statement about a linear *Frobenius equation*, which one views as an analog of a differential equation defined over $\mathbb{P}^1(\mathbb{C})$. Using some "global analysis," (heights, etc.) Taguchi then solves this equation and finishes the proof.

10.4. Meromorphic Continuations of L-functions

In this subsection we will describe the work of Y. Taguchi and D. Wan [TW1], [TW2]. This work insures that, when $\mathbf{A} = \mathbb{F}_r[T]$, L-series of Drinfeld modules (and T-modules, etc.) have *meromorphic* continuations in an extremely general context. As these general L-functions were beyond the scope of what was needed for this volume, we have not mentioned them previously. However, to give the reader an idea of the scope of the results of Taguchi-Wan, we do so now. For more, we refer the reader to [Go4, §3.2 and §3.7]. Thus let Z be an arbitrary scheme of finite type over $\mathrm{Spec}(\mathbf{A})$. Let ψ be a *family of Drinfeld modules over Z*; i.e., there is a line bundle B over Z together with an injection of $\mathbf{A}$ into the $\mathbb{F}_r$-linear endomorphisms of B such that the first coefficient arises from the structure map $Z \to \mathrm{Spec}(\mathbf{A})$ (exactly as when Z is the spectrum of an $\mathbf{A}$-field), and such that the reduction at all the closed points of Z is a Drinfeld module with constant rank on the connected components of Z (see, e.g., [Dr1], [Go8]). Thus without loss of generality we may also assume that Z is connected.

Let $z \in Z$ be a closed point which lies over the prime $\wp$ of $\mathbf{A}$, and let $\mathbb{F}_z$ be the residue field at z. Thus $\mathbb{F}_z$ is a finite extension of $\mathbf{A}/\wp$ of degree $d(z)$. We then set $nz := \wp^{d(z)}$. As in Subsection 4.12, there is a Frobenius morphism F_z of the reduction ψ^z of ψ at z. Moreover, if v a prime of $\mathbf{A}$ not equal to $\wp$ then

$$P_z(u) := \det(1 - F_z u \mid T_v(\psi^z))$$

is a polynomial with $\mathbf{A}$-coefficients and is independent of v. Now let S_∞ be as in Subsection 8.1.

Definition 10.4.1. Let $s \in S_\infty$. We set

$$L(\psi/Z, s) := \prod_{z \in Z \text{ closed}} P_z(nz^{-s})^{-1}.$$

The reader will see directly that Definition 10.4.1 generalizes the definition of L-functions of Drinfeld modules given in Subsection 8.6. The reader will also easily see how to give v-adic versions of the above functions.

In Definition 8.5.12 we introduced essentially algebraic entire functions on S_∞. We now call a function on S_∞ an *essentially algebraic meromorphic function* if and only if it is the quotient of two essentially algebraic entire functions.

Theorem 10.4.2 (Taguchi-Wan). *Let $\mathbf{A} = \mathbb{F}_r[T]$. Then $L(\psi/Z, s)$ is an essentially algebraic meromorphic function.* □

Taguchi and Wan work with certain φ-sheaves. These are "relative" versions of the more elementary φ-sheaves that we introduced in Subsection

10.4. Meromorphic Continuations of L-functions

5.3. They are also related to shtukas and, in fact, are yet another avatar of Drinfeld modules and T-modules, etc. Using these sheaves, one obtains an expression for the L-function $L(\psi/Z, s)$ which is exactly similar to the classical expression for the L-function of an *F-crystal* [K3] of p-adic analysis. The L-function of such crystals is studied via the *Dwork trace formula* [Wan3], of which there are two versions. One version (the multiplicative version) works in characteristic p and allows Taguchi and Wan to establish their results. The reason for the restriction on **A** lies in the need for the authors to establish "overconvergence" properties for a certain character (or a rank one φ-sheaf), which is essentially the second variable y of $s \in S_\infty$. The authors also establish a v-adic version of their results. It is certainly reasonable to expect such results to ultimately hold for arbitrary **A**.

The above results also work when ψ is a family of T-modules or even for the trivial T-module; thus it is most remarkable that the theory of the reduction modulo p of the *classical* Hasse-Weil zeta function is *also* handled by Theorem 10.4.2. In the curve case discussed in this book (e.g., the zeta function of **A** itself) we obtained information about the p-divisibility of class numbers (cf. Subsection 8.14). Now the reduction modulo p of the classical zeta function of a scheme equals the reduction modulo p of its *unit-root* piece and such unit-root zeta-functions are controlled by the *p-adic étale cohomology* of the scheme [Cr1]. Thus it is a very interesting problem to understand just what information is contained in these very general functions. A model of such functions is the following beautiful example of Taguchi and Wan (presented just before §9 of [TW1]). Let Z be the family of affine elliptic curves over $\mathrm{Spec}(\mathbb{F}_r[T])$ given by

$$X_1^2 = X_2(X_2 - 1)(X_2 - T).$$

Then the L-function, $L(Z, s)$ is actually an essentially algebraic entire function on S_∞.

In general, of course, it is very desirable to know that the L-series are actually entire (as in the above example) as opposed to just meromorphic. In [TW2], Taguchi and Wan give sufficient conditions for this to be so. This covers *huge* number of examples of interest. For instance, let L be any finite extension of **k** and let $\mathcal{O}_L$ be the ring of **A**-integers of L. Let ψ be the Drinfeld module over $\mathcal{O}_L$ given by

$$\psi_T := T\tau^0 + \sum_{i=0}^{d-1} a_i \tau^i + \alpha \tau^d,$$

where $\{a_i\} \subset \mathcal{O}_L$ and α is a unit in $\mathcal{O}_L$. (Thus there are *no* bad finite primes for ψ.) Then by [TW2] we conclude that $L(\psi, s)$ is entire, with a similar v-adic version. In general, for an arbitrary Drinfeld module with bad primes, one is still presented with the problem of defining the appropriate Euler factors at the bad primes.

In the papers [TW1] and [TW2] the authors end up with polynomial growth for their estimates of the ∞-adic size of the coefficients of L-series, in agreement with the estimates given in Subsection 8.8. However, as we saw in Subsection 8.24 one can often obtain *exponential* growth, as was used in the "Riemann hypothesis" of that subsection. In a private communication, Taguchi and Wan have told us that they can show such exponential type growth in general (at least for $\mathbf{A} = \mathbb{F}_r[T]$). As such, this is some evidence that the Riemann hypothesis for these characteristic p L-functions *may* hold in great generality.

10.5. The Structure of the A-module of Rational Points

Let L be a field and let Z now be an abelian variety over L. Associated to L and Z we have the abelian group $Z(L)$ of L-rational points of Z. The classical Theorem of Mordell and Weil assures that, under very general hypotheses, (e.g., L is a number field) the group $Z(L)$ is finitely generated.

Now let $\mathbf{A}$ be general and $\mathcal{F}$ an $\mathbf{A}$-field. Let ψ be a Drinfeld module over $\mathcal{F}$; thus the additive group $\mathcal{F}$ becomes an $\mathbf{A}$-module via ψ. We have denoted this module $\psi(\mathcal{F})$. If $\mathcal{F}$ is finite, then, of course, $\psi(\mathcal{F})$ is finite and we have discussed its structure in Subsection 4.12. Suppose now that $L = \mathcal{F}$ is a finite extension of $\mathbf{k}$. It has long been folklore that $\psi(L)$ is *never* finitely generated (as follows, for instance from the remark at the beginning of the proof of Theorem 5 of [Den1]). In fact, we have the following remarkable result [Po3].

Theorem 10.5.1 (Poonen). *The $\mathbf{A}$-module $\psi(L)$ is the direct sum of a finite torsion module and a free $\mathbf{A}$-module of infinite rank.* $\square$

Let M be an $\mathbf{A}$-module. The *rank* of M is the dimension of the $\mathbf{k}$-vector space $M \otimes_\mathbf{A} \mathbf{k}$. The module M is *tame* if every submodule of finite rank is finitely generated as an $\mathbf{A}$-module. (In fact, the reader will easily see that the notion of a "tame module" makes sense for *any* integral domain.) The main step in the proof of Theorem 10.5.1 is establishing that $\psi(L)$ is tame. This is done through the use of local and global heights building on the work of Denis [Den 4].

Let L^{perf} be the perfection of L. Poonen shows that his basic structure theorem also holds for the $\mathbf{A}$-module L^{perf}. This raises a potentially very interesting question as the adjoint ψ^* of ψ is defined over L^{perf}.

Question 10.5.2. Does $\psi^*(L^{\text{perf}})$ have the same structure as $\psi(L^{\text{perf}})$ as $\mathbf{A}$-module?

As of this writing, it is not known what to expect in answer to 10.5.2.

Finally, as pointed out in [Po3], Theorem 10.5.1 is exactly similar to what is known about the structure of k^*, as $\mathbb{Z}$-module, where k is a number field.

10.6. Log-algebraicity and Special Points

Up to now, the results we have established on characteristic p L-series have concerned concepts, such as the class group, which come from characteristic 0. Recently, in [A8], G. Anderson made a first breakthrough in establishing that these L-series *also* contain important information about *characteristic p structures* such as the **A**-module of rational points discussed in our last subsection.

Before describing some of Anderson's results, let us recall the standard formula

$$\exp\left(-\sum_{n=1}^{\infty}\frac{z^n}{n}\right) = 1 - z \qquad (10.6.1)$$

from freshman calculus. This equation allows one to relate the group of classical cyclotomic units (units of the form $\frac{1-\zeta^a}{1-\zeta^b}$ for some root of unity ζ) to special values at $s=1$ of classical L-series. In the notation of [A8], the series $\sum \frac{z^n}{n}$ is *log-algebraic*.

Now let R be a Dedekind domain and let M be an R-module. Let $N \subseteq M$ be an R-submodule. The *divisible closure of N in M*, $\sqrt{N}$, is the set of all $m \in M$ with $r \cdot m \in N$ for some nonzero $r \in R$; clearly $\sqrt{N}$ is a submodule of M. If M is tame (cf. Subsection 10.5) then $\sqrt{N}$ is finitely generated whenever N is; thus $\sqrt{N}/N$ is a *finite* R-module.

One implication of Equation 10.6.1, and the results of classical number theory, is that the cyclotomic units are of finite index in the group of *all* units; thus the full unit group is precisely the divisible closure of the group of cyclotomic units. Return now to the function field setting. Anderson is able to establish a massive generalization of Equation 10.6.1 for certain rank one Drinfeld modules; a result which had its genesis in some very striking examples given in [Th5]. When $\mathbf{A} = \mathbb{F}_r[T]$, this allows him to define an analog (the module of *special points*) of the group of cyclotomic units *for the Carlitz module*. Thus, turning the above procedure around, the full module of "units" for the Carlitz module can be *defined* as the divisible closure of the **A**-module of special points!

More precisely, let $\mathbf{A} = \mathbb{F}_r[T]$ and let m be a nonnegative integer. We define the power series $S_m(t,z)$ by

$$S_m(t,z) := \sum_{i=0}^{\infty} \sum_{\substack{a \in A \\ a \text{ monic}}} \frac{1}{D_i}\left(\frac{C_a(t)^m}{a}\right)^{r^i} z^{r^i + \deg a}.$$

Let $e_C(z)$ be the exponential of the Carlitz module and let $m \geq 1$. Then $\ell_m(x) \colon \mathbf{K}/\mathbf{A} \to \mathbf{C}_\infty$ is defined by

$$\ell_m(x) := \sum_{\substack{a \in A \\ a \text{ monic}}} \frac{e_C(ax\xi)^m}{a},$$

where ξ is the period of the Carlitz module (which is defined up to an element in $\mathbb{F}_r^*$). We set $\ell_0 \colon \mathbf{K}/\mathbf{A} \to \mathbf{C}_\infty$ to be the constant $\sum_{\substack{a \in \mathbf{A} \\ a \text{ monic}}} \frac{1}{a}$. In [A8] the following result is established.

Theorem 10.6.2 (Anderson). 1. *The power series $S_m(t,z)$ actually belongs to $\mathbf{A}[t,z]$; i.e., is a polynomial in t and z with $\mathbf{A}$-coefficients.*
2. *For all $x \in \mathbf{K}$ we have*
$$e_C(\ell_m(x)) = S_m(e_C(\xi x), 1).$$
3. *We have*
$$\frac{S_m(t,z)}{t^m} = \sum_{\substack{a \in \mathbf{A} \\ a \text{ monic}}} a^{m-1} z^{\deg a},$$
for $m > 0$. □

The proof of Theorem 10.6.2.1 involves putting a norm on the coefficients of $S_m(t,z)$ for which these coefficients are *both* discrete and tend to 0. Thus it may be considered a higher dimensional analog of the proof of the more elementary Proposition 8.5.13.

Now let $f \in \mathbf{A}$ be a fixed irreducible and let $\mathbf{k}(f)$ be the extension of $\mathbf{k}$ obtained by adjoining the f-division points of the Carlitz module as in Section 7. Let $\mathcal{O} = \mathcal{O}(f)$ be the ring of $\mathbf{A}$-integers of $\mathbf{k}(f)$. The *module of special points* $\mathcal{S} = \mathcal{S}(f)$ is the $\mathbf{A}$-span (under the *Carlitz module*) of the $\mathcal{O}$-valued points
$$\{S_m(e_C(\xi b/f), 1)\}$$
where m is a nonnegative integer and $0 \neq b \in \mathbf{A}/(f)$. It is shown in [A8] that $\mathcal{S}$ is *finitely generated* (and the rank is computed). This is accomplished by relating $\mathcal{S}$ to special values of characteristic p L-series!

Thus by Poonen's result Theorem 10.5.1, we see that $\mathcal{S}' := \sqrt{\mathcal{S}}/\mathcal{S}$ is a finite $\mathbf{A}$-module. By analogy with the classical index computations of cyclotomic number fields, we see that the Euler-Poincaré characteristic $h(\mathcal{S}')$ (= the Fitting ideal) of $\mathcal{S}'$ may be viewed as an $\mathbf{A}$-*analog of the class number*.

Problem 10.6.3. Show that f acts invertibly on $\mathcal{S}'$.

Equivalently, show $f \nmid h(\mathcal{S}')$. With a little thought, the reader will see that this problem is almost *exactly* analogous to the classical *Kummer-Vandiver Conjecture*, $p \nmid h(p)^+$, (in its unit-index formulation)! So we have yet another analog of the Kummer-Vandiver phenomena in function fields (see Problem 8.22.7). In fact, in 8.14.5.3 we defined the class number $h(f)^-$ in analogy to the classical $h(p)^-$; then using special values of $\zeta_{\mathbb{F}_r[T]}(s)$, Theorem 8.14.4 and 10.6.2.3, Anderson shows that Problem 10.6.3 has an affirmative solution

10.6. Log-algebraicity and Special Points

when $p \nmid h(f)^-$. This is very similar to the classical formula of Kummer $p \nmid h(p)^- \Rightarrow p \nmid h(p)^+$. How all of these different versions of the Kummer-Vandiver phenomenon are related (IF they are related) remains to be worked out. Still it is evident that the techniques now being developed will allow us to peer very deeply into the arithmetic of Drinfeld modules in the near future.

Indeed, the techniques of [A8] present us with many interesting questions. In particular, Anderson defines certain *root numbers* which come up in his L-function calculations. Although these numbers are not at all obviously related to Thakur's Gauss sums (as given in Subsection 9.8) one would expect that they should be intimately connected. In fact, this now seems to be true — see the recent paper [Zh2] which contains both positive results and very intriguing examples. This may lead to insight into the digit expansion procedure used to define Γ-functions, Gauss sums, etc. Moreover, since we have seen in Subsection 9.8 that Gauss sums are intimately related to *arithmetic* gamma functions, this new work hints at a connection between such gamma functions and L-series (see also the discussion at the end of Subsection 9.9). Only further research will allow us to judge the full extent of these connections.

However, inspired by [Zh2], Anderson has given a set of generators of $\mathcal{S}$ which we now present. To begin with, as above, we will *not* be in the "two T's" setup; so we will not distinguish between $\mathbf{C}_\infty$ and C_∞ (see Subsection 7.11). Let $\omega = \omega_f$ be the Teichmüller character associated to (f) as in Subsection 8.11. Let W be the Witt ring of $\mathbf{A}/(f)$ which we regard as being in a fixed algebraic closure of $\mathbb{Q}_p$; thus $\omega \colon \mathbf{A}/(f)^* \to W^*$. By reducing ω modulo p and injecting into $\mathbf{C}_\infty$ we obtain a $\mathbf{C}_\infty^*$ valued character on $\mathbf{A}/(f)$ which we still denote ω; a moment's thought will convince the reader that this character arises from an $\mathbb{F}_r$-algebra injection of $\mathbf{A}/(f)$ into $\mathbf{C}_\infty$. Thus we can use this character, as χ_0, to define basic Gauss-Thakur sums as in Definition 9.8.1, and thus general Gauss-Thakur sums $g(\chi)$ (as in the discussion after 9.8.1).

Let $d = \deg f$ and let i be an integer with $1 \leq i \leq r^d - 2$. Let $\chi := \omega^i$, and let $a \in \mathbf{A}/(f)$. We set

$$\operatorname{sp}(\chi, a) := e_C \left(\sum_{j=0}^{d-1} \omega^{r^j}(a) g(\chi^{r^j}) L(\chi^{r^j}, 1) \right) \in \mathcal{O}.$$

Anderson then shows that $\mathcal{S}$ is generated as an $\mathbf{A}$-module (under C) by the set $\{1\} \cup \{\operatorname{sp}(\chi, a)\}$, where we take all χ and a as above. One should note that this set of generators is *independent* of the choice of ω, though the indexing may depend on this choice.

From the point of view of the "two T's," the above construction is quite remarkable. Indeed, in the definition of $\operatorname{sp}(\chi, a)$ we multiply a Gauss-Thakur sum, which arises when T is a scalar, with an L-value, which arises when T is an operator. It would be very interesting to give a natural reinterpretation of this construction from the "two T's" point of view.

Anderson's result is based on the following characterization of Gauss-Thakur sums. Let $\zeta \in \mathbf{C}_\infty$ be a primitive $(r^d - 1)$-st root of unity and let $\mathbf{A}' := \mathbf{A}[\zeta]$. Let $\mathcal{O}$ be as above and set $\mathcal{O}' := \mathcal{O}[\zeta]$. Let $G := \mathrm{Gal}(\mathbf{k}(f)/\mathbf{k}) \simeq (\mathbf{A}/(f))^*$. As $\mathbf{k}(f)/\mathbf{k}$ is totally ramified above (f), G is canonically isomorphic to the Galois group of $\mathbf{k}(f)[\zeta]$ over $\mathbf{k}[\zeta]$, and we do not distinguish the two. The Gauss-Thakur sums are clearly elements of $\mathcal{O}'$. Let $a \in \mathbf{A}$ be prime to f and let $\sigma_a \in G$ be the corresponding automorphism of $\mathcal{O}'/\mathbf{A}'$. We then have

$$\sigma_a(g(\chi)) = \chi(a)g(\chi).$$

Indeed, this is elementary for the basic Gauss-Thakur sums, and follows for the rest via the digit expansion procedure used to define $g(\chi)$.

Now note that $r^d - 1 \equiv -1 \pmod{p}$. As such, we can decompose $\mathcal{O}'$ into isotypic components under G; we let $\mathcal{O}'(\chi)$ be the component associated to χ. It is easy to see that $\mathcal{O}'(\chi)$ is free of rank 1 over $\mathbf{A}'$, as $\mathbf{A}'$ is still a principal ideal domain.

Theorem 10.6.4. *$\mathcal{O}'(\chi)$ is generated by $g(\chi)$ as an $\mathbf{A}'$-module.*

Theorem 10.6.4 is proved by computing the divisor of a generator of $\mathcal{O}'(\chi)$ and then noting that this is *precisely* the divisor of $g(\chi)$. In particular, 10.6.4 *characterizes* $g(\chi)$ up to an $(r^d - 1)$-st root of unity. It would be very interesting to generalize 10.6.4 to all $\mathbf{A}$, as well as to classical Gauss sums where there are some subtleties involved in decomposing into isotypic components.

Finally we leave the reader with the following problem.

Problem 10.6.5. The techniques of this subsection allows us to produce a good analog for the Carlitz module, and $\mathbf{k}(f)$, of the classical group of units modulo cyclotomic units. Use $\mathcal{S}$ and the structure theorem for rational points (Theorem 10.5.1) to give a definition of the *class group* for the Carlitz module and $\mathbf{k}(f)$.

Thus one would want a finite $\mathbf{A}$-module which would be the obstruction to some *local-global* problem for the Carlitz module. Such a construction would obviously have many profound implications.

One possible solution to 10.6.5 has been proposed by B. Poonen. It is a well-known piece of folk-lore in algebraic number theory that the Tate-Shafarevich group of the units *is* the ideal class group. Thus Poonen suggests trying to do an analogous "Tate-Shafarevich" construction for Anderson's special points. However, this seems to require a functorial definition of 'special points' for Drinfeld modules over *both* global and local fields, and, as of this writing, no such definition is available.

References

[Ab1]   S. Abhyankar: Factorizations over finite fields, in *Finite Fields and Applications*, London Math. Soc. Lecture Notes Series **233**, (1996), 1-21.

[AS1]   S. Abhyankar, G. Sundaram: Galois theory of Moore-Carlitz-Drinfeld modules, *C.R. Acad. Sci. Paris* (to appear).

[Ai1]   A. Aiba: Carlitz modules and Galois module structure, *J. Number Theory* **62** (1997), 213-219.

[Al1]   J.-P. Allouche: Transcendence of the Carlitz-Goss gamma function at rational arguments, *J. Number Theory* **60** (1996), 318-328.

[Al2]   J.-P. Allouche: Sur la transcendance de la série formelle Π, in: *Séminaire de Théorie des Nombres de Bordeaux, 1990*, 103-117.

[Al3]   J.-P. Allouche: Fonction zêta de Carlitz et automates finis, in: *Modules de Drinfeld et l'approximation diophantienne* (ed. Y. Hellegouarch) Caen (1990).

[A1]   G. Anderson: t-motives, *Duke Math. J.* **53** (1986), 457–502.

[A2]   G. Anderson: A two-dimensional analogue of Stickelberger's theorem, in: *The Arithmetic of Function Fields*, (eds: D. Goss et al) de Gruyter (1992), 51–77.

[A3]   G. Anderson: Rank one elliptic A-modules and A-harmonic series, *Duke Math. J.* **73** (1994), 491–542.

[A4]   G. Anderson: Torsion points on Jacobians of quotients of Fermat curves and p-adic soliton theory, *Invent. Math.* **118** (1994), 475-492.

[A5]   G. Anderson: On Tate modules of formal t-modules, *Int. Math. Res. Notices* **2** (1993), 41-52.

[A6]   G. Anderson: On a question arising from complex multiplication theory, in: *Galois Representations and Arithmetic Algebraic Geometry*, Advanced Studies in Pure Mathematics **12** (1987), 221-234.

[A7]   G. Anderson: Another look at the index formulas of cyclotomic number theory, *J. Number Theory* **60** (1996), 142-164.

[A8]   G. Anderson: Log-algebraicity of twisted A-harmonic series and special values of L-series in characteristic p, *J. Number Theory* **60** (1996), 165-209.

[A9]   G. Anderson: A double complex for computing the sign-cohomology of the universal ordinary distribution, (preprint).

[AT1]   G. Anderson, D. Thakur: Tensor powers of the Carlitz module and zeta values, *Ann. of Math.* **132** (1990), 159–191.

[An1]   B. Anglès: Sur les anneaux d'endomorphismes des modules de Drinfeld définis sur des corps finis, *C.R. Acad. Sci. Paris* **t. 319**, Série I, (1994), 1237-1240.

[An2]   B. Anglès: On some subrings of Ore polynomials connected with finite Drinfeld modules, *J. Algebra* **181** (1996), 507-522.

[Ar1] E. Artin: Quadratische Körper im Gebiete der höheren Kongruenzen I, II, *Math. Z.* **19** (1924), 153-246 (= *Coll. Papers*, 1-94).

[AM1] M.F. Atiyah, I.G. Macdonald: *Introduction to Commutative Algebra*, Addison-Wesley (1969).

[Ay1] R. Ayoub: Euler and the zeta-function, *Amer. Math. Monthly* **81** (1974) 1067–1086.

[Ba1] S. Bae: On the modular equation for Drinfeld modules of rank 2, *J. Number Theory* **42** (1992), 123-133.

[Ba2] S. Bae: Drinfeld modules with bad reduction over complete local fields, *Bull. Korean Math. Soc.* **32** (1995), 349-357.

[BaS1] S. Bae, S.-G. Hahn: On the rings of integers of cyclotomic function fields, *Bull. Korean Math. Soc.* **29** (1992), 153-163.

[BaK1] S. Bae, P.-L. Kang: On Tate-Drinfeld modules, *Canadian Math. Bull.* **35** (1992), 145-151.

[BaK2] S. Bae, P.-L. Kang: Local isogeny theorem for Drinfeld modules with nonintegral invariants, *Proc. Amer. Math. Soc.* (to appear).

[BaKo1] S. Bae, J.K. Koo: Torsion points of Drinfeld modules, *Canadian Math. Bull.* **38** (1995), 3-10.

[BaKo2] S. Bae, J. K. Koo: On the singular Drinfeld modules of rank 2, *Math. Z.* **210** (1992), 267-275.

[BaKo3] S. Bae, J. K. Koo: Genus theory for function fields, *J. Austral. Math. Soc.* **60** (1996), 301-310.

[BBT1] P.-G. Becker, W.D. Brownawell, R. Tubbs: Gel'fond's theorem for Drinfeld modules, *Mich. Math. J.* **41** (1994), 219-233.

[Be1] G. Bergman: Coproducts and some universal ring constructions, *Trans. Amer. Math. Soc.* **200** (1974), 33–88.

[Ber1] V. Berthé: Automates et valeurs de transcendance du logarithme de Carlitz, *Acta. Arithm.* **66** (1994), 369-390.

[Ber2] V. Berthé: Combinaisons linéaires de $\zeta(s)/\Pi^s$ sur $\mathbb{F}_q(X)$ pour $1 \leq s \leq q-2$, *J. Number Theory* (to appear).

[Berk1] V. Berkovich: The automorphism group of the Drinfeld half-plane, *C.R. Acad. Sci. Paris* **t. 321** Série I (1995), 1127-1132.

[Beu1] F. Beukers: Differential Galois theory, in: *From Number Theory to Physics*, (eds: M. Waldschmidt et al), Springer (1992), 411-439.

[BGR1] S. Bosch, U. Güntzer, R. Remmert: *Non-Archimedean Analysis*, Springer (1984).

[BoO1] N. Boston, D. Ose: Characteristic p representations that are produced by Drinfeld, (preprint).

[Bo1] N. Bourbaki: *Commutative Algebra*, Addison-Wesley (1972).

[Bro1] M.L. Brown: Singular moduli and supersingular moduli of Drinfeld modules, *Invent. Math.* **110** (1992), 419-439.

[Brow1] W.D. Brownawell: Drinfeld exponential and quasi-periodic functions, in: *Advances in Number Theory* (eds: F.Q. Gouvêa, N. Yui) Oxford University Press, Oxford (1993).

[Brow2] W.D. Brownawell: Submodules of products of quasi-periodic modules, *Rocky Mountain J. Math.* **26** (1996), 847-873.

[Bru1] F. Bruhat: *Lectures on Some Aspects of p-adic Analysis*, TATA Inst. for Fund. Res. (1963).

[Bu1] P. Bundschuh: Transzendensmasse in Körpern formaler Laurentreihen, *J. reine angew. Math.* **299-300** (1978), 411-432.

[Ca1] H. Carayol: Variétés de Drinfeld Compactes, d'après Laumon, Rapoport et Stuhler, Séminaire Bourbaki 44ème année **756** 1991-95 *Astérisque* **206** (1992), 369-409.

[C1]   L. Carlitz: On certain functions connected with polynomials in a Galois field, *Duke Math. J.* **1** (1935), 137–168.
[C2]   L. Carlitz: Some special functions over $GF(q,x)$. *Duke Math. J.* **27** (1960), 139–158.
[C3]   L. Carlitz: Finite sums and interpolation formulas over $GF[p^n, x]$, *Duke Math. J.* **15** (1948), 1001–1012.
[C4]   L. Carlitz: A set of polynomials, *Duke Math J.* **6** (1940), 486–504.
[C5]   L. Carlitz: An analogue of the von Staudt-Clausen theorem, *Duke Math. J.* **3** (1937), 503–517.
[C6]   L. Carlitz: An analogue of the von Staudt-Clausen theorem, *Duke Math. J.* **7** (1940), 62–67.
[C7]   L. Carlitz: An analogue of Bernoulli polynomials, *Duke Math. J.* **8** (1941), 405-412.
[C8]   L. Carlitz: The Staudt-Clausen theorem, *Math. Mag.* **34** (1961), 131-146.
[C9]   L. Carlitz: A class of polynomials, *Trans. Amer. Math. Soc.* **43** (1938), 167-182.
[C10] L. Carlitz: Chapter 19 of "The Arithmetic of Polynomials," *Finite Fields and their Applications* **1** (1995), 157-164.
[C11] L. Carlitz: The arithmetic of polynomials in a Galois field, *Proc. Nat. Acad. Sci. (USA)* **17** (1931), 120-122.
[C12] L. Carlitz: The arithmetic of polynomials in a Galois field, *Amer. J. Math.* **54** (1932), 39-50.
[C13] L. Carlitz: On polynomials in a Galois field, *Bull. Amer. Math. Soc.* **38** (1932), 736-744.
[C14] L. Carlitz: The reciprocal of certain types of Hurwitz series, *Duke Math. J.* **9** (1942), 629-642.
[C15] L. Carlitz: Finite sums and interpolation formulas over $GF[p^n, x]$, *Duke Math. J.* **15** (1948), 1001-1012.
[C16] L. Carlitz: Diophantine approximation in fields of characteristic p, *Trans. Amer. Math. Soc.* **72** (1952), 187-208.
[C17] L. Carlitz: Some sums involving polynomials in a Galois field, *Duke Math. J.* **5** (1939), 941-947.
[C18] L. Carlitz: On factorable polynomials in several indeterminates, *Duke Math. J.* **2** (1936), 660-670.
[C∞]   L. Carlitz: Full list of publications related to finite fields, *Finite Fields and their Applications* **1** (1995), 145-151.
[Cha1] R.J. Chapman: Kummer theory and Galois module structure in global function fields, *Math. Z.* **208** (1991), 375-388.
[Cha2] R.J. Chapman: Carlitz modules and normal integral bases, *J. London Math. Soc. (2)* **44** (1991), 250-260.
[CT1] V. Chari, D. Thakur: On the work of V.G. Drinfeld, *Current Science* **59**(24) (1999), 1297-1300.
[CH1] Z. Chen, D. Hayes: Norm and trace of the j-invariants of Drinfeld modules associated to hyperelliptic curves, *J. Number Theory* **63** (1997), 1-11.
[ChF1] L. Cheng, K. Feng: On independence of cyclotomic units in function fields, *Sci. Sinica* **31** (1988), 601-609.
[ChF2] L. Cheng, K. Feng: On the Minkowski unit in function fields, (English) *Chin. Sci. Bull.* **34** (1989), 809-811.
[Ch1] H. Cherif: Mesure d'irrationalité de valeurs de la fonction zêta de Carlitz sur $\mathbb{F}_q[T]$, *C.R. Acad. Sci. Paris Série I* **310** (1990), 23-26.

[ChM1]   H. Cherif, B. de Mathan: Mesure d'irrationalité de la valeur en 1 de la fonction zêta de Carlitz, relative à $\mathbb{F}_2(T)$, *C.R. Acad. Sci. Paris Série I* **305** (1987), 761-763.

[ChM2]   H. Cherif, B. de Mathan: Irrationality measures of Carlitz zeta values in characteristic p, *J. Number Theory* **44** (1993), 260-272.

[Cl1]   R. Clement: The genus field of an algebraic number field, *J. Number Theory* **40** (1992), 359-375.

[Co1]   G. Cornelissen: Sur les zéros des séries d'Eisenstein de poids $q^k - 1$ pour $GL_2(\mathbf{F}_q[T])$, *C.R. Acad. Sci. Paris* t. **321** Série I. (1995), 817-820.

[Co2]   G. Cornelissen: The field of definition of the divisor of Eisenstein series for $\mathbf{F}_q(T)$, (preprint).

[Co3]   G. Cornelissen: Drinfeld modular forms of weight one, *J. Number Theory* (to appear).

[Cr1]   R. Crew: Geometric Iwasawa theory and a conjecture of Katz, in: *Conf. Proc. Canad. Math. Soc. Vol. 7*, (eds: H. Kisilevsky and J. Labute) Amer. Math. Soc. (1987), 37–53.

[CuR1]   C. Curtis, I. Reiner: *Representation Theory of finite Groups and associative Algebras*, Pure and Applied Math. **XI**, Wiley (1966).

[Dam1]   G. Damamme: Transcendence properties of Carlitz zeta-values, in: *The Arithmetic of Function Fields* (eds. D. Goss et al) de Gruyter (1992), 303-311.

[Dam2]   G. Damamme: Irrationalité de $\zeta(s)$ dans le corps des séries formelles $\mathbf{F}_q((1/t))$, *C.R. Math. Acad. Sc. Canada* **9** (1987), 207-212.

[Dam3]   G. Damamme: Transcendance des valeurs de la fonction zêta de Carlitz, in: *Modules de Drinfeld et l'approximation diophantienne* (ed. Y. Hellegouarch) Caen (1990).

[Dam4]   G. Damamme: Quelques propriétés des fonctions L dans le module de Carlitz, *Université de Caen Prépublication* **65** (1995).

[DamH1]   G. Damamme, Y. Hellegouarch: Transcendence of the values of the Carlitz zeta function by Wade's method, *J. Number Theory* **39** (1991), 257-278.

[Da1]   C. David: Supersingular reduction of Drinfeld modules, *Duke Math. J.* **78** (1995), 399-412.

[Da2]   C. David: The average distribution of supersingular Drinfeld modules, *J. Number Theory* **56** (1996), 366-380.

[DaD1]   S. David, L. Denis: Isogénies minimales entre modules de Drinfeld, (preprint).

[Ded1]   R. Dedekind: Abriss einer Theorie der höheren Congruenzen in Bezug auf einer reellen Primzahl–Modulus, *J. Reine Angew. Math.* **54** (1857), 1-26.

[De1]   P. Deligne: Catégories Tannakiennes, in: *The Grothendieck Festschrift*, Vol. III, Birkhaüser (1990), 111-195.

[De2]   P. Deligne: *Equations différentielles à Points Singuliers Réguliers*, Lect. Notes. Math. 163 (1970).

[De3]   P. Deligne: Valeurs de fonctions L et périodes d'intégrales, *Proc. Symp. Pure. Math.* **33** Part II, Amer. Math. Soc. (1979), 313-346.

[DMOS1]   P. Deligne, J.S. Milne, A. Ogus, K. Shih: *Hodge Cycles, Motives, and Shimura Varieties*, Lect. Notes. Math. **900**, Springer (1989).

[DeH1]   P. Deligne, D. Husemöller: Survey of Drinfeld modules, *Contemp. Math.* **67** (1987), 25–91.

[DeG1]   M. Demazure, P. Gabriel: *Groupes Algébriques Tome I*, North-Holland (1970).

[Dene1]   M. Denert: Affine and Projective Orders in Central Simple Algebras over Global Function Fields, Ph.D Thesis, Gent (1987).
[DeneG1] M. Denert, J. van Geel: The class numbers of hereditary orders in non-Eichler algebras over global function fields, *Math. Ann.* **282** (1988), 379-393.
[Den1]   L. Denis: Géométrie Diophantienne sur les Modules de Drinfeld, in: *The Arithmetic of Function Fields* (eds. D. Goss et al) (1992), 285-302.
[Den2]   L. Denis: Théorème de Baker et modules de Drinfeld, *C.R. Acad. Sci. Paris* **t. 311**, Série 1 (1990), 473-475.
[Den3]   L. Denis: Théorème de Baker et modules de Drinfeld, *J. Number Theory* **43** (1993), 203-215.
[Den4]   L. Denis: Hauteurs canoniques et modules de Drinfeld, *Math. Ann.* **294** (1992), 213-223.
[Den5]   L. Denis: Remarques sur la transcendance en caractéristique finie, *C.R. Acad. Sci. du Canada* 4(v.XIV) (1992).
[Den6]   L. Denis: Le théorème de Fermat-Goss, *Trans. Amer. Math. Soc.* **343** (1994), 713-726.
[Den7]   L. Denis: Dérivées d'un module de Drinfeld et transcendance, *Duke Math. J.* **80** (1995), 1-13.
[Den8]   L. Denis: Problèmes diophantiens sur les t-modules, "Publications de Paris 6," *Journées arithmétiques de Bordeaux* **7** (1995), 97-110.
[Den9]   L. Denis: Indépendance algébrique sur le module de Carlitz, *C.R. Acad. Sci. Paris* **t. 317** Série 1, (1993), 913-915.
[Den10]   L. Denis: Indépendance algébrique et exponentielle de Carlitz, *Acta Arithm.* **LXIX.1** (1995), 75-89.
[Den11]   L. Denis: Méthode fonctionnelle pour la transcendance, *Bull. Aust. Math. Soc.* **50** (1994).
[Den12]   L. Denis: Problème de Lehmer en caractéristique finie, *Compositio Math.* **98** (1995), 167-175.
[Den13]   L. Denis: Indépendance algébrique en caractéristique deux, (preprint).
[Den14]   L. Denis: Irrationalité de valeurs associées à l'exponentielle de Carlitz, *Monat. Math.* **123** (1997), 43-51.
[Den15]   L. Denis: Lemmes de multiplicités et T-modules, *Mich. J. Math.* **43** (1996), 67-79.
[Den16]   L. Denis: Un critère de transcendance en caractéristique finie, *J. Algebra* (to appear).
[Den17]   L. Denis: Fonctions exponentielles en diverses caractéristiques, (preprint).
[Den18]   L. Denis: Valeurs transcendantes des fonctions de Bessel-Carlitz, (preprint).
[Deu1]   M. Deuring: *Algebren*, Ergebnisse der Mathematik und ihrer Grenzgebiete **41** (1968).
[Di1]   B. Diarra: Exponentielles en caractéristique p (modules elliptiques ou de Drinfeld), *Sémin. Anal. Univ. Blaise Pascal 1987-88* **22** (1988).
[D-V1]   J. Diaz-Vargas: Riemann hypothesis for $\mathbb{F}_p[T]$, *J. Number Theory* **59** (1996), 313-318.
[Do1]   D. Dorman: On Singular moduli for rank 2 Drinfeld modules, *Compositio Math.* **80** (1991), 235-256.
[Dr1]   V.G. Drinfeld: Elliptic modules, Math. Sbornik **94** (1974), 594–627, English transl.: *Math. U.S.S.R. Sbornik* **23** (1976), 561–592.
[Dr2]   V.G. Drinfeld: Elliptic modules II, Math. *U.S.S.R. Sbornik* **31** (1977), 159–170.

[Dr3]   V.G. Drinfeld: Varieties of modules of F-sheaves, *Funct. Anal. and Appl.* **21** (English transl.) (1987), 107–122.
[Dr4]   V.G. Drinfeld: Commutative subrings of some noncommutative rings, *Funct. Anal. and Appl.* **11** (1974), 11–14.
[Dr5]   V.G. Drinfeld: The proof of Petersson's conjecture for $GL(2)$ over a global field of characteristic p, *Funct. Anal. and Appl.* **22** (1988), 28-43.
[Dr6]   V.G. Drinfeld: Cohomology of compactified manifolds of moduli of F-sheaves of rank 2, *J. Sov. Math.* **46** (1989), 1789-1821.
[Dr7]   V.G. Drinfeld: Two-dimensional ℓ-adic representations of the fundamental group of a curve over a finite field and automorphic forms on $GL(2)$, *Amer. J. Math.* **105** (1983), 85-114.
[Dr8]   V.G. Drinfeld: Coverings of p-adic symmetric domains, *Funct. Anal. Appl.* **10** (1976), 107-115.
[Dro1] M. Drocur: Untersuchungen zu Hecke-Operatoren auf Drinfeld'schen Modulformem, (preprint).
[Dub1] N. Dubovitskaya: Transcendency of analytic parameters associated to elliptic modules, *Mat. Sbornik*, **127** (1985), 131-141.
[DH1]   D. Dummit, D. Hayes: Rank-one Drinfeld modules of elliptic curves, *Math. Comp.* **62** (1994), 875-883.
[Du1]   D. Dummit: Genus two hyperelliptic Drinfeld modules over $\mathbb{F}_2$, in: *The Arithmetic of Function Fields*, (eds. D. Goss et al) (1992), 117-129.
[DGS1] B. Dwork, G. Gerotto, F. Sullivan: *An Introduction to G-functions*, Ann. of Math. Studies **133**, Princeton Univ. Press (1994).
[EH1]   G. Effinger, D. Hayes: A complete solution to the polynomial 3-primes problem, *Bull. Amer. Math. Soc.* **24** (1991), 363-369.
[EH2]   G. Effinger, D. Hayes: *Additive Number Theory of Polynomials over a Finite Field*, Oxford Univ. Press (1991).
[E1]   N. Elkies: Linearized algebra and finite groups of Lie type, I: Linear, symplectic and unitary groups, (preprint).
[FD1]   B. Farb, R. K. Dennis: *Noncommutative Algebra*, GTM **144** Springer (1993).
[Fe1]   K. Feng: A note on irregular prime polynomials in cyclotomic function field theory, *J. Number Theory* **22** (1986), 240–245.
[Fe2]   K. Feng: Class number "parity" for cyclic function fields, in: *The Arithmetic of Function Fields* (eds. D. Goss et al) (1992), 103-116.
[Fe3]   K. Feng: Zeta function, class number and cyclotomic units of cyclotomic function fields, in: *Zeta Functions in Geometry*, Advanced Studies in Pure Math. **21** (1992), 141-152.
[FeG1] K. Feng, W. Gao: Bernoulli-Goss polynomials and the class number of cyclotomic function fields, *Sci. Sinsia* **32** (1989), 1257-1263.
[FeX1] K. Feng, F. Xu: Kolyvagin's "Euler systems" in cyclotomic function fields, *J. Number Theory* **57** (1996), 114-121.
[FeY1] K. Feng, L. Yin: Maximal independent systems of units in cyclotomic function fields, *Sci. China Ser A* **34** (1991), 908-919.
[FlK1] Y. Flicker, D. Kazhdan: Geometric Ramanujan conjecture and Drinfeld reciprocity law, in: *Number Theory, Trace Formulas and Discrete Groups*, Academic Press (1989), 201-218.
[FP1]   J. Fresnel, M. van der Put: *Géométrie Analytique Rigide et Applications*, Progr. Math. **18**, Birkhäuser (1981).
[GR1]   S. Galovich, M. Rosen: The Class number of cyclotomic function fields, *J. Number Theory* **13** (1981), 363–375.
[GR2]   S. Galovich, M. Rosen: Units and class groups in cyclotomic function fields, *J. Number Theory* **14** (1982), 156–184.

[GR3]   S. Galovich, M. Rosen: Distributions on rational function fields, *Math. Ann.* **256** (1981), 549-560.
[Gei1] J.M. Geijsel: *Transcendence in Fields of Finite Characteristic Mathematical Centre Tracts* 91 , Mathematisch Centrum, Amsterdam (1979).
[Ge1]   E.-U. Gekeler: On power sums of polynomials over finite fields, *J. Number Theory* **30** (1988), 11–26.
[Ge2]   E.-U. Gekeler: *Drinfeld Modular Curves*, Lect. Notes in Math., **1231**, Springer (1986).
[Ge3]   E.-U. Gekeler: On finite Drinfeld modules, *J. Algebra* **141** (1991), 187–203.
[Ge4]   E.-U. Gekeler: On regularity of small primes in function fields, *J. Number Theory* **34** (1990), 114–127.
[Ge5]   E.-U. Gekeler: Some new identities for Bernoulli-Carlitz numbers, *J. Number Theory* **33** (1989), 209-219.
[Ge6]   E.-U. Gekeler: Drinfeld-Moduln und modulare Formen über rationalen Funktionenkörpern, *Bonner Math. Schriften* **119** (1980).
[Ge7]   E.-U. Gekeler: Zur Arithmetik von Drinfeld-Moduln, *Math. Ann.* **262** (1983), 167-182.
[Ge8]   E.-U. Gekeler: Modulare Einheiten für Funktionenkörper, *J. reine angew. Math.* **348** (1984), 94-115.
[Ge9]   E.-U. Gekeler: Le genre des courbes modulaires de Drinfeld, *C.R. Acad. Sc. Paris* **t. 300** (1985), 647-650.
[Ge10] E.-U. Gekeler: A Product expansion for the discriminant function of Drinfeld modules of rank two: *J. Number Theory* **21** (1985), 135-140.
[Ge11] E.-U. Gekeler: Über Drinfeld'sche Modulkurven vom Hecke-Typ, *Compositio Math.* **57** (1986), 219-236.
[Ge12] E.-U. Gekeler: Automorphe Formen über $\mathbf{F}_q(T)$ mit kleinem Führer, *Abh. Math. Sem. Univ. Hamburg* **55** (1985), 111-146.
[Ge13] E.-U. Gekeler: Cohomologie des sous-groupes S-arithmétiques de $GL(2)$ sur un corps de fonctions, *Astérisque* **147-148** (1987), 285-289.
[Ge14] E.-U. Gekeler: Méthodes analytiques rigides dans la théorie arithmétique des corps de fonctions, in: *Groupe d'Etude d'Analyse Ultramétrique 1986,* Université P. et M. Curie, Paris (1988), 47-54.
[Ge15] E.-U. Gekeler: Compactification des Schémas de Modules de Drinfeld, in: *Séminaire de Théorie des Nombres de Bordeaux 1985/86*, exp. 16.
[Ge16] E.-U. Gekeler: Compactification of Drinfeld modular schemes, in: *Proceedings of a Conference on p-adic Analysis, Hengelhoef (1986)*, (eds. N. De Grande-De Kimpe, L. Van Hamme) Publ. Universiteit Brussel, Belgium (1986).
[Ge17] E.-U. Gekeler: On the coefficients of Drinfeld modular forms, *Invent. Math.* **93** (1988), 667-700.
[Ge18] E.-U. Gekeler: Quasi-periodic functions and Drinfeld modular forms, *Compositio Math.* **69** (1989), 277-293.
[Ge19] E.-U. Gekeler: On the de Rham isomorphism for Drinfeld modules. *J. reine angew. Math.* **401**, (1989), 188-208.
[Ge20] E.-U. Gekeler: Sur les classes d'idéaux des ordres de certains corps gauches, *C.R. Acad. Sc. Paris* **t. 309** (1989), 577-580.
[Ge21] E.-U. Gekeler: de Rham cohomology for Drinfeld modules, in: *Séminaire de Théorie des Nombres 1988/89* (ed. C. Goldstein) Progress in Mathematics, Birkhäuser (1991).
[Ge22] E.-U. Gekeler: Sur la géométrie de certaines algèbres de quaternions, *Séminaire de Théorie des Nombres de Bordeaux* **2** (1990), 143-153.

[Ge23] E.-U. Gekeler: de Rham cohomology and the Gauss-Manin connection for Drinfeld modules, *Conference on p-adic Analysis, Trento 1989*, Lect. Notes Math. **1454** Springer.

[Ge24] E.-U. Gekeler: Représentations galoisiennes associées aux modules de Drinfeld (d'après Taguchi), in: *Modules de Drinfeld et l'approximation diophantienne* (ed. Y. Hellegouarch) Caen (1990).

[Ge25] E.-U. Gekeler: On the arithmetic of some division algebras, *Comment. Math. Helv.* **67** (1992), 316-333.

[Ge26] E.-U. Gekeler: Moduli for Drinfeld modules, in: *The Arithmetic of Function Fields* (eds: D. Goss et al) de Gruyter (1992), 153-170.

[Ge27] E.-U. Gekeler: Arithmetic of global function fields related to Drinfeld modules, in: *Proc. 7th KAIST Conference on Algebra and Topology* Taejon, Korea (1992), 17-50.

[Ge28] E.-U. Gekeler: Analytical construction of Weil curves over function fields, *J. de Theorie des Nombres de Bordeaux* **7** (1995), 27-49.

[Ge29] E.-U. Gekeler: Improper Eisenstein series on Bruhat-Tits trees, *Manuscripta Math.* **86** (1995), 367-391.

[Ge30] E.-U. Gekeler: A remark on the dimensions of spaces of modular forms, *Archiv d. Math.* **65** (1995), 530-533.

[Ge31] E.-U. Gekeler: On the Drinfeld discriminant function, *Compositio Math.* **106**, (1997), 181-202

[Ge32] E.-U. Gekeler: Highly ramified pencils of elliptic curves in characteristic two, *Duke Math. J.* **89** (1997), 95-107.

[Ge33] E.-U. Gekeler: Jacquet-Langlands theory over K and relations with elliptic curves, in: *Drinfeld modules, modular schemes and applications, Proc. of a workshop held in Alden-Biesen/Belgium 1996*, (eds: E.-U. Gekeler et al) World Scientific (1997).

[Ge34] E.-U. Gekeler: On the cuspidal divisor class group of a Drinfeld modular curve, (preprint).

[GH1] E.-U. Gekeler, B.H. Snyder: Drinfeld modules over finite fields, in: *Drinfeld modules, modular schemes and applications, Proc. of a workshop held in Alden-Biesen/Belgium 1996*, (eds: E.-U. Gekeler et al) World Scientific (1997).

[GN1] E.-U. Gekeler, U. Nonnengardt: Fundamental domains of some arithmetic groups over function fields, *Internat. J. Math.* **6** (1995), 689-708.

[GeR1] E.-U. Gekeler, M. Reversat: Some results on the Jacobians of Drinfeld modular curves, in: *The Arithmetic of Function Fields*, (eds: D. Goss et al) de Gruyter (1992), 209-226.

[GeR2] E.-U. Gekeler, M. Reversat: Jacobians of Drinfeld modular curves, *J. reine angew. Math.* **476** (1996), 27-93.

[GerP1] L. Gerritzen, M. van der Put: *Schottky Groups and Mumford Curves*, Lect. Notes Math. **817** Springer (1980).

[GK1] R. Gold, H, Kisilevsky: On geometric $\mathbb{Z}_p$-extensions of function fields, *Manuscripta Math.* **62** (1988), 145-161.

[Go1] D. Goss: A short introduction to rigid analytic spaces, in: *The Arithmetic of Function Fields* (eds: D. Goss et al) de Gruyter (1992), 131–141.

[Go2] D. Goss: von-Staudt for $\mathbb{F}_q[T]$, *Duke Math. J.* **45** (1978), 885-910.

[Go3] D. Goss: Modular forms for $\mathbb{F}_r[T]$, *J. reine angew. Math.* **317** (1980), 16-39.

[Go4] D. Goss: L-series of t-motives and Drinfeld modules, in: *The Arithmetic of Function Fields* (eds. D. Goss et al) de Gruyter (1992), 313–402.

[Go5] D. Goss: Drinfeld modules: Cohomology and special functions, In: *Proc. Symp. Pure Math.* **55** part 2, Amer. Math. Soc. (1994), 309–362.

[Go6] D. Goss: Kummer and Herbrand criterion in the arithmetic of function fields, *Duke Math. J.* **49** (1982), 377–384.

[Go7] D. Goss: On the holomorphy of certain non-abelian L-series, *Math. Ann.* **272** (1985), 1–9.

[Go8] D. Goss: π-adic Eisenstein series for function fields, *Compositio Math.* **41** (1980), 3–38.

[Go9] D. Goss: Fourier series, measures and divided power series in the theory of function fields, *K-theory* **1** (1989), 533–555.

[Go10] D. Goss: A formal Mellin transform in the theory of function fields, *Trans. Amer. Math. Soc.* **327** (1991), 567–582.

[Go11] D. Goss: Harmonic analysis and the flow of a Drinfeld module, *J. Algebra* **146** (1992), 219–241.

[Go12] D. Goss: The Γ-function in the arithmetic of function fields, *Duke Math. J.* **56** (1988), 163–191.

[Go13] D. Goss: The Γ-ideal and special zeta-values, *Duke Math. J.* **47** (1980), 345–364.

[Go14] D. Goss: The algebraist's upper half plane, *Bull. Amer. Math. Soc.* **2**(3) (1980), 391–415.

[Go15] D. Goss: v-adic zeta functions, L-series and measures for function fields, *Invent. Math.* **55** (1979), 107–116.

[Go16] D. Goss: On a new type of L-function associated to algebraic curves over finite fields, *Pacific J. Math.* **105** (1983), 143-181.

[Go17] D. Goss: The arithmetic of function fields 2: The 'cyclotomic' theory, *J. Algebra* **81** (1983), 107-149.

[Go18] D. Goss: On a Fermat equation arising in the arithmetic theory of function fields, *Math. Ann.* **261** (1982), 269-286.

[Go19] D. Goss: The theory of totally-real function fields, *Contemp. Math.* **55** Part II, Amer. Math. Soc. (1986), 449-477.

[Go20] D. Goss: Units and class groups in the arithmetic of function fields, *Bull. Amer. Math. Soc.* **13** (1985), 131-132.

[Go21] D. Goss: Analogies between global fields, *Conf. Proc. Canad. Math. Soc. Vol. 7*, (eds: H. Kisilevsky and J. Labute) Amer. Math. Soc. (1987), 83-114.

[Go22] D. Goss: Report on transcendency in the theory of function fields, *Number Theory, New York 1985-88*, (eds. D.V. Chudnovsky et al) Lect. Notes Math. **1388**, Springer (1989), 59-63.

[Go23] D. Goss: L-series of Grössencharakters of type A_0 for function fields, *p-adic Methods in Number Theory and Algebraic Geometry* (eds. A. Adolphson, M. Tretkoff) Contemp. Math. **133** Amer. Math. Soc. (1992), 119-139.

[Go24] D. Goss: Some integrals attached to modular forms in the theory of function fields, in: *The Arithmetic of Function Fields* (eds: D. Goss et al) de Gruyter (1992), 227-251.

[Go25] D. Goss: The adjoint of the Carlitz module and Fermat's last theorem, *Finite Fields and Applications* **1** (1995), 165-188.

[GS1] D. Goss, W. Sinnott: Special values of Artin L-series, *Math. Ann.* **275** (1986), 529–537.

[GS2] D. Goss, W. Sinnott: Class groups of function fields, *Duke Math. J.* **52** (1985), 507–516.

[GrR1] B. Gross, M. Rosen: Fourier series and the special values of L-functions, *Advances Math.* **69** (1988), 1-31.

[GuSh1] L. Guo, L. Shu: Class numbers of cyclotomic function fields, (preprint).

[Ham1] Y. Hamahata: Tensor products of Drinfeld modules and v-adic representations, *Manuscripta Math.* **79** (1993), 307-327.

[HardK1] G. Harder, D. Kazhdan: Automorphic forms on $GL(2)$ over function fields (after V.G. Drinfeld), in: *Automorphic Forms, Representations and L-functions,* (eds. A. Borel, W. Casselman) Proc. Symp. Pure Math. **XXXIII** part 2, Amer. Math. Soc. (1979), 357-379.

[Har1] R. Hartshorne: *Algebraic Geometry,* Springer (1977).

[Ha1] D. Hayes: Explicit class field theory in global function fields, in: *Studies in Algebra and Number Theory,* Advances in Math. **16** (1980), 173-217.

[Ha2] D. Hayes: A brief introduction to Drinfeld modules, in: *The Arithmetic of Function Fields* (eds. D. Goss et al) de Gruyter (1992), 1-32.

[Ha3] D. Hayes: Stickelberger elements in functions fields, *Compositio Math.* **55** (1985), 209-239.

[Ha4] D. Hayes: Explicit class field theory for rational function fields, *Trans. Amer. Math. Soc.* **189** (1974), 77-91.

[Ha5] D. Hayes: Analytic class number formulas in function fields, *Invent. Math.* **65** (1981), 49-69.

[Ha6] D. Hayes: Elliptic units in function fields, in: *Number Theory Related to Fermat's Last Theorem,* Birkhäuser, Boston (1982), 321-340.

[Ha7] D. Hayes: Real quadratic function fields, in: *Number Theory, Canadian Mathematical Society Proceedings,* **7** (eds. H. Kisilevsky, J. Labute) Amer. Math. Soc. (1987), 203-236.

[Ha8] D. Hayes: The refined p-adic abelian Stark conjecture in function fields, *Invent. Math.* **94** (1989), 505-527.

[Ha9] D. Hayes: Brumer elements over a real quadratic base field, *Expositiones. Math.* **8** (1990), 137-184.

[Ha10] D. Hayes: On the reduction of rank-one Drinfeld modules, *Math. Comp.* **57** (1991), 339-349.

[Ha11] D. Hayes: Hecke characters and Eisenstein reciprocity in function fields, *J. Number Theory* **43** (1993), 251-292.

[Ha12] D. Hayes: Introduction to chapter 19 of the "Arithmetic of Polynomials," *Finite Fields and their Applications* **1**, 157-163.

[He1] Y. Hellegouarch: Galois calculus and Carlitz exponentials, in: *The Arithmetic of Function Fields,* (eds: D. Goss et al), de Gruyter (1992), 33-50.

[He2] Y. Hellegouarch: Généralisation de l'exponentielle de Drinfeld, in: *Modules de Drinfeld et l'approximation diophantienne* (ed. Y. Hellegouarch) Caen (1990).

[He3] Y. Hellegouarch: Modules de Drinfeld généralisés, in: *Approximations diophantiennes et Nombres transcendants, Luminy 1990* (ed. P. Phillipon) de Gruyter (1992).

[He4] Y. Hellegouarch: Une généralisation d'un critère de de Mathan, *C.R. Acad. Sci. Paris* **t. 321**, Série I. (1995), 677-680.

[He5] Y. Hellegouarch: Un analogue d'un théorème d'Euler, *C.R. Acad. Sci. Paris* **t. 313** Série I. (1991), 155-158.

[HeRe1] Y. Hellegouarch, F. Recher: Relèvement de modules de Drinfeld en caractéristique zéro, *C.R. Math. Rep. Acad. Canada* **XV** (1993), 167-172.

[HeRe2] Y. Hellegouarch, F. Recher: Generalized t-modules, *J. Algebra* **187** (1997), 323-372.
[Hen1] G. Henniart: φ-spaces and Dieudonné F_x-modules (following Drinfeld), Appendix to [LRS1], *Invent. Math.* **113** (1993) 327-335.
[H1] E. Hille: *Lectures on Ordinary Differential Equations*, Addison-Wesley (1969).
[HY1] L.-C. Hsia J. Yu: On singular moduli of Drinfeld modules in characteristic two, (preprint).
[Hsu1] C.-N. Hsu: On Artin's conjecture for the Carlitz module, *Compositio Math.* **106** (1997), 247-266.
[Hsu2] C.-N. Hsu: Some results on finite Drinfeld modules, (to appear).
[Hsu3] C.-N. Hsu: On Carmichael polynomials, (preprint).
[Hsu4] C.-N. Hsu: A large sieve inequality for rational function fields, *J. Number Theory* **58** (1966), 267-287.
[Hsu5] C.-N. Hsu: The distribution of irreducible polynomials in $\mathbb{F}_q[t]$, *J. Number Theory* **61** (1996), 85-96.
[Hub1] A. Huber: On the Parshin-Beilinson adeles for schemes, *Abh. Math. Sem. Univ. Hamburg* **61** (1991), 249-273.
[Hu1] J. Humphreys: *Linear Algebraic Groups*, Springer (1975).
[I1] E. L. Ince: *Ordinary Differential Equations*, Dover (1956).
[Is1] M. Ishibashi: Effective version of the Tschebotareff density theorem in function fields over finite fields, *Bull. London Math. Soc.* **24** (1992), 52-56.
[IS1] K. Ireland, R. Small: A note on Bernoulli-Goss polynomials, *Canad. Math. Bull.* **27** (1984), 178-184.
[IS2] K. Ireland, R. Small: Class numbers of cyclotomic function fields, *Math. Comp.* **46** (1986), 337-340.
[Jac1] N. Jacobson: *Basic Algebra II*, Freeman (1980).
[Jar1] M. Jarden: The Čebotarev density theorem for function fields: An elementary approach, *Math. Ann.* **261** (1982), 467-475.
[Jo1] K. Joshi: A family of étale coverings of the affine line, *J. Number Theory* **59** (1996), 414-418.
[Kap1] I. Kaplansky: *An Introduction to Differential Algebra*, Hermann (1957).
[Ka1] M. Kapranov: A higher-dimensional generalization of the Goss zeta-function, *J. Number Theory* **50** (1995), 365–375.
[Ka2] M. Kapranov: On cuspidal divisors on the modular varieties of elliptic modules, *Math. USSR Izvestiya* **30** (1988), 533-547.
[Kat1] K. Kato: Generalized class field theory, *Proc. I.C.M. 1990*, Amer. Math. Soc. (1991), 419–428.
[K1] N. Katz: *Exponential Sums and Differential Equations*, Princeton Univ. Press (1990).
[K2] N. Katz: On the calculation of some differential Galois groups, *Invent. Math.* **87** (1987), 13–61.
[K3] N. Katz: Travaux de Dwork, *Séminaire Bourbaki* 24e année, 1971/72 **409** (1972), 167-200.
[Ka1] D. Kazhdan: An introduction to Drinfeld's "shtuka," in: *Corvallis Conference on Automorphic Forms, Representations and L-functions* (eds. A. Borel, W. Casselman) Proc. Symp. Pure Math. **XXXIII** part 2, Amer. Math. Soc. (1979), 347-356.
[KidMu1] M. Kida, N. Murabayashi: Cyclotomic function fields with divisor class number one, *Tokyo J. Math.* **14** (1991), 45-56.
[Ki1] H. Kisilevsky: Multiplicative independence in function fields, *J. Number Theory* **44** (1993), 352-355.

[Kob1] N. Koblitz: *p-adic Numbers, p-adic Analysis and Zeta-functions*, 2nd Edition, Springer GTM **58**, (1984).
[Kor1] H. Kornblum: Über die Primfunktionen in einer arithmetischen Progression, *Math. Z.* **5** (1919), 100-111.
[L1] S. Lang: *Algebra* (Second Edition), Addison-Wesley (1984).
[L2] S. Lang: *Cyclotomic Fields*, Springer (1978).
[La1] G. Laumon: *Cohomology of Drinfeld Modular Varieties, Part I: Geometry, counting of points and local harmonic analysis*, Cambridge Studies in advanced mathematics **41**, Cambridge University Press (1996).
[LRS1] G. Laumon, M. Rapoport, U. Stuhler: $\mathcal{D}$-elliptic sheaves and the Langlands correspondence, *Invent. Math.* **113** (1993), 217-338.
[Lee1] H. Lee: Power sums of polynomials in a Galois field, *Duke Math. J.* **10** (1943), 277-292.
[LMQ1] J. Leitzel, M. Madan, C. Queen: On congruence function fields of class number one, *J. Number Theory* **7** (1975), 11–27.
[Le1] H. W. Lenstra, Jr.: On Artin's conjecture and Euclid's algorithm in global fields, *Invent. Math.* **42** (1977), 201–224.
[Ma1] A. Magid: *Lectures on Differential Galois Theory*, Univ. Lect. Ser. **7**, Amer. Math. Soc. (1994).
[Mat1] B. de Mathan: Approximations diophantiennes dans un corps local, *Bull. Soc. Math. France Mém.* **21** (1970).
[Mat2] B. de Mathan: Approximation exponents for algebraic functions in positive characteristic, *Acta Arithm.* **LX**(4) (1992), 359-370.
[Mat3] B. de Mathan: Exposants d'approximation en caractéristique p, in: *Modules de Drinfeld et l'approximation diophantienne* (ed. Y. Hellegouarch) Caen (1990).
[Mat4] B. de Mathan: Irrationality measures and transcendence in positive characteristic, *J. Number Theory* **54** (1995), 93-112.
[Mau1] V. Mauduit: Carmichael-Carlitz polynomials and Fermat-Carlitz quotients, (preprint).
[Mau2] V. Mauduit: Quotients de Fermat-Carlitz, *C.R. Acad. Sci. Paris* **t. 321** Série I. (1995), 1139-1141.
[Mau3] V. Mauduit: Symbole de Newton. Symbole de Jacobi-Carltz, *C.R. Acad. Sci. Paris* **t. 324** Série I. (1997), 1-6.
[MY1] M. Mendès France, J. Yao: Transcendence and the Carlitz-Goss Gamma Function, *J. Number Theory* **63** (1997), 396-402.
[M1] E. H. Moore: A two-fold generalization of Fermat's theorem, *Bull. Amer. Math. Soc.* **2** (1896), 189-199.
[Mu1] D. Mumford: *Abelian Varieties* Oxford Univ. Press (1970).
[Mu2] D. Mumford: An algebro-geometric construction of commuting operators and solutions to the Toda lattice equation, KdV equation and related nonlinear equations, in: *International Symposium on Algebraic Geometry, Kyoto (1977)* (ed: M. Nagata), Tokyo:Kinokuniya (1978), 115–153.
[Mu3] D. Mumford: *Geometric Invariant Theory*, Springer (1965).
[Mus1] G.A. Mustafin, Non-Archimedean uniformization, *Math. USSR Sbornik* **34** (1978), 187-214.
[NX1] H. Niederreiter, C. Xing: Modules de Drinfeld et courbes algébriques ayant beaucoup de points rationnels, *C.R. Acad. Sci. Paris* **t. 322** Série I. (1996), 651-654.
[O1] O. Ore: On a special class of polynomials, *Trans. Amer. Math. Soc.* **35** (1933), 559–584.

[O2]	O. Ore: Theory of non-commutative polynomials, *Ann. of Math.* **34** (1933), 480–508.
[Ok1]	S. Okada: Kummer's theory for function fields, *J. Number Theory* **38** (1991), 212-215.
[Ok2]	S. Okada: Analogies of Dedekind sums in function fields, *Mem. Gifu. Teach. Coll* **24** (1989), 11-16.
[Os1]	D. Ose: Deformations for function fields, (preprint).
[Ou1]	H. Oukhaba: Groups of elliptic units in global function fields, in: *The Arithmetic of Function Fields* (eds. D. Goss et al), de Gruyter (1992), 87-102.
[Ou2]	H. Oukhaba: On discriminant functions assoctiated to Drinfeld modules of rank 1, *Prépublication de l'Institut Fourier* **188** (1991).
[Pa1]	A.A. Panchishkin: Algorithmes rapides pour la factorisation des nombres et des polynômes, tests de primalité, courbes elliptiques et modules de Drinfeld, *Séminaire de Théorie des Nombres, Université de Caen,* Exposé du 09.10.92.
[PaPo1]	A.A. Panchishkin, I. Potemine: An algorithm for the factorization of polynomials using elliptic modules, (Russian) *Abstracts of Reports of the Conference "Constructive Methods and Algorithms in Number Theory," Minsk, 1989,* p. 117.
[Ph1]	P. Philippon: Une approche méthodique pour la transcendance et l'indépendance algébrique de valeurs de fonctions analytiques, (preprint).
[Pi1]	R. Pink: On compactification of Drinfeld moduli schemes, in: *Moduli Spaces, Galois Representations and L-functions* (Japanese, Kyoto 1993, 1994) RIMS Kokyuroku **884** (1994), 178-183.
[Pi2]	R. Pink: The Mumford-Tate conjecture for Drinfeld modules, Preprint Mannheim, (Sept. 1996), 26 S., in: *Publ. RIMS, Kyoto University* **33** (to appear).
[Pi3]	R. Pink: Compact subgroups of linear algebraic groups, Preprint Mannheim, (August 1996), 58 S.
[Pi4]	R. Pink: ℓ-adic algebraic monodromy groups, cocharacters, and the Mumford-Tate conjecture, Preprint Mannheim, (March 1997).
[Pi5]	R. Pink: Hodge structures over function fields, (in preparation).
[Pi6]	R. Pink: Motives and Hodge structures over function fields, Arbeitsgung talk, Bonn (June, 1997).
[Po1]	B. Poonen: Rigidity and semi-invariants in Drinfeld modules, *J. Number Theory* **55** (1995), 181-196.
[Po2]	B. Poonen: Fractional power series and pairings on Drinfeld modules, *J. Amer. Math. Soc.* **9** (1996), 783-812.
[Po3]	B. Poonen: Local height functions and the Mordell-Weil theorem for Drinfeld modules, *Compositio Math.* **97** (1995), 349-368.
[Po4]	B. Poonen: Torsion in rank 1 Drinfeld modules and the uniform boundedness conjecture, (to appear *Math. Ann.* **308** (1997)).
[Po5]	B. Poonen: Drinfeld modules with no supersingular primes, (preprint).
[Pot1]	I. Potemine: A variant of the Riemann hypothesis for elliptic modules, (Russian) *Abstracts of the Conference "Constructive Methods and Algortihms in Number Theory," Minsk* (1989), p. 126.
[Pot2]	I. Potemine: J-invariant et schémas grossiers des Modules de Drinfeld, *Séminaire de Théorie de Nombres de Caen,* Fascicule de l'année 1994-95, 15p.

[Pot3] I. Potemine: The Hasse theorem for Drinfeld modules over finite fields and local characteristic polynomials of Drinfeld modules with potential good reduction, *Prépublication de L'institut Fourier* **303** (1995).

[Pot4] I. Potemine: Minimal terminal Q-factorial models of Drinfeld coarse moduli schemes, (preprint).

[Pot5] I. Potemine: Complex multiplication of higher rank Drinfeld modules. Hilbert clsss fields of imaginary function fields, (preprint).

[Q1] H.-G. Quebbemann: Cyclotomic Goppa codes, *IEEE Trans. Inf. Theory* **34** (1988), 1317-1320.

[Ra1] W. Raskind: Abelian class field theory of arithmetic schemes, in: *Proceedings Symposia Pure Mathematics* **58**, *Part I*, Amer. Math. Soc. (1995), 85–187.

[Rec1] F. Recher: Propriétés de transcendance de séries formelles provenant de l'exponentielle de Carlitz,*C.R. Acad. Sci. Paris* Série I. **t. 315** (1992), 245-250.

[Re1] I. Reiner: *Maximal Orders*, Academic Press (1975).

[Rev1] M. Reversat: Lecture on rigid geometry, in: *The Arithmetic of Function Fields* (eds. D. Goss et al) de Gruyter (1992), 143-151.

[Rev2] M. Reversat: Sur les revêtements de Schottky des courbes modulaires de Drinfeld, *Archiv d. Math.* **66** (1996), 378-387.

[Ri1] P. Ribenboim: 13 *Lectures on Fermat's Last Theorem,* Springer, (1979).

[Ro1] A. Robert: *Elliptic Curves*, Lect. Notes Math. **326**, Springer (1973).

[Ros1] M. Rosen: Ambiguous divisor classes in function fields, *J. Number Theory* **9** (1977), 160-174.

[Ros2] M. Rosen: The Hilbert class field in function fields, *Expos. Math.* **5** (1987), 365-378.

[Ros3] M. Rosen: Number theory in function fields, (preprint).

[Ros4] M. Rosen: A note on the relative class number in function fields, *Proc. Amer. Math. Soc.* **125** (1997), 1299-1303.

[RS1] M. Rubinstein, P. Sarnak: Chebyshev's bias, *J. Experimental Math.* **3** (1994), 173-197.

[Ru1] I. Rust: Arithmetically defined representations of groups of type $SL(2, \mathbb{F}_q)$, (preprint).

[RuS1] I. Rust, O. Scheja: A guide to constructive class field theory over global function fields, in: *Drinfeld modules, modular schemes and applications, Proc. of a workshop held in Alden-Biesen/Belgium 1996*, (eds: E.-U. Gekeler et al) World Scientific (1997).

[Sa1] T. Satoh: On duality over a certain divided power series algebra with positive characteristic, *Manuscripta Math.* **92** (1997), 153-172.

[SU1] T. Satoh, Y. Uchino: Function field modular forms and higher derivation, (preprint).

[Sc1] W. Scharlau: *Quadratic and Hermitian Forms*, Grundlehren der Mathematischen Wissenschaften **270** Springer (1985).

[Sch1] D. Schilling: *The Theory of Valuations*, Amer. Math. Soc. (1950).

[Schu1] F. Schultheis: Carlitz-Kummer function fields, *J. Number Theory* **36** (1990), 133-144.

[Schw1] A. Schweizer: On the Drinfeld modular polynomial $\Phi_T(X,Y)$, *J. Number Theory* **52** (1995), 53-68.

[Schw2] A. Schweizer: Hyperelliptic Drinfeld modular curves, in: *Drinfeld modules, modular schemes and applications, Proc. of a workshop held in Alden-Biesen/Belgium 1996*, (eds: E.-U. Gekeler et al) World Scientific (1997).

References

[Schw3]   A. Schweizer: On singular and supersingular invariants of Drinfeld modules, *Ann. Fac. Sci. Toulouse Math.* (to appear).

[Schw4]   A. Schweizer: Modular automorphisms of the Drinfeld modular curves $X_0(\mathfrak{N})$, Proceedings of the Journées Arithmétiques 1995, *Collectanea Mathematica* **XLVIII** Barcelona (1997), 209-216.

[SegW1]   G. Segal, G. Wilson: Loop groups and equations of KdV type, *Publ. IHES* **61** (1985), 5-65.

[Se1]   J.-P. Serre: *Corps Locaux*, Hermann (1968).

[Se2]   J.-P. Serre: Local class field theory, in: *Algebraic Number Theory*, (eds: J. W. S. Cassels, A. Fröhlich) Thompson (1967), 129-161 (= *Coll. Papers*, 75).

[Se3]   J.-P. Serre: Propriétés galoisiennes des points d'ordre fini des courbes elliptiques, *Invent. Math.* **15** (1972), 259-331 (= *Coll. Papers*, 94).

[Se4]   J.-P. Serre: *Abelian ℓ-adic Representations and Elliptic Curves*, Benjamin (1968).

[Se5]   J.-P. Serre: *Représentations Linéaires des Groupes Finis*, Hermann (1971).

[Se6]   J.-P. Serre: Une "formule de masse" pour les extensions totalement ramifiées de degré donné d'un corps local, *C. R. Acad. Sci. Paris* **286**, série A (1978), 1031-1036 (= *Coll. Papers*, 115).

[Se7]   J.-P. Serre: *Lie Algebras and Lie Groups*, Benjamin (1965); Lect. Notes in Math. **1500** (1992).

[Sh1]   S. Shatz: Group schemes, formal groups, and p-divisible groups, in: *Arithmetic Geometry*, (eds: G. Cornell, J. Silverman) Springer (1986), 29-78.

[Shu1]   L. Shu: Kummer's criterion over global function fields, *J. Number Theory* **49** (1994), 319-359.

[Shu2]   L. Shu: Class number formulas over global function fields, *J. Number Theory* **48** (1994), 133-161.

[Shu3]   L. Shu: A relative class number formula for global function fields, (preprint).

[Shu4]   L. Shu: Narrow ray class fields and partial zeta-functions, (preprint)

[Sil1]   J. Silverman: *The Arithmetic of Elliptic Curves*, Springer (1985).

[Si1]   S. Sinha: Deligne's reciprocity for function fields, *J. Number Theory* **63** (1997), 65-88.

[Si2]   S. Sinha: Periods of t-motives and transcendence, *Duke Math. J.* **88** (1997), 465-535.

[Sp1]   S. Spencer: Transcendental numbers over certain function fields, *Duke Math. J.* **19** (1952), 93-105.

[St1]   U. Stuhler: p-adic homogeneous spaces and moduli problems, *Math. Z.* **192** (1986), 491-540.

[Ta1]   J. Tate: Number theoretic background, in: *Proceedings Symposia in Pure Mathematics* **33** *Part 2*, Amer. Math. Soc. (1979), 3-26.

[Tag1]   Y. Taguchi: Semi-simplicity of the Galois representations attached to Drinfeld modules of "infinite characteristics", *J. Number Theory* **44** (1993), 292-314.

[Tag2]   Y. Taguchi: Semi-simplicity of the Galois representations attached to Drinfeld modules of "finite characteristics," *Duke Math. J.* **62** (1991), 593-599.

[Tag3]   Y. Taguchi: Ramifications arising from Drinfeld modules, in: *The Arithmetic of Function Fields*, (eds: D. Goss et al), de Gruyter (1992), 171-187.

[Tag4]　　Y. Taguchi: A duality for finite t-modules, *J. Math. Sci., Univ. Tokyo*, **2** (1995), 563-588.

[Tag5]　　Y. Taguchi: On the π-adic theory — Galois cohomology, *Proc. Japan Acad.* **68A** (1992), 214-218.

[Tag6]　　Y. Taguchi: Regular singularity for Drinfeld modules, *Intl. J. Math.* **5** (1994), 595-608.

[Tag7]　　Y. Taguchi: The Tate conjecture for t-motives, *Proc. Amer. Math. Soc.* **123** (1995), 3285-3287.

[Tag8]　　Y. Taguchi: On φ-modules, *J. Number Theory* **60** (1996), 124-141.

[Tag9]　　Y. Taguchi: Analogy between Drinfeld modules and $\mathcal{D}$-modules, in: *RIMS Kokyuroku* (Japanese), (Proc. RIMS) **844** (1993), 190-196.

[TN1]　　Y. Taguchi, Y. Nakkajima: A generalization of the Chowla-Selberg formula, *J. reine angew. Math.* **419** (1991), 119-124.

[TW1]　　Y. Taguchi, D. Wan: L-functions of φ-sheaves and Drinfeld modules, *J. Amer. Math. Soc.* **9** (1996), 755-781.

[TW2]　　Y. Taguchi, D. Wan: Entireness of L-functions of φ-sheaves on affine complete intersections, *J. Number Theory* **63** (1997), 170-179.

[Tak1]　　T. Takahashi: Good reduction of elliptic modules, *J. Math. Soc. Japan* **34** (3) (1982), 475–487.

[Tam1]　　A. Tamagawa: The Eisenstein quotient of the Jacobian variety of a Drinfeld modular curve, *Publ. RIMS, Kyoto Univ.* **31** (1995), 203–246.

[Tam2]　　A. Tamagawa: The Tate conjecture for A-premotives, (preprint).

[Tam3]　　A. Tamagawa: Generalization of Anderson's t-motives and Tate conjecture, *RIMS Kokyuroku* (Proc. RIMS) **884** (1994), 154-159.

[Tam4]　　A. Tamagawa: The Tate conjecture and the semisimplicity conjecture for t-modules, *RIMS Kokyuroku* (Proc. RIMS) **925** (1995), 89-94.

[Tam5]　　A. Tamagawa: The Grothendieck conjecture for affine curves, (preprint.)

[Tam6]　　A. Tamagawa: Ramification of the Galois representation on the pro-l fundamental group of an algebraic curve, *RIMS Kokyuroku* **884** (1994), 54–57.

[Tan1]　　H. Tanuma: A Higher-dimensional analogue of Carlitz-Drinfeld theory, *Proc. Japan Acad.* **71** (1995), 72-74.

[Te1]　　J. Teitelbaum: Rigid analytic modular forms: An integral transform approach, in: *The Arithmetic of Function Fields* (eds. D. Goss et al) de Gruyter (1992), 189-207.

[Te2]　　J. Teitelbaum: The Poisson kernel for Drinfeld modular curves, *J. Amer. Math. Soc.* **4** (1991), 491-511.

[Te3]　　J. Teitelbaum: Modular symbols for $\mathbf{F}_q(T)$, *Duke Math. J.* **68** (1992), 271-295.

[Th1]　　D. Thakur: Shtukas and Jacobi sums, *Invent. Math.* **111** (1993), 557–570.

[Th2]　　D. Thakur: Hypergeometric functions for function fields, *Finite Fields and Their Applications* **1**, (1995), 219-231.

[Th3]　　D. Thakur: Zeta measure associated to $\mathbb{F}_q[T]$, *J. Number Theory* **35** (1990), 1-17.

[Th4]　　D. Thakur: On characteristic p zeta functions, *Compositio Math.* **99** (1995), 231-247.

[Th5]　　D. Thakur: Drinfeld modules and arithmetic in the function fields, *Int. Math. Res. Notices* **9** (1992), 185-197.

[Th6]　　D. Thakur: Gamma functions for function fields and Drinfeld modules, *Ann. of Math.* **134** (1991), 25-64.

[Th7]   D. Thakur: On Gamma functions for function fields, *The Arithmetic of Function Fields*, (eds. D. Goss et al) de Gruyter (1992), 75-86.
[Th8]   D. Thakur: Transcendence of gamma values for $\mathbb{F}_q[T]$, *Ann. of Math.* **144** (1996), 181-188.
[Th9]   D. Thakur: Gauss sums for $\mathbb{F}_q[T]$, *Invent. Math.* **94** (1988), 105-112.
[Th10]   D. Thakur: Gauss sums for function fields, *J. Number Theory* **37** (1991), 242–252.
[Th11]   D. Thakur (appendix by J.F. Voloch): Behavior of function field Gauss sums at infinity, *Bull. London Math. Soc.* **25** (1993), 417-426.
[Th12]   D. Thakur: Exponential and continued fractions, *J. Number Theory* **59** (1996), 248-261.
[Th13]   D. Thakur: Automata-style proof of Voloch's transcendence result, *J. Number Theory* **58** (1996), 60-63.
[Th14]   D. Thakur: An alternative approach to solitons for $\mathbf{F}_q[T]$, (preprint).
[Th15]   D. Thakur: Number fields and function fields (zeta and gamma functions at all primes), in: *Proceedings of a Conference on p-adic Analysis, Hengelhoef (1986)*, (eds. N. De Grande-De Kimpe, L. Van Hamme) Publ. Universiteit Brussel, Belgium (1986), 149-157.
[Th16]   D. Thakur: Gross-Koblitz formula for function fields, in: *Proceedings of the Congress on p-adic Analysis, Trento, May 1989* (eds. F. Baldassarri et al) Lect. Notes Math. **1454**, Springer p. 396.
[Th17]   D. Thakur: Analogies between integers and polynomials — Huzurbazar Memorial Lectures, Bombay Mathematical Colloquium, *Bulletin* **7**(3) (1990), 77-89.
[Th18]   D. Thakur: Iwasawa theory and cyclotomic function fields, in: *Proceedings of the Conference on Arithmetic Geometry*, Cont. Math. **174** (eds. J. Jones, N. Childress) Amer. Math. Soc. (1994), 157-165.
[Thi1]   A. Thiery: Indépendance algébrique des périodes et quasi-périodes d'un module de Drinfeld, in: *The Arithmetic of Function Fields* (eds. D. Goss et al) (1992), 265-284.
[Thi2]   A. Thiery: Transcendance de quelques valeurs de la fonction gamma pour les corps de fonctions, *C.R. Acad. Sci. Paris* Série I **t. 314** (1992), 973-976.
[Thi3]   A. Thiery: Théorème de Lindemann-Weierstraßpour les modules de Drinfeld, *Compositio Math.* **95** (1995), 1-42.
[Thi4]   A. Thiery: $\mathbb{F}_q$-linear Galois theory, *J. London Math. Soc.* **53** (1996), 441-454.
[Tho1]   E. Thomas: On the zeta-function for function fields over $\mathbb{F}_p$, *Pacific J. Math.* **107** (1983), 251-256.
[Ti1]   U. Tipp: On groups acting freely on a tree, (preprint).
[Ti2]   U. Tipp: Local height pairings of Heegner points on Drinfeld modular curves, in: *Drinfeld modules, modular schemes and applications, Proc. of a workshop held in Alden-Biesen/Belgium 1996*, (eds: E.-U. Gekeler et al) World Scientific (1997).
[Tu1]   R. Tubbs: Analytic subgroups of t-modules, *Trans. Amer. Math. Soc.* **349** (1997), 2605-2617.
[V1]   J.F. Voloch: Diophantine approximation in positive characteristic, *Period. Math. Hungar.* **19** (1988), 217-225.
[V2]   J.F. Voloch: Transcendence of elliptic modular functions in characteristic p, *J. Number Theory* **58** (1996), 55-59.
[Wad1]   L. Wade: Certain quantities transcendental over $GF(p^n, x)$, *Duke Math. J.* **8** (1941), 707-729.

[Wad2]   L. Wade: Certain quantities transcendental over $GF(p^n,x)$, *Duke Math. J.* **10** (1943), 587-594.
[Wad3]   L. Wade: Two types of function field transcendental numbers, *Duke Math. J.* **11** (1944), 755-758.
[Wad4]   L. Wade: Remarks on the Carlitz ψ-functions, *Duke Math. J.* **13** (1946), 71-78.
[Wad5]   L. Wade: Transcendence properties of the Carlitz ψ-functions, *Duke Math. J.* **13** (1946), 79-85.
[Wag1]   C. Wagner: Interpolation series for continuous functions on π-adic completions of $GF(q,x)$, *Acta Arithm.* **17** (1971), 389–406.
[Wag2]   C. Wagner: Differentiability in local fields of prime characteristic, *Duke Math. J.* **41** (1974), 285-290.
[Wag3]   C. Wagner: Linear operators in local Fields of prime characteristic, *J. reine angew. Math.* **251** (1971), 153-160.
[Wag4]   C. Wagner: Interpolation series in local fields of prime characteristic, *Duke Math. J.* **39** (1972), 203-210.
[Wag5]   C. Wagner: On the factorization of some polynomial analogues of binomial coefficients, *Arch. Math.* **24** (1973), 50-52.
[Wag6]   C. Wagner: Linear pseudo-polynomials over $GF[q,x]$, *Arch. Math.* **25** (1974), 385-390.
[Wag7]   C. Wagner: Polynomials over $GF[q,x]$ with integer valued differences, *Arch. Math.* **27** (1976), 495-501.
[Wal1]   M. Waldschmidt: Transcendence problems connected with Drinfeld modules, *Istanbul Üniv. Fen. Fak. Mat. Der.* **49** (1990), 57-75.
[Wan1]   D. Wan: On the Riemann hypothesis for the characteristic p zeta function, *J. Number Theory* **58** (1996), 196-212.
[Wan2]   D. Wan: Heights and zeta functions in function fields, in: *The Arithmetic of Function Fields* (eds. D. Goss et al) de Gruyter (1992), 455-463.
[Wan3]   D. Wan: Meromorphic continuation of L-functions of p-adic representations, *Ann. of Math.* **143** (1996), 469-498.
[Wan4]   D. Wan: Poles of zeta functions of affine complete intersections. (preprint).
[Wan5]   D. Wan: Global zeta functions over number fields and function fields, in: *Analytic Number Theory: Proceedings of a Conference in Honor of Heini Halberstam*, vol. 2, (eds: B.C. Berndt et al), Birkhäuser (1996), 767-775.
[Wan6]   D. Wan: L-functions of algebraic varieties over finite fields: rationality, meromorphy and entireness, *Proceedings of the Third International Conference on Finite Fields* Cambridge Univ. Press (1996), 379-393.
[WY1]   T.-Y. Wang, J. Yu: On class number relations over function fields, (preprint).
[Wa1]   L. Washington: *Introduction to Cyclotomic Fields*, Springer (1982).
[We1]   A. Weil: *Basic Number Theory*, Die Grundlehren der Mathematischen Wissenschaften **144** Springer (1967).
[We2]   A. Weil: Sur les fonctions algébriques à corps de constantes fini, *C.R. Acad. Sci. Paris* **210** (1940), 592-594 (= *Coll. Papers*, [1940b]).
[We3]   A. Weil: Number of solutions of equations over finite fields, *Bull. Amer. Math. Soc.* **55** (1949), 497-508 (= *Coll. Papers*, [1949b]).
[We4]   A. Weil: *Courbes Algébriques et Variétés Abéliennes*, Hermann (1971).
[Wi1]   A. Wiles: Modular elliptic curves and Fermat's last theorem, *Ann. of Math.* **142** (1995), 443-551.

[Wo1]　　S.S. Woo: Extensions of Drinfeld modules of rank 2 by the Carlitz module, *Bull. Korean Math. Soc.* **32** (1995), 251-257.

[XZ1]　　F.Xu, J. Zhao: "Euler systems" in global function fields, (preprint).

[Yi1]　　L. Yin: Index-class number formulas over global function fields, *Compositio Math.* (to appear).

[Yi2]　　L. Yin: On the index of the cyclotomic units in characteristic p and its applications, *J. Number Theory* **63** (1997), 302-324.

[Yu1]　　Jing Yu: Transcendence and Drinfeld modules, *Invent. Math.* **83** (1986), 507-517.

[Yu2]　　Jing Yu: Transcendence in finite characteristic, in: *The Arithmetic of Function Fields* (eds: D. Goss et al), de Gruyter (1992), 253-264.

[Yu3]　　Jing Yu: Transcendence and special zeta-values in characteristic p, *Ann. of Math.* **134** (1991), 1-23.

[Yu4]　　Jing Yu: Transcendental numbers arising from Drinfeld modules, *Mathematika* **30** (1983), 61-66.

[Yu5]　　Jing Yu: Transcendence theory over function fields, *Duke Math. J.* **52** (1985), 517-527.

[Yu6]　　Jing Yu: A six exponentials theorem in finite characteristic, *Math. Ann.* **272** (1985), 91-98.

[Yu7]　　Jing Yu: Transcendence and Drinfeld modules: several variables, *Duke Math. J.* **58** (1989), 559-575.

[Yu8]　　Jing Yu: On periods and quasi-periods of Drinfeld modules, *Compositio Math.* **74** (1990), 235-245.

[Yu9]　　Jing Yu: Analytic homomorphisms into Drinfeld modules, *Ann. of Math.* **145** (1997), 215-233.

[Yu10]　　Jing Yu: Irrationality of lattices in finite characteristic, *Mathematika* **29** (1982), 227-230.

[Yuyu1]　　Jing Yu, Jiu-Kang Yu: A note on a geometric analogue of Ankeny-Artin-Chowla's conjecture, (preprint).

[Yub1]　　Jiu-Kang Yu: Isogenies of Drinfeld modules over finite fields, *J. Number Theory* **54** (1995), 161-171.

[Yub2]　　Jiu-Kang Yu: A class number relation for function fields, *J. Number Theory*, **54** (1995), 318-340.

[Zh1]　　J. Zhao: On decomposition law in Carlitz-Kummer function fields, *Chin. Sci. Bull.* **39** (1994).

[Zh2]　　J. Zhao: On root numbers connected with special values of L-functions over $\mathbb{F}_q(T)$, *J. Number Theory*, **62** (1997), 307-321.

Index

C 53
$CH_0(X)$ 348
D_i 44, 173, 366
$D_{i,v}$ 370
I^s 239
$L(\psi, s)$ 258
$L(\rho, s)$ 270
$L[T, \tau]$ 144
L_i 44, 173
M_I^d 194
S_∞ 237
– higher dimensional version 346
$S_{\sigma,v}$ 244
T-modules
– $H_1(E)$ 164
– abelian 147
– coordinate systems 161
– definition 145
– dimension 145
– morphisms 145
– pure 149
– – weight 151
– rank 147
– tensor product 157
– torsion points 151
– uniformization 159
– – tensor products 172
T-motives 144
– $H^1(M)$ 164
– $H_v^1(M)$ 154
– – duality with Tate module 154
– abelian 147
– morphisms 144
– pure 149
– – weight 151
– rank 147
– tensor product 156
$[i]$ 44, 172
Γ-ideals 360
$\Gamma(x)$ 382
$\Gamma(x, y)$ 382

$\Gamma_v(y)$ 371
$\Pi(i)$ 349
$\Pi(x, y)$ 382
$\Pi_\infty(y)$ 361
$\Pi_v(y)$ 371
α^s 237
β^{sv} 244
A-Hurwitz series 352
$\langle x \rangle$ 221
– higher dimensional version 345
ω_σ 271
$\omega_{\sigma,L}$ 274
$\phi(\mathcal{L})$ 70
$\phi[I]$ (=I-division group scheme) 70
τ 3
$\widetilde{\Pi(i)}(i)$ 360
$\xi(\Lambda)$ 223
$\zeta_{\mathcal{O}_L}(s)$ 258
$_2\Gamma_w(y)$ 376
$_2\Pi_w(y)$ 376
$g(x, y)$ 382
$g_v(x, y)$ 387
$k\{\tau\}$ 3
n-dimensional local fields 342
v-adic measures 315
– associated divided power series 316, 318, 319
– Dirac measure 317, 318
v-adic representations 255
– strictly compatible families 256
– – associated L-series 257, 258
A 63
A-field 69
A-magic numbers 309
$\mathbf{C}_\infty$ 63
H 195
$\mathbf{H}^+$ 202
K 63
T 298
V 240
k 63

$\mathbf{k}(I)$ 206
$\mathbf{k}(I)^+$ 207
$\mathcal{F}(n)$ 182
$\mathcal{O}(L)$ 207
$\mathcal{O}^+$ 203

Abel, N. 21–23
Additive polynomials 1, 3
– semi-invariants of 31
– adjoints 15
– analogies with differential operators 20–25, 129
– classification 4
Allouche, J.-P. 366
Anderson, G. 137, 142, 163, 165, 171, 172, 174, 176, 230, 231, 283, 284, 296, 312, 334, 391–393, 397–400
Artin conjecture
– p-adic 290
– arithmetic 285
– classical 323
– geometric 285
Artin, E. 61, 218, 284, 297

Beilinson, A. 343
Bernoulli elements 284
– arithmetic 284
– geometric 298
Bernoulli-Carlitz numbers 353, 354

Carlitz exponential 51
– division values 54
Carlitz factorial 349, 350
– definition 349
Carlitz logarithm 56
– order of convergence 57
Carlitz module 53
– adjoint 62
– division points 54
– tensor powers 157–158, 172–174
Carlitz multiexponential 174
Carlitz multilogarithm 174
Carlitz period see Carlitz exponential
Carlitz, L. 46, 51, 52, 55, 57–59, 274, 290, 292, 310, 311, 314, 315, 329, 349, 350, 353, 354
Central simple algebras 95
– Centralizer Theorem 97
– Brauer group 97
– – invariants 100–101
– cyclic algebras 99
– reduced norm and trace 98
– Schur index 101

– Skolem-Noether Theorem 96
– splitting fields 97
– Wedderburn Theorem 95
Chow group 348
Coleman, R. 163
Criterion for cyclicity
– arithmetic 285
– geometric 307
Cyclotomic units 210, 305
– classical 397

Deligne's Theorem (=the Weil Conjectures) 148
Deligne, P. 148, 198
Denis, L. 366, 389, 390, 396
Deuring, M. 100
Diaz-Vargas, J. 326, 329
Differential Galois groups 24, 25
Digit expansion construction 314, 349, 377, 381, 399, 400
Divided power series 316, 319, 385
Drinfeld divisors 229
Drinfeld modules
– L-series of families 394
– fields of definition 201
– analytic uniformization 76
– associated L-series 258
– – trivial zeroes 289
– associated adjoints 130
– – Tate module duality 135
– associated height 73
– associated rank 71
– associated shtukas 191
– definition 69
– endomorphism rings 80–85
– equivalence with lattices over $\mathbf{C}_\infty$ 76
– fields of definition 202
– isogenies 70
– level I structures 194
– Mordell-Weil Theorem 396
– morphisms 70
– semi-invariants 118
– Tate conjecture 393
– Tate modules 91
– Weil numbers 109–110
Drinfeld, V.G. 63, 66, 89, 179, 184, 192, 193, 195–197, 207, 228

Elkies, N. 126, 129
Entire functions on S_∞ 248
– essentially algebraic 253
– – importance of 254

Index   423

Euler, L.   290, 292, 295, 297
Exponentiation of ideals   239
– v-adic version   244
Exponentiation of positive elements
   237

Faltings, G.   92, 156
Fermat equations
– arithmetic   389
– geometric   389

Gabber, O.   129
Galovich, S.   305
Gamma functions
– arithmetic   399
– – v-adic interpolations   371, 376
– – Chowla-Selberg Theorem   366
– – definition   365
– – functional equations   368–370
– – Gross-Koblitz Theorem   380
– – period relations   363
– geometric
– – v-adic interpolations   387
– – Chowla-Selberg Theorem   386
– – definitions   382, 386
– – Deligne reciprocity   390–392
– – functional equations   382–384
Gamma-ideals   see Γ-ideals
Gauss sums
– basic Gauss sums   376
– characterization   400
– general Gauss sums   377, 399
Gauss-Thakur sums   see Gauss sums
Gekeler, E.-U.   112–114, 218, 223, 303, 304, 312, 334, 365
Graded modules   181
Graded rings   179

Hasse, H.   284
Hayes, D.   85, 88, 193, 207, 210, 212, 215, 223, 290, 295
Hayes-modules   199
Hurwitz series   see **A**-Hurwitz series
Hyperderivatives   385, 386

Iwasawa theory   289
– main conjecture   289, 290

Jacobi sums   378

Kapranov, M.   341–343, 346, 347
Katz, N.   33
Koblitz, N.   392

Kummer, E.   304
Kummer-Vandiver conjecture
– function field class group analog
   313
– classical version   284
– function field special point analog
   398

Lagrange, J.   23
Lattice
– special rank one   223
Lattice in $\mathbf{C}_\infty$ 65
– associated Drinfeld module   67
– associated exponential   65
Log-algebraicity   397

Magic numbers   see **A**-magic numbers
Moore determinant   5, 45
– analogies with Wronskian   21–23
– calculation   8
Mumford, D.   179

Newton polygon   39
– simple   332

Ogus, A.   392
Okada, S.   305
Ore condition   93
Ore, O.   9, 15, 18, 30–33

Parshin, A.   342
Picard-Vessiot extension   24
Poonen, B.   116, 122, 124, 126, 134, 135, 396, 398, 400
Positive elements   198
– higher dimensional version   344

Regulator groups   302
Riemann hypothesis
– classical   323
– possible characteristic p analog   340, 396
Rosen, M.   23, 103, 241, 305
Rubin, K.   85

Scattering matrices   175
Serre, J.-P.   181, 214, 324
sgn-normalized rank one Drinfeld
   modules   199
Shtuka functions   229
– applications to exponentials   229
– applications to logarithms   231
Shtukas   191

- applications to rank 1 Drinfeld modules   227–233
- description of torsion points   186
- poles   191
- zeroes   191

Shu, L.   208, 299, 303, 306, 359
Siegel, C.   290, 292, 295
Sign functions   197
- higher dimensional version   344
- twisted   197

Sinha, S.   382, 391, 392
Sinnott, W.   289, 335, 350
Solitons   391

Taguchi, Y.   92, 93, 321, 393–396
Takahashi, T.   90, 212
Tamagawa, A.   116, 117, 120, 122, 393
Teichmüller character   271
- generalized   271

Thakur, D.   114, 174, 241, 292, 296, 316, 329, 338, 349, 359, 362, 363, 366, 368, 369, 371, 372, 374, 376–382, 386, 391, 399
Thiery, A.   366
Trivial zeroes   277

Value field   240, 242
Vector bundles and locally free sheaves   138

Wade, L.   52, 79, 366
Wagner, C.   315
Wan, D.   326, 332, 334, 340, 394–396
Weil, A.   100, 218, 275, 284–286, 289, 297, 322

Yin, L.   306
Yu, J.   78, 296, 366
Yu, J.-K.   84